PRINCIPLES OF FLIGHT SIMULATION

Aerospace Series List

Handbook of Space Technology	Ley, Wittmann & Hallmann	April 2009
Surrogate Modelling in Engineering Design: A Practical Guide	Forrester, Sobester & Keane	August 2008
Aircraft Performance Theory and Practice for Pilots	Swatton	August 2008
Aircraft Systems, 3rd Edition	Moir & Seabridge	
Introduction to Aircraft Aeroelasticity and Loads	Wright & Cooper	December 2007
Stability and Control of Aircraft Systems	Langton	September 2006
Military Avionics Systems	Moir & Seabridge	February 2006
Design and Development of Aircraft Systems	Moir & Seabridge	June 2004
Aircraft Loading and Structural Layout	Howe	May 2004
Aircraft Display Systems	Jukes	December 2003
Civil Avionics Systems	Moir & Seabridge	December 2002

PRINCIPLES OF FLIGHT SIMULATION

David Allerton
Department of Automatic Control and Systems Engineering
The University of Sheffield

A John Wiley and Sons, Ltd., Publication

This edition first published 2009

© 2009, John Wiley & Sons, Ltd

Registered office
John Wiley & Sons Ltd, The Atrium, Southern Gate, Chichester, West Sussex, PO19 8SQ, United Kingdom

For details of our global editorial offices, for customer services and for information about how to apply for permission to reuse the copyright material in this book please see our website at www.wiley.com.

The right of the author to be identified as the author of this work has been asserted in accordance with the Copyright, Designs and Patents Act 1988.

All rights reserved. No part of this publication may be reproduced, stored in a retrieval system, or transmitted, in any form or by any means, electronic, mechanical, photocopying, recording or otherwise, except as permitted by the UK Copyright, Designs and Patents Act 1988, without the prior permission of the publisher.

Wiley also publishes its books in a variety of electronic formats. Some content that appears in print may not be available in electronic books.

Designations used by companies to distinguish their products are often claimed as trademarks. All brand names and product names used in this book are trade names, service marks, trademarks or registered trademarks of their respective owners. The publisher is not associated with any product or vendor mentioned in this book. This publication is designed to provide accurate and authoritative information in regard to the subject matter covered. It is sold on the understanding that the publisher is not engaged in rendering professional services. If professional advice or other expert assistance is required, the services of a competent professional should be sought.

A catalogue record for this book is available from the British Library.

ISBN: 978-0-470-75436-8 (H/B)

Typeset in 10/12 Times by Laserwords Private Limited, Chennai, India
Printed and bound in Great Britain by CPI Antony Rowe, Chippenham, Wiltshire

For Clare, Fergus and Patrick – the best team I ever played for.

Contents

About the Author   xiii

Preface   xv

Glossary   xvii

1 Introduction   1
1.1 Historical Perspective   1
    1.1.1 The First 40 Years of Flight 1905–1945   1
    1.1.2 Analogue Computing, 1945–1965   3
    1.1.3 Digital Computing, 1965–1985   5
    1.1.4 The Microelectronics Revolution, 1985–present   6
1.2 The Case for Simulation   9
    1.2.1 Safety   9
    1.2.2 Financial Benefits   10
    1.2.3 Training Transfer   11
    1.2.4 Engineering Flight Simulation   13
1.3 The Changing Role of Simulation   14
1.4 The Organization of a Flight Simulator   16
    1.4.1 Equations of Motion   16
    1.4.2 Aerodynamic Model   17
    1.4.3 Engine Model   18
    1.4.4 Data Acquisition   18
    1.4.5 Gear Model   19
    1.4.6 Weather Model   19
    1.4.7 Visual System   20
    1.4.8 Sound System   21
    1.4.9 Motion System   21
    1.4.10 Control Loading   22
    1.4.11 Instrument Displays   23
    1.4.12 Navigation Systems   23
    1.4.13 Maintenance   24
1.5 The Concept of Real-time Simulation   24
1.6 Pilot Cues   27
    1.6.1 Visual Cueing   28
    1.6.2 Motion Cueing   29
1.7 Training versus Simulation   30
1.8 Examples of Simulation   32
    1.8.1 Commercial Flight Training   32

	1.8.2	Military Flight Training	34
	1.8.3	Ab Initio Flight Training	34
	1.8.4	Land Vehicle Simulators	34
	1.8.5	Engineering Flight Simulators	35
	1.8.6	Aptitude Testing	36
	1.8.7	Computer-based Training	36
	1.8.8	Maintenance Training	37
	References		37

2	**Principles of Modelling**	**41**
2.1	Modelling Concepts	41
2.2	Newtonian Mechanics	43
2.3	Axes Systems	51
2.4	Differential Equations	53
2.5	Numerical Integration	56
	2.5.1 Approximation Methods	56
	2.5.2 First-order Methods	58
	2.5.3 Higher-order Methods	59
2.6	Real-time Computing	63
2.7	Data Acquisition	67
	2.7.1 Data Transmission	67
	2.7.2 Data Acquisition	69
2.8	Flight Data	74
2.9	Interpolation	77
2.10	Distributed Systems	82
2.11	A Real-time Protocol	91
2.12	Problems in Modelling	92
	References	96

3	**Aircraft Dynamics**	**97**
3.1	Principles of Flight Modelling	97
3.2	The Atmosphere	98
3.3	Forces	100
	3.3.1 Aerodynamic Lift	100
	3.3.2 Aerodynamic Side force	104
	3.3.3 Aerodynamic Drag	105
	3.3.4 Propulsive Forces	106
	3.3.5 Gravitational Force	107
3.4	Moments	107
	3.4.1 Static Stability	109
	3.4.2 Aerodynamic Moments	111
	3.4.3 Aerodynamic Derivatives	113
3.5	Axes Systems	114
	3.5.1 The Body Frame	115
	3.5.2 Stability Axes	117
	3.5.3 Wind Axes	117
	3.5.4 Inertial Axes	118
	3.5.5 Transformation between Axes	118
	3.5.6 Earth-centred Earth-fixed (ECEF) Frame	119
	3.5.7 Latitude and Longitude	122
3.6	Quaternions	122

3.7	Equations of Motion	124
3.8	Propulsion	127
	3.8.1 Piston Engines	128
	3.8.2 Jet Engines	136
3.9	The Landing Gear	138
3.10	The Equations Collected	143
3.11	The Equations Revisited – Long Range Navigation	148
	3.11.1 Coriolis Acceleration	150
	References	154
4	**Simulation of Flight Control Systems**	**157**
4.1	The Laplace Transform	157
4.2	Simulation of Transfer Functions	161
4.3	PID Control Systems	163
4.4	Trimming	169
4.5	Aircraft Flight Control Systems	171
4.6	The Turn Coordinator and the Yaw Damper	172
4.7	The Auto-throttle	176
4.8	Vertical Speed Management	179
4.9	Altitude Hold	182
4.10	Heading Hold	185
4.11	Localizer Tracking	189
4.12	Auto-land Systems	191
4.13	Flight Management Systems	195
	References	201
5	**Aircraft Displays**	**203**
5.1	Principles of Display Systems	203
5.2	Line Drawing	205
5.3	Character Generation	211
5.4	2D Graphics Operations	214
5.5	Textures	216
5.6	OpenGL®	219
5.7	Simulation of Aircraft Instruments	227
5.8	Simulation of EFIS Displays	235
	5.8.1 Attitude Indicator	237
	5.8.2 Altimeter	239
	5.8.3 Airspeed Indicator	240
	5.8.4 Compass Card	241
5.9	Head-up Displays	242
	References	246
6	**Simulation of Aircraft Navigation Systems**	**247**
6.1	Principles of Navigation	247
6.2	Navigation Computations	250
6.3	Map Projections	252
6.4	Primary Flight Information	254
	6.4.1 Attitude Indicator	254
	6.4.2 Altimeter	255
	6.4.3 Airspeed Indicator	255

	6.4.4	Compass	255
	6.4.5	Vertical Speed Indicator	255
	6.4.6	Turn Indicator	255
	6.4.7	Slip Ball	255
6.5	Automatic Direction Finding (ADF)	255
6.6	VHF Omnidirectional Range (VOR)	257
6.7	Distance Measuring Equipment (DME)	258
6.8	Instrument Landing Systems (ILS)	259
6.9	The Flight Director	260
6.10	Inertial Navigation Systems	263
	6.10.1	Axes	264
	6.10.2	INS Equations	264
	6.10.3	INS Error Model	268
	6.10.4	Validation of the INS Model	272
6.11	Global Positioning Systems	274
	References	282
	Further Reading	283

7 Model Validation — 285
7.1	Simulator Qualification and Approval	285
7.2	Model Validation Methods	288
	7.2.1	Cockpit Geometry	291
	7.2.2	Static Tests	291
	7.2.3	Open-loop Tests	294
	7.2.4	Closed-loop Tests	294
7.3	Latency	298
7.4	Performance Analysis	305
7.5	Longitudinal Dynamics	312
7.6	Lateral Dynamics	323
7.7	Model Validation in Perspective	328
	References	329

8 Visual Systems — 331
8.1	Background	331
8.2	The Visual System Pipeline	332
8.3	3D Graphics Operations	336
8.4	Real-time Image Generation	343
	8.4.1	A Rudimentary Real-time Wire Frame IG System	343
	8.4.2	An OpenGL Real-time IG System	347
	8.4.3	An OpenGL Real-time Textured IG System	350
	8.4.4	An OpenSceneGraph IG System	352
8.5	Visual Database Management	364
8.6	Projection Systems	370
8.7	Problems in Visual Systems	374
	References	376

9 The Instructor Station — 377
9.1	Education, Training and Instruction	377
9.2	Part-task Training and Computer-based Training	378
9.3	The Role of the Instructor	379

9.4	Designing the User Interface		380
	9.4.1	Human Factors	382
	9.4.2	Classification of User Operations	383
	9.4.3	Structure of the User Interface	384
	9.4.4	User Input Selections	388
	9.4.5	Instructor Commands	394
9.5	Real-time Interaction		398
9.6	Map Displays		404
9.7	Flight Data Recording		409
9.8	Scripting		413
	References		421
10	**Motion Systems**		**423**
10.1	Motion or No Motion?		423
10.2	Physiological Aspects of Motion		425
10.3	Actuator Configurations		428
10.4	Equations of Motion		432
10.5	Implementation of a Motion System		436
10.6	Hydraulic Actuation		443
10.7	Modelling Hydraulic Actuators		447
10.8	Limitations of Motion Systems		451
10.9	Future Motion Systems		453
	References		454

Index   **457**

About the Author

David Allerton obtained a BSc in Computer Systems Engineering from Rugby College of Engineering Technology in 1972 and a Postgraduate Certificate in Education (PGCE) in physical education from Loughborough College of Education in 1973. He obtained his PhD from the University of Cambridge in 1977 for research on parallel computing before joining Marconi Space and Defence Systems as a Principal Engineer developing software for embedded systems. He was appointed as a Lecturer in the Department of Electronics at the University of Southampton in 1981 and was promoted to a Senior Lectureship in 1987. He moved to the College of Aeronautics at Cranfield University as Professor of Avionics in 1991, establishing the Department of Avionics. In 2002, he was appointed to the Chair in Computer Systems Engineering at the University of Sheffield.

Professor Allerton has developed five flight simulators at the universities of Southampton, Cranfield and Sheffield, and is a member and past-Chairman of the Royal Aeronautical Society's Flight Simulation Group. He has also served on the UK Foresight Panel for Defence and Aerospace and National Advisory Committees for avionics and also for synthetic environments. In 1998, he was awarded £750,000 by the Higher Education Funding Council for England (HEFCE) to establish a research centre in flight simulation at Cranfield University. He was Director of the annual short course in flight simulation at Cranfield University from 1992 until 2001. He is a Fellow of the Institution of Engineering Technology and a Fellow of the Royal Aeronautical Society and is a Chartered Engineer.

His research interests include computer architecture, real-time software, computer graphics, air-traffic management, flight simulation, avionics and operating systems. He holds a private pilot licence with an IMC rating and represents Yorkshire at tennis (over 55).

Preface

I was lucky – I was a schoolboy in the 1960s. In those days, we mended our own punctures and learnt to take a motorcycle engine apart from first principles. Later on, as a student in the 1970s, I worked on the early computers. They came with circuit diagrams and if they had a fault, as they often did, we laid the schematics out on the bench and fixed it. It was open-season for initiative – we designed our own operating systems, invented our own computer languages and wrote our own compilers. We knew hexadecimal, could read binary paper tapes and wrote interrupt service routines. The computers were slow, so we had to design efficient algorithms and because there was very little memory, efficient organization of data was paramount.

So, in the early 1980s, when we developed microprocessor systems, graphics cards and array processors, building a flight simulator seemed a simple and natural progression. At the time, the sum total of my aeronautical knowledge was that the pilot sat at the front of the aeroplane and the air stewardess sat at the back. In fact, talking to several pilots, even the validity of this assumption turns out to be somewhat flawed. Nevertheless, I embarked on the design of a flight simulator and, some twenty years on, having built five simulators and written over 250,000 lines of code, that practical knowledge forms the basis of this book.

I make no apology for being an engineer. I feel strongly that engineering is an applied discipline and that the subject is learnt by understanding the theory and then applying it to problems[1]. Flight simulation brings together mathematics, computer science, electronics, mechanics and control theory. In other words, flight simulation is an application of systems engineering and it provides a fertile playground to enable undergraduate and postgraduate students to develop their understanding and practice new skills in these subjects. The book focuses on software and algorithms because these are the activities that underpin flight simulation. We were not daunted by machine code programming of the early computers of the 1970s and likewise, students in the twenty-first century should not be intimidated by the complexity and diversity of software needed for a flight simulator.

By the very nature of the subject, the book is wide ranging, which lends itself to criticism that the topics are not covered in sufficient depth. However, the book tries to balance breadth and depth, providing the majority of the software needed to construct a flight simulator, or to develop simulator modules. The flight simulator covered in this book emphasizes modular design, at the level of a distributed network of computers and also at the level of software modules. The aim is to produce a system of 'Lego' bricks, where a module can easily be removed or improved, for example, to develop an aircraft model, or a display or a flight control system.

The first chapter provides the historical background and reviews the trends and concepts behind flight simulation, emphasizing its role in flight training. The general principles behind systems modelling are introduced in Chapter 2 to provide the background for aircraft dynamics covered in Chapter 3, which formulates the equations of motion used in a modern flight simulator. The systems perspective is explored further in Chapter 4, where the design methods used in the flight control systems found in modern aircraft are outlined. Chapter 5 concentrates on the computer graphics used in simulator displays and the use of OpenGL in real-time graphics applications.

---

[1] 'That which we must learn to do, we learn by doing' – Aristotle.

With the increasing role of avionic systems in civil and military aircraft, Chapter 6 addresses the modelling and simulation of aircraft navigation systems, particularly satellite navigation and inertial navigation systems. A major part of simulator development is the validation of a simulator; the methods which underpin the simulator approval processes are introduced in Chapter 7. The computer graphics covered in Chapter 5 is extended to the 3D real-time graphics used in simulator image generators in Chapter 8, where OpenGL and OpenSceneGraph are used to illustrate the techniques used in real-time rendering. Chapter 9 focuses on the user interface needed for an instructor station and ways to provide more effective training and evaluation. The final chapter outlines the equations needed for modern motion platforms, the algorithms used to replicate motion and the inherent limitations of these methods. With such a wide ranging objective, certain topics have had to be omitted and sound generation, control loading, electrical actuation, interfacing and the human factors of flight training are, for the most part, not addressed.

Although flight simulation is not taught as a subject in most higher education organizations, it can provide a catalyst for the teaching of flight mechanics, flight dynamics and avionics and afford opportunities to apply computer graphics, electronics, electrical engineering and control engineering to a research vehicle. The book brings together the specialist disciplines of flight simulation to provide a primer for the newcomer to simulation or flight simulator users. It should provide a springboard to enable colleges and universities to build their own flight simulator, to support project work and to provide an adjunct for teaching at both undergraduate and postgraduate levels. Increasingly, simulation is being used in other industries and the book aims to outline the principles and provide illustrations and sample code for the practitioner faced with the task of developing a simulator.

Much of the software described or illustrated in the book and most of the examples are taken from working software running in the flight simulator at the University of Sheffield, based on the software I have developed over the years. All of the software in the book can be downloaded in source form from www.wiley.com/go/allerton.

I am particularly fortunate in all the help I have received from colleagues and students. At the University of Southampton, Ed Zalsuka developed hardware far in advance of its time that formed the basis of our early simulators, giving a lead of some 10 years in low-cost simulation. Without his help and encouragement, I suspect I would still be working on software for integrated circuits. Dave White (now Chief Scientist at Thales Training and Simulation) introduced me to the fundamentals of flight dynamics. At Cranfield University, Michael Rycroft and John Stollery provided encouragement and facilities to develop my research activities in flight simulation, culminating in the development of a £750,000 research facility funded by the Higher Education Funding Council for England. At the University of Sheffield, David Owens supported the further development of an engineering simulator. Numerous students have helped over the years, particularly Tony Clare with radar modelling and image generation, Stefan Steffanson with OpenGL displays and Huamin Jia with sensor modelling and Sebastien Delmon and Patrick Fayard for their work on Airbus flight control laws. During the writing of the book I have been particularly indebted to Graham Spence, for his help with Linux and networking and the numerous diagrams, images and examples he has provided. He continues to rescue me from software cul-de-sacs I seem to drive into on a regular basis. I am also grateful to Gerhard Serapins from CAE who allowed me to use some of his lecture material on motion systems, including Figures 10.2 and 10.3. As a member of the Flight Simulation Committee of the Royal Aeronautical Society, colleagues on the committee have, without fail, always supported requests for assistance or information and I am very grateful to British Airways, Westland Helicopter, Thales, CAE, Frasca, Colin Wood and Sons for their continued support, in particular, the images provided by Thales, CAE and the Royal Aeronautical Society.

Glossary

AC Advisory Circular
ADF Automatic Direction Finding
AGARD Advisory Group for Aeronautical Research and Development
AIAA American Institute of Aeronautics and Astronautics
API Application Programming Interface
ARINC Aeronautical Radio, Inc
ASCII American Standard Code for Information Interchange
ATC Air Traffic Control
ATG Approval Test Guides
BSD Berkley Software Distribution
BSP Binary-Spaced Partition
CAA Civil Aviation Authority
CAD Computer-Aided Design
CAS Calibrated Airspeed
CBT Computer-Based Training
CEO Chief Executive Officer
CFD Computational Fluid Dynamics
CFR Code of Federal Regulations
CG Centre of Gravity
CMRR Common Mode Rejection Ratio
CRM Crew Resource Management
CRS Course
CRT Cathode Ray Tube

CSMA/CD Carrier Sense Multiple Access – Collision Detection

DC Direct Current

DCM Direction Cosine Matrix

DMA Direct Memory Access

DME Distance Measuring Equipment

DOF Degrees of Freedom

DOP Dilution of Precision

ECEF Earth-Centred, Earth-Fixed

EFIS Electronic Flight Instrument System

EICAS Engine Indicating and Crew Alerting System

EPR Engine Pressure Ratio

ESDU Engineering Sciences Data Unit

FAA Federal Aviation Administration

FAR Federal Aviation Regulation

FCU Flight Control Unit

FFT Fast Fourier Transform

FMS Flight Management System

FSTD Flight Simulation Training Device

GDOP Geometric Dilution of Precision

GPS Global Position System

GPWS Ground Proximity Warning System

GUI Graphical User Interface

HDG Heading

HMD Helmet-Mounted Display

HP Horse Power

HSI Horizontal Situation Indicator

HUD Head-Up Display

IAS Indicated Airspeed

IATA International Air Transport Association

IC Integrated Circuit

ICAO International Civil Aviation Organisation

IG Image Generation

Glossary

ILS Instrument Landing System

IMC Instrument Meteorological Conditions

INS Inertial Navigation System

IOS Instructor Operating Station

IP Internet Protocol

IQTG International Qualification Test Guide

ISA International Standard Atmosphere

ITER Incremental Transfer Effectiveness Ratio

JAA Joint Aviation Authorities

LCD Liquid Crystal Display

LORAN Long range navigation

LVDT Linear Voltage Differential Transformer

MAC Media Access Control

MCDU Multi-purpose Control Display Unit

MCQFS Manual of Criteria for the Qualification of Flight Simulators

NACA National Advisory Committee for Aeronautics

NASA National Aeronautics and Space Administration

NATO North Atlantic Treaty Organization

NDB Non-Directional Beacon

NED North- East- Down

NFD Navigation Flight Display

NOAA National Oceanic and Atmospheric Administration

OAT Outside Air Temperature

OBS Omni Bearing Selector

OSG OpenSceneGraph

OSI Open Systems Interconnection

PAPI Precision Approach Path Indicator

PC Personal Computer

PDOP Position Dilution of Precision

PEP Pilot Eye Point

PFD Primary Flight Display

PID Proportional, Integral, Derivative (control)

QDM Direction (Magnetic)
QFE Pressure relative to Field Elevation
QNH Pressure relative to Nautical Height (sea-level)
RAE Royal Aircraft Establishment
RAF Royal Air Force
RAeS Royal Aeronautical Society
RBI Relative Bearing Indicator
RGB Red- Green-Blue
RMI Radio Magnetic Indicator
RPM Revolutions Per Minute
SGI Silicon Graphics Inc.
SI International system of units
TACAN Tactical Air Navigation
TAS True Airspeed
TCAS Traffic Collision Avoidance System
TCP Transmission Control Protocol
TDMA Time Division Multiple Access
TER Transfer Effectiveness Ratio
TSFC Thrust Specific Fuel Consumption
TTL Transistor-Transistor Logic
UAV Uninhabited Air Vehicle
UDP User Datagram Protocol
UK United Kingdom
US United States
USA United States of America
USAF United States Air Force
USSR Union of Soviet Socialist Republics
VFR Visual Flight Rules
VHF Very High Frequency
VLSI Very Large Scale Integrated (circuits)
VOR VHF Omni-directional Range
VSI Vertical Speed Indicator
ZFT Zero Flight Time

1

Introduction

1.1 Historical Perspective

1.1.1 The First 40 Years of Flight 1905–1945

The aviation pioneers learned to fly by making short 'hops', progressively increasing the length of the 'hop' until actual flight was achieved (Turner, 1913). Training was mostly limited to advice given on the ground. There are a few examples of early training devices, but these were designed to enable pilots to experience the effects of the controls. The Sanders trainer (Haward, 1910), developed in 1910, comprised a cockpit which could be turned into the prevailing wind; if the wind was sufficiently strong, the cockpit would move in response to the pilot's inputs. Similar devices were developed by Walters and Antionette (Adorian et al., 1979) in the same year, where motion of the cockpit was controlled by instructors, as shown in Figure 1.1.

During the First World War, flight training became established in dual-seat aircraft, notably the AVRO 504 which was adopted as a basic trainer by the Royal Air Force (RAF) from 1916 until 1933. The instructor demonstrated manoeuvres, which were practised by the student pilot until a satisfactory standard of proficiency was achieved to 'go-solo', with pilot and instructor communicating via a 'Gosport speaking tube'. Despite the rapid advances in the early years of aeronautics, the handling qualities of these aircraft were poor and many were unforgiving if flown badly. It is claimed that more lives were lost in training[1] than in combat during the First World War (Winter, 1982).

By 1912, Sperry had developed rudimentary autopilot functions incorporating a gyroscope. In a public demonstration in 1914, he flew a Curtiss flying boat 'hands-off' while a mechanic walked along the upper wing surface to the wing tip! By the late 1920s, flight instrumentation enabled pilots to fly safely in cloud and rain, without visual reference to the ground. In instrument meteorological conditions (IMC), an artificial horizon, altimeter, airspeed indicator and a compass enabled pilots to fly by reference to these instruments. Although the flying training syllabus had matured by the 1930s, throughout this period the aircraft was accepted as the natural classroom for flight training, including training for instrument flying, with ground schools providing the theory to support flying training.

In the late 1920s, Edwin Link, who is recognized as the founder of modern day flight simulation, developed a flight training device which enabled a significant part of instrument flying training to

---

[1] In his book *The First of the Few*, Winter claims that 8000 of the 14,166 pilots killed in the First World War died while training in the United Kingdom. In Parliament, on 20 June 1918, the Secretary of State attributed the losses to the lack of discipline of the young pilots.

Principles of Flight Simulation D. J. Allerton
© 2009, John Wiley & Sons, Ltd

Figure 1.1 The Antionette flight training simulator circa 1911 (Courtesy: The Library of Congress)

be conducted in a ground-based trainer (Rolfe and Staples, 1986). Link worked in his father's factory in Binghamton, where they manufactured air-driven pianos and church organs. He had a sound grasp of pneumatics and mechanisms and, having obtained a pilot's licence in 1927, applied his knowledge of engineering to the construction of a flight trainer (Link, 1930), using compressed air to tilt the cockpit and to drive pressure gauges to replicate aircraft instruments. However, Link experienced considerable resistance to his ideas and initially sold his trainer (complete with a slot for coins) to amusement parks.

Advances in aircraft design following the First World War saw a rapid expansion in aviation, including the delivery of freight and mail. The US Post Office subcontracted mail deliveries to the US Army Air Corps. But in the early 1930s, they extended mail delivery to all-weather operations, leading to an alarming increase in fatalities. As a consequence, the US Army Air Corps purchased six flight trainers from Link, specifically for the task of instrument flight training – probably the first time that the value of flight simulation was recognized in flight training.

Link's early contributions to flight training were fundamental:

- He exploited specific engineering technologies to produce a successful flight training device;
- The flight trainer was developed to meet a specific training requirement;
- Although his flight model was simple and inexact, the trainer was effective – it had a positive benefit on pilot training;
- The introduction of the Link trainer established the concept that flying training was not limited to airborne training – effective training could also be undertaken in a ground-based trainer.

The problems of aircraft accidents occurring in instrument flight conditions resurfaced during the Second World War, with aircrews flying long distances in night-time operations and

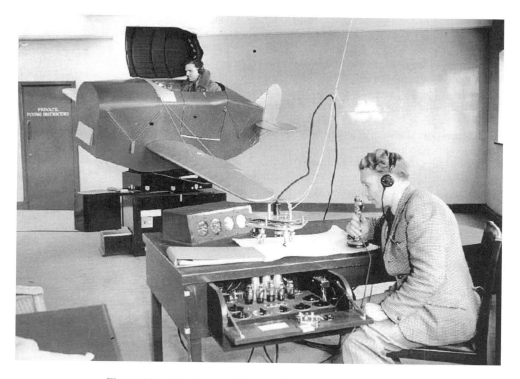

Figure 1.2 A Link trainer (Courtesy: Royal Aeronautical Society)

frequently in bad weather. Under conditions of stress and fatigue, pilots could become disorientated and the ground-based 'Link trainer', shown in Figure 1.2, which was known affectionately as the *Blue Box* (because most were painted blue), enabled pilots to practice in instrument flight conditions; they also provided instructors with a device to assess the instrument flying skills of pilots. Although based on an empirical model, the Link trainer simulated both the environment of an aircraft cockpit and also the dynamic behaviour and performance of an aircraft. Pneumatic actuators were used to inject motion into a small cabin and the instruments were driven to resemble the response and characteristics of an aircraft. In addition, a plotting table was provided so that the instructor could monitor the track flown – the forerunner of the instructor station. Over half a million allied airmen trained on a Link Trainer during the Second World War.

One further development in flight training occurred during the Second World War. As aircraft became more complex to operate, in terms of engine controls, electrics, hydraulics and navigation systems, simulators played an important role in training for the cockpit drills to operate these aircraft and to respond correctly to emergencies. The Silloth trainer (Rolfe and Bolton, 1988), developed by the Royal Air Force, provided a significant amount of training for bomber crews at this time. These flight trainers were fixed-base devices (no motion was provided) – an early example of a part-task trainer, optimized for a specific training application.

1.1.2 Analogue Computing, 1945–1965

Feedback control systems evolved during the Second World War, with servomechanisms being used to stabilize gun turrets for ships and aircraft. The underlying differential equations could be solved by electronic circuits. At the heart of these developments was the operational amplifier, constructed from thermionic valves. By providing a high gain amplifier with resistive feedback, signals could be summed algebraically and more importantly, by providing capacitive feedback, the mathematical operation of integration could be applied to signals. These developments heralded the introduction of the analogue computer. Each amplifier could be configured as a summer or integrator with gains implemented by potentiometers. Short wires were used to connect these elements allowing complex sets of differential equations to be solved in real-time. The simplicity, relative low cost and capability of analogue computers to solve the equations inherent in nuclear power generation, chemical reactions and electrical motor control established analogue computation as a major engineering discipline up to the mid-1960s.

In aeronautics, the analogue computer was used to model the equations of motion of aircraft dynamics (Allen, 1993) as sets of non-linear differential equations, giving the aircraft designer the opportunity to develop advanced control systems in the research laboratory. Indeed, the TSR-2 and Concord programmes made extensive use of analogue computers. Analogue computation offered major advances: it enabled complex differential equations (including non-linear equations) to be solved; it offered real-time simulation; a program could be altered easily by changing the gain of a potentiometer or by patching a few wires. However, this technology also had a number of major drawbacks:

- The valves were unreliable and prone to drift, necessitating regular and complicated calibration;
- The voltage range was limited – the equations had to be scaled manually in order to ensure that the voltage range of the computer was not exceeded, without compromising the resolution of variables used in the computation;
- Multiplication and division of two variables was difficult and slow – mechanical multipliers and inverters were developed, but they lacked resolution;
- Non-linear terms, for example trigonometric functions were difficult to implement – resolvers, with predefined set-points were developed, but these elements were expensive and often limited to a few linear segments.

Although analogue computation made a significant improvement to the fidelity of flight models, advances in motion systems and visual systems were much slower. These early simulators were limited to simple motion drive platforms, often constrained to two or three degrees of freedom and, for the most part, were used to simulate instrument flight conditions. One significant development was the provision of model boards, with very detailed scenery, where a camera was suspended from a small gantry to follow the aircraft flight path over the terrain. The camera was mounted on a gimbal mechanism, allowing the camera to be synchronized with the flight model in pitch, roll and yaw. The camera video output was displayed on a monitor placed directly in front of the pilot, with the model board scaled exactly to the geometry of the pilot eye position. Although very high levels of realism were achieved with detailed model boards, these systems had two pitfalls. First, the pilot could fly off the edge of the model board. One solution was to construct the model board on a flexible base, passing over large rollers, so that the model board effectively repeated. The second limitation was that the camera could collide with the model board, necessitating expensive repairs to the model board and even to the camera. These systems also provided endless opportunities for the prankster, for example, by placing a spider in front of the camera during finals.

There were two significant achievements during this period. First, airlines started to appreciate the benefits of flight simulators, not only to reduce training accidents but also to reduce training costs. Secondly, a significant industry developed in the United Kingdom and the United States to manufacture flight simulators for airlines, notably Redifon and Link Miles in the United Kingdom.

1.1.3 Digital Computing, 1965–1985

While analogue computers were used for real-time applications after the Second World War, digital computers (developed from code-breaking devices) evolved more slowly, initially being introduced into commercial applications, particularly data processing and payroll systems. During this period, the processing speeds advanced, early programming languages were developed and the capacity of storage devices increased. However, it was the development of the transistor in the mid-1960s that initiated the major advances in digital computation. By the 1970s, mini computers were used widely for scientific computing, evolving from the mainframe computers of the 1960s. As the processor speeds increased, sufficient instructions could be executed to enable the equations of motion of an aircraft to be solved at least 15 times per second (15 Hz). At rates below 15 Hz, delays in the computation can be perceived by human pilots, giving rise to unexpected or unusual responses to pilot inputs. The early digital computers achieved this processing rate by using fixed-point arithmetic rather than the slower floating-point arithmetic and by optimizing the code, typically by programming the equations in assembler code. However, by the late 1970s, the mini computers could be programmed in a high-level language, using hardware floating-point processors to achieve an update rate of 50 or 60 Hz. Indeed, several of the simulator companies developed their own processors to meet the very demanding requirement of real-time processing, at this time.

As mini computers advanced in processing speeds, simulator manufacturers applied digital technology to the development of visual systems with two notable achievements. First, with the use of special-purpose hardware, light points defined in a scene could be transformed to the coordinates of a cathode ray tube (CRT). With very fast response video amplifiers, the electron beam of a CRT could be deflected to draw several thousand light points to provide a night or dusk scene. Not only could the light points be rendered accurately, to provide a realistic night-time image, but the intensity of each light point could be modulated to simulate the effects of range or fog. The second advance came with projection. Up to this point, CRT displays were placed in front of the pilot. However, the image lacked realism; the pilot's eyes were focused on the CRT screen, looking at points apparently several miles away. In addition, CRTs were much smaller than the windscreens of transport aircraft. After some experimentation with different lenses (including Fresnel lenses) the collimated projector was introduced into airline simulators. The beams from a CRT reflected from a semi-silvered mirror onto a segment of a spherical mirror, passing back through the semi-silvered mirror to be seen by the pilot. This arrangement afforded two advantages:

- The image was magnified so that it was possible to fill an aircraft window with the image provided by a small CRT. The windows of a commercial transport aircraft are approximately rectangular, allowing the projection system to be mounted on the outside of the simulator cockpit – correction for optical distortion or reversal of the image was provided by adjustment to the CRT beam deflection circuits;
- The light rays from the CRT seen by the pilot were almost parallel, producing an uncanny effect of distance (hence the phrase collimated).

Collimated projection systems were widely used from 1970 to the early 1990s. As monochrome CRTs gave way to colour CRTs to provide daylight images and the scene content increased with faster graphics processors, the collimated projector had sufficient resolution and bandwidth to sustain the 60 Hz frame rate provide by the image generators. However, these projectors had three inherent disadvantages:

- Manufacturing an accurately machined curved mirror was expensive;
- The weight of three or four mirrors added significantly to the (off-axis) load on the motion platform;

- The image was only correct at the pilot eye-point – distortion increased away from the focal point of the projector mirror, to the extent that a pilot would be unable to see anything in the windscreen in front of the other pilot.

1.1.4 The Microelectronics Revolution, 1985–present

Gordon Moore, one of the founders of the Intel Corporation, is widely attributed with encapsulating the advances of the microelectronics revolution in 'Moore's Law' (Moore, 1965). By the 1980s, semiconductor companies were fabricating processors and memories as integrated circuits (ICs). Moore observed that the rate of progress was such that performance doubled with a fixed period of time (and without a concomitant rise in cost), typically 18 months. In other words, every 18 months or so, the capacity of memory chips or the processing speed of processor chips would double, largely driven by a demanding and expanding domestic market for desktop PCs, computer games and audio and video equipment.

By the late 1980s, the standard desktop PC outperformed the mini computers of the 1970s. For flight simulation, the equations of motion could easily be computed at 60 Hz on a single processor. Attention now focused on the development of visual systems. The simulator companies had hitherto developed their own image generators (Barrette, 1986), but with advances in games technologies, they were able to exploit the capability of the graphics ICs. The dusk calligraphic displays were replaced with daylight imagery, with textured surfaces and atmospheric effects rendered at 60 Hz to increase the visual fidelity (Schachter, 1983). The model board was replaced with a digital database of objects to define the terrain, airfield and other geographic features. One further advance was the wide angle projection systems developed in the 1990s. The collimated projectors (Spooner, 1976) were replaced with a light-weight flexible mirror, providing a lateral field of view up to 220° for both pilots, while at the same time significantly reducing the weight of the projection system. The images projected by three (or more) projectors were carefully blended to provide a continuous 'wrap-around' view. The advances in image generation, in terms of the scene detail and image quality, are evident in Figure 1.3, which shows an image taken from a modern civil flight simulator.

Figure 1.3 An image from a civil flight simulator (see Plate 1) Reproduced by permission of Thales

Hydraulics had been introduced to motion platforms in the 1960s. However, by the 1990s, smooth motion could be provided for hydraulic actuators controlled by microprocessors updating the motion system at rates in excess of 500 Hz. Two further advances in motion systems enhanced the fidelity of motion platforms. First, friction at low speeds can induce jerky motion which is detectable on the simulator flight deck. The use of hydrostatic seals largely overcame this problem. The second improvement was to provide sufficient motion cueing in the three linear axes (surge, heave and sway) and the three angular axes (pitch, roll and yaw) of conventional flight. The Stewart platform (Stewart, 1965), with six linear hydraulic actuators reduced some of these limitations. By moving the actuators independently, motion in these different axes could be blended, for example, tilting the platform to simulate surge or raising the platform to replicate heaving motion. By understanding the human balance sensor mechanisms and with the availability of very low friction actuators and high fidelity visual cues, the physical displacement of the hydraulic actuators could be constrained to 'fool' the brain to induce the sensation of motion. In practice, the motion of these actuators is limited to a few metres. Nevertheless, for the range of accelerations in commercial aircraft operations, these platforms provided a remarkably realistic sensation of motion.

As the fidelity of flight simulators increased during this period, airlines realized the benefits to their operation in terms of both increased safety and reduced training costs. Consequently, the operators sought approval from the regulators to replace airborne training with flight simulation. However, the regulators were faced with a proliferation of flight simulation technology and a range of engineering fidelity afforded by these flight simulators. Realizing the need to resolve this situation, the UK Civil Aviation Authority (CAA) and the Federal Aviation Administration (FAA) in the United States, together with the Royal Aeronautical Society (RAeS), arranged a series of meetings leading to the publication of guidelines for the qualification of flight simulators, which were ratified at the RAeS in 1992 (Anon, 1992). These guidelines form the basis of the International Civil Aviation Organisation (ICAO) document 9625 'Manual of Criteria for the Qualification of Flight Simulators (MCQFS)', which enable a flight simulator to be approved for specific training in a training organization. This was a very important milestone in flight simulation:

- It established the international acceptance of flight simulation;
- It enabled training to be conducted in flight simulators to an agreed standard throughout the world;
- It defined several levels of acceptance for flight simulators;
- It enabled flight simulator manufacturers to build flight simulators to comply with a consistent standard of fidelity, defined in engineering terms.

These advances in simulation technology were not limited to the full flight simulators operated by the major airlines. As the complexity of aircraft systems and avionics systems increased, a significant amount of training was needed to operate these systems. Rather than utilize a flight simulator for a specific training activity, part-task trainers were developed, in effect to off-load expensive simulator training to dedicated training devices, optimized for specific tasks. One particular example is the flight management system (FMS). Learning to operate an FMS can require several days of training, depending on the pilot's experience and the aircraft type. However, by simulating an FMS on a laptop computer, flight crews can practise operating an FMS and also be trained and assessed for the cost of a laptop computer and the training software. Part-task trainers are used to train flight crews in many activities which would otherwise be practised in a flight simulator, including navigation procedures, engine-start procedures, aircraft electrical and hydraulic systems operation, FMS operation and radio management procedures. A modern airline flight simulation training device (FSTD) is shown in Figure 1.4, where the panels replicate the aircraft displays and centre pedestal, with additional displays providing course material in the form of diagrams, illustrations and video clips, focusing on the training of the operation of the aircraft systems rather than flying skills. Note that the training may be provided for pairs of students, encouraging crew cooperation at an early stage.

Figure 1.4 A flight simulation training device (FSTD) (see Plate 2) Reproduced by permission of Thales

One other role for flight simulation during this period has been to support engineering design (Allerton, 1999). The cost of flight trials of new equipment and aircraft systems is very expensive and flight simulation has enabled engineers to design and evaluate aircraft systems. Errors eliminated in the design phase are much less costly to rectify than at the flight trials stage. In addition, the simulator provides the designer with much more insight into 'what-if' studies, enabling the designer to compare prototype designs before committing to aircraft equipment. Consequently, the emphasis of flight trials has shifted towards the validation of data derived in simulator studies, rather than expanding the flight envelope. In both the Boeing 777 and the Airbus A380 programmes, engineering simulation made a major contribution to the design of these aircraft.

Since 1985, flight simulation has had a major impact in military flight training. It is arguable that flight simulation was introduced too early into military flight training, giving rise to the view that military flight training was more effective in an aircraft. This argument was further reinforced by the lack of fidelity of flight trainers in the 1960s, giving simulation a bad reputation in many military organizations.

As the cost of training came under strict scrutiny and environmental issues became more prominent, military organizations reviewed the benefits that had been achieved by civil flight training organizations and simulation became an integral part of flying training programmes throughout the world. However, the technology of civil flight training is not appropriate to all military training programmes. In particular, the pilot of a fighter aircraft is not constrained to a forward looking windscreen and during turning manoeuvres, the sustained G-force cannot be replicated with a conventional motion platform. Two advances have, to a limited degree, ameliorated these problems. The motion platform has been replaced with a G-seat in a fixed-base cockpit (White, 1989), to provide more effective training (Ashworth, 1984). As the pilot is subject to G-forces, actuators control the tightness of the harness and provide tactile cues to exert additional pressure on the pilot's body. In addition, the buffet experienced in combat manoeuvres near to the stall limit is provided by vibration of the seat assembly. Two solutions have been developed to provide a pilot with a 360° field of view. In some simulators, the pilot's helmet is instrumented to detect head and eye movement. The image is derived from the eye and head position and is projected directly onto the pilot's eye by a small optical system mounted on the front of the helmet. An alternative approach is the hemispherical dome, used in conjunction with an array of conventional projectors to project the terrain (at relatively low resolution) and several laser projectors to provide high-detail imagery (e.g. aircraft in the scene). The cockpit is placed at the centre of the dome and the projectors are positioned near the base of the cockpit, giving the pilot an all-round view.

One particular area of success in military simulation has been mission rehearsal. By generating the visual database of a combat area (typically derived from satellite mapping), flight crews can rehearse a mission in a flight simulator, even to the extent of applying lessons learnt in the simulator to the mission. Military simulators are also linked via high speed networks to enable flight crews to practise multi-ship missions, either to test tactical manoeuvres against each other or to practise tactics for a specific mission.

1.2 The Case for Simulation

Nowadays, the use of flight simulation in both civil aviation and military training is commonplace and widely accepted (Allerton, 2000). These simulators allow flight crews to practice potentially life-threatening manoeuvres in the relative comfort of a training centre. Similarly, military pilots are able to rehearse complex missions and to practise piloting skills that may be unacceptable in environmental terms (in peace time) or prohibitively expensive, for example, the release of expensive weapons.

1.2.1 Safety

Aviation is underpinned by safety. The primary role of the aviation authorities is to ensure the safety of all aircraft operations, including flight training. As recently as the 1970s, airlines would train with actual aircraft, practising circuits, takeoffs and landings and even simulate system failures. Unfortunately, there were a significant number of training accidents and one major benefit of flight simulation has been to eliminate these accidents.

An airline pilot undergoes two days of training and checking every six months. Quite probably, if pilots had been asked in the 1960s if they would be prepared to renew their licence in a simulator, there would have been universal reluctance to consider such an option. Nowadays, if pilots were asked to undertake this training in an aircraft, there would be a similar rejection of the suggestion.

This is a remarkable transition in a very short time and is simply attributable to the technical advances in modelling, visual systems and motion systems. Flight simulation is accepted by the pilots, operators, unions, regulatory authorities and manufacturers. Indeed, some simulators are qualified for zero flight-time (ZFT) training. With these flight simulators (and an approved flight training organization), all the conversion training is undertaken in a ZFT flight simulator. The first time the pilot will fly the aircraft is with fee-paying passengers on a scheduled flight. Of course, the pilot will be closely monitored by a training captain, but it reflects the advances and acceptance of simulation technology throughout the world.

One further aspect of flight safety is reflected by air accident statistics. Following an accident, the regulatory authorities and airlines will study the report of the air accident investigators and may require airlines to incorporate similar events in their training programmes to enable flight crews to be aware of similar situations. The flight simulator allows pilots to experience a very wide range of flight conditions, so that a pilot will have experienced situations in a simulator within the last six months that most pilots would not normally encounter in their full career. The crew of the Apollo 13 mission attributed their success in coping with the malfunctions on their mission to the many hours spent in a simulator, rehearsing the hundreds of possible flight situations they might encounter.

With the increase in air traffic since the 1970s, flight simulation has made a major contribution to increased flight safety. In particular, flight simulation accelerates experience. In the flight simulator, the instructor can select fog, turbulence, wind-shear or icing conditions in a couple of touch-screen inputs. In normal operations, a pilot might only experience specific wind-shear once in five years. Indeed, with extended operations and four pilots per sector, airline pilots can struggle to complete sufficient manual landings to remain current. In such situations, landings practised in a flight simulator are accepted by the regulatory authorities. One future role for flight simulation may be to maintain the proficiency of pilot handling skills.

In the 1970s, a number of regulators and airlines recognized that a significant proportion of air accidents and incidents were attributable to pilot error, particularly in multi-crew operations, where communications between flight crew or the management of situations broke down. The flight simulator provided the ideal platform to practise crew cooperation procedures and to improve the training of flight crews to respond to hazardous situations. In many training organizations, the flight crew response is recorded and reviewed in debriefing sessions. Lessons learnt from the analysis of cockpit flight data recorders and cockpit voice recorders, following aircraft incidents and accidents, have been incorporated in simulator training programmes. It is probable that the aviation industry has led the world in the use of simulation technology to improve training and safety.

1.2.2 Financial Benefits

A wide body transport aircraft typically has eight flight crews per aircraft and one flight simulator per eight aircraft. With 128 pilots requiring four days flight training per year (four 8-hour shifts), simulation utilization approaches full capacity, without allowing for type conversion programmes for flight crews joining the airline or converting onto type. With airborne operations at least 10 times the cost of simulator operations, the cost of equivalent airborne training would bankrupt an airline. There is a strong case that flight simulation has enabled budget airlines to grow and that, without flight simulation, they would not be able to operate.

The major costs of flight training include:

- The cost of purchasing the flight simulator;
- The cost of building the simulator facility;
- The running costs of the simulator facility (electricity, air conditioning, computer maintenance, spares provision, etc.);
- The staff costs to provide flight training and maintenance of the simulator (instructors, maintenance engineers, administrators, etc.).

The income to a simulator organization comes from selling training time on the simulator. Typically, flight simulators operate 23 hours per day and over 360 days per year. Assuming the simulator and the training organization have the appropriate approvals, the role of the simulator manager is to ensure utilization of the simulator and optimize the income.

In addition to the day-to-day operation of a simulation facility, the procurement of a simulator can also affect the financial return. First, there is a lead-time of the supply of both aircraft and simulators to an airline. An airline will procure specific aircraft for its route structure and then train its pilots to operate these aircraft. Balancing the number of simulators to aircraft types is difficult as route structures can change. Secondly, delays in construction of the building, the delivery of the aircraft or delivery of the simulator can cause a significant loss of revenue. Revenue can be generated by selling spare capacity to other operators. Alternatively, a lack of availability may require pilots travelling to train on simulators operated by other airlines. Thirdly, the long lead times can result in variation of bank rates (for loans) or changes in customer markets.

Clearly, these are tightly balanced financial equations for any simulator operator. However, many of the operational costs are fixed (such as instructor salaries and electrical consumption).

The simulator operator will try to reduce both the purchase cost and the operational cost of the simulator. For the purchase price, the cost of the visual system (the image generators and the projection system) has been as high as 50% of the overall cost of the simulator. However, with the availability of off-the-shelf processors and graphics cards, these costs have dropped dramatically. Although motion systems have been of the order of 10% of the purchase price, as the overall cost has dropped, manufacturers have looked to electrical drive systems to replace the hydraulic systems approved for motion platforms. The operational costs depend on the reliability and availability of the simulator. Simulator slots are often booked several months ahead and loss of a simulator for a few hours in a daily schedule can cause severe disruption to a customer. Reliability of simulator facilities has improved in three ways:

- The simulator sub-systems are much more reliable – there are fewer components and increased redundancy is built into modern microelectronics equipment;
- The modular structure of simulator components enables faulty equipment to be detected and swapped out quickly;
- Simulator systems are continually monitored and checked for variation in performance, allowing preventative maintenance to be undertaken to replace failing components before they actually fail.

For the airline, the choice is to buy or lease a flight simulator. Owning a simulator gives the airline full control of its training programme, a margin of error for operational changes to the airline and spare capacity to generate significant revenue for the airline. The major risk is to ensure revenue earning over a period of 10 years or more, where airlines may change route structures and aircraft types to respond to changing markets. Leasing simulator time can be beneficial, but the airline is dependent on the hourly charging rate of the supplier, the availability of slots, access to other facilities (to keep its aircrew current) and travel costs. Furthermore, if leasing proves impractical, there is a two-year lag to rectify the situation.

1.2.3 *Training Transfer*

The obvious benefit of training on a flight simulator is that time spent training in a simulator can replace time spent training in an aircraft (Caro, 1973). If the simulator is effective (Thompson, 1989), then an hour of training in a simulator could replace an hour, or even several hours training in an aircraft. For example, to practise cross-wind instrument landing system (ILS) approaches in an aircraft requires an acceptable cross-wind and can take at least 15 minutes between approaches. In the simulator, the wind direction and wind speed can be selected and the aircraft can be re-positioned on the approach without flying a circuit or obtaining a landing clearance. Arguably, if the simulator fully replicates the aircraft in this situation, the simulator provides better training than the aircraft.

The term *training transfer* is used to indicate the effectiveness of training in a simulator (Rolfe and Caro, 1982; Caro, 1988; Telfer, 1993). If the number of hours of airborne training needed to achieve a specific level of performance is reduced by synthetic training, the training transfer is positive. Of course, with a poor simulator, an hour spent in the simulator might result in additional time being needed to train in the aircraft, resulting in negative training transfer.

There are three considerations with training transfer. First, how is training transfer measured (Taylor, 1995)? Secondly, are there cost benefits arising from the relative hourly operating costs of the simulator compared with training in an aircraft? Thirdly, does the fidelity (or realism) of a simulator influence the training transfer?

A widely used measure of training transfer (as a percentage) is given by

$$\text{transfer} = \frac{T_c - T_e}{T_c} \times 100 \tag{1.1}$$

where T_c is the amount of airborne time needed by a control group and T_e is the amount of airborne time needed by an experimental group to reach a specific criterion. Clearly, this measurement is independent of the amount of time spent in the simulator. A transfer effectiveness ratio (TER) is more commonly used (Hays, 1992) and is given by

$$\text{TER} = \frac{T_c - T_e}{X_e} \quad (1.2)$$

where X_e is the time spent in the training device. This value equates the number of hours spent in the trainer to equivalent training in an aircraft (for the first X_e hours). One other measurement of training transfer is known as the *incremental transfer effectiveness ratio* (ITER) (Povenmire and Roscoe, 1971). Clearly, after a certain number of hours, the benefit of an extra hour in the training device starts to reduce. In other words, there is a point where (depending on the relative hourly training costs) the cost effectiveness of using a flight simulator is a maximum; thereafter it reduces towards zero. The ITER is given by:

$$\text{ITER} = \frac{Y_{x-\Delta x} - Y_x}{\Delta_x} \quad (1.3)$$

where $Y_{x-\Delta x}$ is the time required to reach a performance criterion by a group having received $x - \Delta_x$ units of training, Y_x is the time required to reach a performance criterion by a group having received Δ_x units of training and Δ_x is the incremental unit of time. In a study of training effectiveness in *ab initio* flight training, Allerton and Ross (1991) obtained the results summarized in Table 1.1.

Assuming that such figures are available to a training organization, then the final column provides insight into the cost effectiveness of training. If the ratio of training costs of the simulator to the aircraft is known, then the cost effectiveness is negative when the value of ITER falls below this ratio.

The issue of fidelity (Hays and Singer, 1989) is very difficult to resolve. The assumption has been that better training is provided by increasing the realism of the simulator. However, this simplification depends very much on the training task. For example, to demonstrate non-directional beacon (NDB) holds with a cross-wind, a desktop animation where the pilot does not actually fly the simulator, but commands heading might prove far more effective than actually training in a full flight simulator. However, operation of an auto-land system in marginal turbulence conditions would probably be best conducted in a flight simulator with good motion and an accurate flight model. There is a view that such issues are best resolved by undertaking a training task analysis. However, there are many examples where effective training has been achieved on very low fidelity equipment and also examples where negative training has occurred with high fidelity equipment.

In order to measure training transfer it is necessary to undertake the simulator training with a group of trainee pilots and compare their performance with a control group who received no simulator training. A number of extensive studies have shown significant transfer of training with the use of synthetic training devices (Waag, 1981; Bell and Waag, 1998). In practice, this option

Table 1.1 Results of a training transfer experiment

Group	Time to first solo (hours)	Simulator training (hours)	Training transfer (%)	ITER
1	14.94	0	–	–
2	11.60	3	22.15	1.10
3	8.24	7	44.70	0.84
4	8.43	11	43.42	– 0.05

may not be practical. One group might justifiably argue that they received poorer training. More significantly, it is necessary to take objective measurements of pilot performance to show the benefits (or not) of simulator training. There are remarkably few instances of such research programmes and for airlines or military organizations, the cost of the experiment would be very high. The corollary of course, is that without such guidelines, nobody is really sure how effective a particular approach to synthetic training is likely to be.

1.2.4 Engineering Flight Simulation

The majority of flight simulators are used for pilot training. However, flight simulation also plays an important role in aircraft design. The major design effort of the modern civil or military aircraft is in the development of systems rather than development of the airframe or engines. Up to the mid-1980s, the testing or validation of aircraft systems would be conducted during flight trials. However, finding a design fault at this stage of the system life cycle is very expensive. In the development of automatic flight control systems, say, a design error could jeopardize the safety of the trials aircraft and detection and isolation of a fault depends on capturing the appropriate data. If these developments are undertaken in a flight simulator, there is no threat to safety, large amounts of data can be acquired and experiments can be quickly modified or repeated.

An engineering flight simulator differs from a training simulator in several ways:

- It only needs sufficient components to undertake the study – irrelevant items can be omitted;
- There is no training element, such as an instructor station;
- There is no requirement to qualify the simulator to meet any specific approvals;
- Data acquisition is very important – the amount and detail of data acquired in an engineering simulator is far higher than in a training simulator.

Design decisions are based on studying the results of trials undertaken in the engineering flight simulator. As the cost of further trials or tests is effectively a computing cost, it is feasible to undertake additional comparative studies or to take prototype designs much further. Of course, the engineering simulator must be based on a detailed and exact model of the actual aircraft and its systems. In practice, this is difficult. If the aircraft is a new combat type, and control systems are being developed which operate near limits of the flight envelope, estimates or approximations used in the modelling may be incorrect. Consequently, airborne flight trials are often used to validate data acquired in simulation studies, where the same inputs are applied to the actual aircraft under the same flight conditions. The data acquired in flight is then scrutinized and compared with data generated in the simulation studies. In some advanced programmes, such as the BAE Systems Typhoon programme, using telemetry, data is transmitted to a ground station where the results can be checked quickly in order to assess if further airborne measurements are required or if the flight test can proceed to the next stage. Otherwise, a lack of airborne data can necessitate an additional airborne trial, increasing the development cost of the programme.

In some aircraft development programmes, particularly for new aircraft, the engineering simulator provides the basis of the software for the subsequent training flight simulator. With the increasing levels of systems integration in modern aircraft, the avionics design teams may require an airframe model and an engine model and similarly, the engine design teams require avionics systems models. Considerable effort is given to the development of modules for an engineering flight simulator and companies developing these modules seek to recover their development costs by charging for data packages. Since early 1990, while the overall cost of flight simulators has dropped, the relative costs of data packages for aerodynamic data, engines data and avionic systems data has increased to a level that has given rise to several international seminars and conferences, focusing on these problems.

One further area where engineering simulators are used is in 'iron-bird' rigs. Aircraft systems, such as undercarriages or braking systems are themselves complex systems, closely integrated with other aircraft systems and connected to the aircraft databuses. In the latter phase of development of an aircraft, the complete set of aircraft systems is constructed (in a large hangar). The airframe dynamics and engine dynamics are simulated but otherwise actual hydraulic and electrical actuation is used to exercise the interface cards, databuses, actuators and sensors. External loads can be applied to actuators to simulate flight conditions and actual aircraft equipment is used for displays, flight computers and avionics. From a systems perspective, the modules of an iron-bird rig are identical to the aircraft modules and extensive tests and fault detection are undertaken to reduce the amount of airborne flight testing.

1.3 The Changing Role of Simulation

Before the mid-1980s, flight simulation was a specialist subject. To achieve the processing necessitated in real-time flight modelling, motion actuation and visual systems, simulator companies developed special-purpose hardware. However, the arrival of the microprocessor and the supporting microelectronic technologies has changed flight simulation in two ways. First, it is possible to model complex non-linear systems using standard off-the-shelf hardware. Secondly, the performance of modern graphics systems simplifies the visualization of simulation.

The flight simulators prior to 1980 focused on replication of the actual aircraft to enable airlines to operate approved flight simulators for their fleet. These simulators cost in excess of $10 million and were designed for one specific aircraft type. The software for these simulators was designed and supported by the simulator manufacturers with the end user (the operator) being dependent on the manufacturer to maintain and support the simulator software.

As computing speeds increased, other industries realized the benefits that simulation could offer. In the field of electronics, IC design moved from the simulation of components at the gate level to the system level, for example, to fabricate an integrated circuit of several hundred thousand gates. Graphics was used in the simulation and visualization of circuit analysis to detect electrical and logical faults prior to fabrication. Modelling packages were used at the block diagram level to simulate mechanical, chemical, nuclear and environmental dynamics. All these developments focused on the provision of powerful design tools to give the designer better insight into system behaviour, often to optimize the system design. For the most part, there was no requirement for real-time performance – there was either no human in-the-loop or alternatively, these tools were used to reduce simulation runs from hours to minutes.

However, it is probably in the aerospace industry where most diversification of simulation technology has occurred. Although the advances in flight simulation have pegged the advances in the related computer technologies, the acceptance of flight simulation has been much slower. There is a sense of 'Catch-22' about flight simulation. Until an organization has used simulation it cannot appreciate the benefits of simulation and while it cannot see the benefits of simulation, it is unlikely to invest in simulation technology. Consequently, it is the reducing cost of simulation technology that has fired the diversification in aerospace.

The financial case for airline simulation has already been outlined. However, if the exacting requirements of airline simulator qualification are applied to other aircraft types, such as regional jets, commuter aircraft, turboprops and general aviation (Allerton, 2002), the case is less convincing. Although the regulatory authorities introduced different levels of qualification, the levels identified by the FAA differ from their European counterparts and have been restricted to fixed wing aircraft, with regulations being extended to rotary wing aircraft in 2001.

The drive for diversification has come from the simulator manufacturers who have tried to match the reducing cost of simulation technology to the markets for simulators. Arguably, this trend has been influenced by the games technologies. For several years, pilots had observed that their children seemed to have better simulation technology in their games than they had in their own training departments. This is understandable – a games company shipping 100,000 games can reduce the end product cost whereas the specialist simulator manufacturer developing a product for a few customers is not able to discount the product or invest in product development at the same level.

The first major development occurred with instrument flight trainers. The availability of computer graphics provided a visual system; stepper motors and servo motors were adapted to replicate aircraft instruments and the equations of motion were solved by a single processor. In addition, mechanical or electrical control loading can be applied to the primary flight controls and an instructor station provides a supervisory role (to set up flight conditions or fail systems) and a monitoring role (to observe tracks and record sessions for debriefing). Such instrument trainers have most of the characteristics of full flight simulators:

- They are real time;
- They are approved for specific training roles;
- They are operated in exactly the same way as the equivalent training aircraft.

In many training organizations, the ground school is distinct from flying training. The ground school lessons are delivered in classrooms and the teaching methods are based on the standard classroom equipment of static overhead projection, whiteboards and blackboards and teaching material that includes hand-outs, notes and textbooks. However, the advances in simulation technology since the mid-1980s have blurred this distinction.

One of the early developments in computer-based training (CBT) was to provide the student with a computer-based textbook. The student could select a page or a topic or was directed through the teaching material according to their rate of learning, typically by their response to questions. In many ways, the early CBT systems were simply one-to-one replacements for conventional learning. However, with the introduction of animation, in the form of computer graphics, more information can be embedded in diagrams or illustrations and more importantly, a student can interact with the computer. For example, an engine-start sequence might require turning on a fuel pump and waiting for the fuel pressure to rise. With a mouse or touch screen, the student can select the appropriate pump and watch the fuel pressure rise.

With the availability of multimedia PC technology, video clips can be combined and merged with animated graphics to illustrate system interactions. More interestingly, if the software for a flight simulator is organized as a re-usable module which can be executed on a workstation, fragments of simulator code can be invoked by CBT software, to operate in exactly the same way that it does in the real-time simulator. With increased display sizes of flat-screens, multiple screens can be combined to show video clips and diagrams in parallel. Although the cost of CBT hardware is low, the major cost is in the development of diagrams and animations. Simple tasks such a detecting mouse actions to select a valve or to show fluid flow in a pipe can require several hours of production. In general, successful use of CBT has been restricted to large programmes, where the cost of CBT development was included in the initial training budget.

Where CBT does offer significant gains is in procedural training. For example, a flight control unit (FCU) may have several selectors, knobs and push switches and interact with other aircraft systems (the aircraft flight control systems and the FMS, in this case). Using a laptop computer, flight crew can practise selections and actually fly a flight plan, where the simulated aircraft responds to FCU inputs and the displays are presented on the laptop screen in the same formats as the aircraft displays. Such an approach to training offers three important advantages:

- The student does not need to learn in a classroom; they can access lesson plans via the Internet, at any time and at any location;
- The result of their exercises can be recorded for assessment by the training organization;

- Students can progress at their own pace; if the lesson plans are carefully structured, the student can be directed back to earlier material to consolidate their understanding or alternatively, where they are progressing quickly, and have demonstrated a good understanding, they can progress faster.

Although the airlines and military organizations have focused on pilot training, recent attention has also been given to maintenance training. Up to the mid-1990s, a large part of maintenance training was conducted with actual aircraft equipment or specially fabricated replicas to enable maintenance engineers to practise the removal and replacement of faulty items. Such training can be very expensive if it takes an aircraft out of service or if mistakes are made with actual equipment, for example, bending the pins of a connector. However, with a synthetic training aid, the student can select the appropriate connector from options displayed on the screen or remove the connector by clicking on the part. A voltmeter probe can then be placed on a specific pin with a simulated voltmeter displaying the voltage. Such systems provide instruction in maintenance procedures and techniques; with increasing dependence on modular hardware in aircraft, technician training concentrates on efficient detection of a faulty module and thorough testing of the replaced unit. The important point to bear in mind is that, just as in the case for pilot training, maintenance trainers are designed to meet a specific training task. If the technician is being trained to tighten nuts with a torque wrench, CBT is not appropriate. However, if the technician is being trained to detect faults when the airline needs to minimize ramp time, then CBT can provide very effective training (and checking).

Flight simulation is moving away from the emphasis on realism in many areas. While it is clearly essential in full flight simulators, in other applications, better understanding may be gained with simplified diagrams, or by answering one of a choice of questions, where the effectiveness of the training is the acquisition of understanding rather than immersion in realistic environments. However, very little research has been conducted into the effect of fidelity in training applications and not surprisingly, with this lack of knowledge in the subject, many organizations opt for realism in training packages.

1.4 The Organization of a Flight Simulator

Baarspul (1990) gives an excellent overview of flight simulation techniques, outlining the technology used in modern flight simulators. The main components of a typical flight simulator are shown in Figure 1.5.

1.4.1 Equations of Motion

The equations of motion are the focal point of all flight simulators. They determine the states of the simulator, taking all the inputs, including pilot controls, winds, aerodynamic terms and engine terms to compute the variables that represent the state of the simulated aircraft, particularly forces, moments, attitude, altitude, heading and velocities. This translation of inputs to outputs depends on the equations of motions used to resolve the linear and rotary motion of the aircraft and includes the aerodynamic data for the aircraft, details of the undercarriage and engine (and possibly propeller) data, usually provided in the form of tables and graphs of data. Note that several of the arrows imply a bi-directional link, for example, several aerodynamic terms are a function of aircraft variables such as angle of attack, Mach number or altitude. In most flight simulators, the equations of motion are updated 50 or 60 times per second, where the forces and moments are computed and applied to the aircraft in less than 1/50th or 1/60th of second.

Introduction

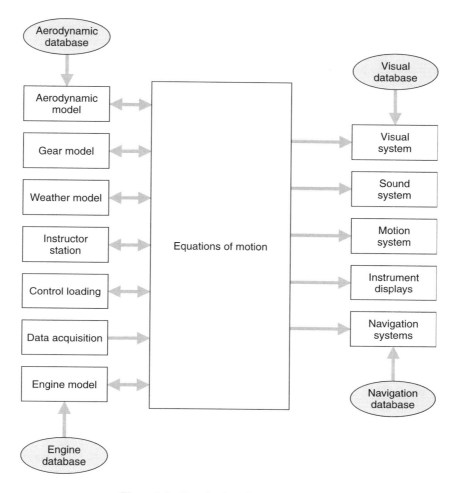

Figure 1.5 Organization of a flight simulator

1.4.2 Aerodynamic Model

The aerodynamic model enables the aerodynamic forces and moments to be computed. For example, the coefficient of lift may be derived as a function of the angle of attack where the specific aerodynamic coefficients are defined in the aerodynamic database. The aerodynamic data is provided as a large (nowadays electronic) database, typically in the form of several thousand graphs of the aerodynamic variables, often as functions of two or three variables. This database is likely to have been obtained from a combination of flight testing, wind tunnel tests and possibly computational fluid dynamics (CFD) analysis. The database will also include a vast amount of validation data to enable the simulator developer to compare the simulator dynamics and performance with actual aircraft data. From the manufacturer's perspective, this data has high commercial value and is usually provided as part of a confidential agreement between the operator, the aircraft manufacturer and the simulator manufacturer.

The aerodynamic model is possibly the most critical element of a flight simulator. An error in modelling the aircraft aerodynamics can lead to a simulation which might fail in the qualification process or be unacceptable to pilots who are experienced on this type of aircraft. Consequently, the aerodynamic data package produced by the aircraft manufacturer is expensive. For this reason, there are very few detailed data packages available in the public domain. Although it is possible to acquire this data from flight trials, the cost of aircraft instrumentation, operation of flight trials, data recording and data analysis can be prohibitive to simulator companies. This is an issue sometimes missed by prospective simulator developers; there is no simple way to acquire this data and just a few errors in the aerodynamic data package can result in an unacceptable aircraft model.

1.4.3 Engine Model

Many of the comments in the previous section also apply to the engine model. The engine manufacturer will undertake extensive tests in engine development. Moreover, the engine manufacturer is likely to be responsible for the engine control systems, so will be using this data to develop an engine model to be used in any control system design activities. In some data packages, the engine dynamics will be provided by the aircraft manufacturer, either because they have undertaken flight tests to identify the engine characteristics or because they have purchased the data from the engine manufacturer for use in the simulator development. Note that the engine data is dependent on the state of the aircraft; implementation of the engine model requires access to variables computed in the flight model.

From the simulator designer's perspective, the engine model is a model of the engine dynamics, rather than a thermodynamic model, which is used to derive engine thrust, fuel flows and engine pressures and rotation speeds. Quite often, the engine model is as detailed as the aerodynamic model. For simulator qualification, the regulatory authority will need to undertake extensive tests of engine failure modes (e.g. surge, stall or total failure) and the engine manufacturer must provide this data. In particular, the engine characteristics change considerably at low speeds and at very low altitude where the engine flow interacts with the ground. For these reasons, there are also very few examples of detailed engine models in the public domain.

1.4.4 Data Acquisition

In a full flight simulator the flight deck is an exact replica of the aircraft flight deck. Indeed, some of the equipment in the simulator may use actual aircraft parts, as it may be more expensive to fabricate a facsimile than to purchase the certified aircraft equipment. In addition to the primary flight controls (control column, rudder pedals, brakes, flap selector, gear selector, etc.), every lever, selector, knob and switch must be interfaced to the appropriate simulator module.

Some of these inputs provide digital data (0 or 1), often referred to as *discrete data*, whereas other inputs are analogue data. The current selection of inputs must be sampled during each frame, converted to an appropriate value and passed to the relevant simulator module. For some sub-systems, for example, a radio panel, these inputs are monitored and computed by a local processor in the module and passed as serial data or parallel data. Nevertheless, a full flight simulator may have several hundred inputs. Typically, the sampling and storage of this data is handled by one (or more) processor(s), dedicated to the data acquisition function.

For analogue data, particularly for the flight controls, data capture is critical to the fidelity of the simulator. The data acquisition software is responsible for minimizing any delay in capturing the data, for ensuring that the resolution of the data is as high as possible and that any signal conditioning or filtering of the data is correctly applied. In practice, data is derived from simulator signals by analogue-to-digital conversion hardware. Although sampling at rates up to 60 Hz is straightforward with modern data acquisition hardware, the typical resolution of sampled data is only 12–16 bits, which is far below the resolution of data used elsewhere in the simulation. In addition, this data is measured by potentiometers that are inherently noisy or linear voltage differential transformers (LVDTs); as the signals are small (in terms of current), they are susceptible to noise and interference from other signals and considerable effort is given to providing a good earth connection and cable shielding for analogue signals in order to minimize any signal distortion from interference.

1.4.5 Gear Model

During taxiing, the undercarriage is in contact with the ground and the interaction between the runway, the tyres, the wheels and the oleos will result in very different dynamics from the aircraft in flight. An additional model is usually provided for ground handling to include the effects of the tyres and the undercarriage assembly. For many simulator applications, the handling of the aircraft during takeoff and around touch down and during the ground roll is critical in terms of fidelity. Many of the emergencies practised by flight crews include incidents prior to rotation (e.g. tyre bursts, engine failures and runway icing) and following touch down (e.g. reverse thrust failure, brake overheat and aquaplaning).

In practice, there are additional state transitions, just prior to takeoff and just after landing, where the aerodynamic contribution to motion is combined with the dynamics of the undercarriage assembly. At this point, the airspeed is such that the pilot has some authority and the aerodynamic forces interact with the undercarriage forces. In addition, ground effect on the aircraft aerodynamics depends on the height of the aircraft above the runway and the speed of the aircraft and this effect must also be included in both the aerodynamic model and the ground handling model. Similarly, high speed turn offs to taxiways can cause tyre scuffing requiring detailed dynamic models of the wheels, brakes and nose wheel steering. There may be a dearth of data for ground handling situations (e.g. a tyre burst) and consequently, elements of the dynamic model may be derived from computed estimates or pilot experience of similar events. It is also important to bear in mind that, for an airline, a simulation sortie may be gate to gate. Pilots have to learn taxiing skills and to manoeuvre a large transport aircraft around an unfamiliar airfield, which can be difficult at night. The consequence to an airline of an unexpected excursion from the runway tarmac to the grass is likely to prove expensive and is best practised in the simulator.

1.4.6 Weather Model

The atmosphere clearly has a major effect on aircraft performance. The equations to compute air pressure, air density and air temperature are well known and straightforward to implement. These terms are used in the flight model and engine model modules. However, other aspects of weather are also modelled in most simulators, particularly wind. Wind has an effect on both navigation but also on the aircraft handling, for example, in cross-wind landings or in turbulence.

Winds also play an important part in flight planning for airline operators, so the wind model must be three dimensional and time varying. Two methods are currently used. The simplest method is to use a wind model to take into account pressure fronts, altitude and location. The alternative approach is to use wind data acquired from agencies that monitor global wind currents; this is generally appropriate as there is no requirement to model current wind patterns in a training exercise.

Atmospheric turbulence also plays an important part in airline training; pilots need to demonstrate that they can handle turbulence and also monitor the interaction of turbulence with automatic landing systems. Normally, turbulence is generated as a set of pseudo random processes, causing perturbation in the three linear aircraft axes. Variation of turbulence components over an airframe can also induce rotational disturbance on the airframe. Two turbulence models are widely used, the Dryden model (Beal, 1993) and the Royal Aircraft Establishment (RAE) model (Tomlinson, 1975) and although they exhibit slightly different characteristics, they have been almost universally adopted.

Following a number of major accident investigations, a better understanding of wind-shear and microburst was gained (partly as a result of simulation studies) and models of both wind-shear and microburst are now included in the training of commercial pilots, to detect the onset of wind-shear and to avoid actions which could lead to an adverse outcome. This is a very good example of flight simulation making a significant contribution to flight safety; the reduction of wind-shear related accidents, since the introduction of mandatory training, has significantly reduced this problem.

The other aspect of weather modelling is to simulate potentially hazardous flight conditions, particularly icing and heavy rain. Airframe icing and runway slush can increase drag during takeoff, affecting takeoff performance, while slush and runway surface water can reduce the effectiveness of braking and steering. In addition to modelling these effects, the conditions for icing should reflect the ambient weather conditions and in the case of ice, snow or rain, these conditions would also be implemented in the visual system.

1.4.7 Visual System

The visual system provides a number of channels of real-time images seen from the pilot eye position. Initially, a database of objects is loaded into the visual system memory. The database may contain fields, airfields, roads, lakes, coastline, vehicles, buildings, trees, forest, vegetations and aircraft. Various standards exist to generate these entities, with OpenFlight™ probably the most widely used format. Each object is reduced to coloured and textured polygons (usually triangles), defined in the coordinate system (and units) of the database. Quite often these objects are arranged in some geometric order so that different levels of detail can be displayed according to the distance from the object.

As the aircraft manoeuvres, the pilot eye position and orientation are computed in the equations of motion and the scene is rendered every frame (typically 60 Hz). Depending on the imaging system, there is a delay between acquiring a new eye position and the pilot seeing the projected image (in addition to any delay in computing the pilot eye position and passing this information to the visual system). This delay, often referred to as *visual latency*, must be kept to a minimum but can extend to three or four frames (four frames at 50 Hz is 80 ms). A value of 100 ms is generally accepted as the worst case allowable latency (depending on the aircraft dynamics) and figures of 20–50 ms are more commonly quoted for current flight simulators.

The quality of the visual system depends very much on the characteristics of the underlying graphics engine. A graphics card will be bounded in terms of the draw rate to render polygons and also the fill rate to texture the polygons. As more scene detail is added, the frame rate may drop below the minimum value for a particular simulator. Similarly, if the polygon count is reduced to increase the rendering rate, the scene detail may reduce to an unacceptable level. These values also depend on:

- The display resolution (in terms of pixels) – if the resolution is increased, the drawing rate is reduced as more pixels need to be drawn per unit area;
- The display refresh rate – if the refresh rate is increased, less time is available per frame for rendering;
- The memory access speed of the graphics frame buffer memory – each pixel must be written to the frame buffer every frame;
- The architecture of the graphics processor – particularly, its interface to the frame buffer memory;
- Any anti-aliasing applied to smooth polygon edges – considerable processing is needed to generate smooth edges.

Each image generator provides a video signal output containing red, green and blue colour components. Each channel is offset according to the projected channel. For example, if there are three channels covering a total of 180°, the forward channel is offset by 0° and the other channels are offset by −60° and +60°. In practice, the projected channels overlap by a few degrees, to avoid any visible gaps in the projected image. Strictly, two pilots seated at the front of an aircraft would not see identical views. In practice, the angular differences are small and are ignored. The image formed by the image generator is generated as a 2D image, in effect for a flat screen. If this image is projected directly onto a curved screen, it will appear distorted. To ensure correct geometry in the projection, as viewed at the pilot eye-point, the distortion of the image is corrected in the projector circuitry. This alignment is set up during installation and is checked regularly during maintenance schedules.

1.4.8 Sound System

The aircraft cockpit or flight deck is a noisy environment. Pilots hear a wide range of sounds including slipstream, engines, sub-systems, air conditioning, ground rumble, actuators, radio chatter, navigation idents, warnings, weapon release and alarms. Although sounds are provided in a simulator to increase fidelity, they are also important cues and they must be consistent with the sounds heard in an aircraft.

Some sounds, such as airspeed or engine revolutions per minute (RPM), vary with flight conditions whereas other sounds give a constant tone (or set of tones), for example, a fire warning. Usually, a separate sound system is provided, taking inputs from other modules, for example, engine RPM from the engine module, slipstream magnitude from the flight model or marker ident from the navigation module.

Generally, two methods are used to generate cockpit sounds. The obvious technique is to record actual aircraft sounds, using a high quality sound recording system placed in the cockpit or flight deck, typically with a microphone suspended from the roof of the cabin. The drawback with this approach is the number of recordings needed to cover all flight conditions. For example, engine sounds vary with airspeed, altitude and engine state (RPM and thrust) and it would be necessary to cover the full range of these variables. The recordings are then accessed every frame to output a small fragment, 1/60th of a second at 60 Hz. The sound signals must also be continuous – any discontinuities between frames would result in interference which would be detectable. Although many commercial sound cards are designed to replay recorded sounds, there are in excess of 50 separate sound components which need to be generated and combined every frame.

The alternative method, and the one more commonly adopted, is to analyse the source of each sound and generate the appropriate waveform. For simple sounds, such as a warning alarm, this is straightforward as these sounds comprise a set of tones. The sound of jet engines, propellers and rotors are more complicated. Nevertheless, the fundamental frequencies of these noise sources can be identified and combined with different forms of white (i.e. random) noise. In effect, each sound consists of several components and for each sound, these components are computed as a function of its primary variables every frame. With earlier sound cards, synthesizer ICs were used but nowadays the sound waveform can be generated by a digital signal processor. One further attraction of this method is that synthetically generated sounds can be compared with actual aircraft sounds to confirm the authenticity of the generated sound, typically by applying a fast Fourier transform (FFT) to the respective sounds and comparing the frequency spectra.

1.4.9 Motion System

As the simulated aircraft is manoeuvred, the pilot will expect to feel the accelerations that would be experienced in actual flight. The accelerations are computed in the flight model and are passed to the motion system. For the standard motion platform comprising six linear hydraulic actuators, each actuator is moved to a new position to try to replicate the accelerations on the pilot's body.

Very fast response actuators are used and these are updated at rates in excess of 500 Hz to eliminate jerkiness in the motion.

Of course, true motion cues cannot be generated. For example, a military aircraft can fly a 4G turn for several minutes whereas the only positive G available from the platform is vertical heave, which is constrained by the length of the hydraulic actuators (typically 2–3 m). Similarly, in a rolling aerobatic manoeuvre the pilot is subject to lateral accelerations and angular moments. The arrangement of the legs of the platform restricts angular motion to 30–40°. However, the human motion sensors and the brain can be fooled. The brain responds to the onset of motion but cannot detect very low rates of motion, so that motion can 'leak away' without the pilot realizing that motion has changed. In addition, if the motion cues are reinforced with strong visual cues by the visual system, the pilot can sense motion that is not actually applied to the platform. In some fixed base simulators, the motion sensed from visual cues can convince pilots that the simulator has a moving platform. Indeed, there are (undocumented) instances where the motion system has been switched off without the flight crew realizing that motion cues had been suppressed.

The motion system contains computers to derive the actuation equations to move the platform to the desired position, together with filters to optimize the platform trajectory and provide appropriate motion for the pilot's balance sensors. Of course, this is a compromise and moreover, different pilots have different perceptions of motion. However, assessment of the motion is a very important aspect of simulator qualification and is checked for critical phases of flight, particularly takeoff, landing and engine failures.

One final and important consideration for the motion platform is safety. The cabin could fall several metres under active control, as a result of a system failure, causing injury to the occupants. Redundant computer systems and extensive monitoring is used to detect unexpected motion, shutting down the hydraulic actuation within a few milliseconds to avoid any injury, if any undesired motion of the platform is detected. One other aspect of safety is that the platform is 4–5 m off the ground. In the case of a fire or emergency, the hydraulic system should lower the platform to its lowest point to allow the flight crew and instructor to escape via ladders.

1.4.10 Control Loading

As an aircraft flies, the slipstream passing over the control surfaces changes the load on these surfaces, particularly the primary flight controls: the elevator, aileron and rudder control surfaces. The force per unit displacement increases (non-linearly) with airspeed and affects the handling of an aircraft. Such effects must also be simulated and this is the function of the control loading system. Control loading is provided by attaching actuators to the flight controls in the simulator so that the actuator provides resistance to motion, typically varying with airspeed.

Prior to 2000, most simulators used hydraulic actuators for control loading. By driving the actuator with a computer system, the control loading system can be programmed to emulate springiness, damping, end stops, backlash and dead bands. The characteristics of hydraulic actuation are so good, that an emulated mechanical end stop feels as though the control column is in contact with a mechanical end stop. Since 2000, advances in electrical motor drives have enabled control loading to be implemented with a single motor drive per channel. Control loading is also an important part of simulator qualification and there have been concerns that residual torques can be detected around the centre position, at very low values of torque where electrical systems lack response. With both hydraulic and electrical control loading systems, the trimming function is simply implemented as a null datum offset for the zero load position. To provide force feedback for the control column or centre stick, the inceptor displacement is sensed using an LVDT or stick force is measured by means of a strain gauge.

Both hydraulic and electrical control loading systems must meet strict safety requirements. A control column, some 12 in. in front a pilot could cause considerable harm if accidentally driven at full power towards the pilot. Similarly, an auto-throttle lever quadrant has the potential to sever a pilot's fingers if moved unexpectedly. Consequently, the applied forces of control loading systems are very closely monitored by separate systems, which can remove any control loading within a few milliseconds.

1.4.11 Instrument Displays

Aircraft instruments cover two eras of flight. Prior to 1980, most aircraft had mechanical instruments. Many of these instruments had complicated mechanisms, including multiple pointers, rotating cards, digital readout and clutch mechanisms. Since 1980, many civil and military aircraft have moved to electronic flight instrument systems, known as *EFIS displays*. These EFIS displays are based on computer graphics with 8-in. ruggedized monitors, typically with the displays updating at least 20 times per second (20 Hz). The graphics hardware in the display has a very fast drawing speed to sustain the frame rate and includes anti-aliasing algorithms to smooth any jagged lines or edges as characters and lines are rendered.

The conventional mechanical displays have been implemented since the Link trainer, initially using mechanically driven pointers or using pneumatics to drive adapted pressure gauges. Nowadays, simulated instruments are driven by servo motors or stepper motors, using specialized mechanisms to provide all the functionality of aircraft instruments, for example, the fail flags of a VOR instrument or the barometric pressure setting for an altimeter. Such instruments are driven by analogue voltage via an I/O card. Many of these instruments contain complex mechanisms to drive several pointers, rolling digits, rotating compass cards and clutch mechanisms for pilot settings. Consequently, with older simulators, these mechanical instruments are often unreliable, requiring regular inspection and maintenance.

With EFIS displays, the simulator manufacturer has the option of using the original aircraft equipment. Alternatively, a modern processor with a low performance graphics card is capable of emulating the EFIS displays found on most aircraft. This option is sometimes referred to as the *stimulate or simulate* debate; either original equipment is stimulated, but the simulator must provide the correct input signals or the equipment is simulated, but then the simulator manufacturer must ensure that all the possible operating modes and functionality have been correctly implemented. In terms of cost, the latter option (in effect, reverse engineering) is generally preferred by simulator developers.

For military aircraft, and more recently for some civil transport aircraft, the simulator may also require a head-up display (HUD). If the original aircraft equipment is used, the appropriate video signal format must be generated for the HUD. Actual HUDs are collimated so that the image appears to be overlaid on the outside world. The optical attenuation is small, so that the external world can be seen through the HUD together with the flight data optically projected via a CRT. One option is to place a small monitor in front of the pilot, combining both 2D graphics (the HUD) with 3D graphics aligned to the projection system. However, such systems can give rise to pilot fatigue as the focal length varies between looking at the HUD and looking at the projected image. This problem can be overcome by projecting the HUD information as a graphical overlay on the visual system; in this case, the pilot simply looks through an open frame representing the HUD frame. This method has the advantage that the pilot does not need to accommodate the HUD information as it is effectively focused at infinity. On the other hand, slight movement of the pilot's head can cause information on the HUD to appear outside the HUD frame.

1.4.12 Navigation Systems

A significant part of flying training covers navigation training. Flight simulation offers two advantages: first, an airborne navigation exercise consumes fuel and secondly, navigation errors in training can be hazardous. Consequently, simulators provide varying degrees of navigation capability. For instrument approaches, VOR, automatic direction finding (ADF) and ILS systems are emulated. This requires integration with radio management panels or an FMS to select receiver frequencies and an up-to-date database of navigation aids. In addition, the errors associated with these systems must be modelled correctly. For example, a VOR operates in the very high frequency (VHF) band and therefore is line-of-sight; it can fail if the aircraft is too low to receive the signal or if the transmitter is obstructed by a hill. Similarly, false glide slope indications, common at some ILS installations, need to be modelled.

For enroute navigation, VOR, inertial navigation systems (INS) and the Global Position System (GPS) are also simulated. In addition to simulation of the correct functionality, it is important to model the error properties of these systems. For example, an INS will drift and exhibit the slow 'Schuler loop' oscillation. Similarly, GPS accuracy varies with the geometric dilution of precision (GDOP) caused by variations in the GPS satellite constellation. In an aircraft, these navigation systems are integrated with the FMS and flight displays (typically via the aircraft databus) and such functionality must be replicated in the simulator, particularly the appropriate behaviour for the relevant failure modes.

For short-range navigation, a 'flat earth' model may be adequate for navigation. Beyond ranges of 100 miles or so, a full spherical earth model is needed for navigation, transforming the aircraft motion to earth latitude and longitude axes. For longer distances, Coriolis acceleration terms (the effect of earth rotation on perceived motion) need to be modelled.

1.4.13 Maintenance

In addition to the systems outlined above, the simulator must be regularly checked to ensure that it is operating within limits. Indeed, for qualified simulators, the operator must keep a log of all scheduled and unscheduled maintenance and repeat diagnostic tests to confirm that the simulator characteristics have not changed significantly, following any maintenance procedures.

Specific tests are conducted for all the sub-system modules. If the module contains a processor or memory, these will be exercised and tested for failures. The visual system and projection system will be checked for alignment, where patterns of rectangles and dots are used to detect drift in the optical systems. Light levels and grey scales are also checked for illumination intensity, together with variation in colour bands for the red, blue and green components, often caused by ageing of the projector bulbs.

For mechanical systems, tests are made for excessive wear or increased friction. Such tests are typically computer-based, where tests are initiated and the results compared with results from previous bench-mark tests. In addition to problems with wear in actuators, sensors may also fail, with ingress of dirt or hydraulic fluid contamination. When sensors or actuators are replaced, extensive diagnostic software is run to check the sensor: that it operates over its full range, that it has been reconnected correctly and that there is no discontinuity or noise on the sensor input.

In airline flight simulators, the simulator usage is constantly monitored and recorded. Usage is an important issue. If, for example, a simulator has been subjected to heavy landings, large jolts can affect the visual system mirror or reduce the life of projector bulbs or motion platform bearings. All faults are logged and at the end of a training session, the instructor can report any malfunctions or anomalies that occurred during the session. Moreover, to guarantee over 95% availability, the airline will have technical support teams, with specialized knowledge of the simulator modules, to provide fast response to any failures or problems and a large inventory of spares to minimize any down time.

1.5 The Concept of Real-time Simulation

In normal everyday life, everything seems to be continuous and instantaneous. The motion of a car along a road is continuous; a ball follows a smooth trajectory through the air; a tree branch waves smoothly in the wind. Of course, we are not strictly solving any equations nor is a tree computing the positions of all of its branches.

Computation is a very different world. We write software programs and the computer executes the instructions of the program. These instructions may involve additions of machine registers, comparing memory locations and jumping from one address in the computer memory to another address. However, each of these instructions takes a finite number of machine cycles and each of these cycles is clocked at the speed of the processor. In other words, for a given computer, a small fragment of code may take several microseconds to execute. This time may also vary depending upon the program and its data. For example, to sort 100 numbers may take much longer than sorting 20 numbers.

Anyone using a computer for office activities is subject to these delays. For example, spell checking a document may take several seconds for a large section of text. The computer has to load its dictionary of words, access the file containing the text and then search for each word in the dictionary. These operations may require several million computer instructions, but generally, a user is not too concerned whether it takes two seconds or three seconds to spell check a document. Of course, if it took five minutes, we might seek an alternative solution (e.g. a better algorithm or a faster computer or a faster disk).

There is one area of office computing where we might notice even small delays. If we move the mouse we expect the cursor to move instantly over the screen, or if we press a key, we expect the appropriate character to appear on the screen. In this case, there is a human in-the-loop and we would expect a response within 1/10th of a second, anything less would be intolerable to the average computer user.

Consider an example where the user moves the mouse around the screen. The computer is, in effect, executing the following code:

```
Check to see if the mouse needs redrawing, and if so
  read the new mouse position
  compute the screen coordinates (x, y) of the mouse
  erase the cursor at its current position
  draw the cursor at (x, y)
```

For each of these actions, a set of computer instructions is executed, but if the computer can execute 10^7 instructions per second, say, then if we only need 10,000 instructions for this code, the software will be executed in 1 ms – from the human perspective, instantaneously. However, on an older computer, the same number of instructions might take 100 ms (or 1/10th of a second) and we might discern a slight jerky motion of the mouse.

Of course, an office computer may be executing other tasks, such as printing a file and therefore, it will need to suspend the printing task frequently to check the mouse for movement. If the mouse handling task is serviced 50 times per second, the cursor will appear to move smoothly in response to the mouse input. However, if the computer is occupied with other tasks, so that the mouse is only serviced twice a second, the mouse will seem unresponsive or slow. Generally, office operating systems provide acceptable mouse management and the mechanism to ensure that the mouse software is activated 50 times per second is part of the operating system and is transparent to the user.

What the operating system does is to discretize time, into very small time steps, so that the mouse management code is guaranteed to be executed 50 times per second. These time steps are smaller than we can discern with our eyes (and hand and brain), giving the mouse what appears to be instantaneous and continuous motion, although it is actually implemented at discrete intervals. Of course, we are all used to such systems; television cameras at a football match capture frames every 1/25th of second, which are transmitted to the television in a house. Because the time step is so small, it appears that the players on the pitch are moving in the normal continuous way that we would expect to see if we sat in the stand at the match.

Exactly the same situation occurs in flight simulation. In an aircraft, the pilot moves the control column. Assuming direct cable linkage to the control surfaces (and ignoring the inertia of the control surface), the elevator moves immediately, causing a perturbation to the aircraft, which is seen as a change in pitch by the pilot who responds with another movement of the control column to correct the pitch attitude. In a flight simulator, the stick position is sampled, the elevator deflection is computed, a new pitch attitude is computed and an image is displayed by the visual system with the new pitch attitude, enabling the pilot to correct the aircraft attitude. The important point is that the overall time for this computation must be sufficiently short so that it appears instantaneous to the pilot. In a modern simulator, these computations must be completed within 1/50th of a second or 20 ms.

This concept is illustrated in Figure 1.6, which depicts 10 frames of a simulation. The arrow for each frame shows the proportion of the frame used in computation of the simulation. If the frame time is sufficiently small, say 1/50th of a second and if the computation in any frame never exceeds the frame time, then the simulation is real-time.

Note that there is an important distinction between real-time computing and fast computing. Although a real-time simulation may require a fast computer, all the computations must be completed within the frame time (Burns and Wellings, 2001), whereas, for a fast computing task, the only metric is the overall time. Some simulation packages imply that generating the code in a more compact form ensures real-time computing. Although this may be true for some applications, in all of the examples covered in this book, the term *real-time* is used to mean that all the computations are solved within one frame during every frame of the simulation.

There is one other important consideration that was touched upon. In addition to the main simulation tasks, the processor may also be required to perform other background tasks. If this additional computing load results in contravening the frame period limit for the simulation task, real-time simulation cannot be sustained. An operating system is therefore a critical part of any real-time simulator. The operating system must guarantee that it will execute the simulation task every frame and never introduce delays that cause the simulation task to exceed its frame limit.

In a safety critical real-time environment, it is necessary to demonstrate that the real-time frame rate can never be violated beyond any reasonable level of doubt (which may be never more than once in 10^7 hours, or once every 1141 years). Although flight simulation software is not safety critical, the real-time constraint must still be fulfilled; usually this is the responsibility of the simulator designer. It is possible to monitor the real-time performance of a flight simulator and record any frame period violations. However, if the frame rate does drop, it is usually apparent to the flight crew as there are observable discontinuities in the visual system, or discernible lags in the aircraft response or even changes of frequency in the sound system outputs.

The simulator designer has, in effect, a time budget to complete all computations within the frame and consequently, tries to exploit as much of the frame time as possible, leaving a small margin for error in these estimates (or for future expansion), particularly as some computation times are data

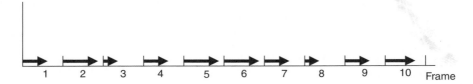

Figure 1.6 Real-time frames

dependent. Given all the constraints on the scene content of the visual system, the processing of the flight model, the engine model, the weather model and so on, it is not uncommon for the frame period to be exceeded occasionally, even for a full flight simulator, particularly as the simulator manufacturer may not have full control over the behaviour of a graphics card under all flight conditions. Nevertheless, ensuring real-time performance, particularly for worst-case conditions is an essential part of system validation and acceptance tests.

1.6 Pilot Cues

A pilot in an aircraft perceives the sensation of flight from the stimulation of the pilot's senses. Of course, these are the same sensors we use in everyday life. They have evolved over millions of years and are optimized for survival. For example, we can perceive the colour, size and distance of a car on a motorway using our visual sensors or run down an escalator using our balance sensors. In flight simulation, these stimuli are referred to as *cues* and, if we are to provide the same sensation of flight in a simulator, it is necessary to provide the same cues as those detected in flight. It is therefore important to understand the cues we use in flight and also the mechanisms in the human body to detect these cues. In addition, if it is not possible to provide the same cues, an understanding of the sensors of the human body and the priority attached to specific cues in processing these cues, may enable the simulator designer to reduce or simplify the cues that need to be provided in a simulator.

The issue of fidelity was discussed briefly in Section 1.2.3. The AGARD (Advisory Group for Aeronautical Research and Development) Flight Mechanics Panel observed in their report on 'Fidelity of Simulation for Pilot Training' (AGARD, 1980) that 'The degree of realism required of the simulation will depend upon the nature of the training task, the level of experience of the students and the level of proficiency they are required to attain'. In reducing fidelity, care is needed to ensure essential cues are not omitted, while on the other hand, considerable savings can be made if unnecessary cues can be eliminated.

There are four stimuli we perceive as humans:

- Chemical – particularly smell and taste – generally, these do not need to be provided for flight simulation;
- Thermal – temperature – for most simulators, the ambient operating temperature of the flight deck or cockpit is sufficient;
- Photochemical – brightness and colour detected by the eyes;
- Mechanical – these include auditory sensors to detect sounds, vestibular sensors to detect gravity and angular and translational motion, kinaesthetic sensors to detect the position and movement of joints and pressure applied to the body and tactile sensors to detect pressure on the skin.

In addition to these sensors, the brain provides a fusion function, combining and processing information to provide the pilot with an understanding of position, orientation, speed and acceleration in the environment. This is an important consideration for the flight simulator designer; it may allow one stimulus to be substituted for another. For example, motion is sensed by cues provided by the visual system (the eyes) and by the motion platform (the vestibular system) offering a potential trade-off of visual cues against motion cues. These stimuli also have different paths through the nervous system of the human body and necessitate different amounts of processing by the brain. Consequently, the brain is able to compensate for these delays. However, if these delays are different in a simulator (e.g. the response of instruments, the visual system or the motion platform to a pilot input), then this can lead to an unacceptable 'feel' of the simulator and in extreme cases, even lead to anxiety resulting in sweating and nausea (the so-called simulator sickness).

1.6.1 Visual Cueing

Visual cues (Buffet, 1980) are sensed by the eye. With two eyes, binocular vision provides further information of size and distance. The characteristics of the human eye are summarized in Table 1.2.

Note that, for a projection system with a lateral field of view of 180°, acuity of 1 arc minute corresponds to 1 part in 10,800. Assuming that three projectors are used, with each channel providing 1000 pixels laterally (allowing for overlap), then current image generators and projectors need to increase resolution by a factor of almost four, to achieve images that correspond to the acuity of normal vision, implying an increase by a factor of 16 of the pixel rendering rate.

The external world seen by a human observer covers the region within touch (1 m) of the observer, the immediate vicinity (up to 30 m), where there is considerable detail and the likelihood of dynamic objects and the far field (beyond 30 m), which is remote and provides a stable 3D environment. The human observer acquires different information in these three regions to perceive images, distance and movement (Farmer et al., 1999).

We perceive images by recognizing familiar shapes, for example, a tank with a turret and barrel would be recognized over a wide range of distance and in different orientations. The brain is combining colour and knowledge – basing its decision-making on the expected image, not necessarily the actual image. In other words, the context in which an image is generated will affect its credibility. For example, in a desert scene, an acceptable tank might be generated by a few hundred polygons, whereas in an urban scene, more detail may be necessary to provide an acceptable image of the tank.

We perceive depth in several ways. Muscles in the eye are used to focus on an object. Other muscles in the eyes converge both eyes to provide a single image (binocular vision). The brain is also able to process the different images formed on the retinas to provide further depth perception. From knowledge of the external world, occlusion (one object in front of another object) and relative size of objects provide important depth cues. In other words, depth might be emphasized by the positioning of objects in the environment. We are used to seeing perspective cues, such as a railway line converging in the distance to provide depth cues. In a simulator, the arrangement of fields might provide useful depth cues in comparison with a more bland landscape. Also, objects placed on the ground appear higher in the visual field towards the horizon. Accordingly, known objects on the

Table 1.2 Characteristics of the human eye

Attribute	Characteristic	Value
Field of view	The angular range detected by the eye. Colour and detail is best in the central region and attenuates significantly towards the limits.	±90° horizontally 50° above 70° below
Colour	Cones in the eye detect the red, green and blue components in colour. The three types of cone receptors are most sensitive at specific wavelengths of light and overall, in the green–yellow region.	Wavelength 400–700 nm
Resolution	The acuity of the eye allows the human to resolve detail. It is maximum in the 3° band in the centre of the field of view. Towards the limits of the field of view, only movement can be detected.	1 arc minute
Brightness	A very wide range of luminance can be detected from darkness to bright sunlight, although discrimination between different levels is best with dark stimuli and significantly reduced with bright stimuli.	
Contrast	Contrast is the ability to distinguish an object from the background, in terms of brightness or colour. Contrast depends on brightness, reducing in sensitivity at low or high levels of luminance.	

ground can reinforce depth cues. One further effect of depth is atmospheric attenuation, where distant objects appear hazy, providing a further (albeit coarse) estimate of distance.

In the personal space, occlusion, size, convergence and accommodation are most important. In the nearby region (up to 30 m) occlusion, height in the visual field and size are important whereas convergence is negligible. Beyond 30 m, binocular vision reduces (both eyes receive the same image) and height in the visual field, occlusion, size and perspective are dominant.

The brain derives motion from visual cues in three ways: the angular field of an object increases moving towards the object; the appearance of an object may alter (e.g. windows in a building become visible); there is a visual reference to motion over a fixed terrain. In a static image, there is no implied motion of the viewer or the object. However, if the object moves it triggers a succession of receptors in the eye. Alternatively, if the head and eyes follow the target, there are sufficient changes to the background to trigger the receptors.

In flight simulation, with the pilot looking forwards, the visual scene not only moves towards the viewer, but points in the scene move towards the edge of the display, which is known as *visual streaming* (or optical flow). Strictly, this depicts relative motion – the same scene would be seen if the pilot moves towards the runway at 250 kt or if the runway were (somehow) moving towards the pilot at 250 kt. Of course, in an aircraft, the pilot may be moving towards a fixed runway at 250 kt but in the simulator, the pilot is stationary and the scene is rendered so that it appears to be moving towards the pilot at 250 kt. The brain finds it hard to distinguish between these cases and assumes that the pilot must be moving as the scene is static. A similar effect can occur in a railway station – if the adjacent train moves off there is a brief sensation that your train is moving and the other train is stationary. In many simulators, the effect of motion provided by a good visual system (known as *vection*) is compelling and has been used to support arguments to reduce the requirement to use a motion platform to provide motion cues.

The very nature of the simulator environment limits the visual cues that can be provided. For the most part, objects are several miles in the distance reducing the contrast, resolution and colour and losing visual cues from accommodation, convergence and retinal disparity. Often there are few objects to provide cues from occlusion, relative size or linear perspective. In addition, the field of view may be limited by the windscreens.

The simulator designer is therefore constrained in the cues that can be provided while needing to ensure that there is sufficient scene content and image resolution. The other requirement for visual fidelity depends on the application. In a civil simulator, considerable detail is usually provided in airport scenes, so that the approach and takeoff have appropriate visual cues, whereas a military helicopter simulator may need to provide occlusion from groups of trees and vection cues for height and speed estimation from detailed texture of the terrain. In subjective terms, the pilot is expecting to see the dominant visual cues that are experienced in flight and the content of the visual database and the performance of the visual system must match these expectations. Objectively, it is difficult to quantify matching the requirements of the visual system with the performance of the image generators. Clearly, there are some cues that cannot be provided. For example, the binocular vision from looking at a screen placed 3 m in front of the pilot may conflict with the binocular vision cues provided during ground handling.

Remarks: these problems would for example occur in a Formula One racing car simulator, where the driver sees the wheels and suspension within 1 m range, the track in the range 2–5 m and also distant objects beyond 5 m. Yet, rendering tyres so that they appear in the correct position does actually work surprisingly well with a collimated visual projector system in racing car simulators.

1.6.2 Motion Cueing

While it is clear that a motion platform, securely anchored to the ground inside a hangar cannot produce the same forces on the human body as an aircraft that can fly at 600 kt and climb to 60,000 ft, an understanding of the human motion sensing system is essential for two reasons:

- It explains the way the human body detects and responds to motion;
- It may identify limitations with the human sensing systems that enable forces to be mimicked in a way that is indiscernible to the pilot in a flight simulator.

The motion sensors of the human body comprise the vestibular system (Howard, 1986) and the haptic system. The vestibular system detects static and dynamic orientation of the head. It also stabilizes the eyes so that clear vision is achieved during movement of the head. In many ways, the vestibular system is equivalent to a gyro-stabilized inertial platform. The haptic system consists of the pressure and touch sensors over the human body, particularly interaction with the seat, pedals and hand controls.

The vestibular system consists of two sets of sensors, one in each ear, which measure the angular and linear accelerations. The angular accelerations are sensed by the semi-circular canals, organized in three mutually perpendicular planes. At low angular frequency (below 0.1 Hz) they measure acceleration. At high frequency (above 5 Hz) they measure displacement. In the mid range, they measure velocity. Engineers will recognize this behaviour as similar to a second order filter. These are heavily damped systems with time constants of 0.1 s and approximately 11 s (yaw axis), 5 s (pitch axis) and 6 s (roll axis). There is also a minimum value of acceleration below which no motion is detectable. For flight, this threshold is between $0.5°/s^2$ and $2°/s^2$. In other words, someone placed inside a large box with no visual stimulus will be unable to detect that they are being moved about one of the three axes, if the acceleration is below $0.5°/s^2$. If this box is a flight simulator cabin, this observation implies that:

- The pilot can be rotated by applying a signal to the motion platform without being aware of the rotation;
- Initial motion can be leaked away to allow additional acceleration to be applied, also without the pilot being aware of the additional motion.

This threshold is exploited in motion systems to regain some of the limitation on displacement of the motion platforms used in flight simulation.

Whereas the semi-circular canals measure angular acceleration, the otoliths sense linear accelerations. The transfer function is similar to the semi-circular canals but with time constants of 0.66 s and 10 s and a threshold of $0.0004 \, m/s^2$. For combined motion of angular and linear accelerations, the semi-circular canal signals dominate with the otoliths providing measurements of predominantly linear motion.

There is one further advantage in knowing the response of the body's acceleration sensors. If the transfer functions of the motion actuators are known, the desired motion of the platform can be matched as closely as possible to the actual motion. Nevertheless, it is important to note that the traditional platform has, for the most part, been abandoned in military simulation in favour of high quality image generation systems, wide-angle projection systems and the use of G-seats to provide tactile motion cues.

1.7 Training versus Simulation

Flight simulators are used in flight training and it is easy to assume that both terms are synonymous and interchangeable. However, flight training is provided to fulfil a training requirement. The equipment used to provide this training may well include a flight simulator but in this role, the simulator is simply equipment. The simulator, together with an instructor and a training syllabus constitutes the training package. Confusion has arisen because, in some cases, the flight simulator is purely a replacement for the aircraft and arguably this has led in the past to the procurement of flight simulators that were poorly matched to the training requirement and therefore provided poor training.

The first stage of procurement for any flight training programme is to undertake a training requirement analysis. This establishes what is required in the training programme, and also what is not required. On the one hand, if the training equipment is underestimated, it may not provide effective training. On the other hand, if the training equipment is overestimated, the cost of the simulation equipment may be excessive.

An often cited example of an effective trainer is the 'cardboard bomber'. In the early stages of flying, pilots have to learn a range of checks, including pre-flight checks, in-flight checks, shut down checks and emergency checks. Sitting in an ordinary seat, in front of a cardboard facsimile of the cockpit, with no moving parts whatsoever, a pilot can point to an instrument or touch a switch appropriate to each check. Photographs are used so that each item resembles physically the actual equipment and is located in approximately the same place as in the aircraft. However, the training requirement of the equipment is simply to help the pilot to remember the checks. Obviously, such equipment would be inappropriate at later stages of training where actual aircraft instruments or switches are needed. Similarly, the use of actual aircraft equipment would be an unnecessary expense in the initial phase of training.

For a major training programme, the training analysis will be conducted by a specialist team, with a good understanding of flying training, simulation technology, instructional methods and human factors. The team will not include flight instructors who will give the subsequent training or the simulator companies who will manufacture the equipment. For an airline or military training organization, the training will be specified in terms of the desired outcomes (or output) of the training. For example, for an instrument trainer, the requirement might state 'on completion of training, a pilot shall be able to demonstrate an ILS approach with a cross-wind up to the maximum allowable value for the aircraft with one engine failed' as part of the training requirement. The implication of just this single statement is that the simulator requires an ILS instrument, a realistic model of engine failure, a weather model which includes cross-winds and presumably a database of navigation aids including ILS beacons. Note that the requirement omits the engine controls, the details of the weather model or the maximum number of ILS approaches to be provided. Presumably, such information would be given elsewhere in the training requirement.

The training needs analysis team will review the training syllabus, discuss the detailed requirement with the customer (possibly to modify or clarify the requirement), review the technical options (to establish any constraints or parameters) and discuss the training methods used by the instructors. For example, some of the questions to be asked about the ILS simulation might include the following:

- The need for aural cues (e.g. Morse idents or marker idents);
- The accuracy of the simulated ILS;
- The range of the ILS;
- The failure modes of the ILS;
- The method of selecting an ILS frequency;
- The number of ILS channels;
- The failure modes that can be set by an instructor;
- The physical representation of the ILS.

This final consideration might give the option of using a graphics display to represent the ILS instrument rather than a mechanical emulation, leading to the clarification of further issues, including

- The resolution of the ILS display;

- The need for anti-aliasing;
- The update rate of the display.

The important point to bear in mind is that the training requirement extends across the whole scope of the training programme, and even includes non-functional requirements, such as

- Access dimensions;
- Power requirements;
- Emergency lighting;
- Reliability and availability figures;
- Air-conditioning;
- Safety issues.

Once the training requirements are clearly defined, these are passed to prospective manufacturers who will be invited to tender to supply the training equipment. There is often some variability in these requirements. For example, the requirement may simply state the tasks to be trained and the level of skill to be attained, and possibly the time available to attain it, using the training equipment. In this case, the simulator manufacturer can match their equipment to the training requirement, advocating one technology rather than another.

One further issue covered in a training requirement is the method of acceptance. For an airline, their senior pilots are likely to retain close links with the simulator company and go through a formal series of acceptance tests at the factory, prior to shipment and delivery of the simulator, followed by further acceptance tests following the installation.

1.8 Examples of Simulation

1.8.1 Commercial Flight Training

The classical application of flight simulation is the training of airline pilots. The simulator comprises a motion platform, supporting an aircraft cabin with visual systems mounted above the cabin, as shown in Figure 1.7.

Pilots joining an airline from flying schools, general aviation or military organizations are trained on a specific type. Their previous training for the commercial pilot's licence will have covered basic flying skills and instrument flying whereas the simulator training for the airline covers operation of the aircraft type on the company's routes. Pilots converting from one type to another type are trained on the specific operational characteristics of the aircraft type, with the amount of training depending on the similarity between types. Pilots are also checked every six months as part of the regulation of pilot ratings. These checks and recurrent training, lasting two days, cover a wide range of operating conditions and in particular, check that flight crews are able to manage all possible emergency conditions. Because the training is specific to an aircraft, a Level-D full flight simulator is used. The simulator is checked regularly by the regulatory authorities and the fidelity of the flight model, the motion platform and visual system is at the limit of current simulator technology. It is important to appreciate that a poor performance during these simulator exercises can lead to loss of a type rating for a pilot and therefore, the level of fidelity of the simulator is required to be as high as possible. In addition, given the criticality of the simulator availability to the airline, the reliability of the simulator must be very high and this is achieved by the quality of both the simulator equipment and maintenance undertaken by the airline.

In other areas of civil training, the relative cost of training in the aircraft compared with the equivalent training cost in an approved flight simulator is the dominant factor in determining

Introduction

Figure 1.7 A civil full flight simulator (see Plate 3). Reproduced by permission of Thales

the usefulness of simulator training. It is most significant in instrument flight training for flying schools. Although the cost of a flight simulator for instrument training is similar to the cost a light twin-engine aircraft, the hourly operating costs of the simulator are far lower. Moreover, the simulator can accelerate training in comparison with airborne training, where hours are lost to bad weather, aircraft serviceability or obtaining clearances to practise navigation procedures. For example, in an instrument procedures trainer, the student can practise an NDB hold, being re-positioned close to an ADF beacon. As the student progresses, the effect of wind on hold patterns, which is arguably a difficult concept to understand for many students, can be added progressively; such structured training is impractical in airborne training.

The other major advances in civil training have occurred in ground school training. With advances in avionics, flight crews must operate EFIS displays, an FMS, an FCU, weather radar, ground proximity warning systems (GPWSs), traffic collision avoidance systems (TCAS), INS, engine management systems, fuel management systems, electrical systems, hydraulic systems, radio management systems and so on. The use of part-task training devices has off-loaded time that would otherwise be needed on a full flight simulator. There may be no requirement for motion or a visual system and rather than fabricate a facsimile panel, computer generated panels containing knobs, switches and displays can be produced as 'soft' panels. Switch selection or rotation of a knob is provided by a touch screen. As the simulation is computer-based, it can be combined with multimedia inputs for commentary, video, diagrams and animation. In addition, student interaction can be monitored, with the training package designed to fulfil a specific set of training tasks.

1.8.2 Military Flight Training

In military flight training, most basic training is undertaken in the aircraft. Instrument flying training is provided in flight simulators together with system familiarization trainers. However, the major developments in military flight training have occurred in operational flight training, where the cost of airborne training (particularly the deployment of munitions) and safety issues justify the cost of simulation. In these advanced simulation facilities, simulators can be linked together, enabling flight crews to practice mission sorties, with instructors introducing enemy forces (land and airborne) often where these synthetic forces have intelligent decision-making capabilities. Such techniques extend to networking flight simulators over large distances allowing international groups to combine their training resources. In addition, much military simulation involves the training in the use of specialist equipment (particularly radar and mission systems) where it is necessary to combine the pilot training with navigator training or training of mission specialists. In these roles, a mission training simulator can also be connected to a flight training simulator to provide realistic training situations and also to practise crew cooperation.

An increasingly important role for flight simulation is mission rehearsal. For flight crews undertaking a mission into unknown territory, they can practise tactics and mission management prior to the mission. In these simulators, data from satellite mapping is used to generate an accurate visual database and knowledge of enemy air defences provides very realistic training to practise for a mission. In some cases, lessons learnt in the simulator result in changes to the mission tactics.

1.8.3 Ab Initio *Flight Training*

Although there is an understandable desire to get the student pilot into an aircraft in the early stages of a flight training programme, with the reducing cost of simulation technology, there is a case to achieve more effective training by replacing some elements of airborne training with simulator-based training. However, the financial case is not strong; Allerton (2002) outlines the advantages for recurrent training of general aviation pilots but points out that there is no financial incentive for flying schools to invest in simulation technology. A few studies with *ab initio* pilots have shown that the quality of flight training can be improved with part-task training in simulators (Lintern *et al.*, 1990; Allerton and Ross, 1991) but generally, the use of simulation in general aviation training is mostly restricted to instrument training in the training of private and commercial pilots.

1.8.4 Land Vehicle Simulators

For many applications, the case for simulation versus 'live' training is based on the relative training costs. For large vehicles such as trucks, earth moving equipment and buses, the cost of a training simulator is less than the cost of training in the vehicle and the cost of damage to the vehicle and its systems in the case of misuse resulting from inexperience. Such systems also present danger to the public where inexperienced drivers are instructed in urban areas. For example, decision making for overtaking other vehicles, can be assessed for a wide range of pre-programmed situations, allowing the driver to be trained and debriefed in the safety of the training centre. In vehicle training simulators, considerable effort is given to the provision of a wide-angle projection system, computer generated imagery for mirrors and realistic urban scenery and traffic.

Similar ideas extend to the training of train drivers. However, in this case, the visual system is simplified because the train is constrained to move along the track; the only variable is the train speed. Some systems are based on pre-recorded images captured by photography to provide very realistic scenes, which are projected at update rates dependent on the train speed. In such systems, track signals are provided as computer generated overlays. In practice, the capability of modern visual systems provides better training, allowing different track layouts to be modelled; the time of day (or season) can be varied and human entities can be modelled to simulate other trains, different signal patterns, trackside maintenance gangs and commuters on platforms.

Several racing car companies have realized the benefits of simulation. Placing the car cockpit in a wide-angle visual system and providing G-loading cues and vibration with a motion seat, the companies can develop driver interfaces, evaluate suspension system settings or modify gearbox ratios to optimize the behaviour of the car for a specific race track. Moreover, with very limited lap practice times, the simulator familiarizes the driver with the track layout, adverse cambers or braking requirements. One current limitation of these simulators is that the driver experiences high G-loads on the neck muscles resulting from a combination of accelerations on bends and during braking. The obvious health and safety considerations of active head loading in a racing car simulator have (not surprisingly) deterred the development of such cues.

1.8.5 *Engineering Flight Simulators*

For the aircraft designer, the development, installation and testing of complex aircraft systems can require many hours of airborne flight tests, which, depending on the equipment can be expensive and dangerous, for example, in developing a military terrain following system. By providing an engineering flight simulator (Allerton, 1996), where the simulator equipment closely matches the aircraft equipment, the system designer can install, integrate and evaluate aircraft equipment in the simulator. The simulator is still flown with the same exercises as the actual flight trials aircraft. The major gain is that design faults are easier to detect in the simulator and more importantly, that early detection of a fault in a programme can reduce the life cycle costs significantly.

Data is acquired in the engineering simulator to assess the performance and behaviour of the aircraft system. Typically, high performance data acquisition equipment is used to support the system analysis. Analysis of the data is also helped by the provision of visualization tools which enable unexpected behaviour or anomalies in the system response to be identified. Alternatively, these tools can assist the designer to determine optimal operating conditions or to confirm that the system fully meets the design requirements. Often, a set of engineering simulators are used as the system is developed, where each simulator is optimized for the phase of development. Initially, a workstation may be sufficient to prove the design concept. Simple pilot-in-the-loop studies are possible with basic desktop simulators. Once the functional specification is established, the aircraft system is developed in a prototype form with interfaces to other aircraft systems, where it can be flown in an engineering simulator, often referred to as an *iron-bird* rig. Such a simulator may have a rudimentary visual system or no motion, but the cockpit controls, sensors, actuators and databuses are often based on actual aircraft equipment. Following tests in an iron-bird rig, the equipment is subsequently installed in the aircraft. A primary objective of the flight test programme is to validate results of the engineering simulator tests and to identify discrepancies between the aircraft and the model used in the simulator.

The requirements of the development programme define the structure and fidelity of an engineering simulator. For example, during the development of cockpit displays, lighting levels may be tested in a simulator that simulates external lighting conditions from very bright sunlight to dark night to establish that the range of display lighting levels is appropriate. In developing an FMS, there is no need for a motion platform but the pilot interface and controls must be exact. Similarly, for a TCAS, or GPWS system, the visual system database must correlate accurately with the position of traffic or terrain detected by these systems.

An engineering simulator may contain software used on the full flight simulator as it is simpler to reuse this software than develop software adapted for the engineering simulator. Alternatively, software developed in an engineering flight simulator may subsequently be used in training simulators. The engineering simulator is often used as a datum, where updates or revisions may be checked in the engineering simulator before being released for training; software modifications can be checked against baseline models. Engineering simulation has possibly provided the most significant advances in aircraft design in the last 20 years, integrating aerodynamics, flight dynamics and structures within a common design framework.

1.8.6 Aptitude Testing

The qualities needed for a pilot include hand-eye coordination, motor skills, spatial awareness, interpersonal communication skills and judgement and management skills. Many training programmes are expensive and consequently, there are considerable benefits in detecting potential deficiency in pilot skills prior to the commencement of any training. Aptitude testing equipment has been developed to isolate these skills and identify areas where a trainee pilot is likely to encounter learning difficulties. In programmes where there is a surfeit of applicants, failure in aptitude tests can result in an immediate rejection from the course. In other programmes, it may identify possible training problems leading to alternative training routes. The overall aim is to use low-cost aptitude testing to detect faults well before they would be identified in a training programme.

An aptitude testing device typically includes a set of inceptors, for example, joysticks, knobs and buttons and a computer screen. For motor skills, the applicant may track a small target or balance a small beam. The input–output relationships are designed to test the applicant's perception and response, for example, by introducing lag or nonlinearities into the inceptor. For cognitive skills, the applicant's response times to tests are measured. Several of these tests may be designed to assess the applicant's performance in situations of high work load. However, the tasks performed by an airline pilot in a low visibility approach, say, or a military helicopter pilot hovering behind a ship in high winds and heavy sea swell, say, are difficult to replicate in an aptitude test where the applicant may have no flying experience. Although aptitude testing may identify potential problems, and considerable effort has been expended by psychologists in devising effective aptitude tests, the concern is that a problem may not be detected in an aptitude test or that a competent pilot may be lost at this early stage. Of course, it is possible to track the performance of pilots in training and correlate this with their rating in aptitude tests although such analysis is not viable for applicants failing an aptitude test.

1.8.7 Computer-based Training

With the availability of high performance computers in flight training, it is not clear how the training syllabus is best presented. Traditionally pilots have been trained by a combination of ground school and airborne training. The flight simulator provides a link between these two elements of training; a pilot is able to undertake training in a simulator to replace training in an aircraft. CBT extends this concept into the ground school training. Recent trends in computer animation and audio-visual computing enable diagrams, photographs, video clips and sound tracks to be organized as a lesson in front of a screen rather than in the classroom with an instructor.

Consider the engine-start procedures for a modern military aircraft. A pilot must learn the 50 or so tasks so that they can be executed without error in a minimum time. The tasks involve switching on pumps, selecting valves, checks on fuel flow rates, waiting for temperature or pressure rises and detection of fault conditions. CBT is positioned midway between practising these tasks on a flight simulator and remembering checks from a chart. The trainee pilot is presented with a screen which can show fuel selectors and flows; the pilot can use a mouse to turn a selector or switch on a pump, where the switch moves and the flow is displayed in a diagram; the engine instruments can be displayed to indicate pressure or temperature just as they would in the aircraft; any interaction by the pilot is matched to an audio track appropriate to the aircraft systems.

CBT now extends to all phases of flight training for both civil and military organizations, covering engine systems, cockpit drills, navigation systems, radio panel operation and FMS training.

However, the intention is not simply to replicate these aircraft systems on a screen. Rather, the lesson can be organized in a linear and logical manner, presenting information at different levels supported with diagrams, animation, photographs and video clips. There is one further extension to CBT which has underpinned its growth; the lesson plans are designed in a structured way where the trainee's interaction and responses are monitored. Not only is the rate of progress available to the instructor, but the students can learn at their own pace, being directed to previous material where there is a lack of understanding. One further extension of this approach is to make training available via the Internet. For example, a CBT lesson on FMS operation can provide an interactive training programme but if the flight is activated, then a server provides the correct flight equations to stimulate the laptop FMS. CBT has seen an increasing use in both airline and military training where it can provide very effective training for procedural tasks. The main problem with CBT systems is the cost and time to develop material. Ratios of 50 : 1 are quoted, depending on the lesson content where it can take 50 hours to produce a 1-hour lesson.

1.8.8 Maintenance Training

Advances in CBT have extended to maintenance training. Rather than dismantle a complete gas turbine engine, the student can interact with images of the engine, selecting a bolt or hose to remove. Inspection of a filter can show evidence of contamination requiring further disassembly. The order and sequence of removal and installations is presented in a logical sequence and fault finding methods are monitored to check on the student's fault diagnosis. In addition, the instructor can monitor a student's progress and deliberately inject fault conditions and failures. The CBT models provide a degree of dynamic fidelity so that test equipment can be connected interactively where observed readings correspond to the system characteristics.

This approach is particularly useful in training electronics technicians. Test equipment can be simulated so that a voltmeter or oscilloscope will give appropriate readings when connected to test points and any fault conditions are correctly represented by the measured conditions. Such ideas extend to facsimile equipment where, for example, a replica of a circuit board can be provided. Although the circuit board is inert and has no functioning electronics, the position of a probe can be detected and display the appropriate signal. The economic benefit of training with synthetic systems is clear:

- The equipment cannot sustain any damage in the training role;
- The cost of synthetic equipment is significantly less than the equivalent aircraft system;
- Potentially major faults can be injected by the instructor;
- The student's progress in failure detection and repair can be monitored by the computer system and provide a record of progress for the instructor.

With modern aircraft, these concepts of maintenance accord with the aircraft systems found on the aircraft, where faults are recorded in flight and downloaded by maintenance crews who undertake fault detection and correction at the board level to facilitate a fast turnaround of the aircraft. Much of the information provided by the onboard data logging equipment enables maintenance crews to rectify the failure quickly and reliably. Exactly the same system software is used in the maintenance trainer to provide sufficient experience of fault finding in a synthetic environment for maintenance crews.

References

Adorian, P., Staynes, W. and Bolton, M. (1979) The evolution of the flight simulator. *RAeS Conference Fifty Years of Flight Simulation*, London.

AGARD (Advisory Group for Aeronautical Research and Development). (1980) Fidelity of Simulation for Pilot Training, Advisory Report AR-159, Flight Mechanics Panel.

Allen, L. (1993) Evolution of flight simulation. *AIAA Conference Flight Simulation and Technologies, AIAA-93-3545-CP*, Monterey.

Allerton, D.J. (1996) Avionics, systems design and simulation. *Aeronautical Journal*, **100**(1000), 439–448.

Allerton, D.J. (1999) The design of a real-time engineering flight simulator for the rapid prototyping of avionics systems and flight control systems. *Transactions of the Institute of Measurement and Control*, **21**(2/3), 51–62.

Allerton, D.J. (2000) Flight Simulation–past, present and future. *Aeronautical Journal*, **104**(1042), 651–663.

Allerton, D.J. (2002) The case for flight simulation in general aviation. *Aeronautical Journal*, **106**(1065), 607–618.

Allerton, D.J. and Ross, M. (1991) Evaluation of a part-task trainer for Ab Initio pilot training. *RAeS Conference Training Transfer*, London.

Anon. (1992) *International Standards for the Qualification of Airplane Flight Simulators*, The Royal Aeronautical Society, London.

Ashworth, W.R., Mckissick, B.T. and Parrish, R.V. (1984) Effects of motion base and G-seat cueing on simulator pilot performance. NASA Technical Paper, No. 2247, NASA, Langley Research Centre.

Baarspul, M. (1990) A review of flight simulation techniques. *Progress in Aerospace Sciences*, **22**, 1–20.

Barrette, R.E. (1986) Flight simulator visual systems–an overview. *SAE Aerospace Technology Conference*, Long Beach.

Beal, T.R. (1993) Digital simulation of atmospheric turbulence for Dryden and Von Karman models. *Journal of Guidance Control and Dynamics*, **16**(1), 132–138.

Bell, H.H. and Waag, W.L. (1998) Evaluating the effectiveness of flight simulators for training combat skills: a review. *International Journal of Aviation Psychology*, **8**(3), 223–242.

Buffet, A.R. (1980) The visual perception of depth using electro-optical display systems. *Displays*, **2**, 1980.

Burns, A. and Wellings, A. (2001) *Real-time Systems and Programming Languages*, Pearson Education, Harlow.

Caro, P.W. (1973) Aircraft simulators and pilot training. *Human Factors*, **15**, 502–509.

Caro, P. (1988) Flight training and simulation, *Human Factors in Aviation*, Academic Press, San Diego, CA.

Farmer, E., Riemersma, J., Van Rooij, I. and Moraal, J. (1999) *Handbook of Simulator-based Training*, Ashgate.

Haward, M. (1910) The sanders teacher. *Flight*, **10**, 1006–1007.

Hays, R.T. (1992) Flight simulator training effectiveness: a meta-analysis. *Military Psychology*, **4**(2), 63–74.

Hays, R.T. and Singer, M.J. (1989) Simulation fidelity, *Training System Design*, Springer-Verlag, New York.

Howard I.P. (1986) The vestibular system, in *Handbook of Perception and Human Performance: Sensory Processes and Perception* (eds K.R. Boff L. Kauffman and J.P. Thomas), John Wiley & Sons, New York.

Link, E.A. Jr. (1930) U.S. Patent No. 1825462.

Lintern, G., Roscoe, S.N., Koonce, J.M. and Sega, L.D. (1990) Transfer of landing skills in beginning flight training. *Human Factors*, **32**, 319–327.

Moore, G.E. (1965). Cramming more components onto integrated circuits. *Electronics*, **38**(8), 114–117.

Povenmire, H.K. and Roscoe, S.N. (1971) Incremental transfer effectiveness of a ground based general aviation flight trainer. *Human Factors*, **13**(2), 109–116.

Rolfe, J.M. and Bolton, M. (1988) Flight simulation in the Royal Air Force in the Second World War, paper No. 1560. *Aeronautical Journal*, **92**, 315–327.

Rolfe, J.M. and Caro, P.W. (1982) Determining the training effectiveness of flight simulators: some basic issues and practical development. *Applied Ergonomics*, **13**(4), 243–250.

Rolfe, J.M. and Staples, K.J. (1986) *Flight Simulation*, Cambridge University Press, Cambridge, England.

Schachter B.J. (ed.) (1983) *Computer Image Generation*, John Wiley & Sons, New York.

Spooner, A.M. (1976) Collimated displays for flight simulation. *Optical Engineering*, **15**(3), 215–219.

Stewart, D. (1965) A platform with six-degrees-of-freedom. *Proceedings of the Institution of Mechanical Engineers*, **180**, (Part 1, 5), 371–386.

Taylor, H.L. (1985) Training effectiveness of flight simulators as determined by transfer of training. *NATO DRG Panel VIII Symposium*, Brussels.

Telfer R.A. (ed.) (1993) *Aviation Instruction and Training*, Ashgate, Aldershot

Thompson, D.R. (1989) Transfer of training from simulators to operational equipment–are simulators effective? *Journal of Education Technology Systems*, **17**(3), 213–218.

Tomlinson, B.N. (1975) Developments in the Simulation of Atmospheric Turbulence, Royal Aircraft Establishment Technical Memorandum FS-46, Farnborough.

Turner, L.W.F. (1913) *Teaching flying, Flight*, **5**(12), 326–328.

Waag, W.L. (1981) *Training Effectiveness of Visual and Motion Simulation*, AFHRL-TR-79-72, USAF Human Resources Laboratory, William Air Force Base, TX.

White, A.D. (1989) G-seat heave motion cueing for improved handling in helicopters. *AIAA Conferecne Flight Simulation Technologies*, Boston.

Winter, D. (1982) *The First of the Few: Fighter Pilots of the First World War*, University of Georgia Press, Georgia.

2

Principles of Modelling

2.1 Modelling Concepts

Man has been intrigued by models. It could be argued that much of mathematics is an abstract modelling notation to better understand the world[1]. In the nineteenth century, the Victorians constructed elaborate mechanical models to depict the motion of the planets. Meccano and Lego have fascinated generations of children. Model aeroplane and locomotive clubs flourish in many towns. While modelling is often used for recreation and pleasure, it is also used to aid conceptual design. It is hard to believe that the Egyptians constructed the Pyramids or the ancient Britons built Stonehenge without making models to devise or practise construction methods. Wave tanks have been used to understand ship motion and to investigate the stability of ships in different sea states. Nowadays, an architect will produce a scaled model, using a computer-aided design (CAD) package, during the planning stage of a new building. Circuit designers simulate the electrical characteristics of logic gates to ensure that an integrated circuit will function correctly, in order to eliminate design errors prior to the costly fabrication process.

This recent role for simulation, to support engineering design, is increasingly important. Computers are used in the design, validation and assessment of prototype systems, providing insight into the performance and behaviour of a system in the early phase of development. Synthetic design environments enable a range of design options to be analysed; for the designer, 'what-if' studies can be made without committing to any technology or building any hardware or alternatively, comparative designs can be assessed to provide insight into the effectiveness and limitations of a new design. Errors detected in the early stages of design cost significantly less to eradicate than errors found in the latter stages of development.

The use of simulation tools in engineering design differs from the use of simulation in flight training. First, an engineering simulator may not need to operate in real-time; typically a few minutes of flow modelling using a CFD package can take several weeks of computation, yet it still provides valuable insight into the behaviour of an aircraft. Secondly, a more basic simulator may be adequate; for example, the development of a weather radar model may not require a visual system or a motion platform. These simulation tools are often used to prove the feasibility or practicality of a design, based on analysis of the data generated in simulation.

The basis of an engineering simulator is very similar to a training simulator. It must provide an accurate mathematical model of the system (Smith, 1979) it simulates. The model may be based

---

[1] In *Notes of Lectures on Molecular Dynamics and Wave Theory of Light*, Lord Kelvin (arguably one of the greatest physicists on the nineteenth century) wrote, 'I am never content until I have constructed a mechanical model of what I am studying. If I succeed in making one, I understand: otherwise I do not'.

Principles of Flight Simulation   D. J. Allerton
© 2009, John Wiley & Sons, Ltd

on an understanding of the physics of the actual system or be derived from data collected from the system, in order to understand the system characteristics. The aim is to replicate the behaviour of a system in a simulator, so that the set of input and outputs relationships is identical for both the system and its simulation model. In other words, if a specific set of inputs to the system produce a particular set of outputs, the same inputs would generate identical outputs in the simulation.

This replication of the complete behaviour of a system is at the very heart of simulation. The model must match the system behaviour over the full operating range and under all conditions. For many systems, where there may be several hundred inputs and outputs and wide variations in the behaviour of the system throughout its operational envelope, it is far from straightforward to derive a model to fulfil this criteria or to validate the model. For many systems, the behaviour of the system in certain operating states may not be known; there may be insufficient experimental data to understand the behaviour or it may be too dangerous or expensive to acquire data. For example, there is no perceived need to undertake inverted flight in commercial transport aircraft flight trials. Yet, there may be a case to train pilots to cope with handling extreme attitudes caused by major atmospheric disturbances. Without any data or a model of the aircraft dynamics in inverted flight, it is difficult to derive an acceptable model or to confirm its behaviour. If an inaccurate model is subsequently developed it may result in an inappropriate pilot response in the aircraft, while the same action in a simulator is acceptable.

For some systems, it may be possible to obtain data from tests and then develop a model that complies with the system behaviour in these tests. In other words, the designer produces a model that mimics the system behaviour in meeting the same input and output relationships identified by the test data. This situation is particularly valid for discrete event models, where the generation and testing of realistic inputs is important in characterizing the system behaviour. For example, consider a supermarket chain building a new supermarket. How many checkout stations should they provide? If they provide too few, the customers will have to endure long delays at the checkout and may take their custom to a competitor. If too many checkout stations are installed, the cost of providing checkout staff may be high, reducing their profits. It would be impractical to experiment at an actual supermarket. However, a simulation study might provide insight into the problem. The supermarket could commission studies of the customer profiles, in terms of the variations of customer arrival rates during the day, including seasonal and daily variations. From knowledge of sales, they could simulate the time spent shopping by customers and implement an algorithm to replicate the selection of a checkout by customers (possibly the one with the shortest queue). Finally, they would need to simulate the time taken to pass through the checkout, which might vary with the experience of the checkout staff. Clearly, nobody understands the array of human processes occurring during shopping, but these sets of random processes, together with probabilistic constraints, could provide a model to enable the supermarket to optimize their construction.

Alternatively, the designer may have a good grasp of the underpinning science but lack any useful data. For example, in the Apollo space programme, there was no actual data of a landing on the moon and all the emphasis had to be given to modelling the spacecraft dynamics as accurately as possible. Nowadays, the first flight of a new aircraft is likely to have been extensively simulated. Data derived from similar aircraft, offline simulations or wind tunnel tests is used to derive the model and the flight tests are used to validate the model.

One common physical characteristic in many systems is that the state of a system varies with time. For example, the rate at which a cup of tea cools depends on the temperature of the tea in the cup. This system follows Newton's law of cooling which states that the rate of loss of temperature of an object is proportional to the difference between its own temperature and the ambient temperature. In mathematical terms the system can be represented by a simple differential equation:

$$\frac{dT}{dt} = -k(T - T_a) \qquad (2.1)$$

where T is the temperature of the liquid, t is time, k is a positive constant and T_a is the ambient temperature. The solution of this equation shows that the temperature decays exponentially with time. If k and T_a are known, the solution of this equation enables the temperature to be predicted as a function of the initial temperature and time.

Many mathematical techniques are used to solve differential equations and analytic methods have been developed and applied successfully to a wide range of differential equations. However, there are also sets of differential equations which defy analytic solution or where a numerical approximation can provide an acceptable answer. We will see that aircraft dynamics is underpinned by differential equations and that sufficiently accurate and robust numerical methods can be exploited to solve these equations, bearing in mind that the solution may need to be computed in a few milliseconds in a real-time flight simulator.

2.2 Newtonian Mechanics

Newton's laws of motion apply to rigid bodies, including aeroplanes, near the surface of the earth. His second law states that the acceleration of a body is proportional to the force that is applied to the body:

$$f = ma \qquad (2.2)$$

where f is the applied force (N), a is the resultant acceleration (m/s^2) and m is the mass of the body (kg). For example, the force of the oarsmen in a rowing boat, pulling in unison on the oars, will cause a boat at rest to accelerate.

Knowing the applied force, it is straightforward to compute the acceleration. Although the mass may not be constant, if the force and mass can be measured (or predicted), then the acceleration can be computed. Newton's law applies to both linear motion and angular motion. For angular motion, the relationship between angular force (known as *torque*) and angular acceleration is given by:

$$T = Ja \qquad (2.3)$$

where T is the applied torque (Nm), J is the moment of inertia (kg m^2) in the axis of rotation of the body and a is the resultant angular acceleration (radian/s^2). For example, a large electrical motor provides the necessary torque to turn a fairground ride, resulting in an angular acceleration of the ride.

Driving a motor car, force is applied (from the combustion of petrol) to the wheels of the car (via the gearbox and engine) causing the car to accelerate or alternatively, force is applied to the brake pedal (increasing the friction between the frame and the wheels) to cause the car to decelerate. Although we drive a car by the continuous application of forces, we control a car by reference to its speed (velocity) and position. Fortunately, it is straightforward to derive velocity and position from acceleration.

$$v = \int a \, dt \qquad (2.4)$$

$$p = \int v \, dt \qquad (2.5)$$

where v is the velocity, p is the position, t denotes time and $\int$ is the integration operator.

In some cases, it may be possible to integrate the functions a and v directly. In other cases, these functions may not take a standard form and integration is then achieved by a numerical approximation method. Consider a simple example: a cannonball is fired horizontally at 500 m/s

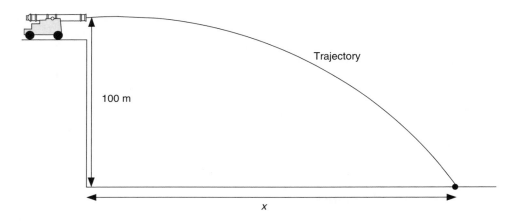

Figure 2.1 Trajectory of a cannonball

from a cliff 100 m above the sea, as shown in Figure 2.1. How far from the cliff will the cannonball impact the sea and how long will it take before it splashes into the sea?

Once the cannonball leaves the barrel, it is only subjected to a gravitational force, causing it to fall while the cannon maintains its horizontal velocity of 500 m/s. Ignoring any frictional forces on the cannonball and assuming that the mass of the cannonball and the gravitational acceleration g remain constant, consider the vertical component of acceleration:

$$v = \int g \, dt = gt + v_0 \tag{2.6}$$

$$p = \int v \, dt = \frac{1}{2} gt^2 + v_0 t + p_0 \tag{2.7}$$

where v_0 is the initial vertical velocity and p_0 is the initial height (referenced to sea level, say). Assuming $v_0 = 0$ (as the cannonball is fired horizontally), $g = 9.81$ m/s^2 and recalling that $p_0 = 100$ m, then

$$t = \sqrt{\frac{2p}{g}} = \sqrt{\frac{200}{9.81}} = 4.52 \text{ s} \tag{2.8}$$

Consider the horizontal component of acceleration.

$$v = \int a \, dt = v_0 \tag{2.9}$$

$$p = \int v \, dt = v_0 t + p_0 \tag{2.10}$$

where a is the acceleration after the cannonball leaves the barrel, $a = 0$ (no further force is applied), $v_0 = 500$ m/s and p_0 can be assumed to be zero, as we are measuring horizontal distance relative to the cliff face.

$$p = v_0 t = 500 \times 4.52 = 2260 \text{ m} \tag{2.11}$$

In this example, simulation of the dynamics of the cannonball is trivial. Knowing the initial conditions, the resultant velocity and position of the cannonball is simply a function of t, the time

Principles of Modelling

of transit. In practice, the forces on the cannonball may vary as a function of several variables, possibly in a non-linear manner. Nevertheless, if we could measure (or estimate) the forces on the cannonball at any instant *and* if we could approximate the integration operation, then we could derive the resultant velocity and position of the cannonball, to a degree of accuracy that it acceptable.

A function of a variable x is shown in Figure 2.2. The derivative of a function is its instantaneous gradient and its integral corresponds to the summation of the function over a specific range, which is given by the area 'under' the function.

To compute the integral of the function over a specific range, the area of each strip can be computed. For a simple approximation, if the strips are assumed to be rectangular, the area of each strip is given by its width multiplied by its height $F(x)$. The accuracy of this approximation to integration increases as the strips become narrower. This form of integration is simple to implement and provided that care is taken to ensure that the intervals are not too coarse, it can provide an acceptable means of performing numerical integration. In Section 2.6, it will be shown that this simple method can be improved. However, for now, the following procedure can be used to implement the integration operation $y = \int x \, dt$

```
void Integrate(float *y, float x) /* 1st order forward Euler */
{
   *y = *y + x * h;
}
```

where h denotes the integration interval.

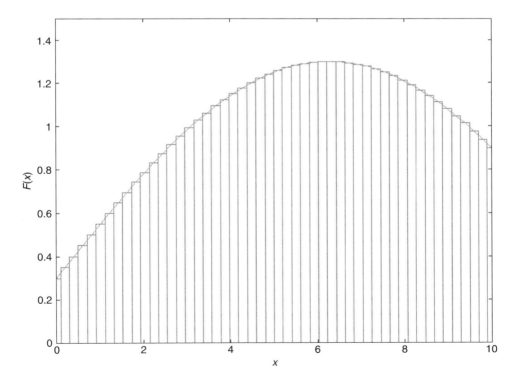

Figure 2.2 Integration of a function of a variable

Returning to the previous example of the cannonball, assume that the integration step is 1 s. The trajectory of the cannonball is computed in Example 2.1.

Example 2.1 Cannonball trajectory simulation

```
/* cannonball-1.c
   dynamic model of a cannonball trajectory
   step length = 1.0 */

#include <stdio.h>
#include <stdlib.h>

#define h    1.0
#define g    9.81

static float Az, Vz, Pz; /* vertical accn (m/s^2), vel (m/s), position (m) */
static float Ax, Vx, Px; /* horizontal accn (m/s^2), vel (m/s), position (m) */
static float t;          /* time (secs) */

void Integrate(float *y, float x);

void Integrate(float *y, float x) /* simple 1st order Euler */
{
    *y = *y + x * h;
}

int main(int argc, char *argv[])
{
    Az = -g;    /* initial vertical accn */
    Vz = 0.0;   /* initial vertical vel */
    Pz = 100.0; /* initial height */
    Ax = 0.0;   /* initial horizontal accn */
    Vx = 500.0; /* initial horizontal vel */
    Px = 0.0;   /* initial horizontal distance */
    t  = 0.0;   /* start from time t=0 */

    do
    {
        printf("%2d %8.2f %8.2f %8.2f %8.2f\n",
                (unsigned int) t, Vz, Pz, Vx, Px);
        t = t + h;
        Integrate(&Vz, Az);
        Integrate(&Pz, Vz);
        Integrate(&Vx, Ax);
        Integrate(&Px, Vx);
    } while (Pz >= 0.0);
}
```

The results are shown in Table 2.1.

Although it is evident that the cannonball splashed into the sea after 4 s, at a distance of over 2000 m from the cliff, with a step length of 1 s the resolution is too coarse to predict the distance travelled or the time of travel to an acceptable accuracy. Repeating the same computation, but reducing the step length from 1 to 0.01 s, the fragment of results shown in Table 2.2 was obtained.

Principles of Modelling

Table 2.1 Cannonball trajectory with a step length of 1.0

T	Vz	Pz	Vx	Px
0	0.00	100.00	500.00	0.00
1	-9.81	90.19	500.00	500.00
2	-19.62	70.57	500.00	1000.00
3	-29.43	41.14	500.00	1500.00
4	-39.24	1.90	500.00	2000.00

Table 2.2 Cannonball trajectory with a step length of 0.01

T	Vz	Pz	Vx	Px
4.45	-43.65	2.65	500.00	2225.00
4.46	-43.75	2.21	500.00	2230.00
4.47	-43.85	1.77	500.00	2235.00
4.48	-43.95	1.33	500.00	2240.00
4.49	-44.05	0.89	500.00	2245.00
4.50	-44.14	0.45	500.00	2250.00
4.51	-44.24	0.01	500.00	2255.00

The improvement in the accuracy of the solution is clear. Figure 2.3 shows the respective trajectories for step lengths of 1 s and 0.01 s. Not only is the final result more accurate, but the trajectory of Figure 2.3b lacks the discontinuity in Figure 2.3a. By implication, the step length can be reduced until the improvement in accuracy is within an acceptable tolerance.

Consider a second example, where the forces on a body are non-linear. A skydiver jumps from an aeroplane – can we determine the altitude and free fall velocity of the skydiver? In the previous example, the cannonball would have continued indefinitely at 500 m/s (until it splashed into the sea). In reality, the friction of the air would slow the cannonball. Similarly, the gravitational force on the skydiver is countered by the air resistance of the body of the skydiver, which increases with the velocity of the skydiver. Eventually, this drag will result in an upward force that balances the downward gravitational force, giving a resultant net force of zero. At this point, no acceleration is applied to the skydiver and the skydiver will maintain a steady-state (or terminal) velocity.

From basic aerodynamics, the drag of the skydiver is given by the following equation:

$$drag = \frac{1}{2}\rho v^2 S C_d \quad (2.12)$$

where ρ is the density of air, v is the velocity, S is the surface area and C_d is the coefficient of drag.

The complexity of the computation should be apparent. First, the density of air ρ changes with altitude. Secondly, the resultant drag varies with the square of velocity. Thirdly, the surface area of the skydiver will change as the skydiver manoeuvres. Finally, the coefficient of drag will depend on the friction of the material of the skydiver's jump suit. These four variables are non-linear so that an explicit solution to the equations would be difficult to derive. An alternative solution is to consider that these non-linear variables are not changing so that the equations are effectively linearized and could be solved as a series of linear equations, where the variables are recomputed over a small range where they are assumed to be constant.

An alternative method (and the one adopted in flight simulation) is to simulate the motion, rather than solving the equations directly. To simplify the problem, assume that $\rho = 1.225$ kg/m³,

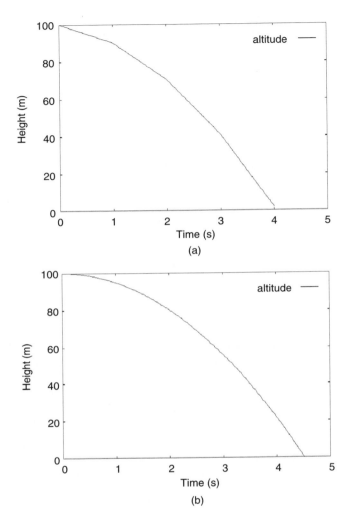

Figure 2.3 (a) Cannonball trajectory, $h = 1.0$. (b) Cannonball trajectory, $h = 0.01$

$S = 1.0 \, \text{m}^2$ and $C_d = 0.05$, so that the only non-linear term is v. The C program in Example 2.2 predicts the motion of the skydiver.

The fragment of the program output in Table 2.3 shows that the skydiver reaches a terminal velocity of 59.66 m/s and, without the help of a parachute, would impact the ground some 54.5 s after jumping from an aeroplane at an altitude of 3000 m (V denotes the velocity in metres per second and H denotes the altitude in metres). Figure 2.4 shows the increasing speed and reducing altitude as a function of time.

If the skydiver deploys a parachute at an altitude of 500 m, say, the coefficient of drag will increase to 1.5. Figure 2.5 shows the altitude of the skydiver as a function of time. In this case, the skydiver impacts the ground 147 s after leaving the aeroplane, but with an impact velocity of only 5 m/s.

Principles of Modelling

Example 2.2 Skydiver simulation

```
/* skydiver-1.c
   dynamic model of a skydiver
   no parachute */

#include <stdio.h>
#include <stdlib.h>

#define h      0.01     /* step length */
#define g      9.81     /* gravity */
#define Rho    1.225    /* air density */
#define S      1.0      /* surface area */
#define Cd     0.43     /* coeff of drag */
#define m      85.0     /* mass of the skydiver (Kg) */

static float Az, Vz, Pz; /* vertical accn (m/s^2), vel (m/s), position (m) */
static float Drag;       /* drag (N) */
static float t;          /* time (s) */

static void Integrate(float *y, float x);

static void Integrate(float *y, float x) /* simple 1st order Euler */
{
    *y = *y+x * h;
}

int main(int argc, char *argv[])
{
    Vz = 0.0;      /* initial vertical vel */
    Pz = 3000.0;   /* initial altitude */
    t  = 0.0;      /* start from time t=0 */

    do
    {
        printf("%8.2f %8.2f %8.2f\n", t, fabs(Vz), Pz);
        t = t+h;
        Drag = 0.5 * Rho * Vz * Vz * S * Cd;
        Az = Drag / m - g;
        Integrate(&Vz, Az);
        Integrate(&Pz, Vz);
    } while (Pz >= 0.0);
}
```

This approach to simulation is very powerful. If we are able to compute the forces on a body, we can then compute the behaviour of the body in terms of its linear and rotational motion. This technique is particularly useful if these terms are non-linear or where the terms are only known at the time of simulation. For the skydiver, aerial manoeuvres will change the skydiver's coefficient of drag. Similarly, for an aircraft, the pilot inputs to the control surfaces and engines, will occur during the simulation, which can be taken into account in computation of the dynamics of the aircraft.

Table 2.3 Skydiver trajectory (no parachute)

T	V	H
54.40	-59.66	5.59
54.41	-59.66	5.00
54.42	-59.66	4.40
54.43	-59.66	3.80
54.44	-59.66	3.21
54.45	-59.66	2.61
54.46	-59.66	2.01
54.47	-59.66	1.42
54.48	-59.66	0.82
54.49	-59.66	0.22
54.50	-59.66	-0.37
54.51	-59.66	-0.97
54.52	-59.66	-1.57
54.53	-59.66	-2.16
54.54	-59.66	-2.76
54.55	-59.66	-3.36
54.56	-59.66	-3.95
54.57	-59.66	-4.55
54.58	-59.66	-5.15
54.59	-59.66	-5.74
54.60	-59.66	-6.34

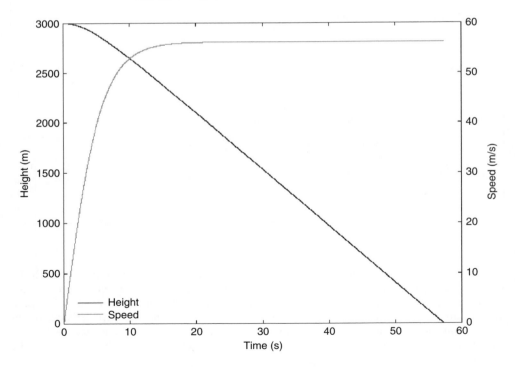

Figure 2.4 Height/speed plot of the skydiver (no parachute)

Principles of Modelling

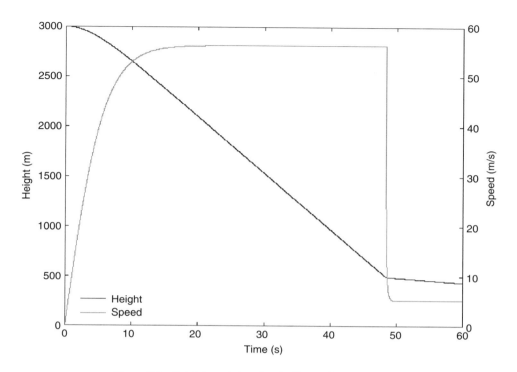

Figure 2.5 Height/speed plot of a skydiver (with a parachute)

2.3 Axes Systems

In simulation, forces occur in a frame of reference where it is convenient to measure or estimate the applied forces whereas the resultant motion may be required in another frame of reference. In the previous example, it was straightforward to compute altitude and velocity of the skydiver relative to the ground. But equally, we could have measured these variables relative to the position of the aircraft or even the distance and bearing from the North Pole. Consider an aircraft taxiing; as the brake or nose-wheel steering is applied to the aircraft, these forces will deform the rubber tyres which are attached to the wheel hubs which, in turn, are attached to oleos attached to the undercarriage assembly attached to the fuselage. These forces will act through the attachment points of the undercarriage assembly leading to motion of the aircraft which needs to be resolved at the centre of gravity of the aircraft to determine the resultant motion of the aircraft. In practice, the effect of these forces may also be needed at the pilot station which is forward of the centre of gravity and, in the case of a transport aircraft, is displaced from the centre line of the fuselage.

In this example, knowledge of the runway friction, the characteristics of the rubber tyres, the brakes and the mass and velocity of the aircraft may provide a good model of the forces applied to an individual aircraft wheel. However, these forces must be transformed through several axis systems in order to compute the longitudinal and lateral forces experienced by a pilot at the front of the fuselage.

Consider an example of a robot in a warehouse, where the robot moves around the warehouse to locate items to be despatched from rows of shelving. The robot steers around the aisles to move to a position (x, y) in order to select an item, where x and y denote coordinates in the frame of the warehouse floor, as shown in Figure 2.6.

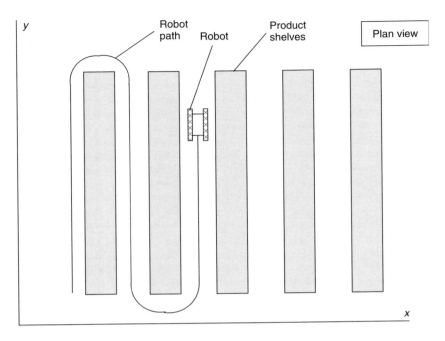

Figure 2.6 Warehouse floor axes

Assume that the speed of rotation of the motors turning the toothed wheels driving the caterpillar tracks can be measured. If both motors turn at equal speed, the robot steers in a straight line with a velocity proportional to the angular rate of rotation of the motors. If one motor turns slower than the other, the robot can steer to the left or right, at a turn rate proportional to the difference in the speed of the motors. At any time, the vehicle will have a forward speed and a turn rate that is a function of the speed of the two motors. In other words, the velocity and turn rate can be measured in the frame of the robot. However, the robot needs to move relative to the frame of the factory floor and therefore, in order to control the robot, it is necessary to know the position and heading of the robot, with respect to the warehouse, as shown (in plan view) in Figure 2.7.

The resultant velocity in the frame of the warehouse floor can be computed by resolving the components of the velocity v in the x and y directions, given by:

$$\dot{x} = v \sin \theta \qquad (2.13)$$

$$\dot{y} = v \cos \theta \qquad (2.14)$$

where θ is the heading of the robot relative to 'north' (the y axis of the warehouse floor) and v is the velocity of the robot. θ is derived from the integral of the turn rate.

Subsequently, the position of the robot can be computed by integrating these velocities to give a coordinate (x, y) in the frame of reference of the floor. This simple example illustrates a technique that is commonly used in the simulation of vehicle dynamics. Forces, accelerations and velocities are determined in the most convenient axis systems (often the vehicle frame) and then resolved in the frame where the vehicle dynamics is most easily understood (often a navigation frame). The main requirement is to establish the rotation of the vehicle frame (usually in three dimensions) with respect to a desired frame of reference. In practice, there may be several frames of reference

Principles of Modelling

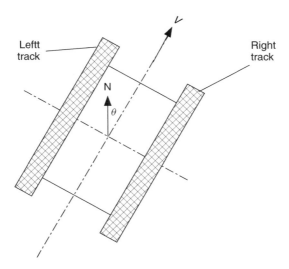

Figure 2.7 Body frame of the robot

but the process outlined above, can be applied through a series of transformations to determine the resultant motion in the frame of an observer.

There is one further important consideration with axes systems, that is, the concept of an inertial frame. We tend to think of the earth as a fixed frame. In fact the earth is a rotating frame, revolving around its axis every 24 hours; it is also rotating about the sun which in turn is moving with respect to distant stars. In flight simulation, it is convenient to consider the earth as a fixed frame of reference. For flights of a relatively short duration or distance, the effect of the earth's rotation can be neglected. However, the Coriolis[2] effect, that is, reference frames rotating with respect to each other should be included in the computation of aircraft dynamics. These corrections are applied in aircraft navigation systems and strictly, should be included in the aircraft equations of motion in a flight simulator.

2.4 Differential Equations

Differential equations occur frequently in science and engineering. If the rate of change of a variable depends on other variables or the rate of change of other variables (their derivatives), these equations can be expressed as a set of differential equations. If a system can be described by a set of differential equations, then analysis of these equations can either predict an exact solution of the system behaviour or provide an insight into the expected behaviour. For example, an electrical circuit consisting of resistors, capacitors and inductors can be expressed as a set of differential equations. If a signal is applied to the circuit, the resultant gain, phase and stability of the circuit can be determined from inspection of these equations. A specific response to a square wave or a sine wave can be predicted. For example, resonance of a car suspension system can be predicted at known frequencies excited by the road surface, or components can be selected to provide sufficient

---

[2] In *Classical Dynamics of Particles and Systems*, Marion J. B. describes a naval engagement near the Falkland Islands in the early stages of the First World War. Although the Coriolis effect had been included by the Royal Navy in their gunnery calculations, the shells missed their targets by 100 yards until the compensation, which normally applied in the northern hemisphere, was reversed.

gain for a radio receiver, or an audio amplifier may become distorted above a certain frequency of input signal.

Not surprisingly, differential equations dominate flight dynamics and a substantial body of theory has been developed over the last 50 years to understand aircraft motion. Of course, pilot inputs are unpredictable – they depend on the information observed by the pilot and the pilot's mental and physical processes. Nevertheless, certain inputs, particularly a step or impulse input, can be represented in mathematical terms and provide insight into the behaviour or handling qualities of an aircraft. If the response is very fast, the aircraft will be agile and manoeuvrable but it may also be very demanding to fly in the presence of disturbances. If the response is slow, the aircraft is sluggish to respond to pilot demands and difficult to fly in situations where changes to the aircraft flight path are needed quickly, for example, in the final stages of an approach. By understanding aircraft behaviour in straight and level flight, in turning, climbing and descending flight and in response to pilot inputs, it is possible to design an aircraft with known and desirable handling characteristics or to design a control system to guarantee specific handling qualities.

By formulating the equations of motion of an aircraft, the longitudinal and lateral dynamics of an aircraft can be simulated by modelling the underlying differential equations. These equations are developed and formulated in Section 3.6 but can be summarized by the following set of equations:

$$\dot{u} = f_u(F_u, m, \theta, \phi, \psi) \tag{2.15}$$

$$\dot{v} = f_v(F_v, m, \theta, \phi, \psi) \tag{2.16}$$

$$\dot{w} = f_w(F_w, m, \theta, \phi, \psi) \tag{2.17}$$

$$u = \int \dot{u}\, dt \tag{2.18}$$

$$v = \int \dot{v}\, dt \tag{2.19}$$

$$w = \int \dot{w}\, dt \tag{2.20}$$

$$\dot{p} = f_p(M_p, J, \theta, \phi, \psi) \tag{2.21}$$

$$\dot{q} = f_q(M_q, J, \theta, \phi, \psi) \tag{2.22}$$

$$\dot{r} = f_r(M_r, J, \theta, \phi, \psi) \tag{2.23}$$

$$p = \int \dot{p}\, dt \tag{2.24}$$

$$q = \int \dot{q}\, dt \tag{2.25}$$

$$r = \int \dot{r}\, dt \tag{2.26}$$

The linear and angular accelerations are computed in the body axes of the aircraft, with the origin at the centre of gravity. The implied orthogonal directions of the u, v and w components are forwards (to the nose), starboard (along the right wing) and downwards (perpendicular to the wings), respectively. Similarly, the p, q and r components refer to angular motion about these axes. The functions f are used to compute the linear and angular accelerations from the forces F and moments M and depend on the aircraft dimensions, mass m, inertias J, the control inputs, air density, airspeed, mach number, angles of incidence and so on. The mass and inertias can be computed from the aircraft weight, fuel, passengers and aircraft dimensions. The aircraft attitude is given by the three angles of orientation: pitch, roll and yaw (θ, ϕ and ψ). The derivation of these equations is described in Chapter 3.

This set of differential equations is solved to determine the linear and angular velocities of the aircraft in the body frame of the aircraft. The terms can then be resolved in an earth axis system to provide meaningful values of speed and position.

$$V_N = t(u,v,w,\theta,\phi,\psi) \tag{2.27}$$

$$V_E = t(u,v,w,\theta,\phi,\psi) \tag{2.28}$$

$$V_d = t(u,v,w,\theta,\phi,\psi) \tag{2.29}$$

$$P_N = \int V_N \, dt \tag{2.30}$$

$$P_E = \int V_E \, dt \tag{2.31}$$

$$P_d = \int V_d \, dt \tag{2.32}$$

$$\dot{\theta} = t(p,q,r) \tag{2.33}$$

$$\dot{\phi} = t(p,q,r) \tag{2.34}$$

$$\dot{\psi} = t(p,q,r) \tag{2.35}$$

$$\theta = \int \dot{\theta} \, dt \tag{2.36}$$

$$\phi = \int \dot{\phi} \, dt \tag{2.37}$$

$$\psi = \int \dot{\psi} \, dt \tag{2.38}$$

where the transformations t provide the appropriate mappings between the axes systems, V denotes the north, east and vertical velocities with respect to the earth and P gives the north and east aircraft position and the aircraft altitude. Similarly, the aircraft attitude (θ, ϕ and ψ) is derived from the body rates p, q and r.

These equations are over-simplified; additional axes systems may be used to formulate these equations (e.g., to include the forces of a helicopter rotor) and the functions, transformations and computations are complex. Furthermore, many of the equations are coupled and the inputs are not known until the simulator is running. Nevertheless, these sets of equations enable the aircraft motion to be solved. Clearly, classical methods of solving differential equations are impractical, as these equations include numerous non-linear terms, which are functions of several variables (which themselves may be non-linear).

The only practical option is to use numerical methods to approximate the integration operation. Although the equations underpinning these sets of differential equations are complex, including lookup tables for non-linear terms, trigonometric terms, rate limits and hysteresis, the real-time computations required for a modern flight simulator are well within the performance of an off-the-shelf PC.

There are two points to consider in selecting a numerical method:

- It must have sufficient resolution and accuracy that any errors it introduces are small in comparison with errors associated with the main terms of the model equations;

- The computations inherent in the numerical methods enable the simulator equations to be solved within the real-time frame rate.

Fortunately, standard numerical integration algorithms meet these requirements and enable the dynamics of aircraft to be easily simulated in high-level programming languages running on modest computer systems.

2.5 Numerical Integration

In the examples given so far, it has been assumed that the mathematical operation of integration[3] can be performed by a simple numerical method. Generally, in solving differential equations, the equations can be formulated as integral equations where either the output of an integration operation is the integral of the input or the input to the integration operation is the derivative of the output. Quite often, integration is shown on a system schematic with the symbol shown in Figure 2.8 where y is the integral of x. Note that an initial value is required for y and the independent variable is time. Normally, limits are not applied to the integration operator – the operation applies from the time of initialization to the current time in real-time simulation. In software, the mathematical shorthand $y = \int x \, dt$ is provided by a mathematical library procedure of the form:

```
integrate(*y, x)
```

Both the integrand x and the integral y are passed to the procedure, although in the case of the variable y, the parameter is a pointer (call by reference) to enable the integral variable to be updated in the procedure.

2.5.1 Approximation Methods

In real-time simulation, the implied differential equation is formed by the interconnection of integrators which simply integrate the input passed to the integration operator. In all numerical analysis, the integral of a variable with respect to time corresponds to the area under the curve representing $x(t)$ as shown in Figure 2.9. The area of each strip is given by its width, which is the sampling period and its height, given by the value of the function at that point in time. Notice that the sampling period in Figure 2.9b will provide a better approximation to the integral value than the coarser sampling period in Figure 2.9a.

In solving differential equations, it is possible that the function $x(t)$ can be mathematically integrated (Griffiths and Smith, 1991 and Murphy et al., 1998). However, in simulation, $x(t)$ is likely to be non-linear; inevitably it will not be in a mathematical form that can be integrated and will vary with flight conditions. Nevertheless, if the function is sampled at rates faster than the frequencies inherent in $x(t)$, it is possible to represent $x(t)$ as a series of discrete points and then

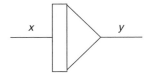

Figure 2.8 An integrator symbol

---

[3] God does not care about our mathematical difficulties. He integrates empirically. – Albert Einstein

Principles of Modelling

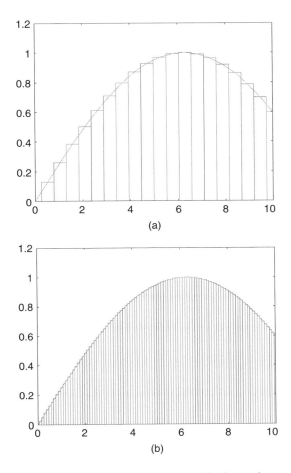

Figure 2.9 (a) Coarse integration step. (b) Fine integration step

compute a polynomial which fits through a number of these points. Integration of a polynomial in t is straightforward but the accuracy of the method depends on the sampling period and also the error in the approximation of the polynomial fit to the curve. This approximation is the basis of the multi-step integration methods used in numerical analysis. A large number of methods have been reported to improve the accuracy of a method or to reduce the computations in the solution (Benyon, 1968).

By choosing an acceptably small step length (or sampling period) and computing a reasonable estimate of the polynomial passing through the current and recent values of $f(x)$, it is possible to provide an integration function to an accuracy that enables the differential equations, which describe a dynamic system, to be solved. The choice of step length depends on the iteration rate of the real-time simulator, typically with values of 20 ms (50 Hz) or 16.6 ms (60 Hz) for flight simulation. In the case of the equations for the motion platform, where the actuator bandwidth may be as high as 400 Hz, a step length of 1 ms (1000 Hz) would be required. The step length is also limited by the processor speed needed for the real-time computations; computing at an iteration rate of 1000 Hz rather than 50 Hz is a 20-fold increase in processing speed. The choice of the order of the polynomial fit depends on the application and the required accuracy. The higher the order of the polynomial, the closer it approximates the function, but the computation of the terms of the polynomial in each integral computation introduces extra computation time into each frame. Consequently, there is a compromise between computation speed and computation accuracy in the selection of a numerical integration algorithm.

2.5.2 First-order Methods

A commonly used first-order method is Euler's forward method. Consider the function $x(t)$ shown in Figure 2.10. If the solution is computed at a fixed rate, the interval between t_n and t_{n+1} is h (for all n), then the value at x_n at time t_n is known and the value x_{n+1} at time t_{n+1} is required. Let f_n be the gradient at t_n, then

$$f_n = \frac{x_{n+1} - x_n}{h} \qquad (2.39)$$

where h is the step length, then

$$x_{n+1} = x_n + h f_n \qquad (2.40)$$

Clearly, this is an approximation; there is an error ε between the true solution at t_{n+1} and the computed value y_{n+1}. It can be shown that for the forward Euler method, this error is of the order

$$\varepsilon_{n+1} \approx \frac{h^2}{2} \qquad (2.41)$$

For the series of inputs to the integrator at time $t_n, t_{n-1}, t_{n-2}, t_{n-3} \ldots \ldots$, we can compute the next output of the integrator (at time t_{n+1}), simply by applying the formula at every step. In other words, the solution is executed frame by frame at the frame rate determined by the step length.

To illustrate this method, consider an example of a simple differential equation with a known solution:

$$y = \frac{dy}{dt} \qquad (2.42)$$

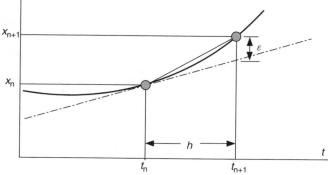

Figure 2.10 First-order Euler integration

Table 2.4 Approximate solution of $y = dy/dt$

x	y	e^t
0	1.0000	1.0000
0.01	1.0100	1.0101
0.02	1.0201	1.0202
0.03	1.0303	1.0305
0.04	1.0406	1.0408
0.05	1.0510	1.0513

The solution is $y = e^t$. Noting that $f_n = dy/dt = y_n$ and setting $h = 0.01$, then

$$y_{n+1} = y_n + hf_n = 1.01 y_n \qquad (2.43)$$

Table 2.4 gives the true value for e^t and the computed solution for y for the first six steps.

Notice how the error accumulates with time, which is of the order 0.0001 (h^2). This is an inherent problem with these methods; the error introduced by the numerical integration algorithm will continue to grow and can only be reduced by decreasing the step length or by moving to a higher order method. Although various improvements have been proposed to Euler's first-order method (Howe, 1989), if numerically accuracy is an issue, higher order integration methods must be used.

2.5.3 Higher-order Methods

Numerous methods have been proposed for the solution of differential equations. However, we will only consider multi-step explicit methods. These are explicit because they are based on previous values, whereas in implicit methods, the predicted value is used in the solution (it occurs on both sides of the equations). Multi-step methods simply 'march' through a solution step-by-step. These methods are particularly useful in simulation and modelling, because information about the states of the model at previous steps may be known and can be used to predict the current state. Although many methods exploit the advantage of a variable step length (to expedite the solution), such methods are impractical in real-time flight simulation, where the frame rate is constant.

There are a number of issues to consider in implementing multi-step methods:

- The methods are based on truncation of the Taylor series, which defines the order of the method; generally higher order methods are more accurate;
- Errors generated in a previous step are included in the computation of the current step and will accumulate. Although this limitation is unacceptable in the solution of an equation over a long period of time, in systems with feedback, numerical errors are attenuated as a result of the feedback mechanism;
- The step length of the method may affect the accuracy of the solution. Clearly, a very coarse integration step can lead to a very inaccurate solution and intuitively, it would seem that the smaller the step, the more accurate is the solution. However, in real-time systems, the solution is usually implemented at an iteration rate which is matched to the time constants of the system;
- The numerical method can introduce instability, which is mostly a function of the step length. Considerable care is needed to ensure that the numerical method chosen to implement a particular model, does not excite instabilities which are not in the model dynamics;

- The solution of the equations is computed to a known resolution, in terms of floating point accuracy. However, rounding errors in the computation of equations derived from series expansions (which are terminated at some implied resolution) can introduce noise into the solution. Generally, rounding errors can be ignored as their effects can be reduced by using extended precision arithmetic, when it is anticipated to be a problem.

The Taylor expansion is given by

$$y(t+h) = y(t) + hy^{(1)}(t) + \frac{1}{2}h^2 y^{(2)}(t) + \frac{1}{6}h^3 y^{(3)}(t) + \frac{1}{24}h^4 y^{(4)}(t) + \ldots \qquad (2.44)$$

where the term $y^{(n)}$ denotes the nth derivative of y.

Euler's forward method is obtained by dropping the terms of second and higher order to obtain

$$y_{n+1} = y_n + hf(t_k, y_k) \qquad (2.45)$$

In general, multi-step methods use only $y_{n-s+1}, y_{n-s+2}, y_{n-s+3} \ldots \ldots y_n$, where s is the order of the method (the number of previous points considered).

For $s = 2$

$$\dot{y}_n = f_n \qquad (2.46)$$

$$\ddot{y} \approx (f_n - f_{n-1})/h \qquad (2.47)$$

Truncating the Taylor series after the second derivative term,

$$y_{n+1} = y_n + h\dot{y}_n + \frac{1}{2}h^2 \ddot{y}_n \qquad (2.48)$$

$$y_{n+1} = y_n + \frac{h}{2}(3f_n - f_{n-1}) \qquad (2.49)$$

This is the well-known second-order Adams–Bashforth method. The integral is simply a function of the current integral value, the current integrand and the previous integrand. The method offers two advantages; first, the error is of the order h^3 in comparison with h^2 for the Euler method and secondly, the method is very simple to implement, requiring four additions, one subtraction and division by two (which can be implemented by an arithmetic shift in integer arithmetic).

The Taylor expansion can be extended to higher order methods, based on difference methods use to fit polynomials that pass through two or more of the previous points. For example, the fourth-order Adams–Bashforth method is given by

$$y_{n+1} = y_n + \frac{h}{24}(55f_n - 59f_{n-1} + 37f_{n-2} - 9f_{n-2}) \qquad (2.50)$$

In numerical methods, the computation of y_{n+1} can be treated as a predictor, enabling f_{n+1} to be computed to improve (or correct) the prediction. If Equation 2.50 is used as a predictor function, then a corrector expression can be applied (Wood, 1992)

$$y_{n+1} = y_n + \frac{h}{720}(251f_{n+1} + 646f_n - 264f_{n-1} + 106f_{n-2} - 19f_{n-3}) \qquad (2.51)$$

Principles of Modelling

The advantage of this method is that the error is of the order h^5 and for small values of h, the truncation error is less than the round-off errors. However, there are two drawbacks. First, there is a considerable increase in computation and secondly, when the computation starts, values for $f_n, f_{n-1}, f_{n-2}, f_{n-3}$ do not exist; a 'starter formula' (usually forward Euler) must be used to generate the first few points, before switching to the Adams–Bashforth method.

The system designer is faced with three compromises in selecting a numerical integration method:

- The speed of the method (if several thousand integrations are executed per second, the impact on processor throughput can be significant);
- The accuracy of the method;
- The stability of the method.

The effect of accuracy can be determined from knowledge of the step length, the truncation error term (given by the order of the method), the system structure and the number of steps to be executed. With improvements in computing speeds, higher order predictor-corrector methods can provide significant improvements in accuracy without affecting the real-time iteration rate (Howe, 1991). Cardullo et al., (1991) provide an overview of the popular numerical integration methods used in real-time simulation.

Stability is much harder to assess. It depends on the system dynamics, the order of the method and the step length (McFarland, 1997). One way to assess the impact of the numerical method in terms of stability is to apply it to a well-conditioned equation and then investigate the limits of the onset of instability. Typically, the first-order equation $\dot{y} = \lambda y$ is used to assess the stability of a selected method.

Consider Euler's method. The discrete form of the equation $\dot{y} = \lambda y$ is given by

$$y_{n+1} = y_n + h\dot{y}_n \qquad (2.52)$$

but $\dot{y}_n = \lambda y_n$, giving

$$y_{n+1} = y_n + h\lambda y_n = y_n(1 + h\lambda) \qquad (2.53)$$

After n steps, starting from y_0,

$$y_{n+1} = y_0(1 + \lambda h)^n \qquad (2.54)$$

In other words, the solution is stable if $\|1 + \lambda h\| < 1$. Assuming that λ can be complex, the stable region is shown in Figure 2.11.

If λ is real, then $h\lambda$ must lie in the interval $(-2, 0)$, or (assuming λ is negative), the step length h must satisfy the constraint $h \leq -2/\lambda$ for Euler's method to be stable. In practical terms, if the fastest time constant of a system is known, then the step length must be chosen to satisfy this constraint. In simulation and modelling, it is possible to simulate a stable system with a coarse step length leading to both inaccuracy and instability. This effect is illustrated by using Euler's forward method on a stable second-order system. Figure 2.12a, b and c shows the response to a step input with a step length of 0.01, 0.1 and 0.3 respectively.

Although the system still appears to be stable in increasing the step length from 0.01 to 0.1, the error in the solution is clearly seen. By further increasing the step length to 0.3, the onset of instability in the solution is evident.

One other method of assessing the stability (rate of growth of errors) in numerical integration algorithms is to solve the harmonic equation $\ddot{y} = -\omega_n y$. This equation can be simulated by two integration operations, as shown in Figure 2.13.

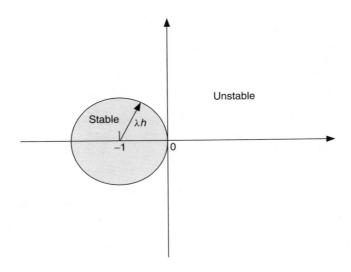

Figure 2.11 Stability of Euler integration

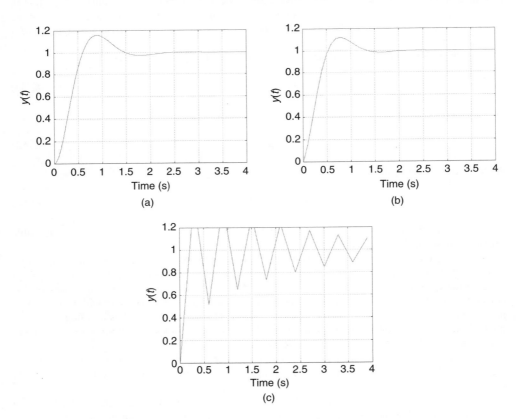

Figure 2.12 Euler integration of a second-order system

Principles of Modelling

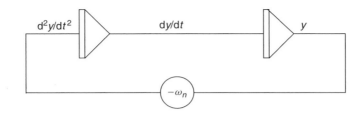

Figure 2.13 Simulation of a harmonic oscillator

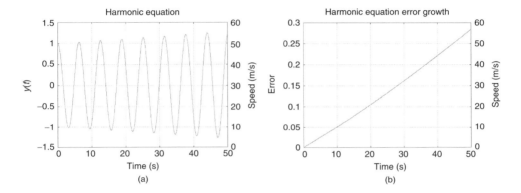

Figure 2.14 Error growth in Euler integration

A perfect oscillator will oscillate at ω_n rad/s at the amplitude given by the initial value of y. Using first-order forward Euler integration, with an initial value $y = 1.0$, the growth of error is apparent in Figure 2.14a, which plots y and Figure 2.14b, which plots the growth of the error, over 50 s.

Clearly, the use of a higher order method will reduce the rate of growth of the error, but cannot eliminate it. If an open-loop simulation is left to run for several hours of simulated time, the errors resulting from numerical integration can become large, to the point where they corrupt the solution. However, it is interesting to observe that such effects do not always arise in flight simulation. For example, if the integration method introduces errors in computing the bank angle, the pilot (or auto-pilot) will continuously correct the bank angle to the desired bank angle, in effect, eliminating the growth of the error.

2.6 Real-time Computing

Programming languages enable the designer to translate an algorithm to executable code. Usually, the main criterion for the software is that it is correct – it produces the correct output to a defined accuracy. However, there is a class of programmes, known as *real-time software* (Burns and Wellings, 2001) where, in addition to correctness, the software must compute its result within a specified time.

Real-time software is not simply fast software. The speed of execution depends on several factors:

- The processor architecture, particularly the clock cycle to extract, decode and execute instructions;
- The complexity of the algorithm being computed, which is likely to be determined by the number of machine instructions to be executed;

- The efficiency of the compilation process to translate a high-level description to machine instructions;
- Any delays in acquiring inputs to the algorithm, or transmitting outputs, particularly data which is read or written via a hardware interface.

For example, in an industrial process, bottles may be filled as they pass along a conveyor belt. The system depositing liquid in each bottle has a fixed period to execute its task. If, for any reason, the processing was delayed, a bottle may only be partially filled. Such a system would have to meet a stringent real-time requirement for the software. During the design phase, any timing constraints will be specified and the software will be thoroughly tested to ensure that it meets it timing requirements, under all possible conditions.

The standard approach in real-time systems is to allocate a fixed period, often known as a *frame*, during which all the computations are executed. These frames are executed repeatedly so that the functions in an application are repeated periodically, at the frame rate. Invariably, the frame rate is matched to the step length used in numerical integration. In effect, the system marches in time, frame-by-frame where the frame period is sufficiently small so that there are no perceived discontinuities in the simulation and any errors introduced by the numerical methods are kept to a minimum.

Modern computers can execute several million instructions per second. However, data transfers involving hardware devices are implemented at much slower speeds. For example, a serial line operating at 9600 baud (9600 bits per second) takes approximately one millisecond to transfer one byte of data. At the receiver, the data arrives every millisecond or so, but the receiver has no knowledge as to when the data will actually arrive – in effect the transfer is asynchronous. When the data arrives, it can be transferred from the hardware registers of the interface card to memory, taking only a few microseconds. One option is for the processor to poll a device until it has received incoming data, or in the case of transmission, to check if the device is ready to transmit more data. However, this method requires the processor to execute instructions waiting for data to arrive before copying the data. Clearly this approach is wasteful of the processing capability of the processor, if it simply waits for a slow device. It would be preferable for the processor to spend its millisecond on more useful computations. This situation is resolved by exploiting the interrupt mechanism provided for most processors, which provides a fundamental part of a real-time system.

In an interrupt driven system, the processor can initiate a transfer with a device but continue with useful computations, executing instructions rather than simply polling the device for completion of a transfer. When the device has data available, the device interrupts the processor. The current state of the processor (the machine registers) is saved in a reserved area (usually a stack) and the processor transfers control to an interrupt service routine allocated to service the interrupting device. In the interrupt service routine, the status of the device is checked, data is copied from the device registers to memory (typically a buffer) and the device may be set for another data transfer. Finally, the processor will return from the interrupt by re-instating the processor state at the time of the interrupt; this is achieved by reloading the machine registers from the region where they were preserved at the time of the interrupt. The processor then resumes the instructions it was executing prior to the interrupt. The important point about this use of interrupts, which are microcoded in the processor architecture, is that the processor does not waste any time waiting for relatively slow devices and the overhead of managing input and output devices is negligible.

A modern operating system allows several processes to execute concurrently. In fact, the processor is switching between processes at relatively high rates in order to achieve this concurrency. The selection of the next process to run is made by a scheduler, which characterizes the operating system. In a real-time operating system, the processes are usually allocated a priority and the scheduler will select the highest priority process that is able to run. Clearly, a process that is waiting for

Principles of Modelling

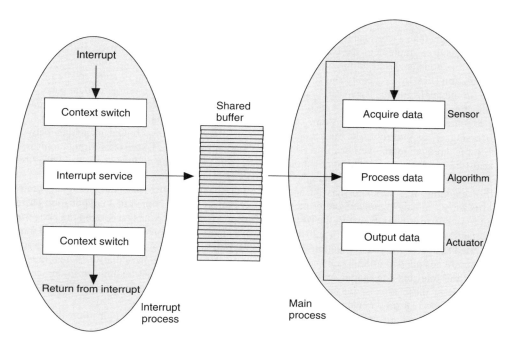

Figure 2.15 Real-time data acquisition

data cannot run and the processor can activate a lower priority process. These scheduling points can occur when an interrupt has been serviced or following a system clock interrupt. There is still a small overhead associated with the processing of interrupts in this way. Each time an interrupt occurs, the machine registers are saved and subsequently re-loaded. If 1000 interrupts occur per second and these context switches take $30\,\mu s$, say, then $30\,ms$ is lost per second (3%), simply in context switching. Figure 2.15 shows a possible arrangement in a real-time operating system. Data is transferred to a buffer in the interrupt service routine, following an interrupt. The main process can access the shared buffer to extract data, without incurring any delay. In practice, care is needed to ensure that the data is not written to the buffer at the same time that another process is accessing the buffer, that there is sufficient space in the buffer to avoid overrunning the buffer region and also that the data in the buffer does not become 'stale'.

One class of real-time programmes involves human operators. The hand–eye coordination response of a human is of the order of a few hundred milliseconds. Any perceived delay between an input and its subsequent actuation (particularly if the human response is in the form of a visual cue) is perceived as an unnatural delay, leading to abrupt changes which are radically different from real life. Consequently, simulations, with an 'operator in-the-loop' must be computed within 20–50 ms to satisfy the operator's sensation of a continuous system. This requirement is particularly critical in flight simulation, where aircraft dynamics, the platform motion and the visual system must be sustained at 50–60 Hz. This value, which corresponds to a computational period less than 20 ms, is a worst-case value. In other words, it is not an average value; it is a hard limit that must never be exceeded. In many systems, if the period is exceeded by a small amount, a complete frame may be lost, in this case reducing the iteration rate from 50 to 25 Hz for a short time until the desired frame rate is recovered.

In many real-time systems, the application software runs under an operating system. This has the advantage that a consistent interface is provided for the system functions and the designer does not have to implement these functions in the application software. However, if the operating system functions are used in combination with the application software, they must not extend the computation time by exceeding the frame limit. Any call to the operating system adds an explicit delay to the application software. In practice, these calls are avoided in the time critical parts of the application software. But there are also aspects of operating system behaviour where computations are executed by the operating system, independent of the application software. For example, the processing of an interrupt, the subsequent buffering of data and any context switching between processes amounts to an overhead imposed on application software. The only solution is to provide a real-time system, where the application software can exert some influence on the behaviour, ordering and prioritization of operating system tasks.

For example, if a simulator has 32 analogue input channels and each analogue input requires 100 µs for the analogue conversion to interrupt the computer and be copied to an application buffer, then 3.2 ms is lost every frame to the overhead of responding to interrupts, accessing the converted value and copying it to memory. In a 20 ms frame, this amounts to approximately 15% of each frame dedicated to operating system functions rather than user applications. In practice, real-time operating systems are designed to minimize this overhead; some architectures are particularly efficient in terms of handling interrupts and input–output interfaces can reduce additional loading on the processor. Nevertheless, these effects and any other delays imposed by an operating system, need to be included in the real-time performance analysis of a simulation.

One further cause of delay occurs in the application software. In some algorithms, the number of iterations depends on data values computed at run-time. For example, a series expansion may terminate after five terms or six terms depending upon a simulation variable. Consequently, there may be a wide variation in the frame time. Another example occurs in flight simulation, where an aerodynamic model is used while the aircraft is airborne whereas a mechanical dynamics model is used while the aircraft is taxiing on the runway. If there is a transition between these models, there may be a jump in processing time during landing. If fact, both models might be executed in the transitory phase during touch down when one or more of the wheels is in contact with the ground, where both the flight model and the undercarriage software are active.

In this latter case, the addition of execution times for each function must still not exceed the frame limit. In practice, although a real-time flight simulator may use a 20 ms frame, say, a conservative margin is normally allocated to ensure that no one frame exceeds the frame limit, for example, to reduce the computation time to 15 ms. Such constraints require very detailed and extensive timing analysis to ensure that no frame exceeds the fixed frame period for all the possible states and constraints of the simulated system. Figure 2.16 shows a typical frame loading for a real-time flight simulator. Although the 20 ms frame is never exceeded, the margin is only 3.5 ms, based on analysis of 1000 frames (50 s).

In the case where the frame period is exceeded, the designer has very limited choices:

- To increase the processor performance;
- To simplify the complexity of the model;
- To attempt to optimize the software.

Generally, it is not feasible in real-time simulation to allow processes to exceed the frame period for a few frames, in the hope that the simulation will catch up within a few frames, when less processing is used (although such techniques are used in computer games). Nevertheless, frames are closely monitored in most real-time simulators, to monitor processor loading and to provide maintenance engineers with valuable diagnostics in the event of unexpected overload.

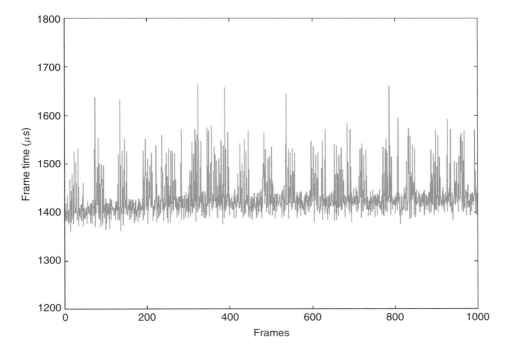

Figure 2.16 Frame activity

2.7 Data Acquisition

In a flight simulator, there are numerous levers, knobs, switches and selectors which enable the flight crew to operate the simulator in the same way that they operate an aircraft. The position and state of these inputs are measured as analogue or digital inputs, typically as a voltage, and are read into a computer via a hardware interface. Similarly, there are many outputs in a simulator, which drive actuators, lamps, stick shakers and possibly flight instruments, where the simulator computer writes the output signal to the hardware interface of the equipment.

Data is transferred to or from simulator equipment via input/output (I/O) cards, which provide both digital and analogue signals. Such cards usually provide fixed voltage ranges and additional signal conditioning is provided to interface the equipment to I/O cards. Although variables are stored in memory as floating point numbers, typically with a resolution of 32 or 64 bits, I/O data is transferred as integer values, typically with a resolution of 10–16 bits and converted to a floating point value for use in the flight model. However, it is important to appreciate that the resolution of the acquired data is several orders of magnitude less than the program variables.

2.7.1 Data Transmission

For digital outputs, most I/O cards are only able to drive a few milliamps, usually at transistor-transistor logic (TTL) voltage levels (0–5 V). Simple amplifier circuits provide the necessary current output and appropriate voltage range. They also usually provide circuit protection to protect the I/O card from inadvertent 'shorting' at the equipment.

Analogue output is more complicated. For digital to analogue (D/A) conversion, an I/O card will provide a specific voltage range (or a set of fixed ranges) and a given resolution. For example, a 12-bit 5-V D/A may provide 0 V for the digital value 000000000000 and 5 V for the digital (binary) value 111111111111, with a resolution of 1.22 mV. From the user's perspective, the 12-bit D/A output 0–5 V would correspond to unsigned integer values in the range 0–4095. Clearly, the scaling between a variable in an application and the D/A value is the responsibility of the programmer. The least significant bit of this D/A is 0.00122 V. Each bit is, in effect, a binary weighting, doubling at each stage from the least significant bit to the most significant bit, so that a 12-bit word with only the most significant bit set corresponds to 2.5 V. For D/A, resistors can be arranged so that they contribute a binary-weighted value, with the voltages of each stage being added, as shown in Figure 2.17, for a 10-bit D/A. The resistor chain comprises binary-weighted resistors. The switches S_0 to S_9 either ground the D/A input or provide a current V_{ref}/R via each resistor R_0 to R_9, which is summed by the operational amplifier.

The output is valid as soon as the inputs have settled and the conversion time depends only on the slew rate of the operational amplifier and the switching time of the switches, providing very fast digital to analogue conversion (of the order a few hundred nanoseconds). Although a circuit of this form is easy to construct, the resistance of each stage must have an accuracy (or tolerance) equal to the least significant bit. In practice, it would be very difficult to calibrate the circuit to meet 0.1% accuracy (needed for a 10-bit D/A) for all resistors. The overall accuracy depends on the accuracy of the reference voltage, the tolerance of the resistors and the impedance of the switches (which can usually be neglected).

For high-resolution D/As, this simple circuit cannot provide sufficient accuracy and an alternative circuit, known as the *R-2R ladder network* is used, consisting only of resistor values R and $2R$, as shown in Figure 2.18. The equivalent circuit is shown in Figure 2.19.

In this circuit, each voltage source is applied to two resistors $2R$ and R, giving a resistance of $3R$ and therefore a current of $V/3R$, so that $i_1 = i_2 = i_3 = i_4$ in all loops. The current i_4 is divided equally on either side of node n_4, so that node 4 supplies $i_4/2$ to node 3, $i_4/4$ to node 2 and $i_4/8$ to node 1 and so on. The total current in R_L is (for this simple four-stage circuit) given

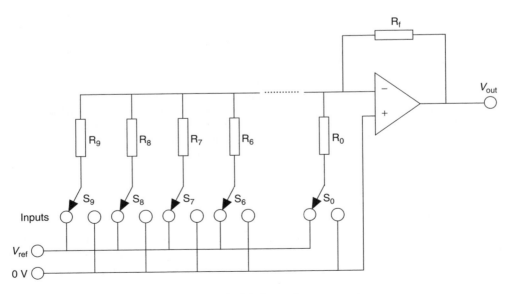

Figure 2.17 Ten-bit digital to analogue conversion

Principles of Modelling

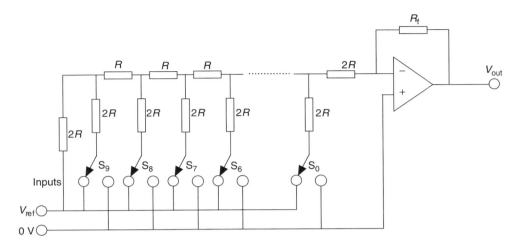

Figure 2.18 Ladder network D/A

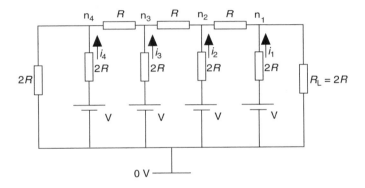

Figure 2.19 Equivalent circuit of a ladder network

by $i_4/2^4 + i_3/2^3 + i_2/2^2 + i_1/2^1$. The output is therefore a binary-weighted output, but in contrast to the weighted resistor network, only two resistance values are required. These resistors can be manufactured to a high tolerance giving excellent linearity for D/A output.

2.7.2 Data Acquisition

The primary inputs of a flight simulator are the flight controls; for a civil aircraft a control column provides elevator and aileron inputs; for a military aircraft a centre stick or side stick provides these inputs. The rudder is controlled by means of rudder pedals and a throttle quadrant provides the engine control inputs, which depend on the type of engine and, for piston engine and turboprop aircraft, the type of propeller. Most aircraft are provided with toe brake systems, a parking brake and some means of nose-wheel steering. For helicopters, the primary pilot inputs are the cyclic control (a centre stick in front of the pilot) and a collective lever. Most of the primary controls also have some form of trimming mechanism to reduce loading on the controls for the flight crew. For fixed wing aircraft, inputs are also provided for flap selection and to lower and raise the undercarriage.

For each of these controls, a transducer converts the control input displacement to a voltage. These analogue inputs provide the positions of the control surfaces, engine settings, system selections for electrics and hydraulics and so on. The typical airline simulator will have several hundred inputs, excluding knobs, selectors and switches for radio panels, navigation panels and engine start panels.

There are two types of input: digital and analogue. A digital input (also known as a *discrete*) can take one of several specific states. For example, a switch is either on or off, or up or down; a five-way selector can only be in one of five positions. In these cases, the input takes one of a range of fixed values, which can be represented as a digital value. For analogue inputs, the input is continuous over a range of position. For example, a control column may operate over a range of $\pm 25°$ or an engine lever may have a movement of $60°$. In this case, the position of the inceptor is measured by a transducer that converts position to a voltage rather than discrete values.

Discrete inputs are relatively simple inputs in a simulator. A reference voltage is usually applied to digital (discrete) inputs, say 5 V for TTL logic. Consider a simple switch with only two states (on or off) given by 0 V or 5 V, which is applied to a digital interface card. In the closed state, the switch grounds the input. In the open state, the switch is not connected and the input floats to 5 V, as shown in Figure 2.20, which illustrates one possible form of digital input.

This scheme has the additional advantage that, if the switch output is accidentally grounded, no electrical damage will occur to the circuit. The value of the resistor R is chosen so that the input current is matched to the digital input circuit of the computer.

For analogue inputs, the analogue voltage provided by a resistive or inductive transducer is fed to an analogue to digital converter (A/D) which provides a digital equivalent value of the applied voltage to a resolution given by the number of digital bits used in the conversion. The flexibility of the operational amplifier (Horowitz and Hill, 1989) provides the basic functions of signal conditioning, amplification and conversion. The schematic of an operational amplifier is shown in Figure 2.21.

The gain of the operational amplifier is

$$\frac{V_{out}}{V_{in}} = -\frac{R_2}{R_1} \qquad (2.55)$$

For a transducer with an output voltage in the range 0–2 V, say, and a computer system with analogue inputs in the range 0–10 V, R_1 and R_2 can be chosen to provide a gain of 5 to use the full

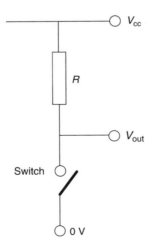

Figure 2.20 Digital switch input

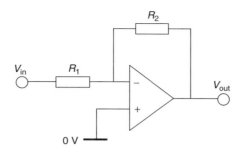

Figure 2.21 An operational amplifier

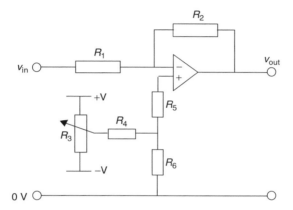

Figure 2.22 Bias compensation for an operational amplifier

resolution of the analogue to digital conversion interface. Similarly if a bias is required to shift an input in the range 2–3 V, it is not sufficient to simply modify the gain, it is also necessary to adjust the bias of the input signal. The bias changes the signal to the range 0–1 V and a gain of 10 ensures the signal is in the range 0–10 V. A typical bias compensation circuit is shown in Figure 2.22.

The variable resistor R_3, in conjunction with the resistor values R_4, R_5 and R_6, can be used with a calibration voltage to provide sufficient bias to shift the input voltage to an appropriate range for the A/D, combined with the gain selection afforded by resistors R_1 and R_2. Such compensation is commonplace in a flight simulator. For example, movement may be sensed by a circular potentiometer. If the mechanical linkage is such that the wiper arm only traverses a proportion of the potentiometer, the input can still be measured over the full operating range of an A/D to ensure that no resolution is lost.

The A/D devices that have been developed for real-time applications are designed to meet the following requirements:

- Speed of conversion – depending on the method of conversion, the analogue input is sampled and held during the conversion, which may take from a few nanoseconds to several milliseconds;
- Sampling rate – although this value depends on the conversion speed, the sampling speed must meet the iteration (or sampling) rate of a simulation. For example, if 32 analogue channels are sampled at 100 Hz, the actual sampling rate of the A/D is 3200 samples per second, approximately 3 ms per sample;

- Resolution – the A/D will provide an input of n bits, where n is typically 8 – 16 bits. For an A/D in the range 0–10 V, a 10–bit A/D will have 1024 (2^{10}) discrete levels giving a resolution of approximately 10 mV. Generally, the higher the resolution, the longer the conversion time. 12-bit A/D devices are widely available and resolutions better than 14 bits are expensive and potentially slow;
- Linearity – an A/D should have a linear conversion over its full range, otherwise compensation will be required following the conversion;
- Common mode rejection ratio (CMRR) – often the input reference and the ground reference are applied as differential inputs. If both inputs are identical, the amplifier output should be zero, independent of the magnitude of the input signals. In practice, if a sine wave is applied to both input terminals of the input stage of an operational amplifier, some residual component is detectable at the output. The CMRR of an amplifier gives a measure of its immunity to interference. For example, mains pickup may be induced in the reference signal and the ground plane of an amplifier. With a good CMRR, this noise is effectively eliminated.

In practice, considerable care is needed to design analogue interface circuits. They must have high impedance to avoid attenuation of the input signal, signal conditioning to ensure the maximum resolution of conversion and good immunity to interference from other equipment.

Three methods are commonly used for analogue to digital conversion. The simplest form is serial conversion, as shown in Figure 2.23. The comparator is simply a high-gain amplifier, which saturates to $+V$ if one input is greater than the other input, otherwise it saturates to $-V$. The counter initially holds zero. All the time the analogue voltage generated by the D/A (of the counter output) is less than the reference voltage, the counter increments at the clocking speed of the counter. As soon as the D/A output exceeds the reference voltage, the comparator changes state, indicating that the conversion is complete. When this occurs, the value in the counter corresponds to the analogue input and the counter is read to provide the converted value. The main drawback of this method is that the conversion time depends on the counter clocking speed and also the magnitude of the reference voltage. Consequently, serial A/D conversion is slow. Nevertheless, it is simple, requiring only a counter, a D/A and a comparator.

An alternative method is to search for the analogue value, changing 1 bit at a time, known as *successive approximation*. Assume a 10-bit converter is used to represent analogue voltages in the range 0–5 V, so that the binary value 0000000000 represents 0 V and 1111111111 represents 5 V. Values in the range 0–2.5 V will have a 0 in the most significant bit, whereas values in the range 2.5–5 V will have a 1 in the most significant bit. Bits can be introduced in 10 cycles (rather than 1024 cycles), either adding a 0 or a 1 at each bit, according to the comparator output. The trial

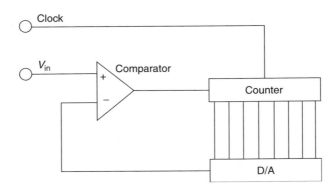

Figure 2.23   Simple analogue to digital conversion

value is initially set to 1000000000. If the comparator output is true, the input is greater than this value, otherwise the current bit is cleared. Logic control circuitry produces a trial bit for the 10th, 9th, 8th ... 1st stage successively, where each trial bit is set or cleared at the clocking rate. The advantage of this method is that the conversion is completed in n cycles for n-bit conversion, independent of the magnitude of the sampled value. Note the importance of a sample-hold circuit as the least significant bit is introduced during the final cycle of the conversion. If the analogue input signal is allowed to change, the A/D will still complete its conversion but may give a totally erroneous value.

The fastest form of A/D conversion is known as the *parallel or flash conversion method*, where a comparator is used for each bit, as shown in Figure 2.24, to illustrate a 3-bit converter. Note

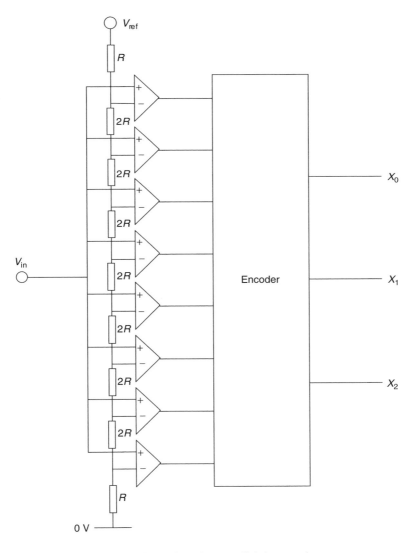

Figure 2.24 Flash analogue to digital conversion

than 2^n comparators are required for an n-bit converter. The resistor chain corresponds to 15/16, 13/16, 11/16, 9/16, 7/16, 5/16, 3/16 and 1/16 of V_{ref}, respectively. The voltage entered into the encoder is a binary-weighted value representing the sampled value. An encoder converts 2^n bits to its corresponding n bits. Note that this method requires 4096 comparators for 12-bit A/D and very accurate resistors but the conversion rate is reduced to the settling time of the comparators. Consequently, flash A/D is used in very high performance signal acquisition applications. In simulation, a sampling rate of 50 Hz for 32 channels requires an overall A/D conversion speed of 640 Hz, approximately 1500 µs per sample, which is well within the conversion time of a successive approximation method.

Considerable attention is given to digital and analogue I/O in flight simulators. The devices selected must provide linearity, sufficient resolution and fast conversion times. Problems can occur as a result of poor grounding of analogue signals; to avoid this, differential analogue inputs are used where the ground reference and signal are provided for each connection. Care is also needed to avoid interference, although generally this results from mains interference or induced noise from electrical machinery. Cross-talk, resulting in interference between wires is rare; as the signal frequencies in flight simulation applications are sufficiently low, this form of interference is unlikely to be encountered. Some I/O devices provide autonomous data transfers, requiring only initialization; the transfers are then implemented independent of the processor. Alternatively, with low-cost embedded systems, it is straightforward to provide dedicated I/O systems, with I/O data transferred to the simulator computers via a network.

2.8 Flight Data

In flight simulation, much of the data is provided in the form of graphical or tabular data. Sometimes, this is because it is the most convenient or efficient form to represent, store and organize the data. Alternatively, the data may be non-linear or functions of several variables and a graphical format simplifies the method to communicate the data. Data may be acquired at regular points or alternatively, for a limited set of points where it was convenient (or safe) to record the values. However, during simulation, it is necessary to compute values which will probably not correspond with the acquired data and it is necessary to interpolate between data points or extrapolate from a set of data points. Clearly, the processes of interpolation and extrapolation can introduce errors in the computation of numeric values and care has to be given to ensure that errors generated in these processes do not affect the solution of the equations or the behaviour of the simulated system.

Often data will be provided as a set of data points. For example, the values in Table 2.5 were recorded for a component of the $dC_L/d\alpha$ lift curve for a fighter aircraft in the subsonic range (below Mach 1). The resultant graph from these 12 points is shown in Figure 2.25.

Data represented in this form has three advantages:

- It is compact – the function is represented as 12 coordinate pairs;
- Interpolation between individual points is straightforward;
- The function passes through the specific tabulated points.

However, this data also exhibits several disadvantages:

- It is necessary to search for a specific line segment as a function of M in order to find the end points of a segment;
- At each tabulated point there is a discontinuity in the function which is probably not representative of the actual system and may introduce dynamics (resulting from an abrupt change) which is inappropriate.

Principles of Modelling

Table 2.5 dC_L/da Data points

M	x(M)
0.0	3.3
0.5	3.4
0.6	3.5
0.7	3.65
0.8	3.85
0.85	4.15
0.875	4.45
0.89	4.65
0.91	4.65
0.92	4.5
0.95	4.0
1.0	3.65

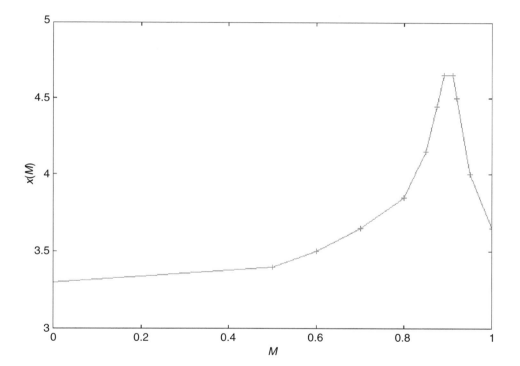

Figure 2.25 A simple function shown with linear line segments

An alternative approach is to use line-fitting methods to obtain a polynomial which is close to the points given in the table. Applying a first-order linear least squares fit for these tabulated values gives the polynomial

$$x(M) = 3.082 + 1.197M \tag{2.56}$$

A second-order linear least squares fit for the same data gives the polynomial

$$x(M) = 3.175 + 0.505M + 0.672M^2 \qquad (2.57)$$

Both polynomial approximations are shown in Figure 2.26; it is clear that the polynomials neither resemble the data characteristics nor pass through any of the actual data points.

Another approach is to compute a high-order polynomial. However, a least squares fit using a 12th-order polynomial is still unlikely to match the data closely and a polynomial fit devised to pass through all the data points (e.g. a Chebyshev polynomial) is likely to contain oscillatory characteristics that are not evident in the data.

Bearing in mind that data acquired experimentally (e.g. from flight trials) is probably valid at the acquired data points (within the resolution of the sensor measurement) and moreover, that an experimentalist capturing the data may have knowledge of the system to justify taking extra points in specific ranges or omitting data points where the characteristic is predictable, then a smooth line passing through all the points with no apparent discontinuity is potentially a better approximation. This approach is similar to the plastic 'flexicurves' that were widely used by draughtsmen and engineers for curve fitting in the days when graphs were plotted on paper rather than computer screens.

The most commonly used curve fitting method to provide a smooth continuous curve passing through all data points is spline fitting. The examples to generate lookup tables from flight data used in this book are based on the spline fitting algorithms in Press *et al.*, (2007). Figure 2.27 shows a spline fit to the data in Figure 2.25 generating 100 points for the range 0–1.0, which shows that the curve is smooth and continuous while also passing through all the data points.

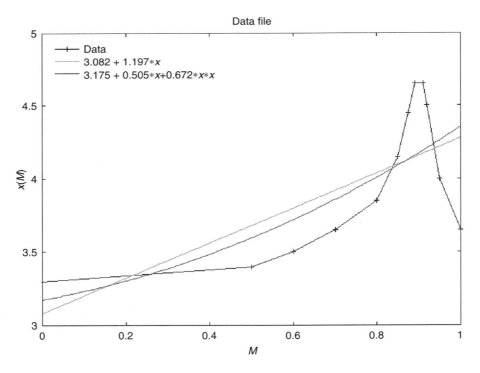

Figure 2.26 First and second-order linear least squares fit to the data points

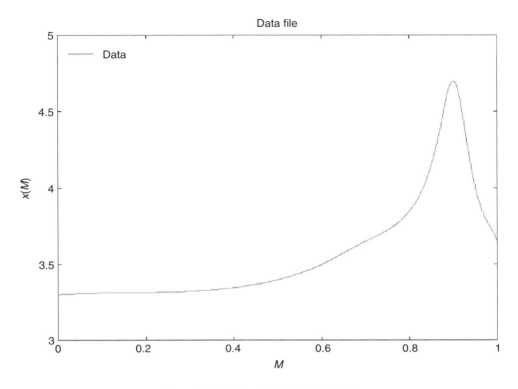

Figure 2.27 Spline fit to the data points

Cubic splines are based on computing a third-order polynomial to fit between three adjacent points, repeating this computation for sets of three points across the range. However, there is a constraint that the gradients (and possibly the second derivative) of these polynomials must be equal at the points where they join. A cubic spline reduces to $N - 2$ linear equations in N unknowns, where N is the number of points to represent the curve. For the natural cubic spline, the two extra conditions are $\ddot{y}_1 = 0$ and $\ddot{y}_N = 0$. In other words, the gradient is not changing at the start and end points – a reasonable assumption for data where the second derivative is slowly changing.

The lookup tables are generated offline so the performance of the spline computation is not an issue. However, it should be appreciated that the spline fit can produce unexpected distortion, particularly if the graph is represented by insufficient points, where the algorithm effectively forces the solution to achieve connecting gradients at the data points in the table. Nevertheless, spline fitting techniques are widely used to generate lookup tables for non-linear functions where the 'reasonableness' of the method outweighs the effect of any numerical errors it may introduce.

2.9 Interpolation

Although a lookup table may contain several hundred points, it is still necessary to compute values between any two points in the table. In other words, an interpolation function is needed to compute values which are not necessarily directly accessible from the table. For example, the data generated by a spline fit to the data given in Table 2.5 are tabulated as an array of floating point values as follows:

```
float f1tab[] = {
3.3000, 3.3019, 3.3036, 3.3051, 3.3065, 3.3076, 3.3087, 3.3096, 3.3104, 3.3111,
3.3116, 3.3121, 3.3126, 3.3129, 3.3132, 3.3135, 3.3138, 3.3141, 3.3143, 3.3146,
3.3149, 3.3153, 3.3157, 3.3161, 3.3167, 3.3173, 3.3181, 3.3189, 3.3199, 3.3211,
3.3224, 3.3238, 3.3255, 3.3273, 3.3293, 3.3316, 3.3341, 3.3368, 3.3398, 3.3430,
3.3465, 3.3504, 3.3545, 3.3589, 3.3637, 3.3688, 3.3743, 3.3801, 3.3864, 3.3930,
3.4000, 3.4074, 3.4153, 3.4237, 3.4327, 3.4422, 3.4524, 3.4632, 3.4747, 3.4870,
3.5000, 3.5138, 3.5284, 3.5434, 3.5588, 3.5744, 3.5900, 3.6056, 3.6208, 3.6357,
3.6500, 3.6637, 3.6772, 3.6912, 3.7062, 3.7228, 3.7416, 3.7631, 3.7880, 3.8167,
3.8500, 3.8887, 3.9353, 3.9925, 4.0631, 4.1500, 4.2554, 4.3802, 4.5233, 4.6500,
4.7010, 4.6500, 4.5000, 4.3211, 4.1476, 4.0000, 3.8930, 3.8184, 3.7622, 3.7107,
3.650};
```

The table contains 100 values in the range 0–1.0, giving an interval of 0.01. Clearly, it is assumed that values outside this range will not be required, although it would be prudent to either detect such occurrences or to provide sensible default values in such a case. Alternatively, extrapolation methods can be used to predict values outside the range from values in the table. For 2D interpolation, we need to return the value $f1(M)$ for any value of M in the range 0–1.0.

A fragment of the curve is shown in Figure 2.28. From the trigonometry of similar triangles, in triangle ABC

$$\frac{p}{q} = \frac{r}{\Delta x} \qquad (2.58)$$

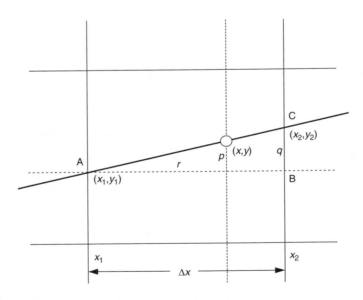

Figure 2.28 2D interpolation

Principles of Modelling

But $p = y - y_1$, $q = y_2 - y_1$ and $r = x - x_1$ which yields

$$\frac{y - y_1}{y_2 - y_1} = \frac{x - x_1}{\Delta x} \text{ giving}$$

$$y = y_1 + \frac{(x - x_1)(y_2 - y_1)}{\Delta x} \qquad (2.59)$$

Δx is easily computed from the overall range and the number of values in the table. x_1 is the nearest x value in the table below x, which allows the offset in the table to be computed to locate y_1. y_2 is the adjacent value above y_1 in the table.

The C code procedure for 2D interpolation is shown in Example 2.3.

Example 2.3 2D interpolation

```
float tabval(float t[], float x, float xmin, float xmax, unsigned int size)
{
    unsigned int p;
    float dx;
    float x1;
    float y1, y2;

    dx = (xmax - xmin) / (float) size;
    p = (unsigned int) ((x - xmin) / dx);
    if (p >= size) {
        return t[size];
    }
    else {
        x1 = xmin + (float) p * dx;
        y1 = t[p];
        y2 = t[p+1];
        return y1 + (x - x1) * (y2 - y1) / dx;
    }
}
```

For example, to access the value in *fltab* for a mach number of 0.345, the following code could be executed:

```
v = tabval(fltab, 0.345, 0.0, 1.0, 100);
```

The same concept can be readily extended to functions of two variables, where 3D interpolation is used to compute $y = f(x, z)$. Values of x over a specific range are stored as a set of values for specific values of z, to represent the graph shown in Figure 2.29.

Consider the case for the computation of engine coefficient of thrust C_T for a piston engine which is a function of the blade angle β and the advance ratio J. For each value of β, C_T values are stored for values of J in the range 0–2.2 in steps of 0.2. Nine sets of data are stored for values

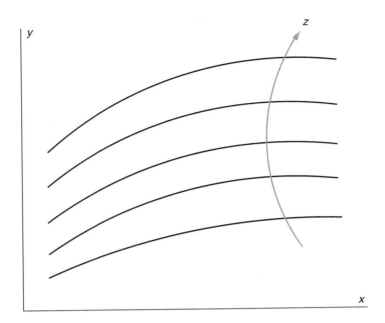

Figure 2.29 3D interpolation

of β in the range 0–40° in steps of 5°. A two-dimensional array representing these values is shown below.

```
float CtTable[] = {
/*
    0.0     0.2     0.4     0.6     0.8     1.0
    1.2     1.4     1.6     1.8     2.0     2.2  J */
    0.000,  0.000,  0.000,  0.000,  0.000,  0.000,
    0.000,  0.000,  0.000,  0.000,  0.000,  0.000,   /*  0 */
    0.000,  0.000,  0.000,  0.000,  0.000,  0.000,
    0.000,  0.000,  0.000,  0.000,  0.000,  0.000,   /*  5 */
    0.000,  0.000,  0.000,  0.000,  0.000,  0.000,
    0.000,  0.000,  0.000,  0.000,  0.000,  0.000,   /* 10 */
    0.092,  0.084,  0.065,  0.037,  0.007, -0.022,
   -0.050, -0.080, -0.108, -0.137, -0.167, -0.196,   /* 15 */
    0.099,  0.096,  0.089,  0.064,  0.036,  0.006,
   -0.023, -0.054, -0.084, -0.115, -0.144, -0.175,   /* 20 */
    0.118,  0.114,  0.110,  0.095,  0.071,  0.043,
    0.013, -0.017, -0.045, -0.073, -0.102, -0.131,   /* 25 */
    0.115,  0.110,  0.106,  0.100,  0.101,  0.080,
    0.053,  0.024, -0.007, -0.038, -0.068, -0.098,   /* 30 */
    0.139,  0.132,  0.126,  0.119,  0.113,  0.108,
    0.089,  0.065,  0.038,  0.010, -0.020, -0.048,   /* 35 */
    0.141,  0.135,  0.128,  0.123,  0.120,  0.114,
    0.112,  0.101,  0.080,  0.057,  0.030,  0.005};  /* 40 */
```

Each row contains C_T values for J in the range 0–2.2 and the nine rows cover the range of β from 0 to 40°. Two additional arrays define the spacing for both J and β.

```
float BetaSpacing[] =
{0.0, 5.0, 10.0, 15.0, 20.0, 25.0, 30.0, 35.0, 40.0};

float JSpacing[12] =
{0.0, 0.2, 0.4, 0.6, 0.8, 1.0, 1.2, 1.4, 1.6, 1.8, 2.0, 2.2};
```

The method is simply an extension of 2D linear interpolation. For given values of J and β, the nearest tabulated values below J and β are located. The end points are first interpolated for β and then the interpolated value for J is located along this line. The attraction of these methods is their simplicity; only a few computations are required to compute these interpolated values. One method to validate the interpolation algorithm is to regenerate these terms, for example, varying J by 0.001 and β by 0.001. Any errors in the continuity of the method should be apparent and specific values can be checked for accuracy. It is also necessary to check that the method is correct near the boundary values and that values outside the given ranges do not lead to run-time errors or produce spurious values. Finally, in simulation it is important to bear in mind that the accuracy of data acquisition may be less than the accuracy of the interpolation methods. Often data measurements acquired experimentally are accurate to 1%, negating the potential benefits of using higher order numerical methods.

The code in Example 2.4 performs 3D interpolation, assuming 0.2 spacing for J and 5.0 spacing for β. The code to check for a valid range for J and β is omitted for brevity.

Example 2.4 3D interpolation

```
static float Cval(float beta, float j, float CTable[])
{
  unsigned int n1, p1, p2, q1;
  float beta1, x1, y1, y2, y3, y4, y5, y6;

  n1 = (unsigned int) (beta / 5.0);
  if (n1 > 7) {
    n1 = 7;
  }
  p1 = n1 * 12;
  p2 = p1 + 12;
  q1 = (unsigned int) (j / 0.2);
  if (q1 > 10) {
    q1 = 10;
  }
  beta1 = BetaSpacing[n1];
  x1 = JSpacing[q1];
  y1 = CTable[p1 + q1];
  y2 = CTable[p2 + q1];
  y4 = CTable[p1 + q1 + 1];
  y5 = CTable[p2 + q1 + 1];
  y3 = y1 + (beta - beta1) * (y2 - y1) / 5.0;
  y6 = y4 + (beta - beta1) * (y5 - y4) / 5.0;
  return y3 - (j - x1) * (y3 - y6) / 0.2;
}
```

2.10 Distributed Systems

For airline flight simulators up to the 1980s, the computers were housed in a large air-conditioned computer room, with thick bundles of cables between the computers and the simulator. With computing resources at a premium, each computer performed a dedicated set of functions, typically with one computer system for the flight model, another for the motion system and additional image generators for the visual system.

Arguably, this was an early form of distributed computing. The flight model outputs to the motion platform and visual system were transmitted as analogue voltages. If the pitch range of $\pm 90°$ is given by ± 10 V, a resolution of 1 mV corresponds to $0.009°$. If the vertical field of view is $40°$ and corresponds to 1000 pixels, say, then each pixel subtends $0.04°$ vertically and smooth angular motion can be achieved. However, if position is passed in the same way, corresponding to ± 10 km, then 1 mV corresponds to 1 m, with the likelihood of 'jerky' motion near the runway.

However, if the information is transmitted digitally between computers, these problems can be avoided, as values are transmitted as floating-point numbers between computers. Prior to the availability of low-cost network interfaces, three options were available for inter-processor communication:

- Shared or reflective memory – a bank of memory is provided which can be accessed by two or more computer systems; this memory is seen in the hardware memory space of each processor connected to the shared memory. While this approach ensures high-speed data transfers, the drawback of this system is the number of address and data lines needed to connect a processor to the shared memory and the need for logic to ensure atomic access to the shared memory to avoid data being read while it is changing;
- Parallel data lines between computers – 8 or 16 parallel data lines provide fast data rates but at the expense of the cabling for data and control (handshake) lines;
- Serial data lines – although only two wires may be needed, the data rates are typically limited to 19,200 bits per second over a distance of up to 20 m.

The third option seems attractive; RS-232 and RS-422 serial interfaces are widely available. However, 19,200 bits per second is approximately 2000 bytes per second. At 50 frames per second, with 4 bytes per variable, this transfer speed corresponds to 10 variables per frame. Consequently, communications between computers require significantly high data rates for real-time simulation.

Fortunately, as the cost of computers fell and processing speeds increased, serial network interfaces were developed for local networking, particularly Ethernet. The benefits of these advances are twofold: first, data rates of 10 Mbits per second are commonplace and secondly, simulator functions can be programmed across an array of interconnected processors rather than allocated to a single high-speed processor.

Since the 1990s, most real-time simulators have exploited the distributed architecture afforded by low-cost networks and workstations. For the simulator designer, the software can be partitioned across an array of processors according to the structure of the simulator software and the performance of the processors. Data is shared between processors by transmitting packets over a dedicated local network. The software is programmed in a high-level language, each processor has an operating system to support software functions and software libraries have been developed to support standard network data transfers.

In a local area network, integrity of the data transfers is paramount and the protocols provided for a network achieve this by re-sending packets if errors are detected or if a packet transfer is not acknowledged within a short time. However, in a real-time system, delays resulting from retried packet transmissions would be unacceptable. All packet transfers must be completed during each fixed frame and these transfers must be guaranteed to meet any timing constraints because, in a distributed system, a node depends on the other nodes to provide data that is timely and consistent. Any variation in the periodicity of the data transfers and the time taken to transfer data may affect the real-time performance and violate the strict timing constraints imposed by a real-time system.

Data transfers in a real-time system must be deterministic. That is to say, the time between sending the data from a node and it being taken by another node is known *a priori* – it is predictable and invariant. However, the action of generating a packet and simply transmitting it directly over the network can lead to indeterministic transfers. The most common problem is that two nodes transmit simultaneously leading to a 'collision' of bus traffic. Typically, a collision is resolved by nodes re-transmitting their packets at a later time. However, subsequent transfers may lead to further collisions, further increasing latency in transfers, violating the timing constraints of a real-time system.

To ensure deterministic transfers in a real-time network, all nodes must comply with a protocol, that is, the set of rules that are adhered to by all the nodes in the network. If the protocol is followed, only one node transmits at a time, avoiding collisions and guaranteeing that the delays incurred in the transfer depend only on the bus characteristics.

Three forms of deterministic protocols are used in real-time networks:

- Master-slave transfers – one node is allocated as a master node. It sends a packet to a slave node which is then allowed to transfer its packets. When these transfers are completed, the slave sends an acknowledgement packet to the master. The master then sends a packet to another node and the cycle repeats until all slave nodes have transmitted their packets. There is a significant overhead in terms of the packet transfers to activate the slave node transfers and to acknowledge their completion, reducing the bandwidth available for data transfers between nodes;
- Token passing – a node only transmits data when it holds a token. On completion of transfers by a node, the token is passed (as a packet) to another node which then has exclusive access to the bus. Each node has a list of the schedule of transfers, passing the token to the next node in the schedule after completing its data transfers. The protocol assumes that all nodes respond immediately (otherwise, delays in one node can accumulate for the other nodes) and depends on all nodes cooperating;
- Time division multiple access (TDMA) – within each frame, a unique time slot is allocated to each node. A node can only transmit during its slot and the time taken for the transfers must not exceed the duration of the slot. Such schemes depend on accurate synchronization of the local clocks of the slave nodes. In practice, a margin has to be built in to ensure that nodes only transfer data within their slots, resulting in some loss of overall bandwidth.

There are many other forms of protocols used in real-time networks and many network architectures have evolved to meet the wide range of demands in real-time computing. In real-time simulation, the criterion is that the protocol is deterministic, reliable and has sufficient bandwidth to accommodate the number of variables transferred between the simulator computers each frame.

The Open Systems Interconnection (OSI) model has seven layers defining network protocol requirements, as shown in Figure 2.30. Its use is widespread as it is followed by all network system designers and is used in the majority of networks throughout the world. The adherence to this standard provides a very high degree of interoperability across platforms and portability of software. Layers 4–7 provide end-to-end file transfer across the world. By necessity, these protocols require re-try mechanisms for loss or corruption of data, acknowledgement of received packets and detection of unresponsive nodes. These protocols are machine independent and provide network addressing to transfer data between computers across continents. However, the complexity of these protocols and the flexibility they must provide restrict their use to non real-time applications. In other words, these protocols are non-deterministic.

The lower levels 1–3 can provide deterministic protocols, while still affording the advantages of node addressing, standardization, interoperability and portability of application software. Level 1

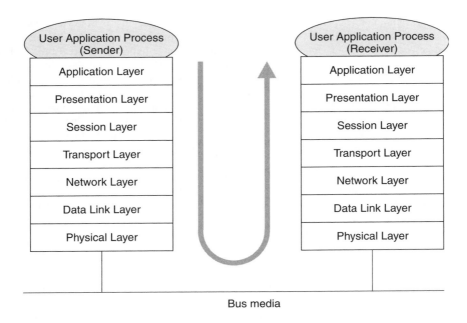

Figure 2.30 Open systems interconnection model

is the physical layer which defines the bus signal formats, timing and hardware interface. Level 2 is the Data Link layer which specifies packet formats, the method of access to the network, the synchronization of packet transfers and error checking. Level 3 is the Network layer, which provides node addressing, the switching and routing of packets and congestion control.

From the simulator designer's perspective, the protocol must not add a significant overhead to network transfers; it must be reliable and provide a straightforward interface to the application software. The protocol used in the simulator covered in this book is based on the User Datagram Protocol (UDP). UDP transfers require an extra 28 bytes in the Ethernet packet for the protocol information, excluding the 32-bit Media Access Control (MAC) packet checksum to detect errors during transmission. Most operating systems provide network libraries which support UDP transfers. Although one criticism of UDP is that the order of packet transmission cannot be guaranteed, for small dedicated networks, such events never occur. UDP provides a simple node addressing method based on the Internet Protocol (IP) address. Each node in the network is allocated a unique IP address of the form $a.b.c.d$ where a, b, c and d denote a value in the range 0–255. The inclusion of a 32-bit checksum means that the receiver has a probability of *not* detecting an error in the packet of the order 2^{-32} or 1 in 2328 million. However, there is a limitation in that the error is only detected at the receiver – the transmitting node is oblivious of the error unless an extra layer of acknowledgement is added, which defeats the benefit of using UDP. In practice, the error rate is likely to be so low that the effect can be ignored in real-time simulation.

Although a large number of bus technologies have been developed for real-time applications, Ethernet offers significant advantages for real-time simulation using UDP transfers:

- The technology is mature, reliable and low cost;
- In addition to the UDP checksum, Ethernet includes a 32-bit hardware checksum;
- A single Ethernet packet can contain between 46 and 1500 bytes of data;
- Ethernet supports data rates of 10 Mbits per second and 100 Mbits per second;

- Ethernet cards are compliant with the OSI model, providing a simple and consistent interface for applications.

At 10 Mbits per second, the data rate of bus transfers is of the order 1 MB per second or 5000 floating point variables per frame (at 50 Hz). Even if only 10% of the frame is allocated to transfers, some 500 variables can be transferred between nodes each frame.

The disadvantage of Ethernet is that strictly it is not a deterministic medium. The hardware in an Ethernet interface uses a Carrier Sense Multiple Access Collision Detection (CSMA/CD) protocol. The carrier used in data transmission is detected. If the bus is quiet then a node can attempt to transfer its packet (the multiple access term implies than any node can attempt to gain access to the bus at any time). In the case where two nodes transmit simultaneously (within one bit period), this event is detected by the hardware interface, the transmission is stopped and the interface waits for a short random time before it attempts to re-transmit its packet. The random delay applied by each node following a collision reduces the likelihood of the same collision re-occurring. The network integrity is based on these random re-try intervals and is therefore non-deterministic. In theory, two colliding nodes could repeatedly select the same random delay indefinitely. In practice, this would never occur. Nevertheless, any collision will lead to additional network transfers, adding latency to the transfers so that deterministic data transfers cannot be assured for a real-time network. However, if the transfers are sequenced so that only one node ever transmits at any time, these transfers are contention-free, no collisions will occur and the resultant data transfers will be deterministic.

Ethernet offers one further advantage that is attractive in real-time applications; multicast data transfers are supported, allowing a data packet to be broadcast to all nodes on the network. In the case where one node needs to transfer the same packet to five nodes, say, it can broadcast the packet and the five nodes needing the data can read the packet from the network. Other nodes, not needing the data can discard the packet. In real-time simulation, multicast transfers can reduce significantly the number of transfers per frame. For example, in a flight simulator, the aircraft pitch attitude is used in the flight model equations, the motion platform dynamics, for the aircraft instrument displays and also by the visual system.

For a dedicated real-time system, the use of IP addresses is important. The nodes on the network are likely to be addressed in some logical manner in the simulator algorithms. There is a need to map between the IP addresses allocated to the nodes on the network and the logical addresses used in the simulation. However, if the simulator designer is able to allocate the IP addresses, these can be given numeric values corresponding to the logical addresses. Most operating systems allow the node IP addresses to be statically allocated. For example, in a system of five nodes, if the following IP addresses are allocated: 192.168.1.1, 192.168.1.2, 192.168.1.3, 192.168.1.4 and 192.168.1.5, then these addresses can correspond to nodes 1–5. For this group of nodes, the broadcast address is 192.168.1.255. A node can inspect the IP address of an incoming packet and effectively route the packet to the appropriate software requiring the specific packet. Of course, this convention assumes a dedicated network, specific to a real-time application. If this network was connected to a local area network, all incoming UDP packets would need to be checked and of course, contention for the network would inhibit real-time access.

In fact, there is a protocol below the UDP known as the *Media Access Control* (MAC) layer. Each Ethernet card has a unique 6-byte physical address and the source and destination MAC addresses are included in the packet. However, this arrangement would require that any real-time application software is dependent on specific hardware addresses and that the transfers would need to be implemented as 'raw sockets'. Most operating systems prohibit raw socket transfers as they make the computer system vulnerable to attacks. Consequently, UDP transfers are often used in real-time applications:

- The bandwidth overhead is minimal, whereas protocols such as transmission control protocol/internet protocol (TCP/IP) introduce significant latency for real-time applications;

- The interface between the application software and the operating system complies with a published standard and is straightforward;
- UDP transfers facilitate the implementation of deterministic protocols for real-time applications.

At the level of the operating system, data is formed as a packet in the memory of the computer and copied to the memory of another computer by the network controllers, as shown in Figure 2.31.

The application software forms the data in the packet in memory. The network interface is programmed to copy this data from memory to the databus, using direct memory access (DMA) independent of the processor. The encoding of the data from parallel to serial is managed by the network controller. The interface can interrupt the processor to indicate completion of the transfers. Similarly at the receiver, the data is copied from the databus as serial data to memory using DMA. The location of the packet in memory and the form of the transfer is defined by programming the interface card. When the transfer is completed, the processor is interrupted to indicate that a new packet is available. In addition, the status of the transfer can also be accessed from the interface.

From the user's perspective, the operating system will perform many of these tasks, reducing the task of reading and writing packets to the allocation of memory for the packet and the construction of the data packet in memory, as shown in Figure 2.32, which outlines the division of responsibility between the application process and the operating system.

In stage 1, the application writes the data to be transferred into the packet, which is allocated in main memory. In stage 2, the operating system is requested to transfer the data. In stage 3, the operating system will form the packet as a valid UDP, say, adding any protocol headers and a checksum. The network hardware interface is then accessed in stage 4 to initiate the transfer. The completion of the transfer is likely to be signalled as an interrupt (stage 5) and the application software can subsequently check the result of the transfer (stage 6). The point to stress is that the user application simply forms a data packet, independent of any protocol, then issues a request to the operating system to transfer the packet, but complying with a specific protocol and is able to monitor the outcome of the transfer.

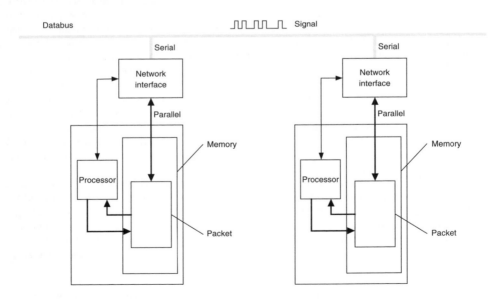

Figure 2.31 Packet transfers

Principles of Modelling

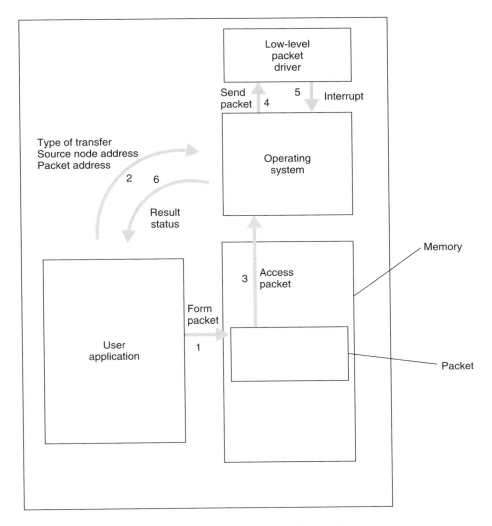

Figure 2.32 Operating system packet transfers

Programs 2.1 and 2.2 show the repeated transfers of UDP packets broadcast between two nodes, using the broadcast IP addresses 192.168.1.255. The network constants and functions are defined in the system-wide header files *types.h, socket.h* and *netinet.h*. Three functions are used to send a packet. The function *socket* opens a socket between the user application and the operating system. Packets are transferred in *sendudp.c* using the function *sendto* and on completion of the transfers, *close* is used to shut down the socket connection. The symmetry of the code used to read packets in *readudp.c* is clear. A socket is opened and closed in the same way, but the function *revcfrom* is used to read packets. The status of these system calls is checked after each call – a negative value indicates a problem. Note the simplicity of transferring data between nodes, which is accomplished in some 70–80 lines of C code. Notice also that the structure and content of the data buffer is left to the user to define, in this example it is simply an array of 200 integers. In *sendpkt.c*, the values

12345678 and 87654321 are written into the first two words of the packet, simply to illustrate how data is passed in the packet.

Additional code is needed (using the function *setsockop*t) to obtain permission to broadcast a packet (necessary in Linux systems). The data structure *s_addr* is used to define the socket information needed in the transfers. In addition, the procedure *bind* in *readpkt.c* is called to bind the socket and the packet descriptor. In this example, a dot is printed after every 50 packets are transmitted or received.

The arguments of the *socket* call define that the transfers are network transfers (AF_INET) using the UDP protocol (SOCK_DGRAM). Prior to any transfer of data, the system data structure *s_addr* is initialized to indicate the type of network transfer (AF_INET), the destination IP address (192.168.1.2) and the port to be used (12345). For each socket, an application can use any number of ports and data transferred using a specific port address will only be read by another node using the same port address. The function *sendto* has to provide the socket (s), the address of the data to be transferred (*&Buffer*), the number of bytes to be transferred (*sizeof*(*Buffer*)), a pointer to the socket descriptor *s_addr* and a variable containing the size of *s_addr*.

The code to receive UDPs is very similar; the function *revcfrom* also provides the socket (s), the address of the buffer to receive the data (*&Buffer*), the number of bytes expected (*sizeof*(*Buffer*)), a pointer to the data structure *s_addr* and a variable containing the size of *s_addr*. The status of *recvfrom* is checked, to detect a checksum error, or if an incorrect number of bytes was received or if any other error occurred during the transfer.

This software interface is common across Linux, Unix and Windows platforms. It is largely based on the Berkley Software Distribution (BSD) package, which although not an international standard has been widely adopted by operating system developers. With the exception of naming conventions for the header files, there is a high degree of interoperability of hardware and portability of software. This simple example works on Windows and Linux platforms. If a network interface card was changed, the software would not need to be modified. Of course, while delegating the responsibility for management of the packet transfers to an operating system simplifies the user application, any overhead imposed by the operating system must be included in any timing analysis of the data transfers in a real-time system.

Although there is no specific requirement to use UDP transfers for real-time flight simulation, the simple software interface, small overhead and interoperability it affords make a strong case for its use. Ethernet can provide the packet transfers in terms of throughput, reliability and low cost to enable deterministic protocols to be implemented. The main criteria for the selection of the protocol in a real-time system are that the transfers are deterministic, the bus bandwidth is sustained (so that network transfers only occupy a small proportion of any frame) and that the application software can easily read and write packets to other nodes on the network. Ethernet, UDPs and a token passing protocol fully meet these requirements and have been used extensively in the flight simulator covered in this book.

Program 2.1 sendudp.c – a program to broadcast UDPs

```
/* program to broadcast UDPs */

#include <sys/types.h>
#include <sys/socket.h>
#include <netinet/in.h>
#include <stdio.h>
#include <termios.h>

#define DEFAULT_PORT 12345
#define DEFAULT_GROUP "192.168.1.255"
```

```c
int main(int argc, char **argv)
{   int Buffer[200];
    int port = DEFAULT_PORT;
    int retval;
    int fromlen;
    int pktcount = 0;
    int permission = 1;
    struct sockaddr_in s_addr;
    int s, addr_len;

    s = socket(AF_INET, SOCK_DGRAM, IPPROTO_UDP); /* open the socket */
    if (s < 0)
    {   perror("socket");
        exit(1);
    }

                /* get permission to broadcast */
    if (setsockopt(s, SOL_SOCKET, SO_BROADCAST, &permission,
        sizeof(permission)) < 0)
    {   perror("permission");
        exit(2);
    }

    printf("ready to write\n");

    s_addr.sin_addr.s_addr = inet_addr(DEFAULT_GROUP);
    s_addr.sin_family = AF_INET;
    s_addr.sin_port = htons(DEFAULT_PORT);
    addr_len = sizeof(s_addr);

    while(1)
    {
        Buffer[0] = 12345678;          /* set arbitrary values in the buffer */
        Buffer[1] = 87654321;

        retval = sendto(s, &Buffer, sizeof(Buffer), 0,      /* send the packet */
                        (struct sockaddr *)&s_addr, addr_len);
        if (retval < 0)
        {   perror("send");
            exit(3);
        }

        pktcount = pktcount + 1;   /* log every 50th packet */
        if (pktcount >= 50)
        {   pktcount = 0;
            printf(".");
            fflush(stdout);
        }
    }

    close(s); /* close the socket */
    return 0;
}
```

Program 2.2 readudp.c – a program to read UDPs

```c
/* program to read UDPs */

#include <sys/types.h>
#include <sys/socket.h>
#include <netinet/in.h>
#include <stdio.h>
#include <termios.h>

#define DEFAULT_PORT 12345
#define DEFAULT_GROUP "255.255.255.255"

int main(int argc, char **argv)
{   int Buffer[200];
    int port = DEFAULT_PORT;
    int retval;
    int fromlen;
    int pktcount;
    struct sockaddr_in s_addr;
    int s;
    int addr_len;
    int permission = 1;

    addr_len = sizeof(s_addr);

    s_addr.sin_family = AF_INET;
    s_addr.sin_addr.s_addr = htonl(INADDR_ANY);
    s_addr.sin_port = htons(DEFAULT_PORT);

    s = socket(AF_INET, SOCK_DGRAM, 0); /* open the socket */
    if (s < 0)
    {   perror("socket");
        exit(1);
    }

        /* get permission to read broadcast packets */
    if (setsockopt(s, SOL_SOCKET, SO_BROADCAST, &permission,
        sizeof(permission)) < 0)
    {   perror("permission");
        exit(2);
    }

        /* bind the socket to its structure */
    if (bind(s, (struct sockaddr *) &s_addr, sizeof(s_addr)) < 0)
    {   perror("bind");
        exit(2);
    }

    printf("ready to read\n");

    while(1)
    {
        retval = recvfrom(s, Buffer, sizeof(Buffer), 0,    /* get a packet */
                          (struct sockaddr *) &s_addr, &addr_len);
```

```
            if (retval < 0)
            {   perror("recvfrom");
                exit(4);
            }

            pktcount = pktcount + 1; /* log every 50th packet */
            if (pktcount >= 50)
            {   pktcount = 0;
                printf(".");
                fflush(stdout);
            }
        }

        close(s);  /* close the socket */
        return 0;

}
```

2.11 A Real-time Protocol

The flight simulator covered in this book runs on seven computers, as shown in Figure 2.33. The packet transfers are based on the UDP transfers described in Section 2.10. The nodes (computers) are assigned logical numbers 1–7, with corresponding IP addresses 192.168.1.1–192.168.1.7. Node 1 acquires digital and analogue inputs and broadcasts a packet containing I/O data. Node 2 solves the equations of motion and broadcasts a packet containing aerodynamic and engine data. Node 3 solves the navigation equations and broadcasts a packet containing navigation and avionics data. Node 4 is the instructor station and broadcasts the instructor commands and scripting information. Although it is possible to limit the instructor station transmissions to frames where the instructor issues a command, transmitting a packet for every frame simplifies the protocol considerably. Nodes 5, 6 and 7 are the image generators; they are passive and do not broadcast any packets.

Node 1 sends its packet and waits for packets from nodes 2, 3 and 4. Node 2 waits for a packet from node 1, sends its packet and waits for packets from nodes 3 and 4. Node 3 waits for packets from nodes 1 and 2, sends its packet and waits for a packet from node 4. Node 4 waits for packets from nodes 1, 2 and 3 and then sends its packet. The visual system nodes (5, 6 and 7) wait for a packet from node 2, which contains sufficient information to render the frame. They are able to perform their respective image generation tasks, before checking the information in the packets from nodes 1, 3 and 4, for example, if the fogging level has changed.

The main requirement is to complete the transfers at the start of the frame, allowing as much time as possible for processing at each node. As the incoming packets are buffered, they can be

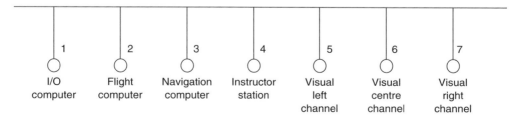

Figure 2.33 Simulator architecture

No.	Time	Source	Destination	Protocol	Info	
363	1.800500	192.168.1.3	192.168.1.255	UDP	Source port: 43118	Destination port: 12345
364	1.800979	192.168.1.4	192.168.1.255	UDP	Source port: 45108	Destination port: 12345
365	1.820191	192.168.1.1	192.168.1.255	UDP	Source port: 12345	Destination port: 12345
366	1.820434	192.168.1.2	192.168.1.255	UDP	Source port: 56789	Destination port: 12345
367	1.820677	192.168.1.3	192.168.1.255	UDP	Source port: 43118	Destination port: 12345
368	1.820920	192.168.1.4	192.168.1.255	UDP	Source port: 45108	Destination port: 12345
369	1.840376	192.168.1.1	192.168.1.255	UDP	Source port: 12345	Destination port: 12345
370	1.840379	192.168.1.2	192.168.1.255	UDP	Source port: 56789	Destination port: 12345
371	1.840618	192.168.1.3	192.168.1.255	UDP	Source port: 43118	Destination port: 12345
372	1.840861	192.168.1.4	192.168.1.255	UDP	Source port: 45108	Destination port: 12345
373	1.860323	192.168.1.1	192.168.1.255	UDP	Source port: 12345	Destination port: 12345
374	1.860559	192.168.1.2	192.168.1.255	UDP	Source port: 56789	Destination port: 12345
375	1.860563	192.168.1.3	192.168.1.255	UDP	Source port: 43118	Destination port: 12345
376	1.861045	192.168.1.4	192.168.1.255	UDP	Source port: 45108	Destination port: 12345
377	1.880258	192.168.1.1	192.168.1.255	UDP	Source port: 12345	Destination port: 12345
378	1.880500	192.168.1.2	192.168.1.255	UDP	Source port: 56789	Destination port: 12345
379	1.880743	192.168.1.3	192.168.1.255	UDP	Source port: 43118	Destination port: 12345
380	1.880985	192.168.1.4	192.168.1.255	UDP	Source port: 45108	Destination port: 12345
381	1.900198	192.168.1.1	192.168.1.255	UDP	Source port: 12345	Destination port: 12345
382	1.900439	192.168.1.2	192.168.1.255	UDP	Source port: 56789	Destination port: 12345
383	1.900684	192.168.1.3	192.168.1.255	UDP	Source port: 43118	Destination port: 12345
384	1.900927	192.168.1.4	192.168.1.255	UDP	Source port: 45108	Destination port: 12345
385	1.920391	192.168.1.1	192.168.1.255	UDP	Source port: 12345	Destination port: 12345
386	1.920405	192.168.1.2	192.168.1.255	UDP	Source port: 56789	Destination port: 12345
387	1.920624	192.168.1.3	192.168.1.255	UDP	Source port: 43118	Destination port: 12345
388	1.920867	192.168.1.4	192.168.1.255	UDP	Source port: 45108	Destination port: 12345
389	1.940322	192.168.1.1	192.168.1.255	UDP	Source port: 12345	Destination port: 12345
390	1.940565	192.168.1.2	192.168.1.255	UDP	Source port: 56789	Destination port: 12345
391	1.940568	192.168.1.3	192.168.1.255	UDP	Source port: 43118	Destination port: 12345

Figure 2.34 UDP transfers monitored by a network analyser

accessed at any time during the frame. In addition, node 1, the I/O computer runs at 50 Hz (20 ms frame), providing a start-of-frame synchronisation for the other nodes and maintaining accurate frame timing.

Figure 2.34 is taken from a network analyser monitoring the network traffic during an autoland approach. The fragment shows packet transfers 363–390, where the transmission time is given the left-hand column. Note that the frames are transmitted at 20 ms intervals (50 Hz frame rate) and that the order of transfers in each frame is 192.168.1.1, 192.168.1.2, 192.168.1.3 and 192.168.1.4. These transfers are typically completed in less than 750 µs, implying that the transfers occupy less than 1 ms (5%) at the start of the frame (with 100 Mbit/s Ethernet). Note also that the multicast broadcast address is 192.168.1.255 (the destination address) and that the destination port is 12345.

The pseudo-code (on Page 94–95) summarises the packets transfers for each frame and the initial values of the node states.

2.12 Problems in Modelling

We had already seen in Section 2.6 that the selection of inappropriate numerical methods can introduce inaccuracy and possibly instability in the simulation of dynamic systems. Generally, these effects can be minimized if the time constants of a system are known, selecting an appropriate order of the method and an appropriate step length for numerical integration. The step length determines the frame rate of the simulator, that is to say, the rate at which the solution is computed, for example, 50 times per second. However, this value is bounded. If the step length is small, the number of frames to be computed per second may overload the processing capacity of the processor. For a coarse step length, in addition to the problems of accuracy and stability, it introduces one further problem. At iteration rates below 20 Hz, discontinuities are just perceptible to the human operator; an apparently continuous system exhibits jerky motion. With perceivable delays, the human operator will over react to input stimuli in a simulator, behaving in a way that is very different from interaction with the actual system. Although the model may be accurate, the errors arise from the coarse step length. In applications where the step length is increased to reduce the computing load, care is needed to ensure that the frame rate does not drop below 20 Hz, for practical purposes.

A simulation is only as good as its model and any model depends on the data used to create the model. While simple models can be created from an understanding of the underlying physics, more complex models are derived from physical models or measurements recorded on the actual system. Quite often, a physical model is based on assumptions and simplifications. For example, a wind tunnel model of an aircraft will operate at very different values of Reynolds number (or airspeed) in comparison with the actual aircraft. The extrapolation of data derived in a wind tunnel experiment to the aircraft data may introduce inconsistencies. Quite often, it is difficult to take a large number of measurements to cover the full operating range of a dynamic system and approximations must be made between measured points, introducing further inaccuracy. It may even be impracticable to obtain measurements over the operating regime of a dynamic system, for example, at the flight envelope of an aircraft near its structural limits. Data may need to be interpolated between measured points or extrapolated from a series of data points. Also, measurements may depend on several variables and it is impractical to take measurements to cover the full range of all the dependent variables.

One further source of error arises from sensor errors. A sensor may not be linear, the resolution of its measurement may be limited or the transducer providing the measurement may not be calibrated correctly. Some measurements can be taken in steady-state conditions, whereas others have to be derived from a continuous measurement. For example, in a wind tunnel, the aircraft can be fixed at a specific angle of attack, the wind speed set to a given value with the control surface deflected to a measured angle. A load balance can measure the force on the model at this one setting of these three variables. Alternatively, it would be difficult to take an accurate measurement in a trials aircraft, flying at a specific airspeed, attitude and elevator deflection. The alternative is to take a series of input and output measurements but at fixed values of airspeed and altitude. In practice, it is difficult to fix specific variables in actual flight and flight trials are designed to minimize any variation in these forms of measurements. Furthermore, a parameter such as force can be measured in a wind tunnel, whereas it has to be derived from other measurements in a flight trial.

Consequently, the acquisition of data using aircraft flight trials is very expensive. The aircraft must be equipped with accelerometers, gyros and recording equipment and many hundreds of flight hours may be needed to acquire all the data. Moreover, the data then needs to be analysed and processed to provide the data in a form that can be used in a simulator. Not surprisingly, the data package for a civil transport aircraft may cost several million dollars.

There are other techniques to estimate the terms needed to model aircraft dynamics. In CFD, if the boundary conditions can be determined for a flow of liquid or gas, then the subsequent flow can be determined in terms of the pressure, temperature and flow rate. The aircraft surface becomes one of the boundaries of the flow and for given flight conditions, the analysis of the flow can be used to determine lift and drag on an aircraft. The solution of the flow used in CFD methods can require very large amounts of processing, typically with a few seconds of flow data needing more than a day to compute with very fast processors. Although CFD methods are some 1000 times too slow for real-time simulation, they are able to compute aircraft dynamics with high accuracy, particularly in non-linear operating regions of the flight envelope. Their further advantage is that, in addition to the CFD software, they only require details of the aircraft 3D geometry to solve the equations. In practice, CFD methods are sensitive to the grid structures used to solve the flow equations.

An alternative approach to the CFD methods used to derive aircraft data is to use explicit equations based on the geometry and matched to experimental measurements, as used by Mitchell (1973) and Smetana (1984). For example, by comparing the drag curves of numerous aerofoils, a good prediction of drag can be deduced from the principle dimensions of the aerofoil. Such methods are used in the Engineering Sciences Data series (ESDU, 2006). However, it must be remembered that these are approximation techniques and that the equations used may be simplified owing to lack of empirical data.

There is one further aspect of modelling data that is easy to overlook. Once a model has been developed, it needs to be validated. Most simulator data packages will provide extensive data to check simulator data against aircraft data, typically in the form of performance graphs or plots of the response of an aircraft to a specific input. It may be straightforward to replicate pilot responses

Node	1	2	3	4

Code executed each frame	formpkt()
	sendpkt()
	repeat
	  p = getpkt()
	  if p = 2
	    copypkt(2)
	    pkt2found = true
	  endif
	  if p = 3
	    copypkt(3)
	    pkt3found = true
	  endif
	  if p = 4
	    copypkt(4)
	    pkt4found = true
	    respond to IOS
	  endif
		repeat
		  p = getpkt()
		  if p = 1
		    copypkt(1)
		    pkt1found = true
		  endif
		until pkt1found
		
		pkt1found = false
		formpkt()
		sendpkt()
		repeat
		  p = getpkt()
		  if p = 3
		    copypkt(3)
		    pkt3found = true
			repeat
			  p = getpkt()
			  if p = 1
			    copypkt(1)
			    pkt1found = true
			  endif
			  if p = 2
			    copypkt(2)
			    pkt2found = true
			  endif
			until pkt1found &
			     pkt2found
			
			pkt1found = false
			pkt2found = false
			formpkt()
			sendpkt()
				repeat
				  p = getpkt()
				  if p = 1
				    copypkt(1)
				    pkt1found = true
				  endif
				  if p = 2
				    copypkt(2)
				    pkt2found = true
				  endif
				  if p = 3
				    copypkt(3)
				    pkt3found = true
				  endif
				until pkt1found &
				     pkt2found &
				     pkt3found

Node	1	2	3	4

	`until pkt2tfound &`
	`      pkt3found &`
	`      pkt4found`
		`endif`
		`if p = 4`
		`    copypkt(4)`
		`    pkt4found = true`
		`    respond to IOS`
		`endif`
		`until pkt3found &`
		`      pkt4found`
			`repeat`
			`if p = 4`
			`    copypkt(4)`
			`    pkt4found = true`
			`    respond to IOS`
			`endif`
			`until pkt4found`
				`pkt1found = false`
				`pkt2found = false`
				`pkt3found = false`

	`pkt2found = false`
	`pkt3found = false`
	`pkt4found = false`
		`pkt3found = false`
		`pkt4found = false`
			`pkt4found = false`	`formpkt()`
				`sendpkt()`

	execute I/O functions	**execute flight model**	**execute nav functions**	**execute IOS functions**
	wait for end-of-frame			

Initial values	`pkt2found = false`
	`pkt3found = false`
	`pkt4found = false`
		`pkt1found = false`
		`pkt3found = false`
		`pkt4found = false`
			`pkt1found = false`
			`pkt2found = false`
			`pkt4found = false`
				`pkt1found = false`
				`pkt2found = false`
				`pkt3found = false`

in a simulator (but not vice versa) and then the simulated response can be checked against the response from actual flight data. Of course, if the two sets of data correlate well, the simulator designer is confident that the simulator provides an accurate model of the aircraft dynamics. The difficulty arises where the two sets of data do not match. The simulator designer has to locate the source of the inconsistency. However, with several variables depending on other variables, it is very difficult to isolate the true cause of a problem. In particular, it is tempting to 'tweak' the data package to match the aircraft data or to use opinion from pilot assessments of a simulator to adjust the model. The main problem is that changing the model in one place to ensure a good data match, may introduce errors in other areas of the model. This is a sensitive area for commercial airlines. If the simulator manufacturer matches the simulator responses to the corresponding aircraft data, but the airline pilots do not accept that the model has the appropriate handling characteristics, it may not be clear where the change should be implemented. The simulator manufacturer is likely to accept objective validation of the sets of data, whereas the airline may favour the subjective views of its senior pilots during simulator acceptance. For this reason, the regulatory authorities are usually involved in the acceptance tests and validation of an airline flight simulator.

References

Benyon, P.K. (1968) A review of numerical methods for simulation. *Simulation*, **11**(5), 219– 238.

Burns, A. and Wellings, A. (2001) *Real-time Systems and Programming Languages*, Pearson Education.

Cardullo, F.M., Kaczmarck, B. and Waycechowsky B.J. (1991) A comparison of several numerical integration algorithms employed in real-time simulation. AIAA Flight Simulation Technologies Conference, New Orleans.

ESDU (2006) *Computer Program for Estimation of Aerofoil Characteristics at Subcritical Speeds: Lift-curve Slope, Zero-lift Incidence and Pitching Moment, Aerodynamic Centre and Drag Polar Minimum*, Item 06020, Engineering Sciences Data, London.

Griffiths, D.V. and Smith, I.M. (1991), *Numerical Methods for Engineers*, Blackwell Scientific Publications, Oxford.

Horowitz, P. and Hill W. (1989) *The Art of Electronics*, Cambridge University Press, Cambridge.

Howe, R.M. (1989) An improved numerical integration method for flight simulation. AIAA Flight Simulation Technologies Conference, Technical Paper (A89-48376 21-09), Boston, MA, Washington, DC, 310– 316, http://www.aviationsystemsdivision.arc.nasa.gov/publications/hitl/rtsim/index.shtml

Howe, R.M. (1991) A new family of real-time predictor-corrector integration algorithms. *Simulation*, **57**(3), 177– 186.

McFarland, R.E. (1997) *Stability of Discrete Integration Algorithms for a Real-time, Second-order System*, NASA, Ames Research Centre.

Mitchell, C.G.B (1973) *A Computer Programme to Predict the Stability and Control Characteristics of Subsonic Aircraft*, TR 73079, Royal Aircraft Establishment.

Murphy, J., Ridout, D. and Mcshane. B. (1988) *Numerical Analysis, Algorithms and Computation*, Ellis Horwood Ltd., Chichester.

Press, W.H., Teukolsky, S.A. and Vetterling, W.T. (2007) *Numerical Recipes: The Art of Scientific Computing*, Cambridge University Press, Cambridge.

Smetana, F.O. (1984) *Computer-assisted Analysis of Aircraft Performance Stability and Control*, McGraw Hill, New York.

Smith, M.J. (1979) *Mathematical Modelling and Digital Simulation for Engineers and Scientists*, John Wiley & Sons, Ltd, New York.

Wood, M.B. (1992) An Improved Adams-Bashforth Numerical Integration Algorithm, Defence Research Agency, Report No. TMA&P7, Controller HMSO, London.

3

Aircraft Dynamics

3.1 Principles of Flight Modelling

Every airline passenger, who has walked out onto an airport apron and looked up at the airliner they were about to board, must have had a momentary thought, wondering if such a large heavy aircraft would get off the ground, let alone fly several thousand miles. Despite such understandable reservations, the aerofoils of gliders, light aircraft, airships, helicopters, airliners, military jets, Concord and the Space Shuttle all generate sufficient lift for flight. Moreover, these aircraft can be controlled by humans and computers and they are capable of transporting people or freight safely over long distances at relatively high speed.

The principles of mechanics and fluid flows were established by Newton, Bernoulli, Euler, Lagrange and Laplace and then applied to aircraft dynamics by G. H. Bryan, F. W. Lanchester, H. B. Glauret, B. M. Jones and many others within a few years of the first flight by the Wright brothers. It was necessary to understand the aerodynamic forces produced during flight. First, the airframe had to withstand the applied forces. Secondly, an understanding of the aerodynamic forces enabled faster or more efficient aircraft to be developed. Thirdly, the aircraft had to controlled by a pilot, implying an understanding of the dynamics of the aircraft.

Fortunately, for the flight simulator designer, many of the equations, which underpin the derivation of aerodynamic forces, are well understood and can be applied to modelling the dynamics of an aircraft. The purpose of this chapter is to understand how the flight simulator designer can apply these principles of aerodynamics and flight dynamics to develop an accurate model of the behaviour of an aircraft, particularly its handling qualities, stability and performance. Building on the concepts developed in Chapter 2, the equations that underpin aircraft motion will be developed to enable an aircraft model to be written as a software program.

Many of these equations are also used in modelling aircraft dynamics and more detailed derivations are given in Etkin and Reid (1996), Boiffier (1998) and Stevens and Lewis (2003). However, it is important to appreciate that, in addition to the equations of motion, flight models require both the data needed to construct the model and also the data to validate the model. Unfortunately, this data is expensive to acquire and, for operational aircraft, is a valuable asset to manufacturers and simulator companies. Consequently, there are very few detailed flight models in the open literature. Often the data that is in the public domain lacks essential components and the reader is advised to seek out many of the excellent National Aeronautics and Space Administration (NASA) reports such as Hanke's (1971) model of the Boeing 747-100, Nguyen *et al.*'s (1979) model of the F-16A and Heffley and Jewell's (1972) report which contains rudimentary data for several aircraft.

Principles of Flight Simulation D. J. Allerton
© 2009, John Wiley & Sons, Ltd

3.2 The Atmosphere

Aircraft fly in the atmosphere, the gas that surrounds the earth. In this book, we will constrain the flight envelope to the troposphere, which is found up to an altitude of about 11,000 m. Admittedly, space vehicles and high altitude aircraft can fly higher into the stratosphere (and beyond). Although the atmospheric properties can still be derived in the upper regions of the atmosphere, they are beyond the scope of this book.

In the troposphere, air is considered to be a gas, obeying Boyle's law in terms of the relationship between pressure, temperature and density. Three quarters of the mass of the atmosphere is contained in the troposphere and is kept in place by the gravitational field of the earth. It is often assumed that gravity is a constant, but the effects of the shape of the earth and the distribution of the mass of the earth can affect the gravitational forces around the earth, as discussed in more detail in Section 3.3.5. Gravity is also affected by altitude, reducing as altitude increases. The effect of the latter variation is that the geometric height of an aircraft (its altitude) also has a geopotential height to take account of the reduction of gravitational force on the air particles where it is flying. The geopotential altitude is given by

$$h = \frac{R_0 z}{R_0 + z} \qquad (3.1)$$

where h is the geopotential altitude, R_0 is the earth radius (6,356,766 m) and z is the geometric altitude. For example, if $z = 11,000$ m then $h = 10,813$ m; a reduction in altitude of 1.7%. Such errors are an important consideration for the simulator designer. Strictly, the geopotential altitude should be used to calculate temperature, pressure and density. Depending on the required fidelity of the simulator, it may be essential to model such effects or alternatively, these effects can be ignored if they do not have any impact on training. Inevitably, there is a trade-off between the accuracy of a model and the computational complexity of the model and this consideration applies as much in modelling the atmosphere as it does to the modelling of aircraft dynamics.

Temperature varies with altitude, reducing as altitude increases in the troposphere. The rate at which the temperature reduces is known as *temperature lapse* and surprisingly, it is constant throughout the troposphere. Temperature is given by

$$T = T_0 - T_L H \qquad (3.2)$$

where T is the temperature (°K), T_0 is the standard sea level temperature (°K), T_L is the temperature lapse and H is the geopotential altitude (metres). The lapse rate is constant at $0.0065°$ m^{-1}. T_0 is usually taken to be 288.15°K (15 °C). At 11,000 m, the temperature will have dropped by 71.5 to -56.5 °C. This calculation assumes dry air and of course, the temperature at sea level is not always 15 °C. In computing the temperature, the temperature lapse is modified, so that daily variations of temperature at low altitude are included, with temperature reducing to the standard value of 216.65 °K at 11,000 m. Air temperature is particularly important in the modelling of engine performance, having a significant effect on the performance of piston engines and jet engines.

Air pressure results from the mass of the column of air molecules above a unit area at a given altitude. Clearly, air pressure reduces as altitude increases. In the United States Standard Atmosphere Model, 1976 (Anon., 1976), the pressure is computed as follows:

$$P = P_0 e^{\frac{g_0 M_0}{R_s T_L} \ln\left(\frac{T_0}{T}\right)} \qquad (3.3)$$

where P is the pressure (Pa), g_0 is the acceleration of gravity (m/s^2 at a latitude of 45 °), P_0 is the pressure at g_0 (Pa), M_0 is the mean molecular weight of air (28.9644 kg/kmol), R_s is the gas constant

(8314.32 Nm/kmol/°K, T_L is the temperature lapse rate (-0.0065 °/m), T_0 is the temperature at sea level (°K) and T is the temperature (°K).

Knowing the temperature and pressure of a gas, the density calculation is derived from the gas law

$$\rho = \frac{P}{T} \cdot \frac{M_0}{R_s} \tag{3.4}$$

Air density reduces from 1.225 kg/m³ at sea level as altitude increases and is used in the computation of the aerodynamic forces and propulsive forces.

An alternative method of computing these terms is to use lookup tables derived from published tables of atmospheric data, interpolating between tabular values. However, the equations given above are straightforward to implement and are commonly used in simulation.

The speed of sound is also provided in an atmospheric model and depends on temperature

$$a = \sqrt{\gamma T \frac{R_s}{M_0}} \tag{3.5}$$

where a is the speed of sound (m/s) and γ is the ratio of specific heats for ideal diatomic gases ($\gamma = 1.4$). The speed of sound is used in the computation of the Mach number, which is needed for aircraft displays and may also be used to derive aerodynamic coefficients needed for the flight model and the engine model. The Mach number is given by

$$M = \frac{V}{a} \tag{3.6}$$

where M is the Mach number, V is the free stream velocity of the aircraft and a is the speed of sound.

The program in Example 3.1 prints the values of temperature, pressure, density and speed of sound from sea level to 10,800 m with close correspondence to the values given in the appendices of McCormick (1995).

Example 3.1 Atmospheric model

```
/* atmos-1.c
   computation of atmospheric temperature, pressure, density
   based on the NASA implementation of the US Standard Atmosphere 1976
      http://modelweb.gsfc.nasa.gov/atmos/us_standard.html
   only valid for troposphere (sea-level to 36,000 ft) */

#include <stdio.h>
#include <stdlib.h>
#include <math.h>

const double Rs  =    8314.32;  /* Nm/(kmol K), gas constant */
const double M0  =     28.9644; /* kg/kmol, mean molecular weight of air */
const double g0  =      9.80665;/* m/s^2, accn of gravity at 45.5425 deg lat. */
const double r0  = 6356766.0;   /* m, Earth radius at g0 */
const double P0  =  101325.0;   /* Pa, air pressure at g0 */
const double T0  =     288.15;  /* K, standard sea-level temperature */
const double Td  =     273.15;  /* K, 0 degrees C */
```

```
    const double Gamma =      1.40;   /* Ratio of Specific heats for an ideal
                                         diatomic gas */

    int main(int argc, char *argv[])
    {
        double As;      /* constant */
        int Z;          /* geometric height m */
        double H;       /* geopotential height m */
        double P;       /* pressure Pa */
        double T;       /* temperature K */
        double rho;     /* density kg/m^3*/
        double Cs;      /* speed of sound m/s */

        As = g0 * M0 / Rs;

        for (Z=0; Z <= 10800; Z+=300)
        {
            H = r0 * (double) Z / (r0 + (double) Z);
            T = T0 - 0.0065 * H;
            P = P0 * exp(As * log(T0 / T) / (-0.0065));
            rho = (P / T) * (M0 / Rs);
            Cs = sqrt((Gamma * Rs / M0) * T);

            printf("%5d %6.2f %8.4f %8.4f %6.2f\n", Z, T, P, rho, Cs);
        }
    }
```

Generally, an atmospheric model is based on the International Standard Atmosphere (ISA), with sea level conditions of temperature 15 °C, pressure 101,325 Pa and density 1.225 kg/m^3. It is necessary to compensate for any variation in sea level temperature or pressure in simulating the atmospheric conditions that prevail on a daily basis. For example, the take-off performance of a light aircraft is significantly reduced on a hot day. Similarly, barometric pressures used in setting aircraft altimeters and air traffic control (ATC) clearances must also be replicated in a flight simulator, particularly the variation of barometric pressure for QFE, QNH or standard altitude settings. Variations from ISA values are incorporated in outside air temperature gauges and altimeters in simulation. However, it is the absolute altitude of the centre of gravity (cg) of the simulated aircraft that is used in all aerodynamic and propulsive computations, where density, pressure and temperature values are derived from the equations above, modified to match the prevailing sea level conditions.

3.3 Forces

3.3.1 Aerodynamic Lift

The density of air is so small in comparison with the density of an aircraft that it offers little buoyancy. However, if an aircraft is propelled through the air, the path taken by air molecules passing over the aircraft wings may be different from the path taken by air molecules passing under the aircraft wings, resulting in a pressure differential. By careful design of the wing profile, the net lower pressure above the wing is sufficient to produce an upward lifting force on the wings of the aircraft. For a Boeing 747 aircraft accelerating along a runway, this pressure differential will be capable of lifting a 300,000-kg aircraft off the ground at about 160 kt. This lift force will be sustained to enable the aircraft to climb to an altitude 6 miles above the surface of the earth and maintain this altitude.

The generation of a lift force derives from the remarkable properties of an aerofoil and underpins the science of aerodynamics. One method to establish the lifting properties of an aerofoil is to keep the aerofoil stationary in an air flow and this is readily achieved in a wind tunnel. As the air flow in the tunnel passes over the aerofoil, the deflection force can be measured using a spring balance. The early experimental aerodynamicists quickly appreciated that the shape of the aerofoil influences the lift generated by an aerofoil and a very large number of aerofoils have been developed and catalogued. For example, the profile of the National Advisory Committee for Aeronautics (NACA) 6409 aerofoil (9% thick, 40% maximum camber, 6% camber) is shown in Figure 3.1.

The magnitude of the lift force also depends on the velocity of the airstream passing over the aerofoil and the density of the air. It was quickly established that the generated lift varies with the square of airspeed and in direct proportion to air density. However, the remarkable characteristic of any aerofoil is that the magnitude of the lift varies with the incident angle it makes to the wind. This angle is usually referred to as α, the angle of attack, or angle of incidence, as shown in Figure 3.2, where the aerofoil in Figure 3.1 is rotated by $\alpha°$.

The lift derived from the aerofoil increases as α increases. Surprisingly, the lift increases linearly with α up to about 10–15°. For many aerofoils, even with an angle of attack of 0° the aerofoil generates some positive lift force. The lift generated by an aerofoil is given by the well-known equation:

$$\text{lift} = 1/2 \rho V^2 s C_L \tag{3.7}$$

where ρ is the density of air, V is the airspeed, s is the aerofoil area and C_L is the coefficient of lift, which depends on the shape of the aerofoil and α. The coefficient of lift C_L is unique for each aerofoil and usually derived from extensive wind tunnel tests or CFD studies. The characteristic of

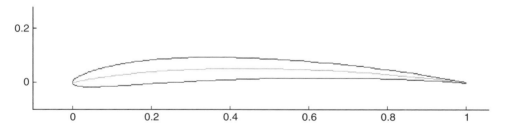

Figure 3.1 NACA 6409 aerofoil profile

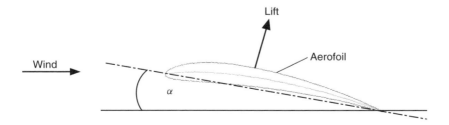

Figure 3.2 Angle of incidence

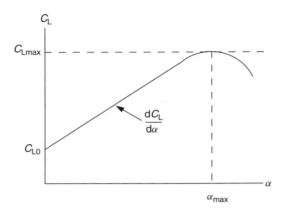

Figure 3.3 Lift curve characteristic

a typical C_L curve, is shown as a function of α in Figure 3.3. The coefficient of lift increases from C_{L0} at $\alpha = 0°$ but reaches a maximum at α_{max} where the coefficient of lift is C_{Lmax}.

After α_{max} the flow of air over the upper surface of the aerofoil becomes turbulent and the coefficient of lift reduces. This condition is known as *stalling*; at a certain value of angle of attack, the airflow will start to break up, often resulting in significant buffeting on the aircraft. If the angle of attack is increased further, the lift will reduce to a very small value and the aircraft will lose altitude very quickly. Such a manoeuvre can be very dangerous near the ground and pilots are taught to detect the onset of a stall and most aircraft are equipped with an aural stall warning device, which is activated a few degrees before α_{max}. There are several important points to appreciate regarding the relationship between lift and angle of attack:

- Stalling depends on the angle of attack rather than airspeed. Although many flight manuals give a stall speed for an aircraft, this is because most aircraft are not equipped with an angle-of-attack gauge. The stall speed is simply the minimum speed at which the aircraft can be flown in level flight before α will exceed α_{max};
- For most aerofoils, the slope of the curve is almost constant up to a few degrees below α_{max};
- The actual characteristic of the lift curve beyond α_{max} is difficult to determine for large values of α. Some experiments have been undertaken with models in wind tunnels but the non-linear aerodynamic characteristics defy analytical solution and are difficult to obtain from experimentation or CFD studies;
- The angle of attack can be negative as well as positive, particularly in inverted flight. For aerobatic aircraft, the aircraft needs to generate almost as much lift inverted as it does in normal flight and the aerofoil will be designed so that the lift curve is almost symmetric.

One further assumption about the lift curve is that the aerofoil shape remains constant in flight. However, aerodynamicists realized that, by adding a flap to the trailing edge or slats to the leading edge, they could improve the lift characteristics, particularly during landing and take-off. Typically, the data provided by a manufacturer will include a set of lift curves for the full range of flap positions, as illustrated in Figure 3.4.

By increasing the flap angle (the pilot lowers the flaps) the lift is increased by a small but significant amount. Figure 3.4 also shows that the maximum value of lift varies for different flap settings (grey dots); not only is C_{Lmax} increased, but the angle of attack at the point of the stall is also increased.

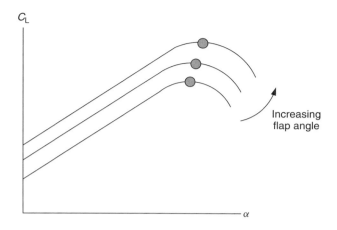

Figure 3.4 Effect of flaps on the lift curve

Although we have only considered a single aerofoil, in practice, the whole aircraft, comprising the fuselage, wings, tail plane, engine nacelles and even the undercarriage may contribute to lift. The tail plane of a Boeing 747 generates more lift than most light aircraft. The data used in a flight simulator to compute C_L will therefore be provided for the whole aircraft, providing the parasitic lift C_{L0} and the induced lift $dC_L/d\alpha$ as a function of Mach number, angle of attack, flap position and possibly other parameters. Depending on the application, data for negative values of α and high incidence values of α (post-stall) will be supplied to appropriate levels of accuracy.

Lift can also vary with altitude close to the ground, where the ground effect of the airflow over the aircraft results in a change of lift caused by the interaction of the flow with the ground. With swept wing aircraft, the air is effectively 'trapped' under the wing at high incidence in the landing phase, acting as a 'cushion of air' below the aircraft. The lift characteristic of a wing is also affected by the material properties of its surface. In the case of icing, the airflow over the wing is altered and there is a significant reduction in lift. Such effects must be accurately modelled in a flight simulator if the simulator is used for flight training in the take-off and landing phase, otherwise errors caused by neglecting such effects could produce unrealistic handling characteristics during training conducted in the flight simulator.

There is one further consideration if the whole aircraft lift is treated as a single entity. Strictly, the lift from the wings, tail and fuselage should be computed separately. In turning flight, the wing into wind will generate more lift than the trailing wing (which may be shielded by the fuselage) so that the angle of attack varies for the two wings, causing one wing to stall before the other. In flight, such an event can lead to a wing drop and quickly develop into a spin. In some simulators, spinning may be an important aspect of training and separate lift computations are performed for each wing. If the whole wing is treated as a single source of lift, it is not possible to simulate spinning, unless some additional component is introduced into the equations of motion to induce spinning artificially at high incidence angles.

Although the angle of attack α has been defined as the angle an aerofoil makes to the wind, the actual angle of attack is computed in a simulator from the forward and downward velocities measured in the body frame of the aircraft, with respect to the cg. The aircraft wing may be offset with respect to the aircraft axes and may also have a twist along the wing, so that the actual angle of attack of the wing has a small angular offset from the airframe axes. In computation of the aerodynamic lift, the actual wing angle of attack, including the additional wing incidence, must be used rather than the aircraft angle of attack.

Although lift cannot be measured directly in flight, in steady-state conditions, with an aircraft flying straight-and-level, the lift must be equal to the mass of the aircraft, giving

$$mg = \tfrac{1}{2}\rho V^2 s C_L \tag{3.8}$$

$$C_L = \frac{2mg}{\rho s V^2} \tag{3.9}$$

There is, therefore, a unique value of C_L for each steady-state value of V for straight-and-level flight. In level flight the angle of attack is the same as the aircraft pitch attitude θ. If V and θ can be measured, the relationship between C_L and α can be derived from flight tests. More commonly, airspeed sensors and angle of attack sensors are used in flight data recording in flight trials.

3.3.2 Aerodynamic Side force

Viewed from above (in plan form), most aircraft are symmetric. However, if the fuselage is not aligned with the direction of flight, it will make an incident angle with the wind, known as *the angle of sideslip*, β. Such conditions can arise if the engines forces about the longitudinal axis are different or if the rudder or an aileron is deflected. In propeller aircraft, side forces may result from the propeller slipstream impacting the aircraft tail. In turboprop aircraft and jet engine aircraft, the high rotational speeds of the engine can induce gyroscopic moments leading to turning motion. Under these conditions, the fuselage acts as a large aerofoil, with an angle of incidence β to the wind, as shown in Figure 3.5. Consequently, lift is generated by the fuselage but in the direction along the wing, causing a lateral force, known as *side force*.

Although the side force generated will be considerably less than the lift force on an aircraft, it is still significant. Generally, pilots try to minimize any side force, partly because the occupants of the aircraft will feel uncomfortable (being pressed against the side of the fuselage) but mainly because the aircraft is flying inefficiently. Side forces also occur in turning flight; as the aircraft is banked, a component of the gravitational force is acting along the wing axis. In flight, a pilot will reduce the side force by use of the rudder (or rudder trim) and the balanced use of the aileron and rudder controls. In an open cockpit aircraft, the pilot will feel the full force of the slipstream when $\beta = 0$. In very early aircraft, sideslip was actually indicated by a small length of string, attached to the airframe in the flow of the slipstream. Subsequently, the slip ball (a ball bearing in a curved

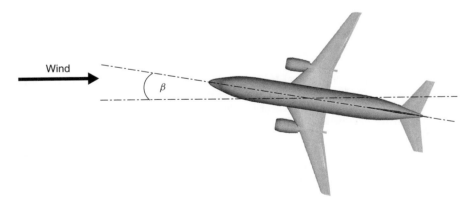

Figure 3.5 Angle of sideslip

glass tube containing a viscous fluid) has provided pilots with a visual indication of sideslip (or more strictly, damped side force).

Side force is given by the following equation

$$\text{side force} = \tfrac{1}{2}\rho V^2 s C_Y \tag{3.10}$$

where C_Y is the coefficient of side force, typically provided as a function of β and rudder input.

The side force term may depend on Mach number, angle of sideslip, flap setting and the shape of fuselage. With flaps, slats and spoilers, any failures of these subsystems may also result in the generation of a side force and therefore, the contribution of these subsystems must also be modelled in order to simulate the effects on side force, resulting from system failures.

3.3.3 Aerodynamic Drag

Based on the explanation of lift in Section 3.3.1, it is tempting to ask why aircraft are not flown close to α_{max} at all times, in order to generate the maximum lift. The answer is simple; while there is a gain in lift from increasing the angle of attack, the surface area in the slipstream also increases, resulting in increased drag on the aircraft. From the earliest days of flight, aerodynamicists were concerned with understanding both lift and drag. If drag increases, the aircraft will lose speed and therefore lose lift, so it was realized that the drag characteristic of an aircraft affects its performance and safety.

Although most aircraft have an elegant aerodynamic shape, the aircraft designer has to compromise between structural integrity and performance. Consequently, just as an aircraft has a unique set of lift curves, so it will have a unique set of drag curves and this data will also be produced by an aircraft manufacturer from instrumented flight trials, wind tunnel tests or CFD studies.

There is also one important difference between the C_L/α curve and the C_D/α curve; for lift the coefficient is linear, whereas for coefficient of drag (for most aircraft) it is quadratic (known as the *drag polar*), up to a few degrees before the stall. Typically, C_D is given by the following equation:

$$C_D = C_{D0} + C_{D1} C_L^2 \tag{3.11}$$

where C_{D0} is the parasitic drag coefficient and C_{D1} is the induced drag coefficient. Past the stall angle α_{max} the lift reduces quickly whereas the drag will continue to increase. This is a particularly important consideration in modelling the stall. The behaviour in the non-linear region beyond the stall varies widely. However, the effect of drag on an aircraft will affect the rate at which airspeed reduces at the onset of stalling. In practice, flight trials data in the post-stall region is difficult to obtain, particularly for civil aircraft.

Aircraft drag is also affected by the flaps. While lowering the flaps will increase lift, it also increases the area of the wing directly in the slipstream, increasing the drag. There are several additional sources of drag:

- Some aircraft have spoilers or speed brakes (surfaces raised on the wing to act as a brake), which increase drag;
- The undercarriage will increase the parasitic drag of an aircraft; if the undercarriage is retractable, the actual drag will vary as the undercarriage is raised or lowered;
- Drag will also vary with sideslip as the area of the aircraft in the slipstream varies with the sideslip angle;
- For propeller aircraft, a windmilling propeller is absorbing energy and is therefore generating additional drag (although this term may be computed as a negative thrust in the propulsion equations).

Like lift, drag also varies with altitude close to the ground, where the ground effect of the airflow over the aircraft changes the coefficient of lift and consequently, the coefficient of drag. The drag characteristic of the airframe is also influenced by the surface properties. In the case of icing, in addition to a reduction in lift, there is a significant increase in drag. The simulator data will also include the effect of drag during taxiing. The drag will vary with the coefficient of friction of the runway surface, the slope of the runway, the temperature and pressure of the tyres and the mechanical friction of the wheels and braking systems. Malfunctions such a tyre burst or brake overheating are also modelled in terms of the changes in drag forces caused by such events. Obviously, data collection for major malfunctions of this form is impractical.

The drag curves supplied for an aircraft will therefore depend on C_{D0}, C_{D1}, Mach number, flap position, β and C_L. Although drag is difficult to measure in flight, wind tunnel models have provided useful approximations in the past and the parasitic drag term can be estimated from the geometric structure of the aircraft. Drag has a major influence on aircraft performance, range and endurance. However, as the drag coefficients C_{D0} and C_{D1} are relatively small, any errors in estimating these terms can introduce relatively large errors in aircraft performance.

3.3.4 Propulsive Forces

The methods used to predict the propulsive forces on an aircraft are covered in Section 3.8. The general assumption is that force is generated along the axial line of an engine. In practice, the hot gases of a jet engine are generated from a turbine rotating at high speed and the direction of thrust may vary with engine RPM. Similarly, for a propeller, each blade is generating lift (perpendicular to the axis of rotation) as the propeller rotates, but the airspeed of a segment of the propeller blade increases with the radius from the propeller hub. Again, the net effect of the generated thrust may not necessarily be along the axial line of the engine/propeller assembly.

For commercial aircraft with reverse thrust mechanisms, further data is needed to predict the thrust when the engine is operating in reverse thrust to retard the aircraft. In addition to the thrust delivered, the delays in reversing the engine 'buckets' and the spool-up times of the engine are critical to the accuracy of the ground handling model in a flight simulator, for example, in simulating an aborted take-off.

In a single-engine aircraft, the engine thrust may not act through the cg of the aircraft and the engine assembly may not necessarily be aligned to the principal axes of the aircraft. Consequently, changes in thrust may induce changes in pitching, rolling or yawing moments. For single-engine propeller aircraft, the airflow from the propeller may result in complex interactions with the airframe. For some aircraft, particularly single-engine turboprops, the airflow over the tail can have a significant effect, inducing large moments immediately after take-off or affecting recovery procedures in spinning. These are both flight regimes where the collection of data is dangerous or where other non-linear aerodynamic effects may be difficult to isolate in captured data.

For multi-engine aircraft, the resultant lines of thrust must be computed and resolved in the principal axes of the aircraft. In practice, the geometric data of engine location and alignment is known and the transformation of engine thrust from the engine axes to the airframe axes is straightforward.

In addition to data for the normal operational envelope of an engine, it is also necessary to model malfunctions. For variable-pitch propellers, the effect of selecting blade fine pitch to feather the propeller must be accurately modelled in terms of the resultant thrust and response of the propeller system. A flight simulator is without doubt the safest place to practice engine failures but if the effects and the response to pilot inputs are not correctly modelled, the outcome could be fatal if similar procedures were followed during an airborne malfunction. For jet engine aircraft, effects of engine surge and stall for multistage compressors, resulting from inappropriate pilot inputs, or high angles of attack or sideslip can result in disturbances to the airflow through the engine and such effects must be correctly modelled. Major malfunctions that are modelled in airline simulators include bird strikes, engine fires and mechanical failures. In recent years, the number of channels monitored by flight data recorders on aircraft and the sampling rates have increased, enabling data from disturbances and incidents to be collected and analysed. Otherwise, the collection of flight data at these extremes of engine operations is either unacceptably dangerous or prohibitively expensive.

3.3.5 Gravitational Force

Strictly, gravity is straightforward to compute. It acts with a standard value through the cg of the aircraft in a direction towards the centre of the earth. In many simulators, this simplification may be acceptable. However, gravity reduces with distance from the centre of the earth, the earth is an oblate sphere rather than a perfect sphere and the gravitational attraction varies slightly over the surface of the earth. If these variations in g are modelled, the local gravitational vector can be computed (typically in the form of lookup tables) at a given latitude, longitude and altitude, based on an international standard (Pavlis *et al.*, 2008).

The earth is ellipsoid in shape, with the semi major axis 6378.2 km and the semi minor axis 6356.8 km. Although this difference is only 21.4 km, it introduces a difficulty in defining the centre of the earth and therefore the latitude of a point in space. If the centre of the earth is defined as the intersection of the equatorial plane and a plane through the poles, then the geocentric latitude is the angle the point makes to this centre. The gravitational latitude is the angle an object will make as it falls, but owing to the distortion of the earth may not be towards the centre. The geodesic latitude is given by the angle made by a perpendicular at the surface of the earth, which also varies with the shape. For the majority of flight simulation applications, gravity is assumed to be a function of latitude and altitude.

3.4 Moments

To climb or descend, a pilot is (in effect) controlling the lift on the aircraft, which is achieved by setting the desired angle of attack, which in turn is accomplished by selecting an appropriate pitch attitude, where the pilot moves the control column to select an elevator angle. In banking the aircraft, which is achieved by movement of the rudder and aileron controls, the lift vector has both vertical and horizontal components, resulting in a change of heading or turning flight. In other words, a pilot selects an aircraft attitude (and power setting) to achieve a desired performance and trajectory. The forces of thrust, lift, drag and side force determine aircraft performance whereas the moments caused by displacing the primary control surfaces produce the aircraft attitude.

In most light aircraft and commercial transport aircraft, the pilot will apply forward or backward motion to the control column to move the elevator on the tail of the aircraft. Similarly, the control column wheel is rotated to move the ailerons on the trailing edge of the wings. For some light aircraft and most military aircraft, movement of a centre stick provides the equivalent actions. For most Airbus aircraft and some fighter aircraft, control inputs are provided by a small side-stick. In all fixed wing aircraft, the pilot presses on a pair of rudder pedals to move the rudder on the tail of the aircraft.

Although aircraft instruments display altitude, heading and airspeed, in visual flight conditions the distant horizon provides an attitude reference. If the fuselage is horizontal, the aircraft has a pitch angle of $0°$, although the aircraft may not be in level flight. If the nose of the aircraft is raised, the aircraft will have a positive pitch angle relative to the horizon. Pitch is an important flight parameter because the pilot is able to control the angle of attack and therefore the aerodynamic lift force by positioning the nose of the aircraft relative to the horizon. Similarly, the angle of bank of the aircraft is also perceived relative to the horizon; if the wings are level (parallel to the horizon), the aircraft has a bank angle of $0°$. The attitude of the aircraft is given by its angles of pitch and roll, relative to the distant horizon seen from the aircraft cockpit. Of course, the horizon may not provide a clear reference line (for example, if there is a distant mountain range), the horizon may be occluded by terrain (for example, flying in a valley or on a runway) or the horizon may not be visible (for example, in foggy or hazy conditions). However, if the horizon is not visible, an instrument known as the *attitude indicator* provides the pilot with a flight instrument to determine the aircraft attitude in pitch and roll.

A magnetic compass provides a heading reference in an aircraft (to determine the yaw angle). Although errors can arise from aircraft accelerations, magnetic interference and poor calibration, a compass enables a pilot to fly the aircraft on a specific heading. In nil wind, the track over the ground and the heading will be the same. However, in the presence of wind, the velocity vector giving the resultant track is derived from the addition of the wind vector and the aircraft velocity vector, given by its airspeed and heading.

In Section 3.3, we saw that if the primary forces on the aircraft can be determined and the aircraft mass is known, then it is straightforward to derive the acceleration, velocity and position of an aircraft. In other words, linear motion can be computed from these forces. Similarly, if the moments and the moments of inertia are known, it is possible to derive the angular accelerations, angular velocities and orientation of the aircraft. In other words, angular motion can be derived from the moments on an aircraft.

A moment is simply a force times its distance from the pivot of rotation. In the case of aircraft motion, this point of rotation is the cg of the aircraft. In practice, the location of the cg depends on the geometry of the airframe, the fuel load, the number (and seating positions) of the occupants, the amount of freight carried and possibly any weapons. Nevertheless, the computation of the position of the cg for a specific aircraft is straightforward (it will be checked by the flight crew before every flight). If the position of the cg is known, then the moment contributed by each control surface (elevator, ailerons and rudder) can be determined from the force applied by each control surface and its distance from the aircraft cg. These aerodynamic terms are provided in a data package or can be derived from wind tunnel tests in the same way that the coefficients of lift, drag and side force are measured.

Consider the pitching moment, as shown in Figure 3.6. Movement of the elevator changes its angle of attack and therefore the lift generated by the elevator. This force, applied at a distance d from the cg produces a pitching moment. If the moment of inertia of the aircraft about the pitching axis is known (it would be included in the data package, as a function of mass and configuration), then

$$\ddot{\theta} = \frac{M}{J} \tag{3.12}$$

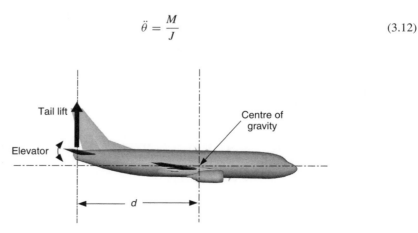

Figure 3.6 Aircraft pitching moment

$$\dot{\theta} = \int \ddot{\theta}\, dt \qquad (3.13)$$

$$\theta = \int \dot{\theta}\, dt \qquad (3.14)$$

where θ is the pitch attitude, M is the moment and J is the moment of inertia. As with any dynamic system, once a body is disturbed by an applied force, its resultant motion results in a countering force (or resistance). We will see in the following sections that the actual derivation of the pitching moment contains many terms in order to be computed accurately, but in simplified terms, the change of position of the elevator control will result in a pitching moment, causing pitching acceleration, a pitch rate and a new angle of pitch.

A similar moment occurs in the roll axis, when the ailerons are applied to bank the aircraft and in the yaw axis, when the rudder is applied to turn the aircraft or to counteract sideslip. In fact, motion in these two axes is coupled; a rolling motion induces sideslip and therefore a yawing moment and similarly, sideslip resulting from application of the rudder induces a rolling moment. In flight training, a pilot is taught to execute these manoeuvres precisely and effectively, in order to minimize any undesired motion. From the simulator designer's perspective, these equations enable the appropriate pitching, rolling and yawing motion to be computed for an aircraft.

3.4.1 Static Stability

A pilot will appreciate the handling qualities of an aircraft in terms of the difficulty experienced in flying the aircraft, particularly in responding to any disturbance. The workload in operating any system is influenced by the stability of the system. For example, balancing a broom stick is difficult, because the system is slightly unstable, whereas steering a motorcar on a motorway is relatively straightforward as the steering system is stable. One definition of stability is the degree to which the system returns to a steady-state if a disturbance is removed, for example, if a window in a house is opened, the house temperature may drop by a few degrees but gradually return to the ambient temperature when the window is closed. In the case of an F-16 aircraft, which is unstable; it would very quickly diverge if the computer control was removed.

Aircraft dynamic stability is reviewed in Chapter 7, but it is also important to appreciate the static stability of an aircraft. If an aircraft is in equilibrium, and is displaced, if the subsequent motion returns the aircraft to its original state, the aircraft is statically stable. On the other hand, if this displacement leads to further displacement away from the equilibrium state, the aircraft in statically unstable. In between these two states, if no further displacement occurs, the aircraft is said to be neutrally stable.

For an analysis of the static stability, the aircraft is assumed to be in a trimmed state, where the aircraft is not accelerating in any axis. This state is referred to as a trimmed state because it implies that no additional input is required from the pilot to maintain this state. For the longitudinal axis, this state is achieved by moving the elevator trim tabs until the aircraft maintains a steady-state flight condition in a 'hands-free' mode (no applied pilot force). However, if the aircraft is then disturbed from this state, the amount of input needed to restore the aircraft to its equilibrium state is an indication of the static stability. Two forms of analysis are undertaken: stick-fixed and stick-free. In a stick-fixed manoeuvre, the control (and the elevator) remains in place during the response, whereas in a stick-free manoeuvre, the elevator is allowed to take up its natural position following the disturbance.

Consider an aircraft in steady straight-and-level flight at an angle of attack α, which is subject to a small disturbance, which changes α and therefore introduces a pitching moment. If the disturbance results in a pitch up and the aircraft response is a pitch down counteracting moment, then the aircraft is statically stable. The condition depends on the $dC_m/d\alpha$ term for the aircraft, as shown in

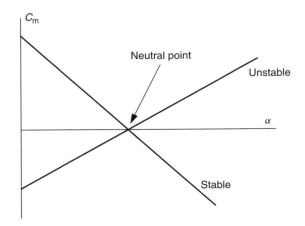

Figure 3.7 Static stability

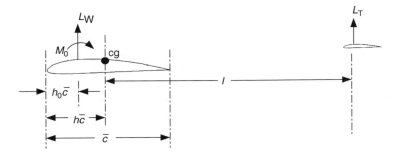

Figure 3.8 Pitching moments

Figure 3.7. If $dC_m/d\alpha < 0$, the aircraft is statically stable. If $dC_m/d\alpha > 0$, the aircraft is statically unstable and if $dC_m/d\alpha = 0$, the aircraft is neutrally stable. For negative values, the greater the magnitude of $dC_m/d\alpha$, the greater is the aircraft stability.

The major terms contributing to the static stability are the lift components provided by the wing and tail, as shown in Figure 3.8.

Summing the moments about the cg,

$$M = M_0 + L(h - h_0)\bar{c} - lL_T$$

where
M_0 is the pitching moment of the wing (and fuselage) about the cg
L is the wing lift
$\bar{c}$ is the wing mean chord
L_T is the tail lift

Expressing these terms as coefficients

$$C_M = C_{M0} + C_L(h - h_0) - C_{LT}\frac{lS_T}{\bar{c}S_W} \qquad (3.15)$$

where
- C_{LW} is the coefficient of lift of the wing
- C_{LT} is the coefficient of lift of the tail
- S_T is the tail area
- S_W is the wing area

The term $lS_T/\bar{c}S_W$ is a constant and can be replaced by $\overline{V}$ which is often referred to as the tail volume coefficient giving

$$C_M = C_{M0} + C_L(h - h_0) - \overline{V} C_{LT} \tag{3.16}$$

In the steady-state, $C_m = 0$.
Differentiating Equation 3.16 with respect to C_L,

$$\frac{dC_M}{dC_L} = (h - h_0) - \overline{V}\frac{dC_{LT}}{dC_L} \tag{3.17}$$

Assuming the aircraft lift is proportional to the angle of incidence α, then the static stability is given by

$$\frac{dC_M}{d\alpha} = a(h - h_0) - \overline{V}\frac{dC_{LT}}{d\alpha} \tag{3.18}$$

where a is the gradient of the aircraft C_L versus α curve.

Moving the aircraft cg aft increases h, increasing the gradient of the $dC_M/d\alpha$ curve and reducing the static stability. The aircraft tail is designed so that $dC_M/d\alpha$ is negative throughout the operating range of the aircraft and over the range of cg positions.

The neutral point is the position where the effect of the cg position is sufficient to give a $dC_M/d\alpha$ value of 0. The static margin is then the distance of the cg from the neutral point, normally expressed as a fraction of the wing mean chord $\bar{c}$. Further analysis of this equation, for both stick-fixed and stick-free conditions is really the domain of the aircraft designer. A small static margin will give light controls and a responsive aircraft but an aircraft that is easily disturbed from steady-state flight. Although a large static margin provides a very stable aircraft, larger tail forces are needed to manoeuvre the aircraft, which may not be acceptable, for example, where a good response is needed in the landing phase. For flight simulation, the aircraft geometry and aerodynamic derivatives are known and the main requirement is to check the aircraft stability over the cg range published for the aircraft. Methods of determining the stick-fixed neutral point in flight tests are covered in most textbooks on aerodynamics. The (simulated) aircraft is flown at different cg positions and the elevator trim angle needed to trim the pitch attitude is measured and plotted against the coefficient of lift C_L. The intersection of these curves coincides with the cg at the neutral point.

3.4.2 Aerodynamic Moments

The three primary controls: the elevator, the ailerons and the rudder are used by the pilot to select the desired aircraft attitude. However, the change of motion can induce additional moments including moments resisting the applied motion. The magnitude of these moments is clearly influenced by the geometry of the aircraft, for example, the size of the elevator surface but also depends on air density and airspeed. A point to bear in mind is that many of these terms are non-linear making analysis of the aircraft motion difficult or requiring linearization of the equations over a relatively small operating region.

The moments are computed in the stability axes and are defined by the following equations for the moments in the roll, pitch and yaw axes, respectively:

$$L_{stab} = 1/2\sigma V_C^2 sb(C_{L\beta} \cdot \beta + C_{L\delta A} \cdot \delta a + C_{L\delta R} \cdot \delta r) + 1/4\sigma Vsb^2(C_{LP} \cdot P + C_{LR} \cdot R) \quad (3.19)$$

$$M_{stab} = 1/2\sigma V_C^2 s\bar{c}(C_{M0} + C_{M\alpha} \cdot \alpha + C_{M\delta E} \cdot \delta e) + 1/4\sigma Vs\bar{c}^2(C_{MQ} \cdot Q + C_{M\dot\alpha} \cdot \dot\alpha) \quad (3.20)$$

$$N_{stab} = 1/2\sigma V_C^2 sb(C_{N\beta} \cdot \beta + C_{N\delta A} \cdot \delta a + C_{N\delta R} \cdot \delta r) + 1/4\sigma Vsb^2(C_{NP} \cdot P + C_{NR} \cdot R) \quad (3.21)$$

where
ρ is air density
V_c is airspeed
α is the angle of attack
β is the angle of sideslip
s is the wing area
b is the wing span
$\bar{c}$ is the wing mean chord
δa is the aileron deflection
δe is the elevator deflection
δr is the rudder deflection

The coefficients C_{ab} are known as the *aerodynamic derivatives*, where the subscript a denotes the axis and the subscript b denotes the source of the moment term. Considerable care is needed in formulating these equations. Generally, the aerodynamic derivatives are non-dimensional, which has the advantage of compatibility between aerodynamic data derived in imperial units and data produced from terms defined in SI units. While this assumption is often valid, aerodynamic data may be generated where the value of wing span b represents the wing semi-span rather than the full wing span. In such cases, a spurious factor of 2 can be introduced into the equations. This situation is difficult to determine unless the method of dimensioning is clearly defined or if there is sufficient test data provided to verify the equations. In some cases, dimensional aerodynamic derivatives are provided and the equivalent non-dimensional derivatives can be computed using the scale factors given in Table 3.1.

The aerodynamic derivatives constitute a major part of the data package for an aircraft. Each derivative may be a function of several variables and data is normally provided in the form of a set of tables for each derivative or its primary components.

An alternative approach is to estimate the aerodynamic derivatives from empirical rules established for aircraft design or to use sources of data to predict these values. For example, Engineering Sciences Data Unit (ESDU) data sheets (ESDU, 1966) provide methods and tables to evaluate aerodynamic derivatives for light aircraft. Similarly, Smetana (1984) provides the sources of data and the methods used to predict these derivatives (including the Fortran programs). Mitchell (1973) also provides methods to estimate derivatives including swept wing fighter aircraft.

Table 3.1 Non-dimensional aerodynamic derivatives scale factors

Dimensional term	Scale factor
$C_{L\beta}$, C_{Lda}, C_{Ldr}, $C_{N\beta}$, C_{Nda}, C_{Ndr}	$1/2\rho V^2 sb$
C_{Lp}, C_{Lr}, C_{Np}, C_{Nr}	$1/4\rho V sb^2$
C_{M0}, $C_{M\alpha}$, C_{Mde}	$1/2\rho V^2 s\bar{c}$
C_{Mq}, $C_{M\dot\alpha}$	$1/4\rho V s\bar{c}^2$

3.4.3 Aerodynamic Derivatives

The validation of a flight model is covered in Chapter 7. If a step input is applied to the elevator (for example, moving the elevator control from its trimmed position backwards by 2 cm for one second, before returning the control to its original position), the resultant oscillatory pitching motion depends on the coefficients in the pitching moment equations. The response in pitch over a few seconds is known as the *short period phugoid*, whereas the response in pitch over a minute or so is known as the *long period phugoid*. Similarly, the response in yaw to a step input applied to the rudder results in an oscillatory yawing motion, known as *Dutch roll*. The phugoid and Dutch roll response provide a unique 'signature' for an aircraft and are used to validate the equations of motion. In addition to the Dutch roll response, an aircraft also exhibits a spiral mode, which is the degree to which a rolling motion induces further rolling motion. These terms also determine the specific handling characteristics of an aircraft and the contribution of the individual aerodynamic derivatives to aircraft handling is summarized in Table 3.2.

One further term that is often included, particularly for propeller driven aircraft, is the additional moment resulting from interaction between the airflow from the propeller and the airframe. This is in addition to the direct moments resulting from the off-axis thrust, particularly in the pitch axis (for example, with under-slung engines) and the yaw axis with multi-engine aircraft. For low-wing single-engine aircraft, increasing power reduces the longitudinal stability but increases the directional stability and elevator effectiveness. A NASA study using a full-scale wind tunnel (Fink and Freeman, 1969) provides a useful source to predict propeller effects for a light twin-engine aircraft. Wolowicz and Yancey (1972) predict the lateral-directional static and dynamic stability characteristics of a twin-engine aeroplane, including power effects. For aerobatic aircraft and turboprops, where the thrust to weight ratio is high, the effects of engine thrust can be significant at low speed (during take-off and landing) and considerable care is needed to model these effects. Similarly, for jet engine aircraft, the gyroscope effects of an engine can induce significant rolling and yawing moments.

Several of these aerodynamic derivatives are dependent on the angle of attack. At high incidence, the airflow over the tail reduces and the effectiveness of the elevator and rudder may reduce significantly. In addition, for jet aircraft, compressibility effects must be taken into account above Mach 0.8. In practice, the accuracy of the aerodynamic data is a major contribution to the fidelity of the flight model. However, with so many derivatives, which are often combinations of half a dozen components, in turn defined as a function of several variables, the scope for error is high and it can be very difficult to isolate an individual variable. For example, if the roll rate is insufficient, the cause may be an error in C_{Lda} or C_{Lp}. If the error is caused by too high damping, although the problem may be reduced by increasing the aileron effectiveness, the modification may still result in undesirable handling qualities. Consequently, considerable care is needed in the methods used to acquire aircraft data, in the implementation of the flight model equations and most particularly, in the rigour of the verification of the flight model.

Most aircraft provide trim controls in pitch and yaw (and sometimes in roll). In the trimmed (non-accelerating) state, a trim control can relieve the force the pilot applies to the control column so that the aircraft maintains the desired pitch attitude with zero pilot force (the 'hands-off' position). These trim controls are usually provided as tabs attached to the respective control surfaces and are controlled by a trim wheel in the cockpit, or in the case of large aircraft, by a small switch, which activates an electric motor to drive the trim tab. For an aircraft, movement of the elevator trim tab alters the mechanical trim datum. In a flight simulator, trimming either alters the elevator position, necessitating an irreversible mechanical trim linkage to move the column or, in the case of a fixed spring control loading, the trim input is added to the column position; in the trimmed state, the elevator control input is zero and the trim input provides an appropriate offset. As the trim effect varies with airspeed, it is usually added to the respective control input rather than as an offset to C_{M0}.

Table 3.2 Aerodynamic derivatives

C_{M0}	The contribution of the distribution of mass of the aircraft to the pitching moment. C_{M0} changes with the centre of gravity (cg) position and the effect of the flaps and undercarriage are often included in this term. The cg position depends on the mass and distribution of the fuel, crew, passengers and luggage and is usually quoted as a percentage of $\bar{c}$.
$C_{M\alpha}$	The major contribution to pitching stability. It determines the natural frequency of the short period phugoid. It also determines the aircraft response to pilot inputs and gusts. The value should be sufficiently large to give an acceptable response to pilot inputs.
C_{Mde}	This term is often referred to as 'elevator effectiveness' or 'elevator power'. It is mainly influenced by the area of the elevator surface and the angular range of movement of the elevator. The value should allow the pilot to control the aircraft safely in an evasive manoeuvre or in response to a disturbance.
C_{Mq}	As the aircraft pitches, resistance to the angular velocity is provided by this term. C_{Mq} is commonly referred to a pitch damping as it dampens the short period phugoid. It provides a major contribution to longitudinal stability and aircraft handling qualities.
$C_{M\dot{\alpha}}$	This derivative also contributes to damping of the short period phugoid.
$C_{L\beta}$	This derivative is known as the *dihedral effect* and counteracts the rolling motion induced by sideslip, which dampens the Dutch roll mode and increases the spiral mode stability.
C_{Lda}	This term is also referred to as 'aileron effectiveness' and depends on the surface area and the range of movement of the ailerons. Roll rate is proportional to aileron deflection. With differential ailerons care is needed to ensure that the data relates to the net contribution of ailerons. Aircraft may have both inboard and outboard ailerons requiring separate data for the contribution of each aileron. In addition, spoilers can also be used to provide roll inputs.
C_{Ldr}	Movement of the rudder induces both a yawing moment and a rolling moment. Generally, this term is small and is often neglected.
C_{Lp}	This term represents resistance to rolling motion as an aircraft banks, providing damping of the Dutch roll mode.
C_{Lr}	As an aircraft yaws, the induced rolling moment is damped out by this term. Its major effect is on the spiral mode and consequently, the term is usually a small positive value.
$C_{N\beta}$	If the aircraft is disturbed in yaw, this term provides a resistance to the yawing motion and consequently, is often referred to as the 'weathercock' derivative and is mostly dependent on the geometry of the aircraft tail. The term determines the natural frequency of the Dutch roll and affects the spiral stability of the aircraft.
C_{Nda}	As aileron input is introduced, the aircraft may also tend to yaw, caused mostly by the drag of the aileron deflections on each wing. (The term adverse yaw is used if this value contributes a yawing moment in the direction opposite to the implied direction of banking). Typically, this value is very small or zero, particularly with differential ailerons.
C_{Ndr}	This term is also referred to as 'rudder effectiveness' and depends on the surface area and range of movement of the rudder. Generally, the rudder is used to counteract sideslip and to coordinate turns, so that the effect of this derivative influences the handing qualities.
C_{Np}	This derivate gives the change in yawing moment resulting from roll. Large positive values increase the Dutch roll damping.
C_{Nr}	This term is the main contributor to damping in yaw. Its effect is to damp the Dutch roll and to increase spiral mode stability.

3.5 Axes Systems

In the previous sections, in discussing the forces and moments applied to an aircraft, references were made to the aircraft axes and also to the earth axes. The aerodynamic, propulsive and gravitational forces on an aircraft, which are typically derived from data packages, are most conveniently computed in the aircraft frame. However, an aircraft may be at 5000 or 20,000 ft; it may be flying level or inverted or banked at 60°. Just knowing the forces and therefore the resultant motion of the aircraft in the aircraft axes is not particularly useful; the attitude, position, altitude, heading and airspeed are needed relative to the earth, for example, to drive the flight instruments in a simulator or to position the aircraft in the visual scene.

Fortunately, only a few axes systems are used in flight simulation and moreover, the transformation of variables between these axes systems is straightforward. It is essential to define conventions for these axes systems, particularly the positive and negative directions for both linear and angular variables. In flight simulation, all axes will be assumed to be orthogonal right hand axes, that is, positive forwards, positive right and positive down.

3.5.1 The Body Frame

The forces, moments, accelerations and velocities are computed in an orthogonal right-handed axis system, centred at the aircraft cg, where x is the forward direction along the fuselage, y is the direction along the starboard wing and z is the direction perpendicular to the underside of the fuselage, as shown in Figure 3.9. The linear forces, accelerations and velocities are positive in these directions. The angular moments and accelerations are positive in the anticlockwise sense looking along the axes from the origin. In other words, pitching up, rolling right wing high and yawing to the left produces positive values. The forces are X, Y and Z; the moments are L, M and N; the linear velocities are u, v and w; the angular velocities are p, q and r about the respective x, y and z axes.

As the propulsive forces are produced by engines (or propellers) attached to the airframe and the aerodynamic forces and moments apply to the control surfaces and airframe structure, it is sensible to resolve and compute these forces in the body frame. Of course, the gravitational forces do not occur in the body frame and, if they are to be combined with forces in the body frame, they also need to be resolved in the body frame. Generally, it is also assumed that the airframe is rigid. However, for airships, missiles subject to high propulsive forces or aircraft with a relatively long thin fuselage, the vehicle body may flex as forces are applied and the appropriate aeroelastic terms must be included in the computation of forces and moments.

The origin of the body frame is at the cg, which depends on the fuel load, passengers, aircraft damage and weapon deployment and the precise location of the aircraft cg must be taken into account in the equations of motion.

The propulsive forces come from propellers or jet engines mounted at the nose, on the wings or on the tail assembly. In the case where an engine or propeller is not mounted to produce thrust

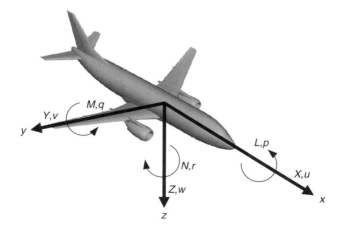

Figure 3.9 Aircraft body axes

in line with the x axis, this alignment will be known, enabling the appropriate components to be resolved in the body axes. Similarly, the moments generated by the engines can be computed from the thrust produced and the distance from an axis, for example, to produce a pitching moment for under-slung engines or a yawing moment for multi-engine aircraft.

The cg of an aircraft follows a trajectory known as the *flight path*. At any instant, the flight path angle γ is the angle the flight path makes with respect to the surface of the earth. It is important to appreciate that the aircraft may not necessarily be pointing in the direction of the flight path. In other words, the pitch attitude θ and the flight path angle γ will differ by the angle of attack and similarly, yaw attitude ψ and the track angle will differ by the angle of sideslip. Figure 3.10 shows the relationship between the pitch and flight path angles; in this example, the aircraft is shown descending and the angles are deliberately exaggerated.

For any flight path angle

$$\gamma = \theta - \alpha \tag{3.22}$$

where α is the aircraft angle of attack.

In the case of level flight, $\gamma = 0$ and $\theta = \alpha$. In climbing flight, the aircraft may be climbing at $5°$ say, with an angle of attack of $8°$, giving a pitch attitude of $13°$. As most aircraft are not fitted with an angle of attack gauge and although the flight path angle could be calculated from the airspeed and vertical rate of climb, pilots select recommended power settings and pitch attitudes to achieve the desired flight path angle.

In simulation, the angle of attack is needed to compute the aerodynamic forces and moments and the pitch attitude in needed in axes transformations, for the artificial horizon instrument display and for the visual system computations. The flight path angle is mostly used in flight control systems. The variables u, v and w, defined in the aircraft body axes, are used in the computation of the overall aircraft speed and the angle of attack, as follows:

$$V_c = \sqrt{u^2 + v^2 + w^2} \tag{3.23}$$

$$\alpha = \tan^{-1}\left(\frac{w}{u}\right) \tag{3.24}$$

where V_c is the speed of the cg of the aircraft along its flight path and α is the angle of attack. For small angle approximations, where $\alpha \approx \tan(\alpha)$

$$\dot{\alpha} = \frac{u\dot{w} - w\dot{u}}{\sqrt{u^2 + w^2}} \tag{3.25}$$

For lateral motion, the aircraft flight path results in a track over the ground. In the presence of wind, or during banked flight, or as a result of rudder inputs or if there is an imbalance in engine

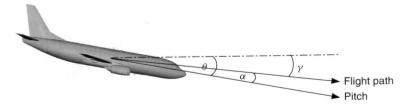

Figure 3.10 Flight path angle

forces, the aircraft will sideslip. In the situation where the angle of sideslip is non-zero, the aircraft will fly a track over the ground that differs from the aircraft heading (or yaw) by the angle of sideslip. For the simulator designer, the angle of sideslip is required for the slip ball display; it is used by flight control systems and is also used in the computation of the aerodynamic forces and moments. The angle of sideslip is computed from the variables u, v and w, defined in the aircraft body axes, as follows:

$$\beta = \tan^{-1}\left(\frac{v}{\sqrt{u^2 + w^2}}\right) \quad (3.26)$$

$$\dot{\beta} = \frac{\dot{v}\sqrt{u^2 + w^2} - v(u\dot{u} + w\dot{w})}{\sqrt{u^2 + w^2}(u^2 + v^2 + w^2)} \quad (3.27)$$

where β is the angle of sideslip.

3.5.2 Stability Axes

Often, the lift and drag forces are the net forces on the aircraft at a specific angle of attack. This is particularly true of data derived from wind tunnel tests, where the vertical and horizontal forces are measured relative to the tunnel floor. Generally, this assumption is also valid for data provided for flight models and a transformation is needed to resolve the aerodynamic forces in the body frame.

In Figure 3.11, the aircraft has an angle of attack α, and the resultant drag in the airframe axes is given by

$$D_x = D \sin \alpha \quad (3.28)$$

where D is the aerodynamic drag. Similarly, there is a component of drag in the z axis and also components of lift and side force in the x and z body axes. For practical purposes, the stability axes can be thought of as the body axes rotated about the y axis by an angle α.

3.5.3 Wind Axes

In the same way that the aircraft makes an angle α to the wind in the vertical plane, so an aircraft makes an angle β to the wind in the lateral plane. Aerodynamic data may be derived in a wind tunnel at different settings of sideslip giving the net forces and moments relative to the wind. In this case, an additional rotation is also required to resolve the aerodynamic forces and moments in the body frame. In practice, aerodynamic data is rarely provided in this form and the effect of sideslip is included directly in computation of the aerodynamic terms. However, where sideslip results from the effect of wind, for example, in the presence of turbulence, the resultant velocity must be transformed into the body frame.

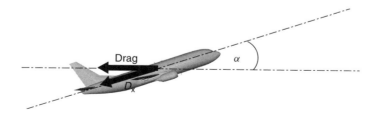

Figure 3.11 Stability axes resolution of drag

3.5.4 Inertial Axes

An aircraft is subject to both internal and external forces. Its subsequent motion could be defined with respect to some fixed point in space. If this point is the origin of a set of axes that are fixed and non-rotating, then this frame is termed an *inertial frame* and strictly, Newton's Laws of Motion are only valid in an inertial frame. In practice, inertial frames are difficult to visualize for aircraft motion, although their concept can help to simplify the computations used in aircraft navigation.

The definition can be relaxed to define an axis system that is only fixed and non-rotating with respect to a moving object. For example, the concrete floor of a building housing the motion platform of a flight simulator is an inertial frame. It is fixed at all times with respect to the platform actuators. The platform actuators and the simulator platform move with respect to a fixed point on the concrete floor.

With some simplification, inertial navigation frames can be used in flight simulation. For example, if a flight simulation is based on a single airfield and the aircraft only flies some 50 miles or so from this airfield, the aircraft motion could be defined relative to the airfield coordinates. In effect, this is a 'flat earth' simplification, ignoring the curvature of the earth. If an aircraft flies on a constant heading between two points of the surface of this flat earth, the flight path is a straight line over the surface of the earth. Whereas, the shortest distance from New York to London is a track along a great circle route, where the aircraft heading would change continuously from New York to London.

The earth can be treated as an inertial frame. However, it rotates with respect to its axes (every 24 hours) and with respect to the sun, approximately every 365 days. There is a further complication; as the earth is spherical, flying at a constant altitude should result in a curved flight path, parallel to the curved surface of the earth, rather than a straight line in space. In addition, the relatively slow turn rate of the earth should also be taken into account, particularly to compute the motion of an aircraft for long distance flights.

Although it is straightforward to compute forces, moments, accelerations and velocities (linear and angular) in body axes, it is also necessary to transform this motion to velocity and position relative to the earth, which can be treated as an inertial frame so that Newton's Laws of Motion apply, enabling aircraft position, velocity and attitude to be referenced to the earth frame.

3.5.5 Transformation between Axes

The transformation of a point or vector can be thought of as either a rotation of the axes or the rotation of the point in the original frame. In either case, the point is transformed from the coordinates of the original frame to coordinates of the new frame. In two dimensions, a vector can be rotated, as shown in Figure 3.12. If the point is rotated in a counter clockwise direction about the origin by an angle θ

$$x' = x \cos \theta - y \sin \theta \qquad (3.29)$$

$$y' = x \sin \theta + y \cos \theta \qquad (3.30)$$

The same equation would also apply if the axes were rotated clockwise by an angle θ, where x' and y' are the coordinates in the rotated frame. If the points (x, y) and (x', y') are treated as vectors, then rotation can be defined by a simple matrix operation:

$$\begin{bmatrix} x' \\ y' \end{bmatrix} = \begin{bmatrix} \cos \theta & -\sin \theta \\ \sin \theta & \cos \theta \end{bmatrix} \begin{bmatrix} x \\ y \end{bmatrix} \qquad (3.31)$$

The method is easily extended to rotations in three dimensions, but the results are simply stated in the following section; formal derivations are given in most textbooks on the subject.

Aircraft Dynamics

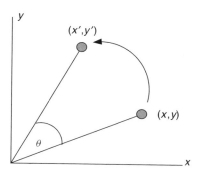

Figure 3.12 Rotation of axes in 2D

3.5.6 Earth-centred Earth-fixed (ECEF) Frame

The ECEF frame has its origin at the centre of the earth. The x and y axes are situated in the equatorial plane; the x axis points from the centre of the earth to the $0°$ meridian of longitude; the y axis points from the centre of the earth to the $90°$ meridian of longitude and the z axis points from the centre of the earth to the (true) north pole. Strictly, to an observer standing on the earth, this frame rotates every 24 hours with respect to the distant stars and Newton's Laws of Motion are not valid. Nevertheless, this frame is very useful in computing aircraft motion with respect to the earth and the complication that the frame is rotating can be resolved to provide an acceptably accurate and rigorous computation of motion.

Aircraft orientation can be defined with respect to this frame, providing three angles θ (pitch), ϕ (roll) and ψ (yaw), known as the *Euler angles*. These three angles define the orientation of one frame relative to another; in this case, the body frame relative to the ECEF frame (and vice versa). These are also the angles recognized by a pilot in aircraft manoeuvres and displayed by aircraft instruments, where the attitude indicator displays pitch and roll and the compass displays yaw (strictly, magnetic heading rather than true heading).

The coordinates of a point in one frame can be transformed by rotations of the three Euler angles to derive the coordinates in the other frame. These operations are not commutative; rotations in a different order will produce a different position, so it is imperative to apply a convention to the order of rotations in all axes transformations. If the axes in the original frame are defined as X, Y and Z, then yaw is a rotation about the Z axis, given by the following transformation:

$$\begin{bmatrix} x \\ y \\ z \end{bmatrix} = \begin{bmatrix} \cos\psi & \sin\psi & 0 \\ -\sin\psi & \cos\psi & 0 \\ 0 & 0 & 1 \end{bmatrix} \begin{bmatrix} X \\ Y \\ Z \end{bmatrix} \quad (3.32)$$

The point $[X\ Y\ Z]^T$ in the original frame is transformed to the point $[x\ y\ z]^T$ in the new frame rotated (in yaw) by an angle ψ.

Similarly, a second rotation in pitch about the Y axis, by an angle θ, is given by the following transformation:

$$\begin{bmatrix} x \\ y \\ z \end{bmatrix} = \begin{bmatrix} \cos\theta & 0 & \sin\theta \\ 0 & 1 & 0 \\ \sin\theta & 0 & \cos\theta \end{bmatrix} \begin{bmatrix} X \\ Y \\ Z \end{bmatrix} \quad (3.33)$$

Finally, a third rotation in roll about the X axis, by an angle ϕ, given by:

$$\begin{bmatrix} x \\ y \\ z \end{bmatrix} = \begin{bmatrix} 1 & 0 & 0 \\ 0 & \cos\phi & \sin\phi \\ 0 & -\sin\phi & \cos\phi \end{bmatrix} \begin{bmatrix} X \\ Y \\ Z \end{bmatrix} \quad (3.34)$$

To combine all three rotations, the transformations in yaw, then pitch, then roll can be described by a single matrix transformation, given by the product of the three matrices in the correct order:

$$\begin{bmatrix} x \\ y \\ z \end{bmatrix} = \begin{bmatrix} \cos\psi & \sin\psi & 0 \\ -\sin\psi & \cos\psi & 0 \\ 0 & 0 & 1 \end{bmatrix} \begin{bmatrix} \cos\theta & 0 & \sin\theta \\ 0 & 1 & 0 \\ \sin\theta & 0 & \cos\theta \end{bmatrix} \begin{bmatrix} 1 & 0 & 0 \\ 0 & \cos\phi & \sin\phi \\ 0 & -\sin\phi & \cos\phi \end{bmatrix} \begin{bmatrix} X \\ Y \\ Z \end{bmatrix} \quad (3.35)$$

Giving

$$\begin{bmatrix} x \\ y \\ z \end{bmatrix} = \begin{bmatrix} \cos\psi\cos\theta & \sin\psi\cos\theta & -\sin\phi \\ \cos\psi\sin\theta\sin\phi - \sin\psi\cos\phi & \cos\psi\cos\phi + \sin\psi\sin\theta\sin\phi & \cos\theta\sin\phi \\ \cos\psi\sin\theta\cos\phi + \sin\psi\sin\phi & \sin\psi\sin\theta\cos\phi - \cos\psi\sin\phi & \cos\theta\cos\phi \end{bmatrix} \begin{bmatrix} X \\ Y \\ Z \end{bmatrix}$$
(3.36)

A matrix to transform between axes is often referred to as a direction cosine matrix (DCM). In this case, the DCM enables a vector in the earth frame $[X\ Y\ Z]^T$ to be computed in the aircraft body frame. It is used in flight dynamics to transform velocities in the body frame to the earth frame. It can also be used to transform earth frame velocities (e.g. wind velocity) to the body frame. It is also used in visual system computations to transform a point (or polygon) defined in earth coordinates to a coordinate frame in the pilot eye frame, aligned to the body frame.

One example where this transformation is used is to compute the gravitational component of force in the body axes of an aircraft. In the earth frame, gravitational force is given by $[0\ 0\ mg]^T$ (where m is the aircraft mass and g is gravitational acceleration); the three components are computed in the body frame as follows:

$$\begin{bmatrix} F_x \\ F_y \\ F_z \end{bmatrix} = \begin{bmatrix} \cos\psi\cos\theta & \sin\psi\cos\theta & -\sin\phi \\ \cos\psi\sin\theta\sin\phi - \sin\psi\cos\phi & \cos\psi\cos\phi + \sin\psi\sin\theta\sin\phi & \cos\theta\sin\phi \\ \cos\psi\sin\theta\cos\phi + \sin\psi\sin\phi & \sin\psi\sin\theta\cos\phi - \cos\psi\sin\phi & \cos\theta\cos\phi \end{bmatrix} \begin{bmatrix} 0 \\ 0 \\ mg \end{bmatrix}$$
(3.37)

giving

$$F_x = -mg\sin\phi \quad (3.38)$$

$$F_y = mg\cos\theta\sin\phi \quad (3.39)$$

$$F_z = mg\cos\theta\cos\phi \quad (3.40)$$

It is important to appreciate the range of the Euler angles:

$$-\pi \leq \psi \leq +\pi \quad (3.41)$$

$$-\pi \leq \phi \leq +\pi \quad (3.42)$$

$$-\pi/2 \leq \theta \leq +\pi/2 \quad (3.43)$$

Aircraft Dynamics

It is not obvious that pitch should be constrained to the range $\pm 90°$. Consider a loop; as the pilot pulls up into the loop, the pitch increases to $80°$, then $85°, 86°, 87°, 88°$ and $89°$. At this point, the aircraft is still pointing forwards, but almost vertically upwards. In the next two degrees of motion, the aircraft passes through $90°$ and subsequently is pointing backwards, but at $89°$ to the horizon. Notice also that during this manoeuvre, as the pitch angle passes through $90°$, the aircraft heading changes from forwards to backwards (an $180°$ instantaneous change in yaw) and also, the aircraft changes from upright to inverted (an $180°$ instantaneous change in roll).

Although at first sight this notation looks cumbersome, it provides a very powerful mechanism to transform a vector from one axis system to another, or between any sets of orthogonal axes. These matrix transformations can be combined as a series of transformations that are represented by a single transformation matrix.

There is one further benefit of this approach; although the reverse transformation implies matrix inversion, this operation is achieved by transposing the transformation matrix.

$$\text{If } \begin{bmatrix} x \\ y \\ z \end{bmatrix} = [A] \begin{bmatrix} X \\ Y \\ Z \end{bmatrix}, \text{ then } \begin{bmatrix} X \\ Y \\ Z \end{bmatrix} = [A]^T \begin{bmatrix} x \\ y \\ z \end{bmatrix} \tag{3.44}$$

For example, the transformation between the body frame velocities (u, v, w) and the earth frame velocities (V_N, V_E, V_d), where V_N is the velocity in the north direction, V_E is the velocity in the east direction and V_d is the velocity in the downward direction, is given by:

$$\begin{bmatrix} V_N \\ V_E \\ V_d \end{bmatrix} = \begin{bmatrix} \cos\psi\cos\theta & \cos\psi\sin\theta\sin\phi - \sin\psi\cos\phi & \cos\psi\sin\theta\cos\phi + \sin\psi\sin\phi \\ \sin\psi\cos\theta & \cos\psi\cos\phi + \sin\psi\sin\theta\sin\phi & \sin\psi\sin\theta\cos\phi - \cos\psi\sin\phi \\ -\sin\theta & \cos\theta\sin\phi & \cos\theta\cos\phi \end{bmatrix} \begin{bmatrix} u \\ v \\ w \end{bmatrix} \tag{3.45}$$

The angular velocities can be transformed between the body and the earth frames using a similar approach, noting that the derivation also uses the Euler angles. The body rates are given by:

$$\begin{bmatrix} p \\ q \\ r \end{bmatrix} = \begin{bmatrix} 1 & 0 & -\sin\theta \\ 0 & \cos\phi & \cos\theta\sin\phi \\ 0 & -\sin\phi & \cos\theta\cos\phi \end{bmatrix} \begin{bmatrix} \dot\phi \\ \dot\theta \\ \dot\psi \end{bmatrix} \tag{3.46}$$

The reverse transformation, to derive the Euler angle rates from the body rates, is given by the transformation:

$$\begin{bmatrix} \dot\phi \\ \dot\theta \\ \dot\psi \end{bmatrix} = \begin{bmatrix} 1 & \sin\phi\tan\theta & \cos\phi\tan\theta \\ 0 & \cos\phi & -\sin\phi \\ 0 & \sin\phi\sec\theta & \cos\phi\sec\theta \end{bmatrix} \begin{bmatrix} p \\ q \\ r \end{bmatrix} \tag{3.47}$$

which expands to the following well-known equations:

$$\dot\phi = p + q\sin\phi\tan\theta + r\cos\phi\tan\theta \tag{3.48}$$

$$\dot\theta = q\cos\phi - r\sin\phi \tag{3.49}$$

$$\dot\psi = q\sin\phi\sec\theta + r\cos\phi\sec\theta \tag{3.50}$$

Notice that these equations contain a singularity when $\theta = -\pi/2$ or $\theta = +\pi/2$, where the terms $\tan\theta$ and $\sec\theta$ are infinite. Such conditions occur in aerobatic manoeuvres where the aircraft loops or climbs at a near vertical angle. Two techniques are used to overcome these problems. The pitch angle can be constrained so that the computation results in a valid floating point number. For example, $\tan(89.5°) = 114.6$ and this value could be substituted in computations when the pitch attitude is between $89.5°$ and $90.5°$. The numerical error introduced by this approximation only occurs at this extreme flight attitude where its effects on the aircraft behaviour may not be apparent. The more rigorous and commonly used method is to use quaternions, which are described in Section 3.6.

3.5.7 Latitude and Longitude

In flight simulation, airfields, geographic features and navigation beacons are defined by their latitude and longitude. Consequently, it is common for aircraft position to be defined by latitude and longitude rather than a position relative to a fixed datum. Assuming that the earth is a perfect sphere and ignoring any variation in gravity over the surface of the earth, the equations to derive latitude rate and longitude rate are given by:

$$\dot\lambda = \frac{V_n}{R+h} \qquad \lambda = \int \dot\lambda \, dt \qquad (3.51)$$

$$\dot\mu = \frac{V_e \sec \lambda}{R+h} \qquad \eta = \int \dot\mu \, dt \qquad (3.52)$$

$$\dot h = -V_d \qquad h = \int \dot h \, dt \qquad (3.53)$$

where λ is latitude, μ is longitude, R is the radius of the earth and h is the altitude above mean sea level. Note the singularity as λ approaches $-90°$ or $+90°$, which occurs flying directly over the north or south poles. Care is needed to cope with this problem, although in practice, few simulators are likely to be used to practice take-offs or landings near the poles.

Care is also needed with the precision of the computation of latitude and longitude. The range of latitude is $\pm 90°$ ($\pm \pi/2$) and the range of longitude is $\pm 180°$ ($\pm \pi$). A standard 32-bit floating point variable typically has 8 bits for the exponent and 24 bits for the mantissa. For numbers in the range used to compute latitude and longitude, this corresponds to an accuracy of approximately 10^{-7}. For an earth circumference of approximately 40×10^6 m, the floating point resolution is of the order 4 m. In other words, the smallest movement of an aircraft may be as high as 4 m, which would result in unacceptably jerky motion in the visual system as an aircraft taxies slowly. To avoid such problems, double precision (64-bit) floating point arithmetic is normally used in the computation of latitude and longitude.

3.6 Quaternions

Quaternions (also known as the *four-parameter method*) are widely used in flight simulation (Fang and Zimmerman, 1969). They enable the Euler angles to be computed but avoid the singularity in the computation of the body rates in Equations 3.48–3.50. The four quaternion parameters e_0, e_1, e_2 and e_3 are defined, in terms of the Euler angles, as follows:

$$e_0 = \cos\frac{\psi}{2} \cos\frac{\theta}{2} \cos\frac{\phi}{2} + \sin\frac{\psi}{2} \sin\frac{\theta}{2} \sin\frac{\phi}{2} \qquad (3.54)$$

$$e_1 = \cos\frac{\psi}{2} \cos\frac{\theta}{2} \sin\frac{\phi}{2} - \sin\frac{\psi}{2} \sin\frac{\theta}{2} \cos\frac{\phi}{2} \qquad (3.55)$$

Aircraft Dynamics

$$e_2 = \cos\frac{\psi}{2}\sin\frac{\theta}{2}\cos\frac{\phi}{2} + \sin\frac{\psi}{2}\cos\frac{\theta}{2}\sin\frac{\phi}{2} \tag{3.56}$$

$$e_3 = \sin\frac{\psi}{2}\cos\frac{\theta}{2}\cos\frac{\phi}{2} - \cos\frac{\psi}{2}\sin\frac{\theta}{2}\sin\frac{\phi}{2} \tag{3.57}$$

These identities can be used to initialize the quaternion values. At all times, the quaternion parameters must satisfy the following constraint:

$$e_0^2 + e_1^2 + e_2^2 + e_3^2 = 1 \tag{3.58}$$

The quaternion parameters are formed from the body rates as follows:

$$\dot{e}_0 = -\frac{1}{2}(e_1 p + e_2 q + e_3 r) \quad e_0 = \int \dot{e}_0' \tag{3.59}$$

$$\dot{e}_1 = \frac{1}{2}(e_0 p + e_2 r - e_3 q) \quad e_1 = \int \dot{e}_1' \tag{3.60}$$

$$\dot{e}_2 = \frac{1}{2}(e_0 q + e_3 p - e_1 r) \quad e_2 = \int \dot{e}_2' \tag{3.61}$$

$$\dot{e}_3 = \frac{1}{2}(e_0 r + e_1 q - e_2 p) \quad e_3 = \int \dot{e}_3' \tag{3.62}$$

The DCM used to transform velocities in the body frame can be defined in terms of the quaternion parameters, as follows:

$$\begin{bmatrix} e_0^2 + e_1^2 - e_2^2 - e_3^2 & 2(e_1 e_2 - e_0 e_3) & 2(e_0 e_2 + e_1 e_3) \\ 2(e_1 e_2 + e_0 e_3) & e_0^2 - e_1^2 + e_2^2 - e_3^2 & 2(e_2 e_3 - e_0 e_1) \\ 2(e_1 e_3 - e_0 e_2) & 2(e_2 e_3 + e_0 e_1) & e_0^2 - e_1^2 - e_2^2 + e_3^2 \end{bmatrix} = \begin{bmatrix} a_{11} & a_{12} & a_{13} \\ a_{21} & a_{22} & a_{23} \\ a_{31} & a_{32} & a_{33} \end{bmatrix} \tag{3.63}$$

Combining elements from the DCM given in terms of the Euler angles enables the Euler angles to be computed:

$$\theta = \sin^{-1}(-a_{31}) \tag{3.64}$$

$$\phi = \tan^{-1}\left(\frac{a_{32}}{a_{33}}\right) \tag{3.65}$$

$$\psi = \tan^{-1}\left(\frac{a_{21}}{a_{11}}\right) \tag{3.66}$$

where a denotes the elements of the DCM. Note that this representation also depends on the numerical accuracy of the integration algorithm and that the magnitude of the quaternion values must be continually adjusted to satisfy the constraint in Equation 3.58, which is applied by modifying the computation of the quaternion parameter rates, as follows:

$$\dot{e}_0 = -\frac{1}{2}(e_1 p + e_2 q + e_3 r) + k\lambda e_0 \tag{3.67}$$

$$\dot{e}_1 = \frac{1}{2}(e_0 p + e_2 r - e_3 q) + k\lambda e_1 \tag{3.68}$$

$$\dot{e}_2 = \frac{1}{2}(e_0 q + e_3 p - e_1 r) + k\lambda e_2 \qquad (3.69)$$

$$\dot{e}_3 = \frac{1}{2}(e_0 r + e_1 q - e_2 p) + k\lambda e_3 \qquad (3.70)$$

where $kh \leq 1, \lambda = 1 - (e_0^2 + e_1^2 + e_2^2 + e_3^2)$ and h is the integration step length (frame rate). This method, known as the *method of algebraic constraint*, where λ is adjusted during each iteration, ensures that the constraint is maintained. Quaternions are widely used in flight simulation, particularly in military simulators where aerobatic manoeuvres are practised. However, care must be taken to initialize the quaternion parameters and to ensure that the numeric constraint outlined above is not violated.

3.7 Equations of Motion

In the previous sections, we have reviewed how the forces and moments on an aircraft can be derived and how terms computed in one axis system can be transformed to a more convenient axis system. In this section, we look at the equations which underpin aircraft motion. In particular, an aircraft has six degrees of freedom, translating and rotating in three linear dimensions and three angular dimensions (Fogarty and Howe, 1969).

Recalling the axes shown in Figure 3.9, the axes conventions and primary terms are summarized in Table 3.3.

The aircraft forces are given by:

$$\sum F_x = \frac{d}{dt}(mu) \qquad (3.71)$$

$$\sum F_y = \frac{d}{dt}(mv) \qquad (3.72)$$

$$\sum F_z = \frac{d}{dt}(mw) \qquad (3.73)$$

The aircraft moments are given by:

$$\sum L = \frac{dh_x}{dt} \qquad (3.74)$$

$$\sum M = \frac{dh_y}{dt} \qquad (3.75)$$

$$\sum N = \frac{dh_z}{dt} \qquad (3.76)$$

Consider an element of mass dm rotating at an angular velocity ϖ, where $\varpi = pi + qj + rk$, as shown in Figure 3.13.

Table 3.3 Axes notation

Axis	Linear velocity	Angular velocity	Moment	Force	Moment of momentum	Moment of inertia
X	U	P	L	F_x	h_x	I_{xx}
Y	V	Q	M	F_y	h_y	I_{yy}
Z	W	R	N	F_z	h_z	I_{zz}

Aircraft Dynamics

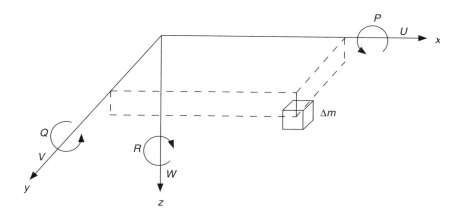

Figure 3.13 Forces on an element of mass

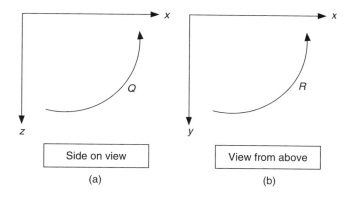

Figure 3.14 Linear velocity components

Consider the velocity in the x direction as shown in Figure 3.14. The pitching rate Q contributes Qz and the yawing rate contributes $-Ry$. Clearly, the roll rate term contributes no velocity along the x axis. Recalling that the x axis is forwards, the y axis is in the direction of the right wing and the z axis is down, then the linear velocity components of Δm in the x, y and z directions, respectively, are given by:

$$\dot{x} = U + Qz - Ry \qquad (3.77)$$

$$\dot{y} = V - Pz + Rx \qquad (3.78)$$

$$\dot{z} = W + Py - Qx \qquad (3.79)$$

Differentiating Equations 3.77–3.79 to derive the accelerations:

$$\ddot{x} = \dot{U} + \dot{Q}z + Q\dot{z} - \dot{R}y - R\dot{y} \qquad (3.80)$$

$$\ddot{y} = \dot{V} - \dot{P}z - P\dot{z} + \dot{R}x + R\dot{x} \qquad (3.81)$$

$$\ddot{z} = \dot{W} + \dot{P}y + P\dot{y} - \dot{Q}x - Q\dot{x} \qquad (3.82)$$

Substituting for $\dot{x}$, $\dot{y}$ and $\dot{z}$ from Equations 3.77–3.79

$$\ddot{x} = \dot{U} + QW + QPy - Q^2x - RV + RPz - R^2x + \dot{Q}z - \dot{R}y \qquad (3.83)$$

$$\ddot{y} = \dot{V} - PW - P^2y - PQx + RU + RQz - R^2y - \dot{P}z + \dot{R}x \qquad (3.84)$$

$$\ddot{z} = \dot{W} + PV - P^2z + PRx - QU - Q^2z + QRy + \dot{P}y - \dot{Q}x \qquad (3.85)$$

The inertia forces of the mass element are given by:

$$F_{xi} = -m\ddot{x} \qquad (3.86)$$

$$F_{yi} = -m\ddot{y} \qquad (3.87)$$

$$F_{zi} = -m\ddot{z} \qquad (3.88)$$

The inertia moments of the mass element are given by:

$$M_{xi} = F_{zi}y - F_{yi}z \qquad (3.89)$$

$$M_{yi} = F_{xi} - F_{zi}x \qquad (3.90)$$

$$M_{zi} = F_{yi}x - F_{xi}y \qquad (3.91)$$

The total inertia forces and moments acting on the aircraft are obtained by summing Equations 3.86–3.91 over the total aircraft mass. However, several terms vanish if the origin is taken as the cg, in particular:

$$\sum \Delta mx = 0 \qquad (3.92)$$

$$\sum \Delta my = 0 \qquad (3.93)$$

$$\sum \Delta mz = 0 \qquad (3.94)$$

$$\sum \Delta mxy = 0 \qquad (3.95)$$

$$\sum \Delta myz = 0 \qquad (3.96)$$

($\Sigma \Delta mxy$ and $\Sigma \Delta myz$ are assumed to be zero if the aircraft is symmetric about the xz plane).
The moments of inertia are defined as:

$$I_{xx} = \sum \Delta m(y^2 + z^2) \qquad (3.97)$$

$$I_{yy} = \sum \Delta m(x^2 + z^2) \qquad (3.98)$$

$$I_{zz} = \sum \Delta m(x^2 + y^2) \qquad (3.99)$$

$$I_{xz} = \sum \Delta mxz \qquad (3.100)$$

Although the moments of inertia could be computed from these summations, as the geometry and material of the aircraft is known, these values are usually determined during aircraft manufacture, from the mass distribution of the aircraft. The forces are determined from Equations 3.83–3.85, eliminating the terms containing mx, my and mz:

$$F_x = m\ddot{x} = m(\dot{U} + QW - RV) \tag{3.101}$$

$$F_y = m\ddot{y} = m(\dot{V} - PW + RU) \tag{3.102}$$

$$F_z = m\ddot{z} = m(\dot{W} + PV - QU) \tag{3.103}$$

Similarly, Equations 3.89–3.91 can be expanded, substituting the terms in Equations 3.83–3.85 to provide the moments:

$$L = I_x \dot{P} + QR(I_z - I_y) - I_{yz}(Q^2 - R^2) - I_{xz}(\dot{R} + PQ) - I_{xy}(\dot{Q} - PR) \tag{3.104}$$

$$M = I_y \dot{Q} + PR(I_x - I_z) + I_{xz}(P^2 - R^2) - I_{yz}(\dot{R} - PQ) - I_{xy}(\dot{P} + QR) \tag{3.105}$$

$$N = I_z \dot{R} + PQ(I_y - I_x) - I_{xy}(P^2 - Q^2) - I_{xz}(\dot{P} - QR) - I_{yz}(\dot{Q} + PR) \tag{3.106}$$

If the aircraft has a symmetrical distribution of mass in the fore and aft plane of geometry, which is the case for most aircraft, then $I_{xz} = 0$ and $I_{yz} = 0$ and Equations 3.104–3.106 reduce to:

$$L = I_x \dot{P} + QR(I_z - I_y) - I_{xz}(\dot{R} + PQ) \tag{3.107}$$

$$M = I_y \dot{Q} + PR(I_x - I_z) + I_{xz}(P^2 - R^2) \tag{3.108}$$

$$N = I_z \dot{R} + PQ(I_y - I_x) - I_{xz}(\dot{P} - QR) \tag{3.109}$$

These equations ignore the gyroscopic effects of propellers, rotor blades and rotating turbines. If h denotes the angular momentum vector, defined in the body axes, then:

$$L = I_x \dot{P} + QR(I_z - I_y) - I_{xz}(\dot{R} + PQ) + Qh_z - Rh_y \tag{3.110}$$

$$M = I_y \dot{Q} + PR(I_x - I_z) + I_{xz}(P^2 - R^2) - Ph_y + Rh_x \tag{3.111}$$

$$N = I_z \dot{R} + PQ(I_y - I_x) - I_{xz}(\dot{P} - QR) + Ph_y - Rh_z \tag{3.112}$$

The forces F_x, F_y and F_z in Equations 3.101–3.103 are computed from the aerodynamic terms in the flight model (based on the aerodynamic data pack) and the propulsive forces provided in the engine model (based on the engine data pack) to derive the linear accelerations $\dot{U}$, $\dot{V}$ and $\dot{W}$ in the body axes. Similarly, the moments L, M and N in Equations 3.110–3.112 are computed from the flight model and engine model to derive the angular accelerations $\dot{P}$, $\dot{Q}$ and $\dot{R}$ in the body axes. The terms $\dot{U}$, $\dot{V}$, $\dot{W}$ and $\dot{P}$, $\dot{Q}$, $\dot{R}$ are then integrated to provide the linear velocities U, V, W and the angular body rates P, Q, R in body axes.

3.8 Propulsion

For fixed wing aircraft, thrust is produced in one of three ways:

- A combination of a piston engine and a propeller;
- A turbojet, based on the gas turbine, including the modern turbofan used in commercial transport aircraft and high performance military aircraft;
- A combination of a turbine and a propeller, known as a *turboprop*.

In addition, propellers are classified as fixed pitch and variable pitch. For light aircraft, it is common for the engine manufacturer to select a propeller where the pitch angle of the propeller blade meets an overall performance requirement. However, for larger aircraft, the propeller efficiency can be improved by changing the propeller blade angle in flight, where the pilot is provided with a lever to set the blade angle.

For a jet engine, comprising a compressor, a combustion chamber and a turbine, air is compressed by an axial flow compressor and passes to the combustor. In the combustion chamber, atomized kerosene is mixed with the compressed air and burnt. This hot gas then passes through the turbine which extracts energy from the gas to drive the compressor. Typically, three quarters of the energy is used to drive the compressor, with the remaining energy providing thrust from the exhaust. For turboprop or turboshaft engines, the compressor drives the propeller. In addition, with an afterburner, fuel can be added to the exhaust gases to increase thrust at the expense of an increase in fuel flow rate. The pilot controls the amount of fuel injected into the combustion chamber to select the desired engine power or thrust. A turbofan aircraft will have engine gauges for engine pressure ratio (EPR) for thrust selection, engine RPM to cross-check engine performance, exhaust gas temperature and possibly fuel flow rate.

For piston engine aircraft, a mixture of fuel and air is injected into each engine cylinder and detonated just after the piston has compressed the fuel-air mixture. The set of pistons are connected to a rotating crankshaft which drives the propeller. The pilot has a throttle lever to control the amount of fuel injected into the engine to select engine power, a mixture lever to select the fuel-to-air ratio (particularly to weaken the ratio for more economical flight in the cruise) and possibly a pitch lever to select the propeller blade angle. Depending on the aircraft, gauges are provided for engine RPM, manifold pressure, cylinder head temperature and exhaust gas temperature. By referring to the pilot's notes for the aircraft, a pilot will select a specific engine RPM and manifold pressure appropriate to the phase of flight. For most light aircraft, the RPM gauge is used to indicate engine power.

Although it is possible to predict propeller performance with a reasonable degree of accuracy, for jet engines and piston engines it is commonplace to base the dynamic model of the engine on data sheets provided by a manufacturer. Often, these are based on performance data enabling the simulator designer to derive a performance model covering the aircraft flight envelope. In addition, the engine (and propeller) will have a dynamic response to pilot inputs and disturbances and these must be modelled accurately, for example, the time for a jet engine to spool up to maximum RPM from flight idle conditions is likely to take several seconds. Because the cost of acquiring the propulsion data for a simulator is very high for the engine and propeller manufacturers, there is very little engine data in the public domain. As the propulsion terms used in modelling flight dynamics are less significant (in terms of simulator fidelity), simplifications are possible, provided that sufficient accuracy is maintained in computing the performance data. In addition to pilot inputs, the temperature and density of the air entering the engine and the Mach number (for jet engines) and airspeed (for propellers) are also needed in the computation of engine thrust.

3.8.1 Piston Engines

As a propeller rotates, each blade makes an angle of attack to the air, generating lift to propel the aircraft forwards. The angle of incidence depends on the twist of the blade and, as each blade is an aerofoil, the lift generated by the propeller depends on the air density and the relative speed of air passing over the aerofoil, which is produced as a result of the propeller speed of rotation. In addition, the propeller thrust also depends on the air through the propeller plane of rotation, given by the aircraft airspeed. The efficiency of a propeller depends on the airspeed and blade angle and for a variable-pitch propeller, a pitch lever is used to control the blade angle in the vertical plane of rotation. In effect, the pitch lever acts like a gearbox, selecting the most efficient operating condition. The pitch lever is used to rotate the blade angle, typically from 15° (fully fine) to 90° (fully coarse). At a blade angle of 90°, where it is orthogonal to the plane of rotation or parallel to the slipstream, the blade is 'feathered' offering minimal drag. Feathering a propeller is necessary in the case of an engine failure in a twin-engine aircraft. The difference in propeller drag can induce fast roll rates if the pilot fails to respond correctly to an engine failure. In order to reduce the asymmetric forces resulting from the engine failure, the pilot selects the feathered position for the pitch lever of the failed engine. A fixed pitch propeller can be modelled by setting the propeller at a specific blade angle.

The engine levers move through an angular range, depending on the mechanical installation of the throttle quadrant. In a flight simulator, this displacement is measured by a potentiometer, either a rotary potentiometer attached to the lever shaft or a linear potentiometer connected to the lever. In simulation, it is more convenient to treat these inceptor inputs as values in the range 0–1 rather than angular displacements, where the fully forward position corresponds to 1.0.

The pitch lever selects engine RPM; although it acts to limit engine RPM, in the normal engine operating range the engine provides sufficient power to maintain the selected RPM. However, with insufficient power, the engine may not be able to provide the selected RPM or at high airspeed, the power absorbed by the propeller may be such that it exceeds the selected RPM. For most piston engines used in light aircraft, the maximum RPM is 2700 RPM and a simple square law is appropriate to derive the selected RPM.

$$n_{max} = 2700 * pl * pl \qquad (3.113)$$

where n_{max} is the maximum RPM and pl is the position of the pitch lever control. In practice, a more accurate model of the relationship between the pitch lever and RPM could be obtained from flight tests.

The manifold pressure is the pressure of the air entering the engine which depends on the pressure altitude (computed in the atmospheric model), the throttle position and the engine RPM. As the engine RPM increases, the engine produces more pressure at the manifold. The manifold pressure also increases with engine lever input. The change in manifold pressure is given by:

$$\Delta mp = (0.04635 * tl - 0.0469) * n \qquad (3.114)$$

where Δmp is the contribution to manifold pressure by the engine, tl is the throttle lever position and n is the engine RPM. At 2700 RPM with the throttle lever fully forwards (1.0), $\Delta mp = 1.5$ in. Mg. Although SI units are mostly used in this book, as performance data for most piston engines is given in imperial units, it is convenient to compute the internal engine terms in imperial units, converting the engine output terms to SI units for use in the flight model. The net manifold pressure is given by:

$$mp = pressure\_altitude + \Delta mp \qquad (3.115)$$

where Δmp is the manifold pressure and *pressure_altitude* is computed in the atmospheric model.

The fuel entering the engine is a mixture of air and fuel from the carburettor (or fuel injection system). As air density reduces with altitude, the fuel-to-air ratio increases. The mixture lever enables the pilot to select an appropriate fuel-to-air mixture ratio. However, if the mixture becomes either too weak or too rich, the engine power output falls rapidly. Most mixture levers have more effect towards the end of their forward travel and a simple approximation is given by:

$$mixture\_lever = 2.0 - ml * ml \qquad (3.116)$$

where ml is the mixture lever position.

The fuel : air ratio is approximated by:

$$far = mixture\_lever * \rho * 0.1 \tag{3.117}$$

where *far* is the fuel-to-air ratio and ρ is air density. If the fuel-to-air ratio is less than 20/1 (corresponding to a very lean mixture of 0.05) or greater than 8/1 (corresponding to a very rich mixture of 0.125), or if there is no fuel or if both magnetos are off, then a power factor of zero is assumed, which will cause the engine to stop (although the propeller may continue to turn, driven by the slipstream). The relationship between the engine power factor and fuel-to-air ratio is non-linear but a standard model can be used for most aviation fuels. A spline fit to the following coordinate pairs gives a close approximation to the graph of power factor *pf* versus fuel-to-air ratio (by weight) based on values given in Stinton (1985). The tabulated values used to simulate the effect of fuel-to-air ratio on engine power for a piston engine are shown in Figure 3.15. Typically, maximum power occurs with a fuel-to-air mixture of 0.08 and best economy occurs at the maximum gradient of the curve, at approximately 0.067.

The engine power is a result of the power obtained from combustion and power loss caused by friction losses and mechanical inefficiency of the engine. Normally, the static horsepower of an engine is quoted as the maximum power an engine can deliver at rest at sea level (maximum pressure attitude). For a model of a 160 HP (horsepower) engine, the static horsepower shp (the power produced by combustion) is computed as:

$$shp = mp * (0.0039 * n - 1.0) \tag{3.118}$$

with a power loss *fpow* proportional to the square of RPM given by:

$$fpow = 0.0413 * n^2 / 2700.0 \tag{3.119}$$

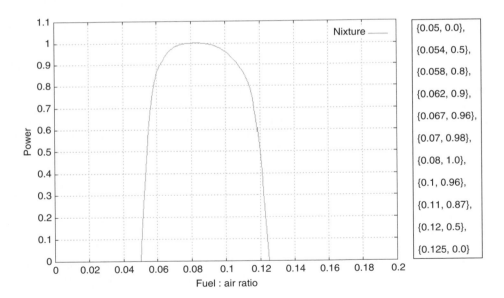

Figure 3.15 Combustion efficiency for a piston engine

where the units of power are horsepower (HP). For safety, two separate magneto circuits are usually provided for aero engines with two sets of spark plugs, with each magneto generating the current for both plugs in each cylinder. If however only one magneto is operational, there is a reduction in power caused by the reduced effectiveness of the combustion, typically reducing the static horsepower by 5%. This effect is noticeable during engine checks, where a 50 RPM drop occurs as the individual magnetos are checked (briefly switched off).

The net engine power *epow* is the difference between the power generated and the power lost and is given by:

$$epow = shp * pf - fpow \qquad (3.120)$$

The fuel flow rate is computed from the computed static horse of the engine, the engine RPM and the carburettor setting controlled by the mixture lever. If the fuel flow reaches a small value (or if the fuel switch is turned to off or the fuel tank is empty), there is insufficient fuel injected into the engine to cause combustion. Based on engine performance graphs, a typical fuel flow rate *fflo* (lb/hr) is given by:

$$fflo = (0.235 * shp + 0.0125 * rpm - 9.69) * ml * 0.000225 \qquad (3.121)$$

The main calculation for the propeller is the computation of the propeller thrust generated and the power required from the propeller. These terms are not straightforward to compute although propeller manufacturers do produce performance charts for their propellers in terms of the blade angle β and the advance ratio J. The advance ratio J depends on the engine RPM and the speed of air through the propeller resulting from the aircraft airspeed and is given by:

$$J = \frac{V}{nD} \qquad (3.122)$$

where V is airspeed, n is the propeller angular velocity and D is the propeller diameter. For a 6 ft diameter propeller:

$$J = 16.9 * 1.944 * V/n \qquad (3.123)$$

where V is the airspeed (m/s) and n is the propeller angular velocity (RPM). Care is needed to avoid numeric overflow at low engine RPM when the engine is starting or stopping. At other times, the propeller speed is unlikely to be rotating at less than 650 RPM (a typical idle speed).

The propeller thrust and power terms depend on the coefficients C_T and C_P, respectively; these are the terms provided by the propeller manufacturer and are given by:

$$C_T = \frac{T}{\rho n^2 D^4} \qquad (3.124)$$

$$C_P = \frac{P}{\rho n^3 D^5} \qquad (3.125)$$

where T is the thrust produced by the propeller, ρ is the air density ratio, n is the propeller angular velocity, D is the propeller diameter and P is the power required by the propeller. By deriving dimensionless terms for C_T and C_P, it is straightforward to compute the thrust T and the power P.

These curves are provided as functions of advance ratio J and the blade angle β, requiring two dimensional interpolation to compute these values from lookup tables generated from these graphs. Typical graphs for the coefficient of thrust C_T and power C_P for a Piper Comanche (PA-30) propeller are given in Figures 3.16 and 3.17, as a function of advance ratio. For both graphs, curves are plotted for blade angle β from values of $0-40°$ at $5°$ intervals.

The propeller power is given by:

$$ppow = \frac{C_P \, n^3}{6430041 * \rho} \qquad (3.126)$$

where $ppow$ is in horsepower, propeller rotation speed n is in RPM and ρ is the density ratio.

The difference between the engine power generated and the power absorbed by the propeller results in a torque which produces an angular acceleration on the propeller, given by:

$$T = \frac{(epow - ppow) * 7120.91}{n} \qquad (3.127)$$

where T is torque. Dividing the torque by the moment of inertia of the propeller provides the angular acceleration on the propeller:

$$\dot{n} = \frac{T}{0.2} \qquad (3.128)$$

where $\dot{n}$ is the angular acceleration of the propeller and the moment of inertia of the propeller is assumed to be $0.2 \, \text{kg m}^2$. In practice, $\dot{n}$ needs to be limited as very high torque could be generated at very low values of RPM.

The angular acceleration of the propeller $\dot{n}$ is also adjusted for three conditions:

- If the engine is stopping, the resistance of the engine and propeller reduces $\dot{n}$;
- If the pitch lever is pulled back into the feather region, the coarse blade angle will cause the propeller to slow down;
- If the engine starter is engaged (before the engine reaches idle RPM), $\dot{n}$ will increase as a result of the additional power provided by the starter motor.

The propeller angular velocity n is derived by integrating the angular acceleration:

$$n = \int \dot{n} \, dt \qquad (3.129)$$

In simulation, the value of n is limited to the range 0–3000 RPM; the propeller cannot turn backwards and under normal operating conditions, the engine is unlikely to exceed 3000 RPM.

For a constant speed propeller, the blade angle is constantly adjusted by a governor mechanism to maintain the selected RPM. In other words, the rate of change of blade angle $\dot{\beta}$ depends on the difference between the actual RPM and the governed RPM and also the angular acceleration of the propeller. Using typical values for a constant speed propeller:

$$\dot{\beta} = \frac{n - n_{\max}}{10} + \frac{\dot{n}}{20} \qquad (3.130)$$

Aircraft Dynamics

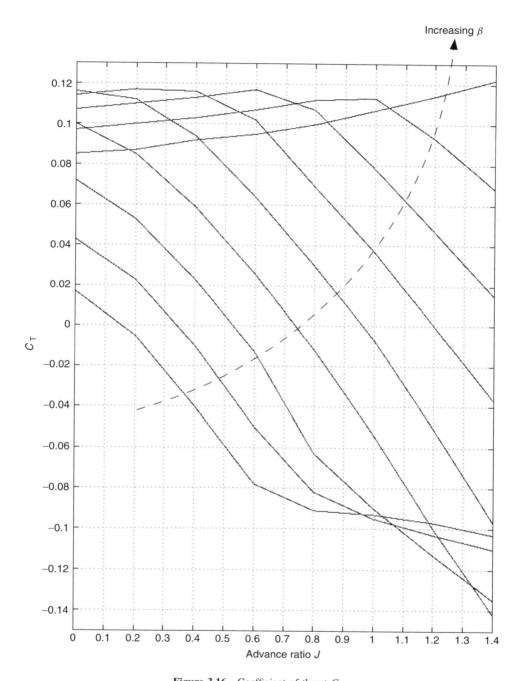

Figure 3.16 Coefficient of thrust C_T

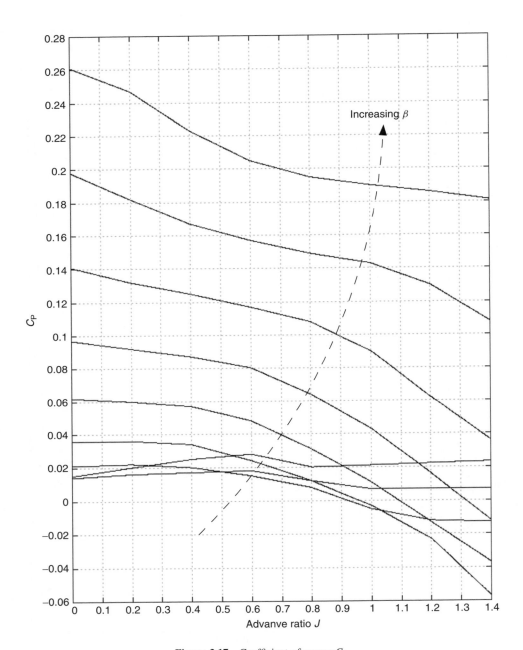

Figure 3.17 Coefficient of power C_P

where $\dot{\beta}$ is the rate of change of blade angle (degrees per second) and n_{max} is the governed RPM (selected by the pitch lever). A reasonable limit for $\dot{\beta}$ is $\pm 50\,°/s$. The blade angle β is computed by integrating the rate of change of blade angle:

$$\beta = \int \dot{\beta}\, dt \qquad (3.131)$$

Finally, the propeller thrust is computed (taking account of the combination of SI and imperial units) as follows:

$$T = \frac{0.0038121 * n^2 * C_T}{\rho} \qquad (3.132)$$

where T is the thrust (N) and ρ is the density ratio. One additional consideration is the ground run. Airflow from the propeller interacts with the ground and there is possibly up to 30% loss in available thrust close to the ground for most light aircraft. In flight, the application of the carburettor heat lever diverts the engine exhaust resulting in a small loss in power and on some aircraft, cowl flaps also have a small effect on engine output. Although these terms are relatively small, they are significant in terms of aircraft performance, particularly in go-around situations and with engine failures.

The inputs to the engine model are the pilot inputs (throttle, pitch and mixture levers), the aircraft airspeed, air density and pressure altitude. The engine model is normally updated at the same frame rate as the aerodynamic equations and provides engine thrust, fuel flow, manifold pressure and engine temperatures (both exhaust gas and cylinder head). Having developed an engine model, it is necessary to check the simulated performance against the aircraft performance data published for the aircraft. In practice, there is very little information in the public domain describing engine or propeller performance and without access to the propeller C_T and C_P curves, it is difficult to estimate propeller performance. Worobel and Mayo (1971) and Worobel (1972) provide a useful description of a method to predict propeller characteristics, including a Fortran programme.

In cases where engine data is not available, an approximation to simplify the computation of thrust is

$$P = TV \qquad (3.133)$$

where P is power, T is thrust and V is airspeed. In other words, if power is assumed to be proportional to the throttle lever position, the thrust can be estimated. Furthermore, as McCormick (1995) observes, engine efficiency reaches a maximum value of 80% around an advance ratio of 0.75 reducing either side of 0.75. Although this is no substitute for actual propeller charts, this approximation (knowing the maximum engine power output) can provide an insight into aircraft performance. For some aircraft engines, there is a further complication in understanding the engine control system. For example, the Aztazou engine in the Jetstream-100 turboprop controls the fuel flow to maintain engine RPM rather than blade angle. In simulating these control systems, where the propeller charts exhibit considerable non-linearities, care is needed in the design of a controller to avoid instability or oscillation in order to replicate the dynamic response of the combined engine and propeller propulsion system.

3.8.2 Jet Engines

For propellers and piston engines, there is sufficient guidance in performance estimation in the literature to make a reasonable estimate of performance. The case is different for jet engines and turboprops and it is difficult to produce a performance model without access to a considerable amount of data provided by the engine manufacturer. This information contains both performance data to derive thrust as a function of Mach number and altitude and also data to predict the dynamic response of the engine to thrust demands. In addition, fuel flow, turbine RPM and engine temperatures must also be provided for engine instruments, which are derived from the engine model parameters.

While a thermodynamic model of the fan, compressor, combustor, turbines and nozzles for a turbofan aircraft can provide accurate performance data, the complexity of the modelling of the gas flows (both axial and rotary) are sufficiently complex that real-time models are impractical. Knowledge of the thrust specific fuel consumption (TSFC), which provides the weight of fuel per hour per unit of thrust (N/hr/N), measured in level flight versus true airspeed provides an estimate of fuel flow rate because thrust matches drag in non-accelerating level flight.

The notes in this chapter are based on data for the Pratt & Whitney JT9D-7A turbofan engine, for two reasons: first, the engine is representative of a high bypass turbofan engine (albeit dated) and secondly, performance data for this engine is provided in both McCormick (1995) and Hanke (1971). Care is needed with engine data because engine power is also used to drive electrical aircraft systems and is also bled for air conditioning. Net thrust for level flight at sea level varies from 37,000 lbs at low speed to 21,000 lbs at Mach 0.6, whereas it remains almost constant at 12,000 lbs at 30,000 ft for speeds in the operating range from Mach 0.3 to Mach 0.95. Considerable care is taken by manufacturers to ensure that engines are operated in the safe region where the engine is unlikely to surge or stall. Engine stall occurs when the local angles of attack of the rotor blades exceed the stall angle, which can happen with gusts, rapid power demands or disruption to the air flow resulting from unusual manoeuvres. Stalling can induce variations in rotation speed throughout the engine leading to surge. In the training of airline flight crews, an engine model should incorporate such effects for training in engine emergencies.

An engine model typically has four inputs:

- The pilot throttle levers;
- The ambient temperature of the engine air intake;
- Mach number;
- Altitude;

and four outputs:

- Thrust (forward and reverse);
- Engine RPM;
- Fuel flow;
- Exhaust gas temperature.

The pilot throttle lever is rotated forward from the idle position for forward thrust. A release lever must be engaged to move the lever into the backward position to select reverse thrust when the aircraft is on the ground (weight on wheels). The linkage and movement of the lever is prone to mechanical backlash and also has a small dead-zone around the flight idle position. A lookup table is used to provide the actual lever position and the throttle position applied to the fuel injection system, as shown in Figure 3.18.

The static EPR is derived from the engine power lever angle and the ambient air temperature, which is obtained from the weather model, as illustrated in Figure 3.19.

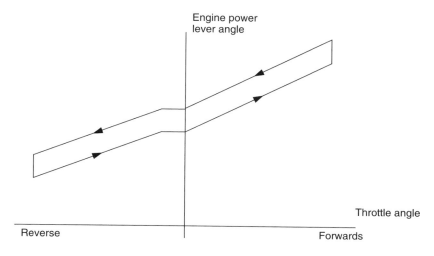

Figure 3.18 Engine power lever angle characteristic

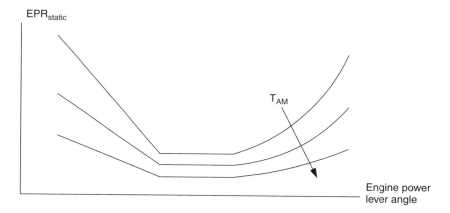

Figure 3.19 Static engine pressure ratio as a function of power lever angle

Clearly, changes to the demanded EPR depend of the moment of inertia of the compressor and turbine rotating at high speed and the dynamic response of the overall system is typically modelled as a first-order lag. In addition, the intake can be affected by the undercarriage and flaps in the flight idle state (typically during landing). The decay speed of the EPR also depends on Mach number and altitude, as shown in Figure 3.20.

Finally, the thrust is computed as a function of EPR, the ambient pressure ratio and Mach number, as shown in Figure 3.21. For reverse thrust, correction for Mach number effects is not normally included.

Two additional terms should be incorporated in the engine model. First, in selecting reverse thrust, there is a delay while the reverse buckets are deployed. During this short period (usually of a fixed duration), the net thrust is modified to take account of the reversal. Secondly, the engines can generate drag when wind milling in the engine-out state and this term is computed for engine failure conditions.

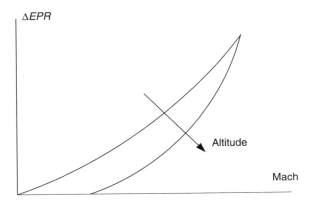

Figure 3.20 Effect of Mach number on EPR

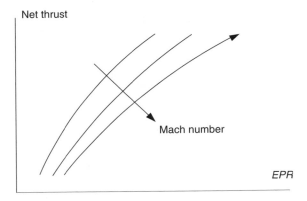

Figure 3.21 Engine thrust as a function of EPR

The manufacturer will also provide validation data to check engine performance at specific true airspeeds and altitude and also to confirm engine response times, for specific atmospheric conditions of temperature and pressure.

3.9 The Landing Gear

As an aircraft taxies, and during the ground roll prior to take-off or following touchdown, the wheels are in contact with the ground. The aircraft may have sufficient airspeed so that the aerodynamic terms still generate forces on the airframe but in addition, forces generated by the undercarriage assembly are also applied to the airframe. Typically, the undercarriage comprises a nose wheel assembly and two main gear assemblies, attached to the fuselage or wings. The nose wheel gear absorbs the load once the main gear assemblies are in contact with the runway and is used for steering. The main gear takes the initial force of impact from contact with the runway and also provides braking (in addition to any reverse thrust provided by the engines). The majority of aircraft accidents occur in the landing phase and consequently, considerable attention is given to the fidelity of the undercarriage model in flight simulation, particularly in civil flight training. It is an area of

Aircraft Dynamics

flight training where its value cannot be overstated. In a simulator, pilots can practise for tyre bursts, brake failures, wet or icy runway conditions, crosswind approaches and microburst without any risk to an aircraft or its flight crew.

In many flight simulators, the landing gear model is almost as detailed as the airframe model. Although several studies have focused on the generic properties of landing gear models (Barnes and Yager, 1985) and (Kruger et al., 1997), specific models will have been derived from taxiing tests and will be included in the data package for the simulator. There are very few detailed landing gear models in the public domain.

Although an undercarriage assembly is a complex mechanical system, it can be represented as a single equivalent oleo, comprising a spring and a damper. The spring must provide sufficient force to support the aircraft between the airframe and the wheels in contact with the ground. The damper provides sufficient force to absorb the download force of a heavy landing, in order to avoid any structural damage to the aircraft (up to a reasonable limit). The location of the (equivalent) gear assemblies will be known and a typical layout is shown in Figure 3.22. Note that the position of the gear locations is given by the centre of the respective wheels and that they are relative to the aircraft cg at the 25% mean wing chord position.

These notes are based on the landing gear model of the Boeing 747 model (Hanke, 1971), which provides the basis for a generic undercarriage model. As the aircraft descends, it will reach a height above the runway where a wheel is in contact with the ground and the respective oleo will be in compression. For each gear assembly, the forces and moments are computed and then transformed to the body frame of the aircraft. The length of each oleo (its compression) is determined from the aircraft position and attitude, as follows:

$$\Delta s_i = \Delta h + x_i \sin \theta - y_i \sin \phi \cos \theta - z_i \cos \phi \cos \theta \qquad (3.134)$$

where the subscript i denotes the respective gear: nose gear $= 1$, left main gear $= 2$ and right main gear $= 3$. Δs_i is the length of the ith oleo and Δh is the height of the aircraft cg above the runway.

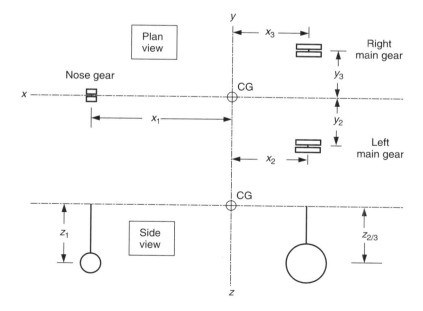

Figure 3.22 Gear locations

In the body frame shown in Figure 3.22, the location of each assembly is given by (x, y, z). The rate of compression of each oleo can also be derived from the aircraft dynamics, as one end of the oleo is, in effect, 'attached' to the runway and the other end is attached to the airframe.

$$\Delta \dot{S}_i = \dot{h} + x_i \dot{\theta} \cos\theta + y_i (\dot{\theta} \sin\phi \sin\theta - \dot{\phi} \cos\phi \cos\theta) + z_i (\dot{\theta} \sin\theta \cos\phi + \dot{\phi} \sin\theta \sin\phi) \quad (3.135)$$

where $\dot{h}$ is the aircraft vertical speed. Note that Δs_i is negative if the oleo is in contact with the ground; in this case, the landing gear forces are combined with the aerodynamic forces on the aircraft frame, otherwise the forces and moments generated by the landing gear are zero. When one or more wheels is in contact with the runway, the pitch and roll angles will be small and small angle approximations can be used ($\sin\theta \approx \theta$ and $\cos\theta \approx 1$).

At each wheel in contact with the ground, there are three forces, as shown in Figure 3.23: F_n the normal force, F_μ the friction force (including braking friction) and F_s the side force, which arises when the plane of rotation of the tyre is not in the direction of the motion of the aircraft (typically during turning).

As the oleo compresses, the spring force increases. The tyre will also compress, providing a reaction in combination with the oleo. Typically, the oleo is gas filled and the compression force is based on the isothermal compression law, giving a square law characteristic in terms of the oleo force per unit displacement, up to the physical limit of the length of the strut. The damping force is proportional to the oleo compression rate (or more typically, to the square of the compression rate). The data for the oleo spring force and damping force is provided by the aircraft manufacturer or the equipment manufacturer. The spring force plus the damping force is the net force in the oleo F_o. It is not difficult to estimate the spring force coefficient or the damping force coefficient; the spring force must be sufficient to support the maximum weight of the aircraft. The damping must be sufficient to absorb a rate of descent up to 2000 ft per minute (without the oleo bottoming) and also should damp out any oscillatory motion caused by the spring force in a few seconds.

Each wheel is subject to friction, comprising breakout friction (at low speed), rolling friction and braking friction. The coefficient of breakout friction applies below approximately 10 kt, reducing from 0.014 at rest to 0 at 10 kt. Generally, a constant is used for the coefficient of rolling friction, where the friction force is given by

$$F_{\text{friction}} = V_g c_{\text{breakout}} + c_{\text{rolling}} F_n \quad (3.136)$$

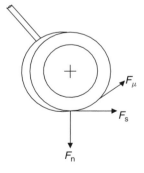

Figure 3.23 Tyre forces

where V_g is the ground speed and $c_{breakout}$ and $c_{rolling}$ are the coefficient of breakout and rolling friction, respectively. Braking friction depends on the braking force applied by the pilot, the coefficient of friction of the brakes and the aircraft mass. The coefficient of friction of the runway surface μ_r limits the braking force that can be applied before the wheel skids. Typical values of μ_r are 0.4 for a dry runway, 0.1 for a wet runway and 0.001 for any icy runway. The net braking force is the addition of the breakout friction, the rolling friction and the braking friction, applied to the main gear wheels, in the opposite direction to the aircraft heading.

The wheel side force occurs when the tyre is not rolling in the direction of motion of the airframe, as shown in Figure 3.24. The tyre track angle β results in a side force perpendicular to the plane of rotation of the wheel. This track angle occurs during turning motion, where the airframe motion and the direction of the wheels are slightly different or, in the case of the nose wheel, when the pilot applies force to the rudder pedals (or tiller) to steer the aircraft. The magnitude of the side force depends on the vertical load applied to the tyre (the force in the oleo), which causes a deflection δ_{t_i} for each tyre. The resultant side force is usually given as a square law in terms of this tyre deflection. The net side force also depends on the ground track angle β and the steering angle and is limited to a maximum side force given by the normal force and the coefficient of sliding friction (typically 0.6). The track angle of each wheel is given by (thdg$_i$ − ttrk$_i$), where thdg$_i$ is the heading angle of the gear assembly. For the main gear, this is the aircraft yaw angle but for the nose gear it is the aircraft yaw angle plus the gear angle resulting from the pilot rudder input. The track angle ttrk$_i$ is the direction of the flight path of the ith gear assembly. In the body frame, the velocity of the gear in the x axis is u; the velocity in the y axis is $v + rx_i$, where r is the rate of yaw in the body frame and x_i is the distance of the gear assembly from the aircraft cg in the x axis. This velocity in the body frame is transformed from the body frame to the Euler frame as the gear direction is also defined with respect to the Euler frame.

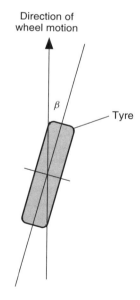

Figure 3.24 Tyre ground track angle

At each gear assembly, the forces are resolved in the body frame axes:

$$F_{x_i} = F_{\mu_i} - F_{n_i}\theta \tag{3.137}$$

$$F_{y_i} = F_{s_i} + F_{n_i}\phi \tag{3.138}$$

$$F_{z_i} = F_{\mu_i}\theta - F_{s_i}\phi + F_{n_i} \tag{3.139}$$

For the nose wheel ($i = 1$), the x-axis force term also includes the nose wheel side force steering component

$$F_{x_1} = F_{\mu_1} - F_{n_1}\theta - F_{s_1}\delta s \tag{3.140}$$

The forces are summed to provide the resultant body frame forces

$$F_x = \sum_{i=1}^{3} F_{x_i} \tag{3.141}$$

$$F_y = \sum_{i=1}^{3} F_{y_i} \tag{3.142}$$

$$F_z = \sum_{i=1}^{3} F_{z_i} \tag{3.143}$$

The moments in the body frame as given by

$$M_x = \sum_{i=1}^{3} (F_{z_i} y_i - F_{y_i} \Delta z_i) \tag{3.144}$$

$$M_y = \sum_{i=1}^{3} (-F_{z_i} x_i - F_{x_i} \Delta z_i) \tag{3.145}$$

$$M_z = \sum_{i=1}^{3} (F_{y_i} x_i - F_{y_i} y_i) \tag{3.146}$$

Where Δz_i is the distance of the ith gear assembly below the cg given by

$$\Delta z_i = z_i + \Delta s_i \tag{3.147}$$

Figure 3.25 shows the results of a simulation where the aircraft (a Boeing 747-100, mass 260,000 kg) is dropped at rest from a height of 2 ft. Figure 3.25a shows the oleo displacements (in inches) for the nose wheel gear, left main gear and right gear assemblies. Note the natural frequency and damping ratio of the landing gear that is evident in Figure 3.25b, which shows the aircraft altitude (above the runway), the vertical speed and the pitch and roll angles. One attraction of using a general model of this form is that the location of the gear assemblies and the oleo spring forces and coefficients of damping can be easily changed for different aircraft.

In practice, much more detailed models are used to include tyre temperature effects, tyre scuffing, tyre bursts, wheel spin-up, low-speed tiller steering and automatic (anti-lock) braking systems. For airline flight training, the ground handling model is very important, providing practice in ground

Aircraft Dynamics

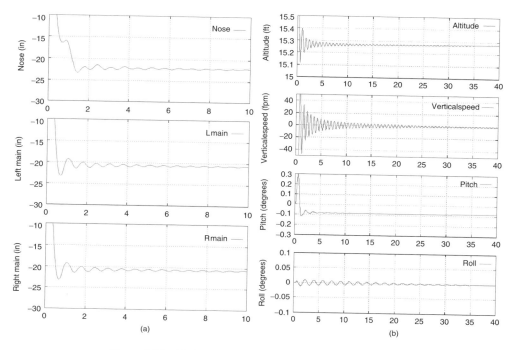

Figure 3.25 (a) Oleo displacement. (b) Landing gear response

manoeuvring, braking techniques, taxiing and particularly a touch down response that it is appropriate to the simulated aircraft. Conversely, a poor ground model can introduce negative transfer of training in a critical area of flight training. For example, a poor ground model could induce over-rotation in the take-off (causing a tail strike) or be too tolerant of a heavy landing, leading to significant problems if similar handling techniques were applied in flying the actual aircraft.

3.10 The Equations Collected

The implementation of the equations of motions is summarized in Figure 3.26. The aerodynamic coefficients are computed to compute the aerodynamic forces and aerodynamic moments. Similarly, the engine forces and moments are computed. These forces and moments are combined to compute the body frame linear and angular accelerations. These accelerations are integrated to form the body frame linear velocities (U, V, W) and the body frame rates (P, Q, R), which are used to compute the quaternion terms needed to compute the DCM to derive the Euler angles. These angles are used to transform from body frame velocities to the earth frame velocities and to compute the aircraft position (latitude, longitude and altitude), which is also needed for the atmospheric model.

The diagram is simplified in three ways: first, the wind components are omitted, secondly, the landing model terms are omitted and thirdly, aerodynamic terms resulting from flap and undercarriage are also omitted.

The equations of motion can be implemented in the following steps:

1. Acquire the inceptor inputs $\delta e, \delta a, \delta r, \delta p, \delta f$ in the range ± 1.0 for signed values and $0-1.0$ for unsigned values.

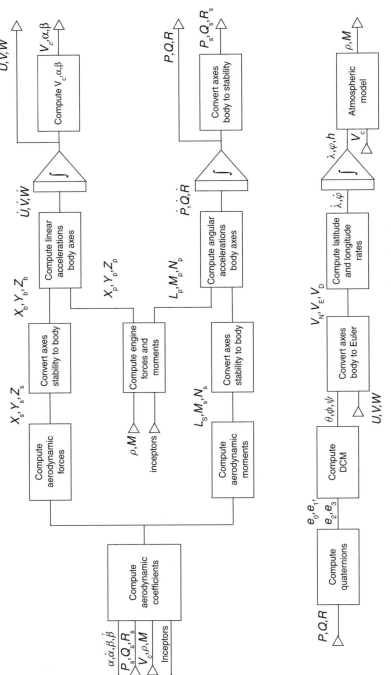

Figure 3.26 Implementation of the equations of motion

Aircraft Dynamics

2. Compute the angles of attack and sideslip

$$V_c = \sqrt{u^2 + v^2 + w^2} \tag{3.148}$$

$$\alpha = \tan^{-1}\left(\frac{w}{u}\right) \tag{3.149}$$

$$\alpha_w = \alpha + \alpha_{\text{offset}} \tag{3.150}$$

$$\dot{\alpha} = \frac{u\dot{w} - w\dot{u}}{\sqrt{u^2 + w^2}} \tag{3.151}$$

$$\beta = \tan^{-1}\left(\frac{v}{\sqrt{u^2 + w^2}}\right) \tag{3.152}$$

$$\dot{\beta} = \frac{\dot{v}\sqrt{u^2 + w^2} - v(u\dot{u} + w\dot{w})}{\sqrt{u^2 + w^2}(u^2 + v^2 + w^2)} \tag{3.153}$$

3. Compute the coefficients of aerodynamic forces $C_L, C_{Lt}, C_D, C_{y\beta}, C_{y\delta r}$
4. Compute the coefficients of aerodynamic moments in pitch $C_{m0}, C_\alpha, C_{\delta e}, C_{mq}, C_{m\dot{\alpha}}$
5. Compute the coefficients of aerodynamic moments in roll $C_{l\beta}, C_{l\delta r}, C_{l\delta a}, C_{lp}, C_{lr}$
6. Compute the coefficients of aerodynamic moments in yaw $C_{n\beta}, C_{n\dot{\beta}}, C_{n\delta r}, C_{n\delta a}, C_{np}, C_{nr}$
7. Compute the body frame forces

$$\text{lift} = \tfrac{1}{2}\rho V_c^2 s C_L \tag{3.154}$$

$$\text{drag} = \tfrac{1}{2}\rho V_c^2 s C_D \tag{3.155}$$

$$\text{side force} = \tfrac{1}{2}\rho V_c^2 s (C_{y\delta r}\delta r + C_{y\beta}\beta) \tag{3.156}$$

8. Compute the engine forces and moments E_x, E_y, E_z and E_L, E_M, E_N
9. Compute the gear forces and moments G_x, G_y, G_z and G_L, G_M, G_N
10. Resolve the body frame forces

$$F_x = \text{lift}\,\sin\alpha - \text{drag}\,\cos\alpha - mg\sin\theta + E_x + G_x \tag{3.157}$$

$$F_y = \text{side force} + mg\sin\phi\cos\theta + E_y + G_y \tag{3.158}$$

$$F_z = -\text{lift}\,\cos\alpha - \text{drag}\sin\alpha + mg\cos\theta\cos\phi + E_z + G_z \tag{3.159}$$

11. Compute the body frame accelerations

$$\dot{u} = \frac{F_x}{m} - qw + rv \tag{3.160}$$

$$\dot{v} = \frac{F_y}{m} - ru + pw \tag{3.161}$$

$$\dot{z} = \frac{F_z}{m} - pv + qu \tag{3.162}$$

12. Compute the body frame aerodynamic velocities

$$u_{\text{aero}} = \int \dot{u}\, dt \tag{3.163}$$

$$v_{\text{aero}} = \int \dot{v} \, dt \tag{3.164}$$

$$w_{\text{aero}} = \int \dot{w} \, dt \tag{3.165}$$

13. Include the wind components (north, east and down)

$$u = u_{\text{aero}} + \text{wind}_N a_{11} + \text{wind}_E a_{21} + \text{wind}_d a_{31} \tag{3.166}$$

$$v = v_{\text{aero}} + \text{wind}_N a_{12} + \text{wind}_E a_{22} + \text{wind}_d a_{32} \tag{3.167}$$

$$w = w_{\text{aero}} + \text{wind}_N a_{13} + \text{wind}_E a_{23} + \text{wind}_d a_{33} \tag{3.168}$$

14. Include the turbulence components

$$u = u + \text{turbulence}_N a_{11} + \text{turbulence}_E a_{21} + \text{turbulence}_d a_{31} \tag{3.169}$$

$$v = v + \text{turbulence}_N a_{12} + \text{turbulence}_E a_{22} + \text{turbulence}_d a_{32} \tag{3.170}$$

$$w = w + \text{turbulence}_N a_{13} + \text{turbulence}_E a_{23} + \text{turbulence}_d a_{33} \tag{3.171}$$

15. Compute the Earth velocities

$$V_N = u a_{11} + v a_{12} + w a_{13} - \text{wind}_N \tag{3.172}$$

$$V_E = u a_{21} + v a_{22} + w a_{23} - \text{wind}_E \tag{3.173}$$

$$V_d = u a_{31} + v a_{32} + w a_{33} - \text{wind}_d \tag{3.174}$$

16. Compute the aircraft latitude and longitude rates

$$\dot{\lambda} = \frac{V_N}{R + h} \tag{3.175}$$

$$\dot{\mu} = \frac{V_E}{\cos \lambda (R + h)} \tag{3.176}$$

17. Compute the aircraft position

$$\lambda = \int \dot{\lambda} \, dt \tag{3.177}$$

$$\mu = \int \dot{\mu} \, dt \tag{3.178}$$

$$h = \int V_d \, dt \tag{3.179}$$

18. Compute the body rates in stability axes

$$p_{\text{stab}} = p \cos \alpha + r \sin \alpha \tag{3.180}$$

$$r_{\text{stab}} = r \cos \alpha - p \sin \alpha \tag{3.181}$$

19. Compute the body frame moments in stability axes

$$M_{stab} = \tfrac{1}{2}\rho V_c^2 s \bar{c}(C_{m0} + C_{m\alpha}\alpha_w + C_{m\delta e}\delta e) + \tfrac{1}{4}\rho V_c s \bar{c}^2(C_{mq}q + C_{m\dot{\alpha}}\dot{\alpha}) \quad (3.182)$$

$$L_{stab} = \tfrac{1}{2}\rho V_c^2 sb(C_{l\beta}\beta + C_{l\delta a}\delta a + C_{l\delta r}\delta r) + \tfrac{1}{4}\rho V_c sb^2(C_{lp}p_{stab} + C_{lr}r_{stab}) \quad (3.183)$$

$$R_{stab} = \tfrac{1}{2}\rho V_c^2 sb(C_{n\beta}\beta + C_{n\delta a}\delta a + C_{n\delta r}\delta r) + \tfrac{1}{4}\rho V_c sb^2(C_{np}p_{stab} + C_{nr}r_{stab}) \quad (3.184)$$

20. Compute the body frame moments in the body frame

$$M = M_{stab} + \text{lift}(cg - 0.25)\bar{c}\cos\alpha + \text{drag}(cg - 0.25)\bar{c}\sin\alpha + E_M + G_M \quad (3.185)$$

$$L = L_{stab}\cos\alpha - R_{stab}\sin\alpha + E_L + G_L \quad (3.186)$$

$$R = R_{stab}\cos\alpha + L_{stab}\sin\alpha - \text{sideforce}(cg - 0.25)\bar{c} + E_N + G_N \quad (3.187)$$

21. Compute the body frame angular accelerations

$$\dot{p} = \frac{L + (I_{yy} - I_{zz})qr + I_{xz}(\dot{r}p + pq)}{I_{xx}} \quad (3.188)$$

$$\dot{q} = \frac{M + (I_{zz} - I_{xx})rp + I_{xz}(r^2 - p^2)}{I_{yy}} \quad (3.189)$$

$$\dot{r} = \frac{R + (I_{xx} - I_{yy})pq + I_{xz}(\dot{p} - qr)}{I_{zz}} \quad (3.190)$$

22. Compute the body rates

$$p = \int \dot{p}\, dt \quad (3.191)$$

$$q = \int \dot{q}\, dt \quad (3.192)$$

$$r = \int \dot{r}\, dt \quad (3.193)$$

23. Compute the quaternions

$$\lambda = 1 - (e_0^2 + e_1^2 + e_2^2 + e_3^2) \quad (3.194)$$

$$\dot{e}_0 = -1/2(e_1 p + e_2 q + e_3 r) + \lambda e_0 \quad (3.195)$$

$$\dot{e}_1 = 1/2(e_0 p + e_2 r - e_3 q) + \lambda e_1 \quad (3.196)$$

$$\dot{e}_2 = 1/2(e_0 q + e_3 p - e_1 r) + \lambda e_2 \quad (3.197)$$

$$\dot{e}_3 = 1/2(e_0 r + e_1 q - e_2 p) + \lambda e_3 \quad (3.198)$$

$$e_0 = \int \dot{e}_0\, dtt \quad (3.199)$$

$$e_1 = \int \dot{e}_1\, dt \quad (3.200)$$

$$e_2 = \int \dot{e}_2 \, dt \qquad (3.201)$$

$$e_3 = \int \dot{e}_2 \, dt \qquad (3.202)$$

24. Compute the DCM

$$a_{11} = e_0^2 + e_1^2 - e_2^2 - e_3^2 \qquad (3.203)$$

$$a_{12} = 2(e_1 e_2 - e_0 e_3) \qquad (3.204)$$

$$a_{13} = 2(e_0 e_2 + e_1 e_3) \qquad (3.205)$$

$$a_{21} = 2(e_1 e_2 + e_0 e_3) \qquad (3.206)$$

$$a_{22} = e_0^2 - e_1^2 + e_2^2 - e_3^2 \qquad (3.207)$$

$$a_{23} = 2(e_2 e_3 - e_0 e_1) \qquad (3.208)$$

$$a_{31} = 2(e_1 e_3 - e_0 e_2) \qquad (3.209)$$

$$a_{32} = 2(e_2 e_3 + e_0 e_1) \qquad (3.210)$$

$$a_{33} = e_0^2 - e_1^2 - e_2^2 + e_3^2 \qquad (3.211)$$

25. Compute the Euler angles

$$\theta = \sin^{-1}(-a_{31}) \qquad (3.212)$$

$$\phi = \tan^{-1}\left(\frac{a_{32}}{a_{33}}\right) \qquad (3.213)$$

$$\psi = \tan^{-1}\left(\frac{a_{21}}{a_{11}}\right) \qquad (3.214)$$

3.11 The Equations Revisited – Long Range Navigation

In the equations of motion covered in the previous sections, the forces and accelerations are formulated in the body frame and the earth is assumed to be a non-rotating inertial frame. However, this simplification introduces two problems: first, the effect of the earth rotation is omitted and secondly, the resultant accelerations are most conveniently computed in the navigation frame, whereas the navigation frame has only been used to represent aircraft velocity and position. In addition, gravitational acceleration has been transformed to the body frame, whereas this acceleration can be most simply treated as an acceleration in the navigation frame. In the following sections, it will be shown that the revisions to the equations of motion to incorporate Coriolis and centripetal accelerations are straightforward.

The equations covered in the previous sections are sometimes referred to as 'flat earth' equations. If an aircraft flew at a constant altitude in a trimmed (non-accelerating) state, it would follow a trajectory parallel to the ground with this representation. In practice, if the aircraft flew in a straight line in inertial space, although its flight path might be tangential to the surface of the earth initially, the aircraft would start to fly away from the earth as a result of centripetal acceleration. In addition, the rotation of the earth has also been ignored in the formulation of the equations of motion.

Although the transformation between north and east velocities to latitude and longitude rates has been covered, it does not include any terms for the Coriolis accelerations resulting from the rotating earth frame.

For the majority of simulation applications, the exclusion of these terms has very little effect on the accuracy of the simulation. Unlike missile guidance, say, navigation is achieved by corrections from visual references or ground-based navigation aids. In the simulation of missile systems or inertial navigation systems, or for simulated flight over several hours, these terms would be significant and would therefore be included in the equations of motion.

There is one further limitation that these additional terms impose. Up to now, it has been convenient to compute the forces and accelerations in the body frame. The resultant velocities are then resolved into the appropriate local frame. However, Coriolis acceleration and centripetal acceleration occurs in the navigation frame, so there are benefits from computing these accelerations and the aerodynamic and gravitational accelerations in the navigation frame. This implementation also simplifies the computation of the gravitational acceleration, which was resolved into body frame components in the previous implementation. In addition, the wind velocity components have been resolved in the body frame, whereas it is more convenient to include them in the computation of the local frame velocity components.

Consider the aircraft illustrated in Figure 3.27. Assume that it is constrained to fly along a meridian of longitude (in the plane of the page) and is also constrained to pitching motion (ignoring the roll and yaw terms). The local navigation frame is positioned at the aircraft cg with the x axis parallel to the surface of the earth.

$$\dot{\theta} = \omega_{yb} - v_x/(R+h) \tag{3.215}$$

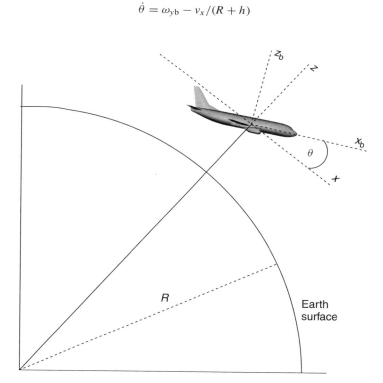

Figure 3.27 Two-dimensional navigation in an earth frame

where

> ω_{yb} is the pitch rate in body axes (q) resulting from the aircraft dynamics
> v_x is the transport rate given by the rotation of the aircraft around the earth
> R is the earth radius
> h is the altitude of the aircraft above the earth surface.

Resolving the aircraft forces into the local frame

$$f_x = f_{xb} \cos\theta + f_{zb} \sin\theta \qquad (3.216)$$

$$f_z = -f_{xb} \sin\theta + f_{zb} \cos\theta \qquad (3.217)$$

The net accelerations in the local frame result from the aerodynamic forces, the gravitational force and the motion of aircraft around the earth. The gravitational acceleration is assumed to act perpendicular to the surface of the earth (towards the centre of the earth).

$$\dot{v}_x = f_x + v_z\dot{\theta} = f_x + v_x v_z/(R+h) \qquad (3.218)$$

$$\dot{v}_z = f_z + g - v_x\dot{\theta} = f_z + g - v_x^2/(R+h) \qquad (3.219)$$

In other words, accelerations computed in the body frame can be resolved into the navigation frame. Although this is a contrived example, it is used to illustrate the computation of motion in different axes, which is readily extended to six degrees of freedom.

In the following sections, three orthogonal axes systems will be used to describe aircraft motion:

- Body axes – with an origin at the aircraft cg, the x axis pointing forwards, the y axis along the starboard wing and the z axis perpendicular to the xy plane, pointing downwards;
- ECEF axes – an inertial reference frame with an origin at the centre of the earth, defined in terms of latitude, longitude and altitude, where one axis passes through the Greenwich meridian and the other axis lies in the equatorial plane;
- The navigation (or local) axes, with the origin on the surface of the earth directly below the aircraft, with the axes continuously realigned to point north, east and downwards, towards the centre of the earth.

The aircraft body frame translates and rotates with respect to the navigation frame. Likewise the navigation frame translates and rotates with respect to the earth frame. Although appearing cumbersome, the transformations between the linear and angular motion are straightforward. The following description is based on notes from Titterton and Weston's (2004) excellent introduction to inertial navigation and McFarland's (1975) description of a kinematic model used at NASA. For reasons of brevity, the equations are stated rather than derived.

3.11.1 Coriolis Acceleration

The theorem of Coriolis applies to two rotating reference frames. Motion in one frame can be computed in the other frame if the relative angular velocity between the two frames is known. As a consequence of the relative motion of the frames, motion of a body defined in one frame may appear to be different (in terms of measurements) in another frame. Such a situation can be illustrated with a playground roundabout. If two children sit on opposite sides of the roundabout and throw a ball to each other, viewed from above (the inertial frame) the ball will follow a straight path. Viewed from the rotating roundabout, the ball will appear to take a curved path.

Aircraft Dynamics

The Coriolis effect should not be confused with relative motion. The earth rotates once every 24 hours and therefore a point on the equator is moving at 1670 km/hr or 464 m/s, relative to the centre of the earth. Someone jumping up and down on a pogo stick on the equator will not suddenly perceive the earth to be moving underneath them at 464 m/s while they are in the air, because the pogo stick also had a velocity of 464 m/s when it left the ground. To the pogo stick user, the earth appears to be stationary.

Consider a light aircraft, flying at 80 kt from Plymouth airport to Glasgow airport in the United Kingdom on a perfect summer's day (i.e. with nil wind). The aircraft flies almost true north (along a line of longitude), taking 4 hours to cover the distance of almost 328 nautical miles between the two airports (assuming nil wind). In inertial space, if the aircraft maintains a constant airspeed, altitude and vertical speed, the net acceleration on the aircraft is zero and therefore, the aircraft trajectory in inertial space will follow a straight line. Viewed from the rotating earth, the flight path would appear curved or if the aircraft latitude and longitude is plotted, the aircraft track will curve to the right.

In a rotating earth frame, the velocity of the surface of the earth at the equator is 1670 km/hr. In an inertial frame, taking off from Plymouth (latitude 50.4 °N), the aircraft has a westward component of 295.7 m/s. Landing at Glasgow (latitude 55.9 °N), the aircraft will have a westward component of 260.0 m/s. Assuming the aircraft flies due north at all times, there is an average (implied) change in eastward velocity of 17.8 m/s, which must have resulted from an acceleration in the navigation frame of the aircraft. If the pilot actually flew true north for 4 hours, the aircraft would miss Glasgow by 256 km. Over relatively short distances, this effect is negligible and a pilot can follow land marks or navigation beacons to maintain an accurate track. Over long distances, GPS provides latitude and longitude position updates and inertial navigation systems compensate for earth rotation. Landing at Glasgow airport, the aircraft would appear to have been subject to a deceleration of 17.8 m/s over 4 hours or 0.001236 m/s². Admittedly, this is only 1/1000th of the gravitational acceleration and, for many applications of flight simulation, the earth can be treated as a fixed non-rotating frame; if flight guidance is based on navigation aids, aircraft tracks plotted on navigation charts will be correct.

However, where a flight is based on course corrections (e.g. missile guidance) or the flight is over several hundred nautical miles, the Coriolis effect will introduce accelerations which should be included in the computation of aircraft motion, in order to be consistent with aircraft navigation. In the general case, the accelerations resulting from a rotating earth frame should be included in the equations of motion.

Coriolis applies when one reference frame rotates relative to another reference frame. Motion defined in one frame can be computed in the other frame in terms of the velocity and position in one frame and the relative rotation between the frames. Consider an inertial frame and an earth-fixed frame. The well-known equation is:

$$v_e = v_i - \omega_{ie} \times r \tag{3.220}$$

where the vector v_i is the velocity in the inertial frame, v_e is the velocity in the earth rotating frame, r is the position vector in the inertial frame, $\times$ denotes the cross product and ω_{ie} is the rotation difference of the two reference frames, given by $\omega_{ie} = [0 \ 0 \ \Omega]^T$, where Ω is the earth rotation rate (1 revolution per 23 hours, 56 minutes and 4 seconds) giving an angular rate of rotation of $\frac{2\pi}{86,164} = 0.00007292$ rad/s. Consider motion in the inertial frame. From Equation 3.221

$$\frac{d^2}{dt^2} r_i = \frac{d}{dt} v_e + \frac{d}{dt} [\omega_{ie} \times r] \tag{3.221}$$

Differentiating Equation 3.220,

$$\frac{d^2}{dt^2} r_i = \frac{d}{dt} v_e + \omega_{ie} \times v_e + \omega_{ie} \times [\omega_{ie} \times r] \tag{3.222}$$

The earth is assumed to rotate at a constant rate, therefore $\frac{d}{dt}(\omega_{ie}) = 0$. The net acceleration on the aircraft is given by the aerodynamic acceleration a, the gravitational acceleration g and the Coriolis term, as follows:

$$\frac{d}{dt} v_e = a - \omega_{ie} \times v_e - \omega_{ie} \times [\omega_{ie} \times r] + g \tag{3.223}$$

The term $\omega_{ie} \times v_e$ is the acceleration resulting from the aircraft moving over a rotating earth and the term $g - \omega_{ie} \times [\omega_{ie} \times r]$ represents the local gravity vector formed from the gravitational acceleration of the earth mass and the centripetal acceleration of the earth rotation. The net acceleration in the inertial frame is

$$\frac{d}{dt} v_e = a - \omega_{ie} \times v_e + g_l \tag{3.224}$$

In the navigation frame

$$\frac{d}{dt} v_e|_n = \frac{d}{dt} v_e|_i - (\omega_{ie} + \omega_{en}) \times v_e \tag{3.225}$$

Substituting $\frac{d}{dt} v_e|_i$ in Equation 3.225, the general form of the aircraft acceleration in the navigation frame is given by

$$\frac{d}{dt} v_e|_n = a - [2\omega_{ie} + \omega_{en}] \times v_e + g_l \tag{3.226}$$

In navigation axes, the acceleration is given by

$$\dot{v}_e^n = C_b^n a^b - [2\omega_{ie}^n + \omega_{en}^n] \times v_e^n + g_l^n \tag{3.227}$$

where C_b^n is the transformation (DCM) from the body frame to the navigation frame and g_l^n is the local gravitational acceleration in the navigation frame.

The equations to derive the aircraft accelerations and moments given in Section 3.10 are revised as follows:

1. Acquire the inceptor inputs, compute the angles of attack and sideslip, compute the aerodynamic forces and moments in the body frame, compute the engine forces and moments, compute the gear forces and moments (Equations 3.149–3.157)
2. Compute the forces in the body frame

$$F_x = \text{lift} \sin \alpha - \text{drag} \cos \alpha + E_x + G_x \tag{3.228}$$

$$F_y = \text{side force} + E_y + G_y \tag{3.229}$$

$$F_z = -\text{lift} \cos \alpha - \text{drag} \sin \alpha + E_z + G_z \tag{3.230}$$

Aircraft Dynamics

3. Resolve the forces in the body frame in the navigation frame

$$F_n = F_x a_{11} + F_y a_{12} + F_z a_{13} \tag{3.231}$$

$$F_e = F_x a_{21} + F_y a_{22} + F_z a_{23} \tag{3.232}$$

$$F_n = F_x a_{31} + F_y a_{32} + F_z a_{33} \tag{3.233}$$

4. Compute the distance of the aircraft from the centre of the earth

$$R = R_0 - P_z \tag{3.234}$$

5. Compute the linear accelerations in the navigation frame

$$\dot{v}_n = \frac{F_n}{m} - 2v_e \Omega \sin\lambda + \frac{v_n v_d - v_e^2 \tan\lambda}{R} \tag{3.235}$$

$$\dot{v}_e = \frac{F_e}{m} + 2v_n \Omega \sin\lambda + 2v_d \Omega \cos\lambda + \frac{v_e v_d - v_n v_e \tan\lambda}{R} \tag{3.236}$$

$$\dot{v}_d = \frac{F_d}{m} + g - 2v_e \Omega \cos\lambda - \frac{v_e^2 + v_n^2}{R} \tag{3.237}$$

6. Compute the linear velocities in the navigation frame

$$v_n = \int \dot{v}_n \, dt \tag{3.238}$$

$$v_e = \int \dot{v}_e \, dt \tag{3.239}$$

$$v_d = \int \dot{v}_d \, dt \tag{3.240}$$

7. Compute the turbulence terms in the navigation frame

$$T_n = T_u a_{11} + T_v a_{21} + T_w a_{31} \tag{3.241}$$

$$T_e = T_u a_{12} + T_v a_{22} + T_w a_{32} \tag{3.242}$$

$$T_d = T_u a_{13} + T_v a_{23} + T_w a_{33} \tag{3.243}$$

8. Compute the relative velocity in the navigation frame (to include wind and turbulence terms)

$$v_{nr} = v_n - \text{wind}_n + T_n \tag{3.244}$$

$$v_{er} = v_e - \text{wind}_e + T_e \tag{3.245}$$

$$v_{dr} = v_d - \text{wind}_d + T_d \tag{3.246}$$

9. Compute the body frame velocities

$$u = v_{nr} a_{11} + v_{er} a_{21} + v_{dr} a_{31} \tag{3.247}$$

$$v = v_{nr}a_{12} + v_{er}a_{22} + v_{dr}a_{32} \qquad (3.248)$$

$$w = v_{nr}a_{13} + v_{er}a_{23} + v_{dr}a_{33} \qquad (3.249)$$

10. Compute the aircraft position (λ, ϕ, h) from Equations 3.176–3.180
11. Compute the body frame moments (L, M, N) and body frame angular accelerations ($\dot{p}, \dot{q}, \dot{r}$) from Equations 3.181–3.191.
12. Compute the earth rotation rates in the navigation frame

$$\omega_p = \Omega \cos \lambda + \frac{v_e}{R} \qquad (3.250)$$

$$\omega_q = \frac{-v_n}{R} \qquad (3.251)$$

$$\omega_r = -\Omega \sin \lambda - \frac{v_e \tan \lambda}{R} \qquad (3.252)$$

13. Include the earth rate terms in the body rates

$$p = p - \omega_p a_{11} + \omega_q a_{21} + \omega_r a_{31} \qquad (3.253)$$

$$q = q - \omega_p a_{12} + \omega_q a_{22} + \omega_r a_{32} \qquad (3.254)$$

$$r = r - \omega_p a_{13} + \omega_q a_{23} + \omega_r a_{33} \qquad (3.255)$$

14. Compute the quaternions (Equations 3.195–3.203)
15. Compute the DCM (Equations 3.204–3.212)
16. Compute the Euler angles (Equations 3.213–3.215)

References

Anon (1976) U.S. Standard Atmosphere, NASA-TM-X-74335.

Barnes, A. and Yager, T. (1985) Simulation of aircraft behaviour on or close to the ground. AGARD Report No. AG-285.

Boiffier, J. L. (1998) *The Dynamics of Flight – The Equations*, John Wiley & Sons, Chichester.

ESDU (1966) *The Equations of Motion of a Rigid Body Aircraft*, Item 67003, Engineering Sciences Data, London.

Etkin, B. and Reid, L. D (1996) *Dynamics of Flight: Stability and Control*, John Wiley & Sons, New York.

Fang, A. C. and Zimmerman, B. G. (1969) Digital Simulation of Rotational Kinematics, NASA TN-D-5302.

Fink, M. P. and Freeman, D. C. (1969) *Full-scale Wind-tunnel Investigation of Static Longitudinal and Lateral Characteristics of a Light Twin-engine Airplane*, NASA TN-D-4983, Langley Research Centre, Hampton.

Fogarty, L. E. and Howe, R. M. (1969) Computer Mechanisation of Six-degree of Freedom Flight Equations, NASA CR-1344.

Hanke, C. R. (1971) The simulation of a large jet transport aircraft, Vol. I: mathematical model, NASA CR-1756, Vol. II: Modelling Data, Boeing Report No. N73-10027, Boeing Company, Wichita, KS.

Heffley, R. K. and Jewell, W. F. (1972) Aircraft Handling Qualities Data, NASA CR-2144.

Kruger, W., Besselink, I., Cowling, D., Doan, D. B., Kortum, W. and Krabacher, W. (1997) Aircraft landing gear dynamics: simulation and control. *Vehicle System Dynamics*, **28**, 119–158.

McCormick, B. (1995) *Aerodynamics, Aeronautics and Flight Mechanics*, John Wiley & Sons, New York

McFarland, R. E. (1975) A Standard Kinematic Model for Flight Simulation at NASA-Ames, NASA-CR-2497.

Mitchell, C. G. B (1973) A computer programme to predict the stability and control characteristics of subsonic aircraft, TR 73079, Royal Aircraft Establishment.

Nguyen, L. T., Ogburn, M. E., Gilbert, W. P., Kibler, K. S, Brown, P. W. and Deal, P. L. (1979) Study of Stall/Post-stall Characteristics of a Fighter Airplane with Relaxed Stability, NASA TP-1583.

Pavlis, N. K., Holmes, S. A., Kenyon, S. C. and Factor, J. K. (2008) *An Earth Gravitational Model to Degree 2160: EGM2008*, General Assembly of the European Geosciences Union, Vienna.

Smetana, F. O. (1984) *Computer-assisted Analysis of Aircraft Performance Stability and Control*, McGraw Hill, New York

Stevens, B. L. and Lewis, F. L. (2003) *Aircraft Control and Simulation*, John Wiley & Sons Hoboken, NJ

Stinton, D. (1985), *The Design of the Aeroplane*, Sheridan House Inc., London

Titterton, D. and Weston, J. (2004), *Strapdown Inertial Technology*, Institution of Engineering and Technology, Stevenage.

Wolowicz, C. H. and Yancey. R. B. (1972) *Lateral-Directional Aerodynamic Characteristics of Light Twin-Engine Propeller-Driven Airplanes*, NASA TN-D-6946, NASA Flight Research Centre, Edwards, CA.

Worobel, R. (1972) *Computer Program User's Manual for Advanced General Aviation Propeller Study*, NASA CR-2066, Hamilton Standard, Windsor Locks, Connecticut.

Worobel, R. and Mayo, M. G. (1971) *Advanced General Aviation Propeller Study*, NASA CR-114399, Hamilton Standard, Windsor Locks, Connecticut.

4

Simulation of Flight Control Systems

4.1 The Laplace Transform

Electrical circuits containing resistors can be decomposed by applying two rules to simplify the network. Resistors in series can be combined by substituting a group of resistors in series with an equivalent resistor, as shown in Figure 4.1. Similarly, resistors configured in parallel can be reduced to an equivalent resistor, as shown in Figure 4.2.

Any combination of resistors can be reduced to sets of resistors in series or parallel, which in turn can be simplified to an equivalent single resistor for each group of resistors. This reduction proceeds until the network of resistors is reduced to a level of complexity where it is straightforward to derive the relationship between the circuit output signal and the input signal applied to the circuit. For example, in Figure 4.3, the circuit gain is given by

$$\frac{V_{out}}{V_{in}} = \frac{R_2}{R_1 + R_2} \tag{4.1}$$

The relationship between the voltage across a resistor and the current through the resistor is given by the well-known Ohm's law

$$v = iR \tag{4.2}$$

where i is the current (amps) through the resistor, v is the voltage (volts) across the resistor and R is the resistance (ohms).

In addition to resistors, electronic circuits may also contain other passive components, including capacitors and inductors. For a capacitor

$$i = \frac{C\,dv}{dt} \text{ or } \int i\,dt = Cv \tag{4.3}$$

where i is the current through the capacitor, v is the voltage across the capacitor and C is the capacitance (farads). Similarly, for an inductor

$$v = L\frac{di}{dt} \tag{4.4}$$

Principles of Flight Simulation D. J. Allerton
© 2009, John Wiley & Sons, Ltd

Figure 4.1 Series resistor networks

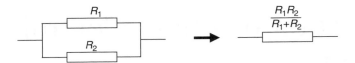

Figure 4.2 Parallel resistor networks

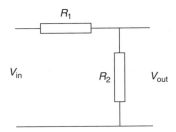

Figure 4.3 Resistor network

where i is the current through the inductor, v is the voltage across the inductor and L is the inductance (henrys).

The voltage across a capacitor is proportional to the integral of the current through the capacitor and for an inductor, the voltage is proportional to the rate of change of current. Circuits can therefore be treated as sets of simultaneous differential equations, in terms of the network currents. The Laplace operator s provides a convenient shorthand to aid in the analysis of circuits, where resistors are represented by an impedance R, capacitors by an impedance $1/sC$ and inductors by an impedance sL. A network of components can then be reduced to a single equivalent impedance, where the relationship between the circuit output and the circuit input is given as a polynomial in s, which is more commonly known as the *transfer function*. For example, consider the RC network, shown in Figure 4.4.

The circuit gain is given by

$$\frac{V_{out}(s)}{V_{in}(s)} = \frac{\frac{1}{sC}}{R + \frac{1}{sC}} = \frac{\frac{1}{RC}}{s + \frac{1}{RC}} = \frac{a}{s+a} \tag{4.5}$$

where $a = 1/RC$

Consider a step input, where $V_{in} = 1$ V. The Laplace transform of a unit step input is $1/s$. Therefore the output $V_{out}(s)$ is given by

$$V_{out}(s) = V_{in}(s) \frac{a}{s+a} = \frac{a}{s(s+a)} \tag{4.6}$$

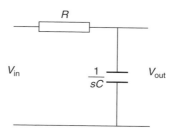

Figure 4.4 RC network

From an inspection of tables of Laplace transforms (Spiegel, 1965), the output signal can be transformed from a polynomial in s to the equivalent signal given as a function of time, giving

$$V_{out}(t) = 1 - e^{-at} \tag{4.7}$$

Notice the convenience of the Laplace transform. The circuit can be defined and manipulated as a polynomial in s, providing a straightforward method to determine the time response.

The Laplace transform also provides insight into the steady-state condition of a system. The steady-state of a system given by $f(s)$ is defined by $sf(s)$ as s tends to zero. In Equation 4.6, the steady-state gain for the step input is given by

$$V_{out}(s) = \frac{as}{s(s+a)} = 1 \tag{4.8}$$

The output tends to unity, which is confirmed by the equivalent time domain expression

$$V_{out}(t) = 1 - e^{-at} \text{ or } V_{out}(t) = 1 \text{ as } t \to \infty \tag{4.9}$$

In other words, from analysis of the circuit, the time response to a step input can be computed as a function of R and C for this basic first-order system. For more complex circuits, the transfer function can be resolved to form a polynomial in s to provide the system time response to an impulse, ramp or step input. Moreover, inspection of the roots of the denominator provides insight into the stability and the time response of a circuit.

Although we have only considered the Laplace transform for electrical circuits, similar methods of simplification and analysis apply to mechanical systems (and also to chemical, thermal, financial and nuclear systems). For mechanical systems, a damper has a transfer function D, a mass has a transfer function $1/sM$ and a spring has a transfer function sK, where D is the coefficient of damping, M is the mass or inertia and k is the spring elasticity. In other words, a mechanical system comprising dampers, masses and springs is equivalent (in Laplace terms) to an electrical circuit of resistors, capacitors and inductors, respectively.

For the systems engineer, the Laplace transform applies not only at the component or network level but it also applies to block diagrams, particularly where there are feed forward and feedback branches. In Figure 4.5, three common forms of block diagram are shown, together with their reduced form, bearing in mind that G_1 and G_2 may be transfer functions in s. Figures 4.5a and b are used to simplify complex block diagrams. Figure 4.5c gives the equivalent of a feedback loop. Having reduced the block diagram to a single transfer function in s, inspection of the roots of the denominator can provide considerable insight into the system response and stability.

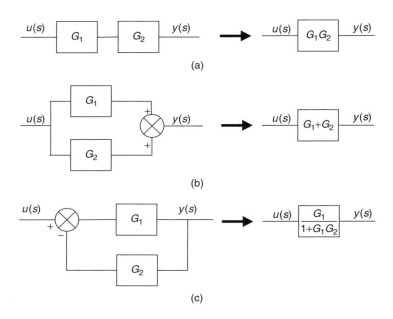

Figure 4.5 Transfer function blocks

In cases where the system is excited by a sinusoidal input, it is very useful to be able to determine the system gain or phase at specific frequencies and also the overall bandwidth of the system, where a reasonable gain can be expected. By substituting $s = j\omega$ in the system transfer function, where $j = \sqrt{-1}$ and ω is frequency (rad/s), the transfer function reduces to a complex number as a function of the frequency ω. The system gain at a frequency ω is given by the magnitude of the complex number and the phase is given by the tangent of the angle represented by the complex number on an Argand diagram (the imaginary component divided by the real component).

For these reasons (and many more), Laplace transforms are widely used in engineering design. It is important to appreciate that they are a mathematical convenience and specific assumptions apply the Laplace transforms. Often, engineers overlook the mathematical rigour of the Laplace transform and the reader is advised to seek out the many text books on Laplace transforms to confirm the benefits they afford and to appreciate the rigour and elegance of the method. However, the mathematics of the Laplace transform is straightforward and they are widely used in the design of control systems to establish the system response, or the system stability or the steady-state gain. Laplace transforms also provide the basis of filter design, in effect shaping the frequency response to provide sufficient gain or attenuation in specific frequency bands.

In flight simulation, control systems are often specified as transfer functions and therefore the simulator designer needs to implement transfer functions in software, so that the input–output relationship of a system is modelled accurately. For the systems engineer, the time response, the frequency response and the transfer function are equivalent and interchangeable. For the reader confronted with the design or implementation of a control system, there are many excellent textbooks on the design of control systems, providing a sound introduction to Laplace transforms and the techniques to analyse system stability or to tune system performance. These topics are outside the scope of this book but in the subsequent sections, it will be assumed that the reader has a sound grasp of transfer functions.

4.2 Simulation of Transfer Functions

The Laplace operator s^n corresponds to the nth derivative of a variable and similarly $1/s^n$ corresponds to the nth integral of a variable. For example, sx is, in effect, shorthand for dx/dt and likewise x/s is equivalent to $\int x \, dt$. Although s is an operator, it can be used as a symbol in manipulating transfer functions, subject to the restrictions that apply to the Laplace transform. Consequently, a transfer function can be rearranged in the form of a polynomial in terms of the operator $1/s$, where each occurrence of $1/s$ is an integration operation.

For example, consider the following transfer function:

$$\frac{y(s)}{x(s)} = \frac{16}{s^2 + 4s + 16} \tag{4.10}$$

Rearranging the transfer function and treating the Laplace operator as an algebraic operator (Ord-Smith and Stephenson, 1975)

$$s^2 y + 4sy + 16y = 16x \tag{4.11}$$

Dividing through by s

$$sy + 4y = \frac{16}{s}(x - y) \tag{4.12}$$

$$y = \frac{1}{s}\left(\frac{16}{s}(x - y) - 4y\right) \tag{4.13}$$

The equivalent block diagram is shown in Figure 4.6. A 20s the term $1/s$ corresponds to an integrator, the transfer function can be implemented by two simple integrations (given in C code):

```
integrate(&t, 16 * (x - y)
integrate(&y, t - 4 * y)
```

where the procedure `integrate(*y, x)` performs numerical integration equivalent to the mathematical operation $y = \int x \, dt$.

Notice that t is computed by the first integrator and then used in the integrand of the second integrator and that y is computed by the second integrator and used in the integrand of the first integrator. Care is needed that the values of x and t used in the integrands are the values at the start of the step and that they are not updated until the end of the step. The advantage of this method is that it is simple and intuitive. However, it does introduce some intermediate variables;

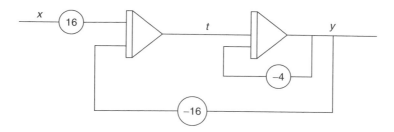

Figure 4.6 Second-order transfer function schematic

in this case, t is the integral of $16(x - y)$. Although the state variables are y and dy/dt, dy/dt is not explicitly computed following the rearrangement of the equations. In addition, it is necessary to set the right-hand integrator to the initial value of y, and the left-hand integrator to the initial value of t.

An alternative approach is to use the Tustin transformation (Franklin et al., 1994) to represent transfer functions in real-time systems, where the Laplace operator s is replaced by the sampled data operator Z, given by the bilinear transformation

$$s = \frac{2}{T}\left(\frac{Z-1}{Z+1}\right) \tag{4.14}$$

where T is the sampling interval.

The transformation of a transfer function in s to a transfer function in Z can be achieved by hand or by using a symbolic manipulation tool to transform the numerator and denominator to polynomials in Z in the following form

$$D(Z) = \frac{y(Z)}{x(Z)} = \frac{b_0 + b_1 Z^{-1} + b_2 Z^{-2} + \cdots}{a_0 + a_1 Z^{-1} + a_2 Z^{-2} + \cdots} \tag{4.15}$$

Rearranging the equation, ignoring the fourth and subsequent terms of the numerator and denominator and treating Z as a difference operator

$$a_0 y_k + a_1 y_{k-1} + a_2 y_{k-2} = b_0 x_k + b_1 x_{k-1} + b_2 x_{k-2} \tag{4.16}$$

$$y_k = \frac{1}{a_0}(-a_1 y_{k-1} - a_2 y_{k-2} + b_0 x_k + b_1 x_{k-1} + b_2 x_{k-2}) \tag{4.17}$$

Using the transfer function given in Equation 4.10, with a sampling interval of 10 ms (100 Hz), $T = 0.01$

$$s = 200\left(\frac{Z-1}{Z+1}\right) \tag{4.18}$$

Rearranging the transfer function

$$\frac{y(Z)}{x(Z)} = \frac{1 + 2Z^{-1} + Z^{-2}}{2551 - 4998Z^{-1} + 2451Z^{-2}} \tag{4.19}$$

Cross-multiplying and substituting each Z operator with its corresponding difference values

$$y_k = (4998 y_{k-1} - 2451 y_{k-2} + x_k + 2x_{k-1} + x_{k-2})/2551 \tag{4.20}$$

The method also requires the formulation of initial conditions, for example, values of y_{k-1}, y_{k-2} and so on at time $t = 0$. This restriction is a potential weakness of the method because a separate starter formula is required. In addition, care is needed with numerical overflow because Tustin transforms can generate a very large dynamic range of coefficients according to the transfer function coefficients. However, Tustin's method does not introduce intermediate variables – the only variables are the input and output variables of the transfer function. Moreover, the method does not require a numerical integration method although it is a difference method based on past values of y and x, so care is needed with the choice of the sampling interval T.

4.3 PID Control Systems

In simulation, it may be necessary to implement a control system. Although the controller may be provided as a transfer function, in systems where the documentation simply implies a controller, but no description of the controller is provided, the simulator designer needs to implement a control system to meet the system requirements. There are numerous textbooks on control system design and a plethora of control design methods and the reader, intent on gaining more insight into control system design, would be well advised to follow the standard texts on the subject. However, the control systems implemented for aircraft are often first- or second-order systems (partly to meet certification requirements) and are amenable to implementation using PID controllers, where PID stands for proportional, integral and derivative control.

Many control systems are required to track an input demand. For example, the temperature in a house should match the temperature set by a thermostat; a car cruise control system is designed so that the speed of the car follows a set speed; an aircraft autopilot system ensures that an aircraft maintains a selected altitude. Many of these systems take the form shown in Figure 4.7, where the output y is chosen by the set point r. In this example, the output y of the system $G(S)$ is required to track the input r and this is achieved by the PID controller $C(s)$.

PID controllers are widely used, partly because they are effective and partly because they are straightforward to design. They are particularly common in systems that exhibit first- or second-order characteristics and for damped stable systems where PID control offers improvements in the system response. The general form of the PID controller is given by

$$C(s) = K_P + \frac{K_I}{s} + K_D s \qquad (4.21)$$

where s is the Laplace operator. The design of an effective control system involves adjusting the coefficients K_P, K_I and K_D, which is referred to as tuning the controller. Generally, increasing K_P and K_I tends to reduce the steady-state errors but may not meet stability criteria, while increasing K_D can improve the system stability. Although the controller has three terms, for some applications, the K_I or K_D terms can be omitted.

The simplest form of controller is a constant K. In this form, the output y is fed back and the system eventually reaches a steady-state condition. The control input applied to $G(s)$ is proportional to the difference between the input r and the output y and consequently, this system is referred to as *proportional* control. The advantage of proportional control is its obvious simplicity. However, depending on the value of K, the controller may introduce instability into the system response; it may not necessarily produce an ideal response or the output may exhibit a steady-state error.

Proportional control can be augmented with both *integral* and *derivative* control. The derivative term implies either that the derivative of y is available (possibly as a state variable) or that it can be obtained by differentiating y. Derivative control may provide a faster response. Both proportional and derivative control may introduce (or fail to eliminate) steady-state errors and integral control can be used to reduce these errors. For integral control, the error term e is integrated. If e is zero, the output y must equal the input r and the integral term has no further effect.

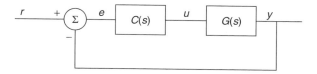

Figure 4.7 A basic PID control system

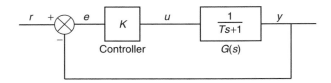

Figure 4.8 Proportional control

Consider proportional control of the first-order system shown in Figure 4.8 where $C(s) = K$. The open-loop response to a unit step input $u(s)$ is given by

$$y(s) = \frac{1}{Ts+1} \qquad (4.22)$$

giving a response $y(t) = 1 - e^{-Tt}$ in the time domain.
The closed-loop transfer function is

$$\frac{y(s)}{r(s)} = \frac{KG}{1+KG} = \frac{\frac{K}{Ts+1}}{1+\frac{K}{Ts+1}} = \frac{K}{Ts+1+K} = \frac{K}{1+K}\frac{1}{\frac{T}{1+K}s+1} \qquad (4.23)$$

The equivalent time domain response is

$$y(t) = \frac{K}{1+K}\left(1 - e^{-(1+K)\frac{t}{T}}\right), \text{ with a steady-state output} \frac{K}{1+K}$$

The closed-loop system is still first order; as K increases, the steady-state error and the time constant reduces, as shown in Figure 4.9, for $T = 1$.

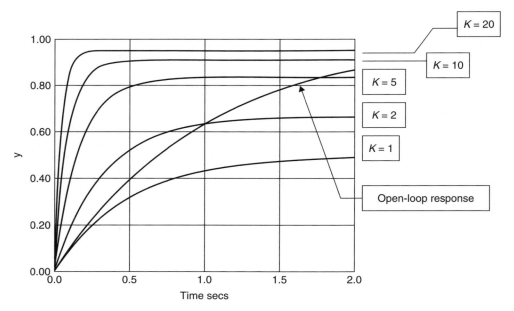

Figure 4.9 Effect of varying the gain in proportional control

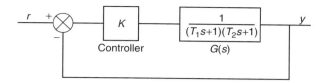

Figure 4.10 Second-order system with proportional control

Notice that although the steady-state error reduces, it is not eliminated. Note also that the response improves with increasing gain. Although it might be tempting to use a very large value of K, in a practical system the range of the input u to the system $G(s)$ will be limited, for example, the fuel flow rate input for a jet engine.

Consider the second-order system, shown in Figure 4.10, where the system transfer function given by

$$G(s) = \frac{1}{(T_1 s + 1)(T_2 s + 1)} \quad (4.24)$$

The closed-loop characteristic equation for a system with gain G and a controller K is given by $1 + KG = 0$, so that

$$\frac{K}{T_1 T_2 s^2 + (T_1 + T_2)s + 1} + 1 = 0 \quad (4.25)$$

Let $T_1 = 1$ and $T_2 = 0.5$, then the characteristic equation is given by

$$s^2 + 3s + 2 + 2K = 0 \quad (4.26)$$

Equating the terms with the characteristic equation for a general second-order system

$$s^2 + 2\zeta \omega_n s + \omega_n^2 = 0 \quad (4.27)$$

if we choose a damping factor $\zeta = 0.6$ (giving an overshoot of approximately 10%), then $2\zeta\omega_n = 3$, $\omega_n = 2.5$ and $\omega_n^2 = 2 + 2K$, giving $K = 2.125$. The response to a unit step is shown in Figure 4.11. Note that the steady-state condition is $\frac{K}{1+K} = 0.68$ and the overshoot is 0.75/0.68 (10%).

The controller can be extended to include integral and derivative control terms and can also be applied to higher order systems, where the characteristic equation is derived for the close loop system and the poles of the system are assigned by careful selection of the coefficients K_P, K_I and K_D to produce the desired response. In practice, PID is often an iterative process and the effect of each of the three terms on the overall system performance is summarized in Table 4.1.

Although a PID controller takes the form

$$C(s) = K_p + \frac{K_I}{s} + K_d s \quad (4.28)$$

For convenience, the controller is often written as

$$C(s) = K_p \left(1 + \frac{1}{T_I s} + T_d s\right) \quad (4.29)$$

In a digital system, the inputs are likely to be sampled at a fixed rate and therefore the controller and the systems are updated at the sampling interval Δt. In other words, the implied integration and differentiation are implemented in the Z domain.

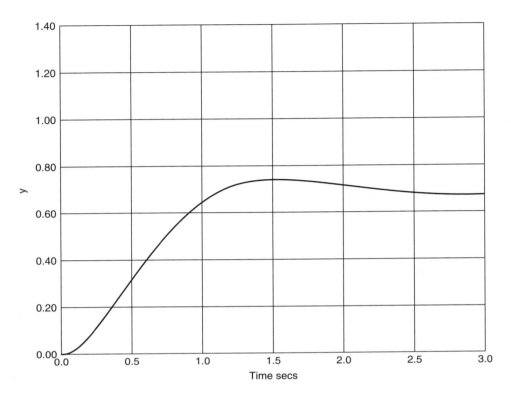

Figure 4.11 Second-order system with damping

Table 4.1 Effect of PID control terms

Control term	Rise time	Overshoot	Settling time	Steady-state error
K_p	Decrease	Increase	Small change	Decrease
K_I	Decrease	Increase	Increase	Eliminate
K_d	Small change	Decrease	Decrease	Small change

The simplest approximation for differentiation is a first-order difference, where

$$\left.\frac{dy}{dt}\right|_n = \frac{y_n - y_{n-1}}{\Delta t} \qquad (4.30)$$

and similarly, for integration

$$\int y(t)\, dt = \sum_{k=1}^{n} y_k \Delta t \qquad (4.31)$$

The discrete form of the PID controller (Bennett, 1994) is given by

$$m_n = K_p \left[e_n + \frac{1}{T_I} \sum_{k=1}^{n} e_k \Delta t + T_d \left(\frac{e_n - e_{n-1}}{\Delta t} \right) \right] \qquad (4.32)$$

Simulation of Flight Control Systems

Let $K_I = K_p(\Delta t / T_s)$ and $K_d = K_p(T_d / \Delta t)$, then Equation 4.32 can be implemented by executing the following code every step, as follows:

```
s = s + e
m = Kp * e + Ki * s + Kd * (e - eold)
eold = e
```

where `s` is the summation of the errors over n steps and `m` is the input to the system.

The simplicity of this code is one of the attractions of a PID controller. Once the values for K_p, K_I and K_d are determined, the controller can be implemented with three lines of code. The summation term `s` and `eold`, which represents the value of `e` at the previous step, must be initialized.

One source of error is that both the integration and differentiation operations are first order. Higher order integration algorithms such as fourth-order Adams–Bashforth are commonly used for the integral term. Higher order approximations for the derivative term use previous values to reduce the effects of noise that occur in sampling using a first-order approximation, for example Bibbero (1977)

$$\frac{dy}{dt} = \frac{1}{6T_s}[y_n + 3y_{n-1} - 3y_{n-2} - y_{n-3}] \tag{4.33}$$

where T_s is the time step. In this case, the approximation requires initial estimates of y for the first four steps.

The value `m` applied to the plant is likely to be limited by practical constraints. For example, an aircraft actuator control system may compute a demand angle of 80°, whereas the actuator may be limited to ±20°. In these situations, the difference between the demanded output (from the control law) and the actual plant output is likely to be different for a relatively long time. If an integral control term is used, the value `s` will increase at each sample time. If the plant finally reaches its desired value, a potentially large value will have accumulated for `s`. As the system has now met its desired input, the control term `m` should be zero, but `s` will continue to demand an increased input while the integral term reduces.

This situation is known as *integrator wind-up* or *integral saturation*, giving a relatively poor response in constrained conditions. Two methods are commonly used to overcome this problem:

1. The integral term is constrained to a fixed range by adding a test on the magnitude of `s` to the PID control law:

```
s = s + e
if (s > smax)
    s = smax
if (s < smin)
    s = smin
```

2. The integrator is inhibited when the control actuator saturates so that the integrator output remains constant while the saturation condition applies.

Both methods constrain the integral term from building up to a large value but they also introduce a problem. As the actuator comes out of saturation, the integral term no longer relates to the current system dynamics (it is simply the value of the integrator term at the time it was inhibited) and therefore, the integral term can introduce a lag under these conditions.

Consider the basic PID expression

```
m = Kp * e + Ki * s + Kd * (e - eold)
```

This expression implies that, in the steady-state, m must be equal to the integral term. In many systems, the steady-state conditions occur when m = 0, so that the integral term would also be zero. However, for systems where m is non-zero in the steady-state, for example an engine control system will maintain a minimum flow even at the minimum value of engine demand, this expression is not valid.

One solution is to modify the equation to add a constant term M which is the manipulated variable at the steady-state point, as follows

```
m = Kp * e + Ki * s + Kd * (e - eold) + M
```

M is effectively the means to set the operating point for the controller. Without M, the integral action will provide compensation but there will be difficulties in changing smoothly without introducing a major disturbance. For example, the operating point may be reset when the system is changed from manual to automatic (e.g. an autopilot function is engaged or disengaged), resulting in a large change in demand.

Normally, it is preferable if these changeovers occur in a 'bumpless' manner, which can be achieved by calculating the value of M for a given steady-state operating point and inserting this new value at the changeover, when the error term (e) is zero. At this point, the integral term is set to zero and the output m is set to M. However, M also depends on the output load and may be incorrect if the load changes. If the error term is not zero, there will be a sudden change in the manipulated variable m due to the proportional action. One method to ensure a valid value for m is to monitor these values for all operator settings (manual control). At the point of change-over, m is copied to mc and e is copied to ec. Either the changeover is made when the error (e) is zero or M is preset to the nominal value and the integral term s is set to the following value:

```
s = mc - Kp * ec - M
```

An alternative method is known as the *velocity method* and is widely used for bumpless transfers. The method derives the change in the value of the manipulated variable rather than its absolute value.
Given the standard form of the PID equation:

$$m(t) = K_p \left(e(t) + \frac{1}{T_I} \int e(t) + T_d \frac{de(t)}{dt} \right) \qquad (4.34)$$

Differentiating this equation

$$\frac{dm(t)}{dt} = K_p \left(\frac{de(t)}{dt} + \frac{1}{T_I} e(t) + T_d \frac{d^2 e(t)}{dt^2} \right) \qquad (4.35)$$

In difference form

$$m_n - m_{n-1} = K_p \left(e_n - e_{n-1} + \frac{\Delta t}{T_I} e_n + \frac{T_d}{\Delta t} [e_n - 2e_{n-1} + e_{n-2}] \right) \qquad (4.36)$$

In terms of previous error values

$$\Delta m_n = K_p \left[\left(1 + \frac{T_s}{T_I} + \frac{T_d}{T_s}\right) e_n - \left(1 - 2\frac{T_d}{T_s}\right) e_{n-1} + \frac{T_d}{T_s} e_{n-2} \right] \quad (4.37)$$

Let

$$K_1 = K_p \left(1 + \frac{T_s}{T_I} + \frac{T_d}{T_s}\right) \quad (4.38)$$

$$K_2 = -\left(1 - 2\frac{T_d}{T_s}\right) \quad (4.39)$$

$$K_3 = \frac{T_d}{T_s} \quad (4.40)$$

then

$$\Delta m_n = K_p [K_1 e_n + K_2 e_{n-1} + K_3 e_{n-2}] \quad (4.41)$$

This value is used to compute m at the time of changeover and ensures a bumpless transfer. However, if the changeover results in a large error, the response may be slow if the integral action is slow (i.e. if T_I is large).

4.4 Trimming

Trimming is an essential part of basic flying. In the steady-state, it ensures that the pilot input forces (elevator, aileron and rudder) are zero. Otherwise, a pilot will be required to apply a continuous force to maintain the aircraft in its steady-state, resulting in an increasing level of fatigue in the pilot's hands or feet. Trim wheels are provided on most aircraft to alter trim tabs on the primary flight controls. Having set the aircraft to the desired attitude and performance, the trim controls enable the aircraft to maintain a specific attitude and performance in the 'hands-off' state. The pilot workload is reduced significantly as small disturbances can be easily corrected by small inputs allowing the aircraft to return to the steady-state condition.

Trimming is also important in flight simulation:

- If the aircraft is restored to a previous situation by the instructor, the aircraft controls and trims may not be in the trimmed state for the restored attitude and performance;
- If any tests are applied, it is normal practice for the aircraft to be trimmed before the test is initiated;
- If a control system is activated, the aircraft model should be re-initialized to a trimmed state to avoid any large initial inputs caused by engaging the control law.

In these situations, rather than requiring the pilot to set the trim controls, the simulation software computes the trimmed state and adjusts the trim controls to ensure the aircraft is trimmed. For example, if a simulator is restored to an approach at 5 nm, with an airspeed of 165 kt, a rate of descent of 900 ft/min, a flap setting of 20° and with the undercarriage down, it is reasonable to assume that the aircraft would be set in the trimmed state a pilot would otherwise have selected.

Generally, automatic trimming (where the trimming is applied by the simulator rather than the pilot) applies to the longitudinal axis and the elevator and engine controls. For a given configuration, it is necessary to compute the aircraft control positions for trimmed flight to achieve a desired airspeed and vertical speed at a specific altitude. An alternative view of trimming is that it is the set of aircraft control positions to ensure non-accelerating flight in the longitudinal axis. In other words, the airspeed is constant (the longitudinal acceleration is zero) and the pitch attitude is constant (the aircraft pitch rate is zero).

However, the computation of these trimmed conditions is not straightforward. Altering the elevator can result in a change of attitude and vertical speed leading to a change in airspeed. Similarly, changing the engine power settings may induce pitching accelerating and longitudinal acceleration. Nevertheless, for a specific performance value, there is a unique power setting and aircraft attitude and pilots are able to set an aircraft performance accurately and efficiently. The equivalent function in a simulator can therefore be implemented by means of a similar iterative process.

As the desired aircraft performance and the aircraft flight path are known, the angle of attack and the thrust must be computed for these values. Of course, in a flight simulator, the aerodynamic derivatives will also be known, allowing the exact simulation equations to be used to compute the unknown trimmed state variables. In effect, the computation of these conditions is a search for the unique values of angle of attack and thrust and a simple iterative search method can be used.

The value of angle of attack will be in the range $\pm 15°$. At the desired angle of attack, the z force in stability axes will be zero. The search involves initially setting the minimum angle of attack to $-15°$ and the maximum angle of attack to $+15°$. Selecting a trial value at the midpoint, the aircraft forces are computed for this trial angle of attack. If the z force is positive, the value must be too high and the lower bound is set to the midpoint value. Otherwise, the trial value is too low and the higher bound is set to the midpoint value. This process repeats until the computed z force is within a small range, for example, less than 1 N. Given the aircraft altitude, airspeed, vertical speed, flap setting and undercarriage position, air density (ρ) is computed as a function of altitude and flight path angle (γ) is given by $\tan^{-1}(V_d/U)$, where V_d is vertical speed and U is airspeed.

The search proceeds as follows:

$\alpha_{min} = -15°$
$\alpha_{max} = 15°$
REPEAT
$\quad \alpha_t = (\alpha_{min} + \alpha_{max})/2.0$
$\quad \theta = \gamma + \alpha_t$
$\quad w = u \tan \alpha_t$
$\quad C_l = $ aerodynamic coefficient of lift
$\quad lift = 0.5\rho(u^2 + w^2)sC_l$
$\quad C_d = $ aerodynamic coefficient of drag
$\quad drag = 0.5\rho(u^2 + w^2)sC_d$
$\quad zforce = -lift \cos \alpha_t - drag \sin \alpha_t + mg \sin \theta$
$\quad if\ (zforce > 0)\ then$
$\qquad \alpha_{min} = \alpha_t$
$\quad else$
$\qquad \alpha_{max} = \alpha_t$
UNTIL($|zforce| < 1$)

terminating when the resultant z force is less than a predefined datum (e.g. 1 N).

In the steady-state, the pitching moment is zero

$$M_{stab} = \frac{1}{2}\sigma V_C^2 s \bar{c}(C_{M0} + C_{M\alpha}\alpha + C_{M\delta E}\delta e) + \frac{1}{4}\sigma V s \bar{c}^2(C_{Mq}q + C_{M\dot{\alpha}}\dot{\alpha}) \quad (4.42)$$

As q and $\dot{\alpha}$ are also zero

$$C_{M0} + C_{M\alpha} \cdot \alpha + C_{M\delta E}\delta e = 0 \quad (4.43)$$

Giving

$$\delta e = \frac{-(C_{M0} + C_{M\alpha}\alpha_w)}{C_{M\delta E}} \quad (4.44)$$

where $\alpha_w = \alpha + wing\ incidence$

Knowing the value of angle of attack in the trimmed state, the thrust is computed by resolving the lift, drag and gravity component in the body frame

$$thrust_t = drag\ \cos\alpha - lift\ \sin\alpha + mg\ \sin\theta \quad (4.45)$$

As the engine model is usually non-linear (and taking care to eliminate the aircraft transient response), an iterative search can be made to compute the engine lever position to produce the trimmed thrust, $thrust_t$.

The search proceeds as follows:

$tp_{min} = 0.2$
$tp_{max} = 1.0$
REPEAT
    $tp = (tp_{min} + tp_{max})/2.0$
    $thrust = engine(tp)$
    if $(thrust < thrust_t)$ then
        $tp_{min} = tp$
    else
        $tp_{max} = tp$
UNTIL$(|thrust - thrust_t| < 1)$

terminating when the computed thrust is within a predefined datum from the trim thrust (e.g. 1 N), where $engine(tp)$ is the aircraft engine model, which computes thrust as a function of airspeed, altitude and engine lever setting tp.

4.5 Aircraft Flight Control Systems

The design of aircraft flight control systems is a combination of flight dynamics and applied control (Cook, 1997). The response and stability of the longitudinal and lateral dynamics of aircraft is covered in Chapter 7. A basic airframe with actuators is often a compromise of aircraft design and consequently the handling of an aircraft cannot be optimized for all flight conditions. However, with the addition of a flight control system, as shown in Figure 4.12, the aircraft response can be changed to provide a more agile aircraft or a more stable aircraft (Stevens and Lewis, 2003).

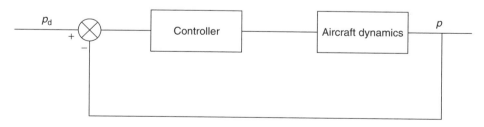

Figure 4.12 A flight control system

A demanded term p_d is compared with the current value of p (where p is typically an aircraft performance parameter, such as altitude, heading or airspeed) and the controller provides the appropriate inputs for the aircraft actuators. In particular, for civil aircraft in controlled airspace, the pilot workload to maintain a flight level, a heading and airspeed in all phases of flight would be very high and a primary role of the flight control systems is to reduce the pilot workload (and also pilot fatigue during long flights). The flight control engineer needs a sound understanding of the aircraft dynamics, so that the controller achieves the desired response and stability (Roskam, 2003) of the aircraft system.

The flight control systems used in aircraft are usually referred to as *autopilot functions*. In addition to the control of the primary aircraft states (altitude, heading and airspeed), an autopilot can also provide coordinated turns, control the rate of climb or descent and acquire tracks to radio beacons. Auto-land systems enable aircraft to land in conditions of very low visibility. In all these situations, the flight control system has full responsibility for aircraft manoeuvres and the role of the flight crew is to select the appropriate inputs and to monitor the autopilot functions closely. For the flight simulator designer, the schematics of the flight control systems may be provided. Alternatively, it may be necessary to implement the autopilot functions that are provided on the aircraft. The following sections outline the design of controllers for standard autopilot functions.

4.6 The Turn Coordinator and the Yaw Damper

Aircraft are turned by banking. Movement of the ailerons causes the aircraft to roll and then turn, where the rate of turn depends on the airspeed and the angle of bank. Pilots are instructed to execute a controlled turn; as the aircraft banks, the lift vector can be considered to rotate about the roll axis, reducing the vertical lift component. To execute a level turn, elevator input is increased slightly on entry to the turn and reduced on exit from the turn to minimize any change of altitude in a turning manoeuvre.

Although the lift vector is perpendicular to the wings in the body frame, if the aircraft banks about the roll axis, a component of gravitational acceleration is now resolved in the body y axis, producing a lateral acceleration or side force. In an aircraft, banking by application of the aileron alone can introduce an undesirable side force, where the occupants of the aircraft experience a lateral motion, similar to the motion felt in a motor car taking a bend. This effect is well known and pilots are taught to coordinate their aileron input with rudder input to reduce the side force. With appropriate rudder application, the side force is zero and the occupants experience the more natural feeling of a vertical force, perpendicular to their seat. The slip ball provides an indication of side force (or sideslip), enabling the pilot to apply appropriate rudder inputs during the turn. Of course, sideslip can also result from propeller slipstream, rudder imbalance, crosswinds and airframe asymmetry. Nevertheless, it is desirable to reduce these effects in turning flight and a turn coordinator control system is provided in transport aircraft to reduce lateral motion during a turn.

Sideslip is measured either from body velocities provided by the inertial reference system or directly from measurements from a side-slip vane on the airframe. The controller is simple. Rudder is applied to reduce the angle of sideslip and as the system drives towards zero, a simple proportional controller is appropriate, as shown in Figure 4.13.

Figure 4.14a shows the aircraft response to an aileron input of 25° for 5 s, resulting in a sideslip of approximately 3°. In this example, the yaw damper is disengaged. Figure 4.14b shows the aircraft response to the same input with a turn coordinator where $K = 20$. The rudder compensation is clearly shown in Figure 4.14b with the excursion in sideslip reducing to 1°. The aircraft in this example is a Boeing 747-100, at 180 kt at 3000 ft, configured with gear up and flaps 20. The rudder input is computed from the angle of sideslip as shown in Example 4.1.

Simulation of Flight Control Systems

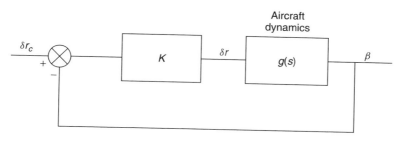

Figure 4.13 Turn coordinator

Example 4.1 Turn coordinator

```
void FCS_TurnCoordinator(float *Rudder)
{
   *Rudder = -Model_Beta * 20.0;
}
```

As the roll and yaw axes of an aircraft are coupled, a yawing input can induce a rolling motion and vice versa. This mode of motion, known as *Dutch roll* occurs in most aircraft and is covered in detail in Chapter 7. If aileron is applied to bank the aircraft, the induced sideslip will introduce a yawing moment, which in turn can excite a rolling moment. In most aircraft, this motion damps out naturally after 30–60 s. However, during this time, the aircraft may yaw laterally by several degrees, which can be uncomfortable for passengers and result in increased workload for the pilot causing undesirable yawing from pilot inputs or as result of a disturbance such as turbulence. For example, in an approach, the pilot is endeavouring to line up with the extended centre line of the runway; any small disturbance or turn manoeuvre can result in a yawing motion, making it difficult to establish the heading accurately and, moreover, the use of the rudder to correct these effects may actually exacerbate this motion. Generally, these effects are undesirable and a yaw damper control system is used to reduce the yawing motion.

The role of the yaw damper is to cancel out this oscillatory motion. On aircraft fitted with a yaw damper, the pilot may be able to disengage the yaw damper and often two yaw damping systems are provided if the aircraft has lower and upper rudder systems. In its simplest form, the yaw rate can be fed back to command rudder input to reduce the yawing rate, driving the yaw rate to zero. However, suppressing the yaw rate is clearly undesirable in situations where the pilot attempts to turn the aircraft by banking. The objective is to suppress the Dutch roll but pass the desired yawing motion and this is provided by a simple filter known as a *wash-out filter*, typically provided by a transfer function of the form

$$\frac{s}{s + 1/T_w} \qquad (4.46)$$

where T_w is the time constant of the wash-out filter, as shown in Figure 4.15.

In Figure 4.16a, the yaw rate disturbance resulting from the Dutch roll is evident; with the yaw damper engaged, the sideslip oscillation is largely suppressed, as shown in Figure 4.16b. In practice, the yaw damper is combined with the turn coordinator to suppress both sideslip and Dutch roll oscillation. The actual yaw damper system will be much more complex in a commercial transport aircraft; it must operate over the full flight envelope and will be a compromise between suppressing the undesirable motion associated with Dutch roll and providing the pilot with sufficient authority to manoeuvre the aircraft in hazardous conditions.

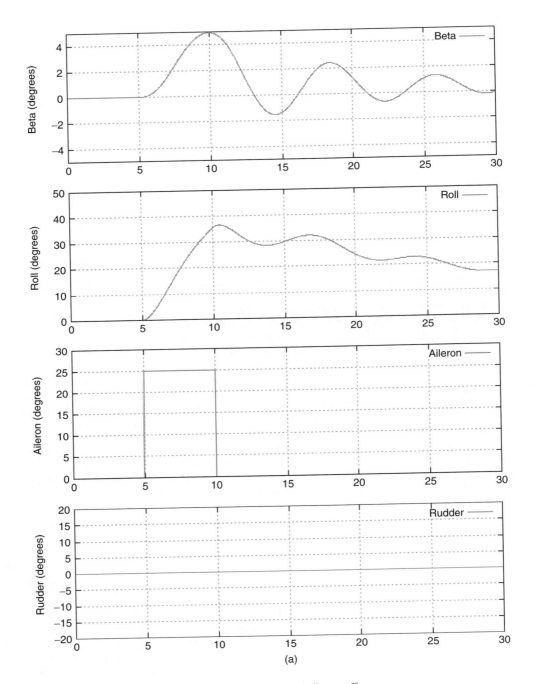

Figure 4.14 (a) Turn coordinator off

Simulation of Flight Control Systems

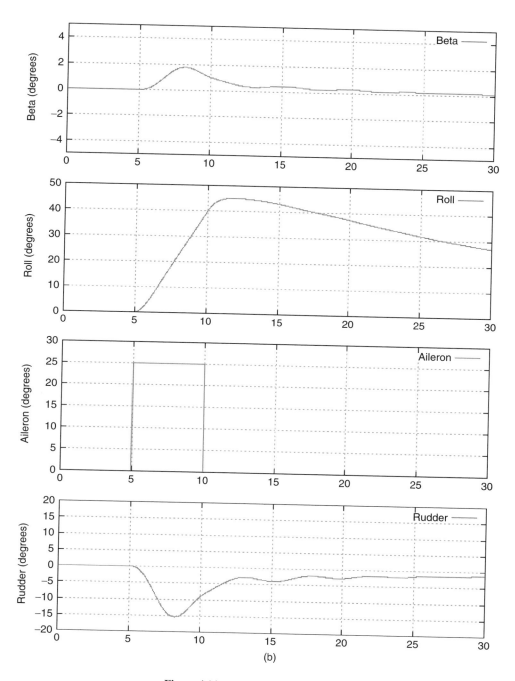

Figure 4.14 (b) Turn coordinator on

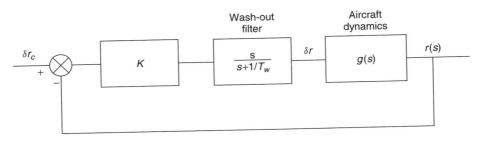

Figure 4.15 A yaw damper

The code to implement the yaw damper to include yaw rate feedback and the wash-out filter is straightforward. The variable ydampint is the integral value of the yaw damper input ydamp given in Example 4.2.

Example 4.2 Yaw damper

```
void FCS_YawDamper(float *Rudder)
{
  Maths_Integrate(&ydampint, ydamp);
  ydamp = Model_R * 20.0 - ydampint / 1.0;
  *Rudder = *Rudder + ydamp;
}
```

4.7 The Auto-throttle

During an approach, the pilot has to maintain the glide slope and localizer. Similarly, during a hold pattern, the pilot maintains heading and altitude. In both situations, small changes in pitch attitude can lead to a significant change in airspeed, requiring the pilot to adjust both the pitch attitude and the power setting. Indeed, a change in pitch results in a change in airspeed and a change in power setting can cause a pitching moment. In order to reduce the pilot workload, an auto-throttle maintains a commanded airspeed. If the pilot lowers the nose of the aircraft, the fuel flow to the engines is reduced to anticipate the possible increase in airspeed and similarly, if the nose of the aircraft is raised, the auto-throttle increases the fuel flow to avoid the loss of airspeed from the climbing manoeuvre. In most phases of flight, there is either a preferred or efficient operating airspeed. For example, the final phase of an approach is flown at a reference speed. The auto-throttle is consequently a very important control system, maintaining the airspeed to within a knot in most phases of flight.

In addition to reducing pilot workload, the auto-throttle also plays an important role in suppressing the long period phugoid motion. Although an aircraft may be disturbed in pitch, by turbulence say, the variation in airspeed that occurs in the subsequent phugoid motion is avoided, reducing significantly the variation in altitude and pitch. The auto-throttle also operates in conjunction with other automatic control modes. For example, in climbing and acquiring a commanded altitude, the airspeed during the climb and in levelling out at the top of the climb is maintained by the auto-throttle.

In a transport aircraft, the engine levers are usually controlled by electrical (or hydraulic) actuators commanded by the auto-throttle. If the actual airspeed is below the commanded airspeed, the throttle is advanced; if the airspeed exceeds the commanded airspeed, the throttle is reduced. However, the time constants associated with a turbofan aircraft are relatively large and consequently, considerable care is needed to design an auto-throttle that is both responsive and stable. A typical auto-throttle control system is shown in Figure 4.17, where dp is the engine power lever position, in the range 0–1.

Simulation of Flight Control Systems

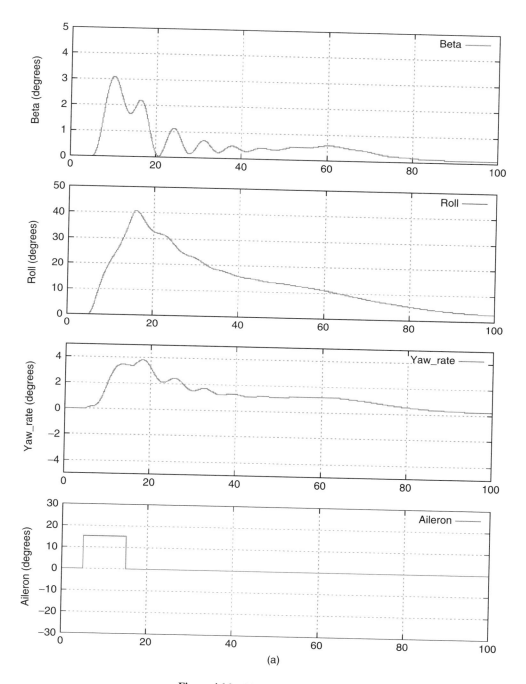

Figure 4.16 (a) Yaw damper off

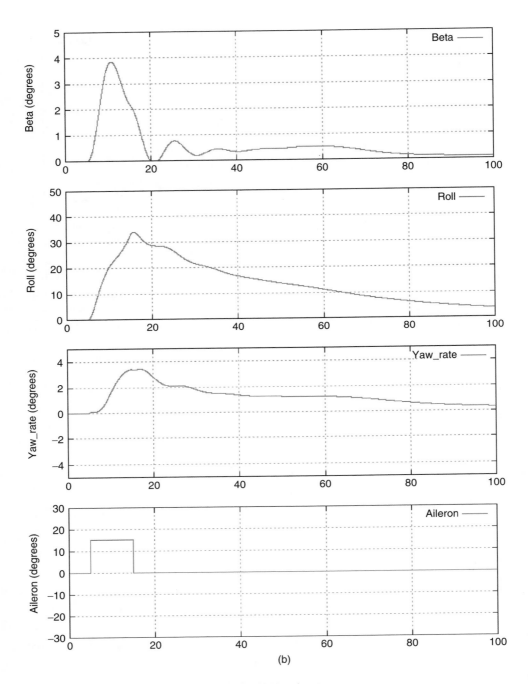

Figure 4.16 (b) Yaw damper on

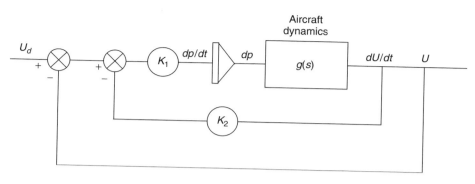

Figure 4.17 Auto-throttle control system

The rate of change of throttle position dp/dt is zero when the airspeed U reaches the reference speed U_d and when the aircraft is no longer accelerating ($dU/dt = 0$). The aircraft response to an increase in demanded airspeed from 240 to 270 kt ($K_1 = 10, K_2 = 2$) is shown in Figure 4.18a. The response to a deceleration from 350 to 250 kt is shown in Figure 4.18b. In both cases, the aircraft altitude was maintained at 3000 ft. A throttle position of 0.2 corresponds to the flight idle position of the engine levers. The selection of K_1 and K_2 depends on the time constant of the engine to respond to power demands.

The auto-throttle procedure in Example 4.3 contains feedback terms for airspeed (u) and rate of change of airspeed (Udot). The computed engine lever position is constrained to the range 0.2 (flight idle) to 1.0 (fully forward).

Example 4.3 Auto-throttle

```
void FCS_SetSpeedHold(float Vref)
{
  APSpeed = Vref;
}

float FCS_SpeedHold(float Vref)
{
  float tp;
  float dv;
  float IAS;

  IAS = Model_U; // * sqrt(Weather_DensityRatio);
  dv = Vref - IAS;
  tp = (dv - Model_UDot * 2.0) * 10.0;
  Maths_Limit(&tp, 0.2, 1.0);
  return tp;
}
```

4.8 Vertical Speed Management

In climbing and descending phases of flight, the vertical speed must be controlled to meet ATC requirements, or to operate the engines efficiently or to be at a given altitude at a specified time. There is one further constraint for commercial transport aircraft; cabin pressurization systems have a limit to the rate at which they can maintain cabin pressure without discomfort to passengers.

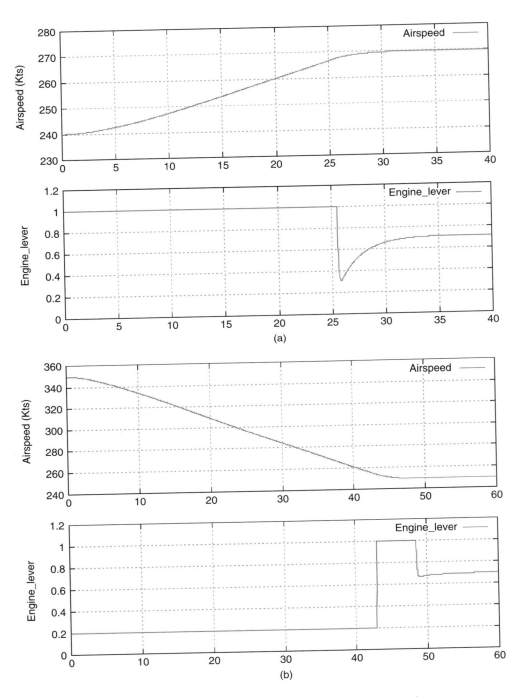

Figure 4.18 (a) Auto-throttle acceleration. (b) Auto-throttle deceleration

Simulation of Flight Control Systems

Most airline passengers will have experienced 'ear-popping' during a descent. Typically, an aircraft will descend at 1000 ft/min from the cruise altitude to the approach segment and a flight plan will include climb and descent rates for each vertical segment of the flight plan, which will be flown at specific configurations (flap settings) and airspeeds. Similarly, the rate of climb depends on the aircraft engine rating, the aircraft mass (particularly fuel) and the outside air temperature. The climb rate also depends on the steady-state angle of attack in the climb – a margin is left to avoid entering the stall region. The angle of attack during the climb will depend on the lift and drag terms, which may be chosen to meet specific aircraft operating requirements.

Although vertical speed corresponds to the aircraft rate of climb dh/dt, where h is altitude, it is more sensible to define the vertical flight path by the flight path angle γ given by

$$\gamma = \theta - \alpha \tag{4.47}$$

For a steady-state climb at a given airspeed there is a unique angle of attack; the flight path angle control law is therefore equivalent to a pitch control system, which can be implemented using a pitch rate control law. If the commanded flight path angle γ_c is the same as the current flight path angle γ, then the pitch angle is the required angle for the commanded flight path angle and the pitch rate must be zero. In other words, a simple proportional plus integral controller can be used for pitch rate control, as shown in Figure 4.19.

The code for the controller is given in Example 4.4. The elevator is limited to ±1.0, the normalized elevator input. In addition, during a turning manoeuvre, the vertical speed (or flight path angle) should be maintained and this is achieved by including the appropriate components of pitch rate and yaw rate in the feedback term to derive e, the input to the PI controller. The feedback pitch rate term q is modified as follows:

$$e = q_c - (q \cos \phi - r \sin \phi) \tag{4.48}$$

in order to account for the bank angle ϕ in a turn (McLean, 1990). For the PI controller, values of $K_p = -50$ and $K_i = -4$ were found to produce a critically damped response. The commanded pitch rate is proportional to the flight path error and is also limited to 1°/s.

Example 4.4 Flight path angle hold

```
float FPAHold(float GammaC)
{
  const float FPA_Kp = -50.0;
  const float FPA_Ki = -4.0;

  float e;
  float Gamma;
  float Qc;
  float de_fcs;

  Gamma = Model_Pitch - Model_Alpha;
  Qc = (GammaC - Gamma) * 0.3;
  Maths_Limit(&Qc, -1.0 / Maths_ONERAD, 1.0 / Maths_ONERAD);
  e = Qc - (Model_Q * cos(Model_Roll) - Model_R * sin(Model_Roll));
  FPA_s = FPA_s + e;
  de_fcs = FPA_Kp * e + FPA_Ki * FPA_s;
  Maths_Limit(&de_fcs, -1.0, 1.0);

  return de_fcs;
}
```

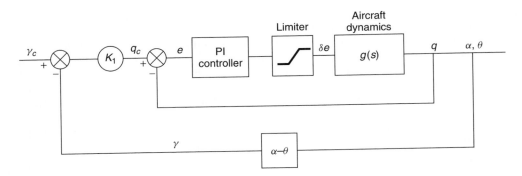

Figure 4.19 Flight path angle controller

The response of the flight path angle controller is shown in Figure 4.20a. In this example, the aircraft is trimmed for level flight at 3000 ft, 180 kt with flaps 20 (gear up) and a flight path angle of 5° is commanded. Note that the change from level flight to an established climb with a flight path angle of 5° is completed within 10 s, with the aircraft established with a climb rate of 1590 ft/min. Figure 4.20b shows the response of the flight path angle controller for the same flight conditions, changing from level flight to a flight path angle of −3°, typically to acquire the glide path during an ILS approach.

4.9 Altitude Hold

In controlled airspace, an aircraft must maintain its assigned altitude to within 50 ft. At 250 kt, a 1° change in pitch attitude will lead to a change of over 400 ft/min in vertical speed. Although commercial pilots are able to fly within such limits, the workload to maintain altitude by manual flying is high and an autopilot with altitude hold is essential. Altitude hold is readily accomplished with a flight path angle controller. Beyond 200 ft, say, of the assigned altitude, the aircraft climbs (or descends) at a standard rate, say, 1000 ft/min. Within 200 ft of the assigned altitude, the climb speed reduces linearly from 1000 ft/min at 200 ft from the commanded altitude to zero at the commanded altitude.

A vertical speed hold system is simply a flight path angle hold system, where the commanded flight path angle is derived from the vertical speed and the ground speed. Strictly, $\gamma_c = \tan\left(\frac{V_{ref}}{V_{gs}}\right)$, where $V_{gs} = \sqrt{V_N^2 + V_E^2}$, but for flight path angles below 10° small angle approximation is used to implement the vertical speed hold control law as shown in Example 4.5.

Example 4.5 Vertical speed hold

```
float FCS_VSpeedHold(float Vref)
{
    return FPAHold(Vref / Model_Vgs);
}
```

The code for altitude hold is simply a computation of the commanded vertical speed as a function of the vertical distance from the reference altitude as shown in Example 4.6. This value is limited to the operational climb speed of the aircraft, in this case ±1000 ft/min.

Simulation of Flight Control Systems

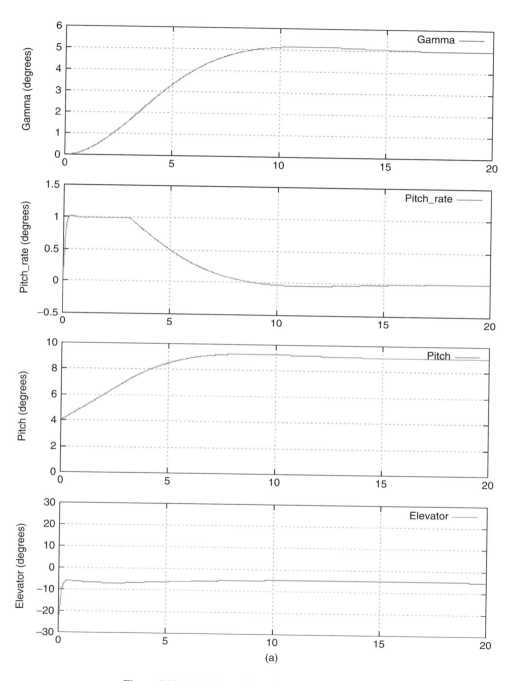

Figure 4.20 (a) Flight path angle controller $-5°$ climb

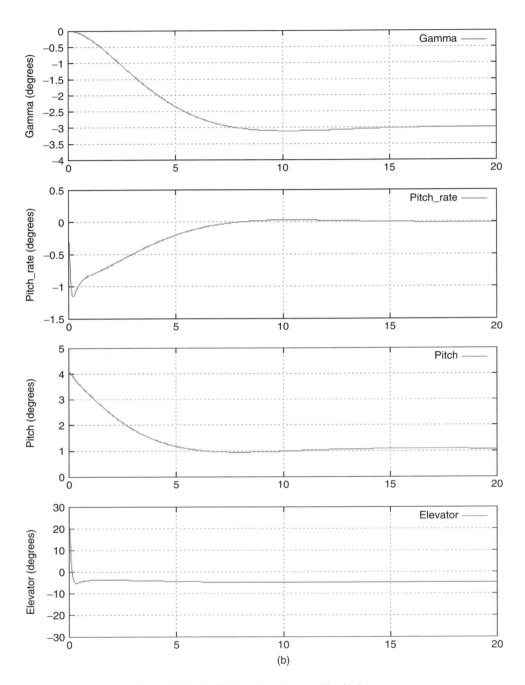

Figure 4.20 (b) Flight path angle controller 3° descent

Example 4.6 Altitude hold

```
float FCS_HeightHold(float Href)
{
    float vs;
    float h;

    h = Model_Pz - Aero_CGHeight;
    vs = -(Href - h) * 0.083333;  /* > 200 ft? */
    if (vs > 5.08) {  /* > 1000 fpm? */
        vs = 5.08;
    } else if (vs < -5.08) {  /* < -1000 fpm? */
        vs = -5.08;
    }
    return FCS_VSpeedHold(vs);
}
```

Figure 4.21a shows a climb from trimmed level flight at 3000–8000 ft at 320 kt. Note the reduction in vertical speed from 1000 ft/min after passing 7800 ft and the slight increase in pitch attitude during the climb (as air density reduces). Figure 4.21b shows a descent from 8000 to 3000 ft at 320 kt.

4.10 Heading Hold

Aircraft direction and aircraft track are controlled by selecting aircraft heading (in degrees magnetic). In the case of track following, if both airspeed and groundspeed measurements are available, the wind component can be computed, enabling a heading to be selected to follow a desired track. In autopilot modes, the maximum bank is typically limited to ±20° in order to meet certification requirements – if higher angles of bank were permissible, turbulence could result in a further increase in bank angle, leading to a high wing load factor that could result in structural damage to the airframe. There is a further constraint on turning for commercial transport aircraft, which is that the maximum turn rate is 3°/s.

The relationship between bank angle and turn rate is given in Clancy (1975)

$$\phi = \frac{\dot{\psi} V}{g} \quad (4.49)$$

where ϕ is the angle of bank, $\dot{\psi}$ is the turn rate, V is the true airspeed and g is gravitational acceleration. Assuming that ϕ is limited at 20°, the maximum turn rate is shown for different airspeeds in Table 4.2.

For light aircraft, standard rate one (3°/s) turns are possible up to 127 kt. For civil transport aircraft, the turn rate reduces as higher speeds, although most turning manoeuvres are limited to small heading changes at waypoints of the flight plan. At lower speeds, for example during hold patterns or manoeuvring in the circuit, where airspeeds in the range 150–180 kt are typical, turn rates reduce to 2°/s or so.

Consequently, when large heading changes are required, an aircraft turns at the maximum bank angle and therefore, the maximum achievable turn rate. As the required heading change reduces, the bank is reduced until the desired heading angle is reached. In other words, a proportional controller can be used for heading hold, where the bank angle is proportional to the difference between the commanded bank angle and the instantaneous bank angle.

The bank angle controller is implemented as a PD controller, as shown in Example 4.7.

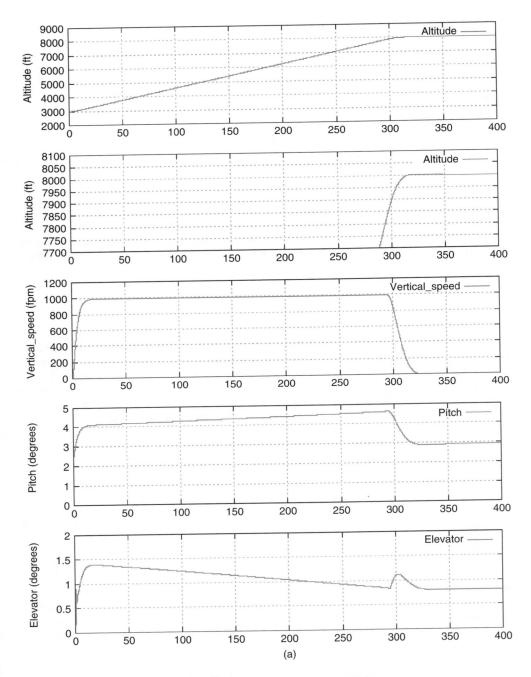

Figure 4.21 (a) Climb from 3000 to 8000 ft

Simulation of Flight Control Systems

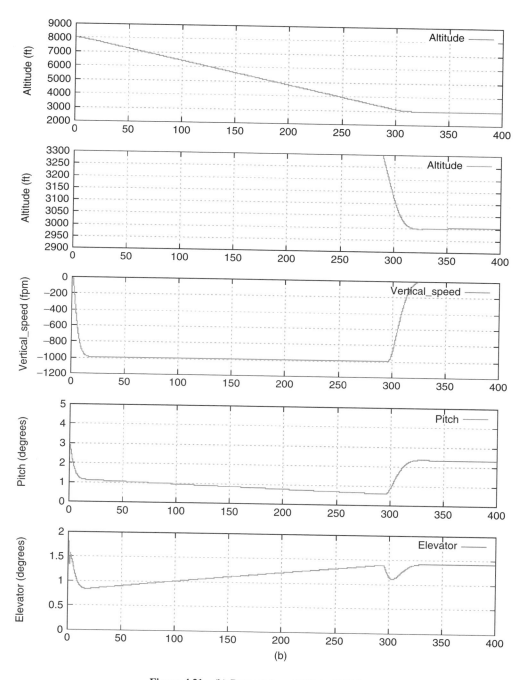

Figure 4.21 (b) Descent from 8000 to 3000 ft

Table 4.2 Turn rate limits

Airspeed (kt)	Turn rate (°/s)
80	4.77
120	3.18
150	2.54
200	1.91
250	1.53
300	1.27
350	1.09
400	0.95
450	0.85

Example 4.7 Bank angle hold

```
float BankAngleHold(float RollC)
{
  const float Bank_K1 = 3.0;
  const float Bank_K2 = 3.0;

  float da;

  Maths_Limit(&RollC, -DEG20, DEG20);
  da = (RollC - Model_Roll) * Bank_K1 - Model_P * Bank_K2;
  Maths_Limit(&da, -1.0, 1.0);
  return da;
}
```

The commanded roll angle `RollC` is limited to the range ±20°. The required aileron displacement (normalized to the range ±1.0) is computed from the difference between the commanded bank and the instantaneous bank angle `Roll` and the roll rate feedback term `P`. The commanded bank angle is computed from the heading hold controller as shown in Example 4.8.

Example 4.8 Heading hold

```
float FCS_HeadingHold(float HdgRef)
{
  float dHdg;
  float trc;

  HdgRef = HdgRef + (float) (AeroLink_NavPkt.MagneticVariation);
  dHdg = HdgRef - Model_Yaw;
  Maths_Normalise(&dHdg);
  trc = dHdg * 0.15; /* 20 deg -> 3 deg/s */
  if (trc > DEG3) {
    trc = DEG3;
  }
  else if (trc < -DEG3) {
    trc = -DEG3;
  }
  return BankAngleHold(trc * Model_U / 9.81);
}
```

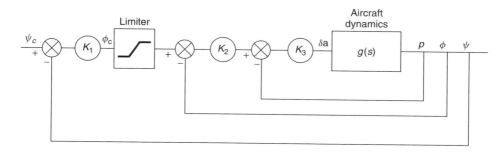

Figure 4.22 Heading hold controller

The heading reference is converted from degrees magnetic to degrees true. The change of heading required is normalized to the range $\pm 180°$. The turn rate required then reduces from the maximum turn rate to zero over the last $20°$ of the turn onto the desired heading. The maximum turn rate is limited to $3°/s$ and similarly, the bank angle is limited to $20°$. The value selected to start reducing the turn rate depends on the roll response of the aircraft. If initiated too early, the turn rate reduces very slowly but if initiated too late, the aircraft can overshoot the commanded heading. A typical heading hold control system is shown in Figure 4.22.

Figure 4.23 shows the aircraft response to a heading change from 30 to $80°$, where $K_1 = 0.15, K_2 = 3$ and $K_3 = 3$. Note the slight overshoot in yaw rate before the constant turn rate is established at the maximum bank angle of $20°$. As the heading control system interacts with the yaw damper, significant non-linearity is introduced into the controller, which is evident in the aileron response. Nevertheless, a stable turn through $50°$ is completed in just over $30\,s$.

In addition to heading control, an aircraft may also follow a selected track, requiring a track angle controller. In practice, this is straightforward because the track angle can be computed from the aircraft heading and the angle of sideslip (or the true airspeed and ground speed). The drift angle to compensate for the wind is readily computed, so that an aircraft heading is commanded, which includes compensation for the drift angle. In cases where the drift angle is estimated, the drift is continuously monitored and corrections are applied to follow a desired track.

4.11 Localizer Tracking

In lateral guidance, in addition to the heading hold function described in Section 4.10, an aircraft will also track a VOR radial (inbound or outbound) or an ILS localizer. In this case, the flight control system computes a track to intercept the localizer at a reasonable angle. If the intercept angle is too shallow, the aircraft will take too long to acquire the localizer track. Alternatively, if the intercept angle is too steep, the aircraft may overshoot the localizer, repeatedly reacquiring the localizer.

Localizer tracking is illustrated in Figure 4.24. The relative bearing α of the aircraft A from the beacon B can be computed from the aircraft position and the beacon coordinates given in the navigation database. The QDM β of the desired track to B is given by the VOR omni-bearing selector (OBS) selection (for VOR tracking) or the runway QDM (for ILS localizer acquisition). The lateral guidance is simply the computation of the aircraft track AP to intercept the localizer, given by $\beta + g(\beta - \alpha)$, where g is an appropriate gain value (typically 6). The aircraft heading (the track less the wind component) is continuously recomputed and passed to the heading-hold subsystem. The objective is to compute a trajectory that provides a smooth intercept, avoiding an overshoot of the localizer. Clearly, if a small value of gain is used, the aircraft may not intercept the localizer before reaching B.

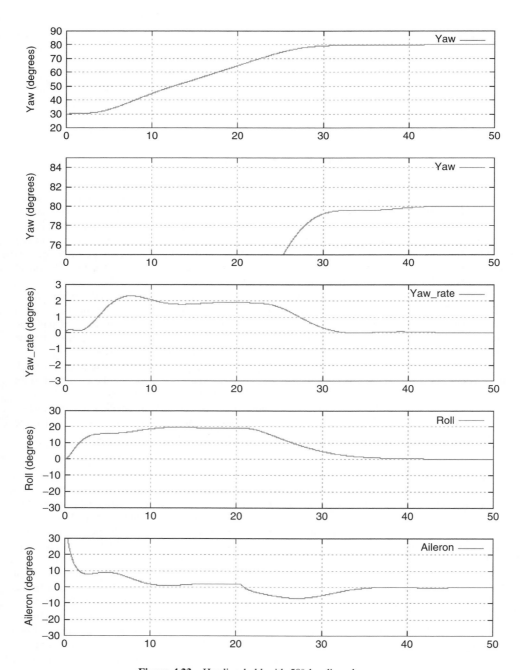

Figure 4.23 Heading hold with 50° heading change

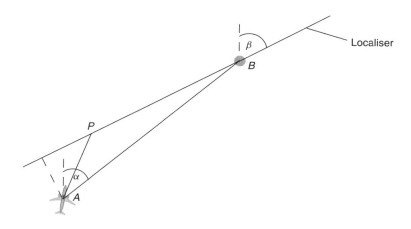

Figure 4.24 Localizer acquisition

Two points should be noted. First, the intercept is continuously recomputed, so that the actual track follows a curved path, reducing the turn rate as the aircraft reaches the localizer. Secondly, the computations are performed with true angles, where the localizer is defined as a magnetic value for aircraft navigation. Figure 4.25 shows a localizer capture of the 080 radial to the Midhurst VOR (MID), starting approximately 25 nm south west of the VOR.

4.12 Auto-land Systems

In an instrument approach, the ILS provides two guidance signals to enable an aircraft to be flown down to 200 ft above the ground. The ILS signals are displayed to the pilot giving information of the lateral angular deviation from the localizer and the vertical angular deviation with respect to the glide slope. At 200 ft, if the runway is in view, the pilot is able complete a visual approach. Otherwise, if the visibility is poor and the runway is not fully visible, then the pilot will either attempt another approach (a go-around) or divert to another airport.

For commercial transport aircraft, this limit of a decision height at 200 ft would be uneconomic for flight operations and auto-land systems are widely used for flight guidance, all the way to touchdown with nil visibility for Cat III-C auto-land systems. For such operations, the ILS receiver and ILS ground transmitter must meet high levels of reliability (typically four independent redundant channels) and continuity and, in addition, the flight crew must be approved for Cat III-C operations. Once engaged, the auto-land system (Hogge, 2004) will capture the glide path and localizer and reduce the rate of descent prior to touchdown (the flare) without intervention of the flight crew, who have a critical system monitoring role during the approach.

In a flight simulator, auto-land functions must be implemented where auto-land operations are required for flight crew training. From the knowledge of the aircraft position and location of the glide slope and localizer transmitters, the lateral and vertical error from the localizer and glide slope, respectively, can be readily computed as angular errors. In fact, the localizer capture for an ILS is very similar to localizer capture for a VOR. In the case of a VOR, the radial is set by means of a course (CRS) pointer whereas for an ILS, the localizer direction is aligned to the centre line of the runway. In addition, the ILS localizer receiver is approximately four times more sensitive than a VOR receiver and generally, is only used up to a limit of 20 miles from touchdown.

The basis of the navigation computations for ILS localizer capture is shown in Figure 4.26 (not to scale).

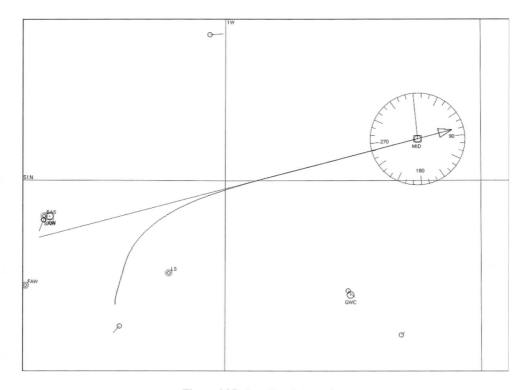

Figure 4.25 Localizer interception

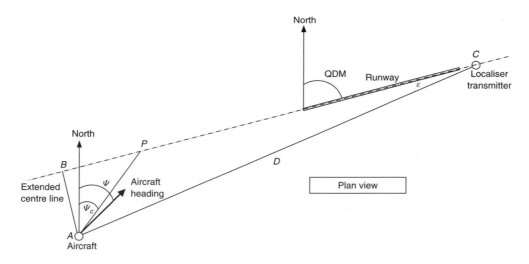

Figure 4.26 ILS localizer capture geometry

The aircraft position and the location of the ILS localizer transmitter are known (the transmitter location is available from the navigation database), enabling the localizer error ε to be computed, to simulate the ILS localizer receiver signal. Typically, the localizer transmitter is positioned about a kilometre from the far end of the runway. The point P can be computed as $\frac{3}{4}$, say, of the distance along BC in triangle ABC. An alternative method, and the one used in the flight simulator software, is to compute the desired heading ψ_c as

```
Runway_QDM-c*Localiser_Error
```

where c is a constant that determines the relative bearing of P from A to give the commanded heading ψ_c to intercept the localizer BC. If this angle of intercept is too coarse, the aircraft will intercept the localizer but fly through it requiring further correction to reacquire the localizer. On the other hand, if the intercept angle is too shallow, the time to acquisition may be excessive. In addition, the angular rate of closure also depends on the airspeed of the aircraft and the distance D from the transmitter.

As the aircraft follows the commanded heading towards the localizer, the angular error ε is continuously recomputed, leading to a change in the position of P and the commanded intercept heading ψ_c, resulting in a gradual curved flight path track until the aircraft is positioned on the localizer with a heading of the runway QDM.

A similar method is used for glide slope capture. The basis of the navigation computations for ILS glide slope capture is shown in Figure 4.27 (not to scale). The glide slope error ε is computed from the aircraft position and the location of the glide slope transmitter, which will be published in a navigation database. Typically, the transmitter is some 300 m from the start of the runway, positioned adjacent to the touchdown markers on the runway. The reference glide slope angle is normally around 3° but depends on the terrain close to the airport or reflections from the terrain. For some city airports, steeper climb slopes are used, for example, London City airport has a 5.5° glide slope.

The point P is computed as a function of the glide slope error ε, which can be computed either from the geometry of triangle ABC, to provide an appropriate flight path angle to intercept the glide slope at P or alternatively, the commanded flight path angle γ_c can be computed as

```
Runway_GSref-c*Glideslope_Error
```

where c is a constant to give a commanded flight path angle γ_c to incept the glide slope. As in the case of localizer acquisition, as ε reduces, the flight path angle also reduces, giving a manageable intercept where the flight path angle γ equals the glide slope reference angle at the point of intercept.

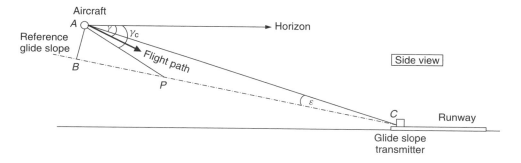

Figure 4.27 ILS glide slope capture geometry

As in the case of the localizer, the angular rate of closure depends on the airspeed and the distance from the runway.

In the steady-state, as the aircraft follows the glide path, and assuming a glide slope of 3°, the rate of descent is $V \sin(3°)$. At 160 kt, say, the vertical speed is 848 ft/min (14 ft/s). Any vertical impact above 200 ft/min is a heavy landing, likely to cause structural damage to the aircraft and possible injury to passengers. At touchdown, the vertical speed needs to reduce to less than 100 ft/min. For light aircraft with a nose wheel, this phase of flight is referred to as the *round out*. At 15 ft or so above the runway, the pilot fully retards the power lever and raises the noise to fly level, in effect to fly a foot or so above the runway with reducing speed. In this attitude, the aircraft touches down with a very small rate of descent just above the stall speed. For large transport aircraft, the manoeuvre is slightly different because the approach is at a much higher airspeed and also because the landing roll can use a large proportion of the overall runway length. There are two further complications. First, the engines may take a few seconds to spool down to flight idle power and, secondly, in increasing the pitch attitude to reduce the rate of descent, it is possible to over rotate so that the aircraft starts to climb. In this state, the aircraft could easily climb to 100 ft or so above the runway, with the engines at flight idle and the airspeed reducing to the stall speed, with a potentially catastrophic outcome. Consequently, the final phase of the approach requires three actions:

1. The power is reduced to flight idle below 50 ft;
2. The localizer tracking gain is reduced to avoid large banking manoeuvres close to the runway;
3. Flight path angle hold is engaged with a very small rate of climb (e.g. $\gamma = 0.005$).

The transition from glide path descent rate to touch down must allow sufficient time to establish the flight path angle hold below 10 ft but without extending significantly the landing distance. Finally, as the wheels touch down, a positive pitch attitude must be maintained so that the front wheel is slowly lowered as the airspeed reduces, to avoid structural damage to the nose wheel.

In the description of auto-land systems above, wind has been conveniently ignored, whereas it has a major impact on automatic landing systems. However, the runway centreline track is known and the wind speed can be computed from airspeed and ground speed (it is also likely to be a known value in the simulator) to determine the drift needed for localizer tracking. If the aircraft is approaching at 160 kt with a 20-kt crosswind, say, the drift angle is approximately 7°. In the transition below 50 ft, two actions are needed. First, the aircraft is yawed so that aircraft is aligned with the runway centre line at touch down and, secondly, the wing into wind is slightly lowered to avoid drift across the runway and also to reduce the likelihood of an engine strike, which may occur in the presence of a gust, (particularly for aircraft with under-slung engines).

The code for the Boeing 747 auto-land system used in the flight simulator covered in this book is given in Example 4.9.

Example 4.9 Auto-land

```
void FCS_Autoland(float *Elevator, float *Aileron)
{
  unsigned int ILSBeacon;
  unsigned int ILSRunway;
  float Qdm;
  float h;
  float hdot;
  float fpa;
  float gserr;
```

```
  if (AeroLink_NavPkt.ILS1.ILSBeacon) {
    Qdm = (float) (AeroLink_NavPkt.ILS1.RunwayQdm);
  }
  else {
    return;
  }

  hdot = -Model_Vd;
  h = -Model_Pz + (float) (AeroLink_NavPkt.GroundLevel) + Aero_CGHeight;
  if (h < 15.0) {
    *Aileron = FCS_HeadingHold(Qdm + 1.0 *
              (float) (AeroLink_NavPkt.ILS1.LocaliserError));
    if (Model_OnTheGround) {
      *Elevator = PitchHold(Model_Pitch * 0.5);
    }
    else {
      *Elevator = FPAHold(-0.005);
    }
  }
  else {
    gserr = (float) (AeroLink_NavPkt.ILS1.GlideSlopeError);
    Maths_Limit(&gserr, -DEG1, DEG1);
    fpa = -DEG3 + 5.0 * gserr;
    if (fpa > 0.0) {
      fpa = 0.0;
    }
    *Aileron = FCS_HeadingHold(Qdm + 10.0 *
              (float) (AeroLink_NavPkt.ILS1.LocaliserError));
    *Elevator = FPAHold(fpa);
  }
}
```

4.13 Flight Management Systems

For a flight of several hundred miles, the flight crew of a modern transport aircraft would have a potentially heavy workload if they were to fly the aircraft manually throughout the flight. In practice, they will select automatic flight control modes to manage the flight and to reduce the pilot workload. Figure 4.28 shows a simulated FCU for an Airbus A340.

The panel enables a pilot to select both display modes and control modes for the aircraft. For example, the navigation display mode can be selected by turning the left of the two grey selectors

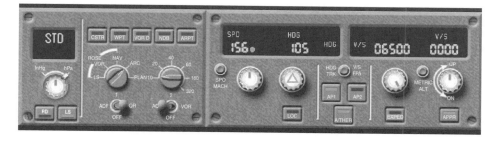

Figure 4.28 Flight control unit

to LS, VOR, NAV, ARC or PLAN or the two switches marked ADF, OFF or VOR can be used to select the signal source used for the navigation display. The panel also includes push switches, for example, the two leftmost switches can select or deselect the ILS display and flight director display on the primary flight display.

The main functions of the FCU are provided by the four knobs on the right-hand panel, to control speed, heading, altitude and vertical speed. By turning a knob to the desired speed or heading, the selected value is displayed in the panel directly above each knob. In addition, each knob can be pulled towards the pilot, giving control to the pilot or pushed away from the pilot, giving control to the flight management system. In the former case, the pilot can, in effect, engage or disengage one of the four flight control modes described in Sections 4.6–4.9. For example, to descend to 3000 ft at 350 kt, the pilot will select an altitude of 3000 ft and an auto-throttle speed of 350 kt and then engage speed and altitude to initiate the descent. During the descent, the commanded speed or altitude can be adjusted by selecting a new value, simply by turning the respective knob. Alternatively, the pilot can select the rate of descent (or climb) rather than an assigned altitude and select a heading (magnetic or true) or a track angle. In addition, illuminated push switches are provided to engage one or both autopilots (AP1 and AP2), to engage the auto-throttle (A/THR), to engage an expedited descent (EXPED), to engage auto-land (APPR) or to track a VOR localizer (LOC).

From the simulator designer's perspective, these functions are provided by the FCU selections, which in turn engage or disengage the automatic control modes covered in this chapter. In some simulators, an actual aircraft FCU may be used, which will need to be stimulated by data via the aircraft databus connection. Some manufacturers provide simulated systems, with a specific protocol for serial communications or Ethernet. Depending on the training application, the FCU (and similar panels) can be simulated as software panels, displayed on a flat-screen panel, where interaction with the switches and knobs is by means of a mouse or tracker ball. Whichever system is used, any change in the operating mode is detected (e.g. a switch is pressed or a knob is turned) and, depending on the action, the appropriate control mode is selected. In simulating these systems, considerable care is needed to ensure that an emulated unit fully complies with the manufacturer's specification of the equipment.

Although the flight crew could fly each segment of a flight plan manually, by selecting the aircraft speed, altitude and heading, this would be a time-consuming (and potentially error prone) activity for the flight crew and a FMS is provided to load, modify and execute a flight plan. The flight plan will include both the lateral flight plan and the vertical plan, given as a set of waypoints from the departure point to the destination point. The flight plan will also contain essential information to manage the flight safely and economically. Of course, the flight crew can override the flight plan at any time or use the FCU to select alternative settings.

For most airlines, the flight plan is generated by computers in the operations centre and provided to flight crews in a format that can be uploaded into the aircraft. Otherwise, the flight crew would need to enter manually the specific details of waypoints, approaches, navigation beacons and so on. The flight plan segments will be generated using knowledge of company routes and aircraft performance to produce speeds and climb rates for efficient (and economical) operation of the aircraft.

Once uploaded by the flight crew, the flight plan is accessed by means of a separate display, known as *multi-purpose control display unit* (*MCDU*). Buttons positioned around the edges of the display screen enable the flight crew to select specific modes, to enter or edit data or simply to confirm (or display) information in the flight plan. The FMS will drive the FCU to select vertical speed, flight path track or airspeed for each segment of the flight plan. An FMS will contain an ARINC-424 compliant database containing radio navaids, waypoints, airports, runways, airways, holding patterns, airport arrival/departure and let-down procedures and company routes. For example, the FMS will look up nearby navaids and select an appropriate VOR frequency to reduce pilot workload. The database will also contain detailed performance data specific to the aircraft and its engines.

Simulation of Flight Control Systems

The primary function of the FMS is to select a speed, heading, climb rate and altitude for each segment of the flight (Spitzer, 2000). In its simplest form, each waypoint is defined by an identifier, a latitude, a longitude, an altitude and an initial airspeed. The FMS extracts waypoint pairs to give the vertical profile and lateral track between two consecutive waypoints. For lateral guidance, the track between two waypoints is the equivalent of a localizer and the navigation method used for localizer tracking can be used to acquire and follow a flight plan track. For vertical guidance, the aircraft either follows the selected vertical speed or the assigned altitude for the next waypoint, using a conventional height-hold autopilot function.

The FMS database and the flight plan will contain detailed information relating to engine performance, airways, ATC restrictions, company routes, climb/descent profiles and also time at waypoints. An aircraft FMS will contain a computer to predict top of descent points, to predict wind effects on the flight plan, to check speed/climb constraints and to implement step climbs. If an actual aircraft FMS is used, it must be integrated with the simulator systems to be provided with the same inputs as the aircraft FMS and its outputs will be fed to the FCU. If the FMS is simulated, in addition to the provision of user interface functions responding to key inputs and the display of FMS pages, the FMS also provides guidance commands to the FCU and implements flight path optimization for effective management of the flight plan.

In the FMS computations, the relative bearing and distance to a waypoint is needed and generally, great circle tracks are computed (which also correspond to radio wave paths). The distance d between two points (λ_1, ϕ_1) and (λ_2, ϕ_2) on the surface of the earth, where λ denotes latitude and ϕ denotes longitude, is given by

$$d = R\sqrt{(\Delta\lambda)^2 \cos\lambda_1 \cos\lambda_2 (\Delta\phi)^2} \tag{4.50}$$

where R is the earth radius and

$$\Delta\lambda = \lambda_2 - \lambda_1 \tag{4.51}$$

$$\Delta\phi = \phi_2 - \phi_1 \tag{4.52}$$

The relative bearing ψ of a point (λ_2, ϕ_2) from a point (λ_1, ϕ_1) is given by

$$x = \sin\lambda_2 - \sin\lambda_1 \cos d \tag{4.53}$$

$$y = \cos\lambda_2 \sin\Delta\phi \cos\lambda_1 \tag{4.54}$$

$$\psi = \tan^{-1}\left(\left|\frac{y}{x}\right|\right) \tag{4.55}$$

The absolute values of x and y are used in the computation of relative bearing because most *arctan* functions are limited to the range $\pm\pi/2$. The actual value of ψ depends on the four possible signs of x and y. If $x < 0$ then $\psi = \pi - \psi$. If $y < 0$ then $\psi = -\psi$. Care is also needed to cater for small values of x, which could lead to numeric overflow. In this situation, $\psi = \pi/2$. In C, the maths function `atan2(y,x)` computes ψ correctly for all four quadrants, whereas `atan(y/x)` is restricted to the range $\pm\pi/2$.

By computing the relative bearing of a waypoint to the next waypoint, and as the location of the successive waypoint is known, the aircraft can intercept and track each segment. For the majority of FMS operations, the aircraft tracks from one waypoint to the next, where the FMS updates the list of waypoints as the aircraft reaches each waypoint. There are four basic variations of tracking used with an FMS, as shown in Figure 4.29. The pilot may need to fly direct to a waypoint further

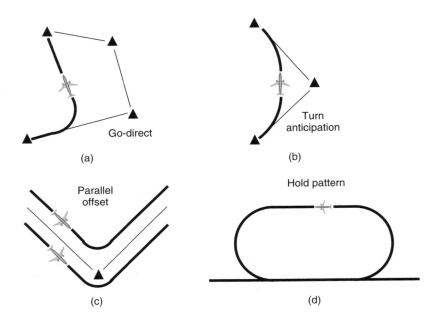

Figure 4.29 FMS functions

ahead in the flight plan, as, shown in Figure 4.29a. More commonly, rather than waiting to turn overhead a waypoint, if the turn onto the next segment is anticipated, the routing is much more efficient, as shown in Figure 4.29b. In the case where two aircraft are flying similar routes, parallel offset tracking is provided to ensure lateral separation, as shown in Figure 4.29c. Finally, standard hold patterns are published for airports and are stored in the FMS database. From the position and orientation of the hold pattern, the FMS provides guidance to enable the aircraft to fly an accurate hold pattern with minimal pilot intervention, as shown in Figure 4.29d.

The tracking method for each segment is similar to the localizer tracking method described in Section 4.11. The next waypoint is treated as a beacon and the bearing of the next waypoint from the current waypoint is the equivalent of the localizer QDM, as shown in Figure 4.30 where the aircraft is required to follow the track from waypoint X to waypoint Y. The bearing of waypoint X

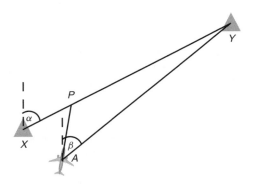

Figure 4.30 Flight plan segment tracking

Simulation of Flight Control Systems

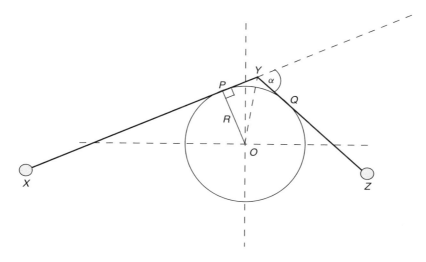

Figure 4.31 Flight plan turn anticipation

from waypoint Y is given by α. The bearing of the aircraft A from waypoint Y is given by β. The localizer error $\delta = \beta - \alpha$. The flight control system is commanded to hold a heading of $\alpha + g\delta$, where g is an appropriate gain to intercept the flight plan segment (typically a value of 6).

The other main tracking requirement is to anticipate the turn to the next segment, to avoid overshooting the waypoint at Y, as shown in Figure 4.31 where the aircraft turns through an angle α to track the next flight plan segment from Y to Z. Ideally, the aircraft will track along the arc from P to Q, rather than overflying the waypoint at Y. The point P (the start of the turning manoeuvre) is computed to initiate the turn to intercept the segment from Y to Z.

Let

$$d = PY \tag{4.56}$$

The relationship between bank angle and turn rate is given by

$$\tan \phi = \frac{\dot{\psi} V}{g} \text{ or} \tag{4.57}$$

$$\dot{\psi}_{max} = \frac{g \tan \phi_{max}}{V} \tag{4.58}$$

where
    ϕ is the bank angle
    V is the airspeed
    $\dot{\psi}$ is the turn rate

For a constant rate of turn

$$V = R\dot{\psi} \text{ or} \tag{4.59}$$

$$R = \frac{V}{\dot{\psi}} \tag{4.60}$$

where

$\dot{\psi}$ is the desired turn rate
R is the radius of the turn

Let angle $PYO = \beta$, then

$$\beta = \frac{\pi - \alpha}{2} \tag{4.61}$$

$$\tan \beta = \frac{R}{d} \tag{4.62}$$

$$d = \frac{V^2}{g \tan(\phi_{max}) \tan \beta} \tag{4.63}$$

For civil transport aircraft, the maximum bank angle (ϕ_{max}) approved for automatic flight is typically 25°. In other words, the distance from the waypoint Y where the capture of the next segment is initiated is a function of the turn angle α and the airspeed V, as given in Equation 4.63. In practice, these computations are more complex; first, the speed and altitude may change for the next segment and, secondly, there is a finite time to initiate and complete the turn, introducing a small variation in turn rate during the turn. In the former case, the computation is simplified if the change of altitude or airspeed is delayed until the turn is completed. An aircraft FMS will contain

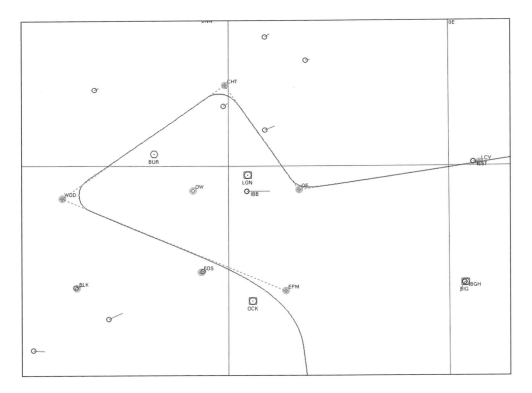

Figure 4.32 An example of turn anticipation in flight plan tracking

detailed performance data for the aircraft, including time to bank, time to accelerate and climb and descent rates at specific airspeeds and altitudes, in order to predict lateral guidance terms.

By way of illustration, the following rudimentary flight plan contains four segments, following a flight plan based on five NDBs in the London area. The resultant aircraft flight path is shown in Figure 4.32. For each waypoint, the columns denote the name, latitude and longitude of the waypoint and the altitude (feet) and airspeed (kt) of the segment starting at the respective waypoint. Note the acquisition of the first segment from an initial position south of the Epsom NDB (EPM).

```
WPT1  N5119.13  W00022.22  3000  200
WPT2  N5127.15  W00052.67  3000  220
WPT3  N5137.09  W00030.53  3500  220
WPT4  N5127.92  W00020.43  4000  220
WPT5  N5130.22  E00004.13  4000  200
```

References

Bennett, S. (1994) *Real-time Computer Control: An Introduction*, Prentice-Hall, Harlow.

Bibbero, R. J. (1977) *Microprocessors in Instruments and Control*, John Wiley & Sons, Ltd, New York.

Clancy, L. J. (1975) *Aerodynamics*, Pitman Publishing Limited, London.

Cook, M. V. (1997) *Flight Dynamics Principles*, Arnold, Amsterdam.

Franklin, G. F., Powell, J. D. and Emani-Naemi, A. (1994) *Feedback Control of Dynamic Systems*, Addison-Wesley, Reading, MA.

Hogge, E. F. (2004) B-737 Linear Autoland Simulink Model, NASA/CR-2004-213021, Langley Research Centre, Virginia.

McLean, D. (1990) *Automatic Flight Control Systems*, Prentice Hall International.

Ord-Smith, R. J. and Stephenson, J. (1975) *Computer Simulation of Continuous Systems*, Cambridge University Press, Cambridge.

Roskam, J. (2003) *Airplane Flight Dynamics and Automatic Flight Controls*, Darcorporation.

Spiegel, M. (1965) *Schaum's Outline of Laplace Transforms*, McGraw Hill, New York.

Spitzer, C. R. (2000) *The Avionics Handbook*, CRC Press Inc.

Stevens, B. L. and Lewis, F. L. (2003) *Aircraft Control and Simulation*, John Wiley and Sons.

5

Aircraft Displays

5.1 Principles of Display Systems

In flight simulators before 1990, aircraft instruments were emulated either by mechanical adaptation of actual instruments or by the use of servo mechanisms (and latterly stepper motor drives) to replicate the function and characteristics of actual aircraft instruments. Nowadays, it is common practice to use computer graphics to simulate aircraft instruments. First, the high degree of realism provided by synthetic displays is acceptable in a training role. Secondly, the cost of developing computer-generated displays is significantly less than the cost of developing electromechanical instruments. Thirdly, flat-screen displays are now commonplace in civil and military aircraft.

As the PC has evolved, computer displays in the office and home have become ubiquitous. The application of computer graphics extends to word processing, spreadsheets, web access and image processing. It has been the domestic demand for PC technology, which has driven the major improvements in performance (both speed and storage) and the concomitant reductions in the cost of computing. For the PC designer, the display is a critical component: it must provide colour, it must have high resolution for clarity and the time to generate an image must be negligible. The requirements for aircraft displays, which have kept pace with the advances in PC display technology, are identical.

Three display technologies span the last 50 years. The CRT was developed for television, initially for monochrome displays, and since the 1960s for colour displays. CRT technology was the ideal solution for early computer displays, where the technology to display TV pictures was adapted to display text and subsequently computer graphics. Liquid crystal displays (LCDs) developed quickly from the small monochrome displays used in watches and calculators to the flat-screen colour displays used in computers (and aircraft) since the 1990s. LCDs afforded significant reductions in weight and power consumption with increased quality and reliability. More recently, plasma screen technology has been exploited for domestic television with the consumer demand for large flat screens driving down the cost.

In a CRT, the glass tube is formed as a screen with a thin neck. The inside of the screen is coated with phosphor (in crystalline form), a thermionic valve is inserted in the neck and air is evacuated from the tube, which is sealed. By applying a large voltage between the cathode and anode of the valve, a stream of electrons is transmitted. Electrons hitting the screen coating cause the phosphor to luminesce at the focal point of the electron beam. Two sets of magnetic deflection coils mounted in the neck of the CRT are used to deflect the beam in the horizontal and vertical directions of the screen. In effect, the beam can be positioned on the screen. If the beam is switched on, it will create a series of dots as it moves. If the beam is turned off, no light is produced. However,

---

Principles of Flight Simulation D. J. Allerton
© 2009, John Wiley & Sons, Ltd

the phosphor only luminesces for a short time (typically 10–50 ms); therefore, the screen must be repeatedly refreshed to generate a permanent image.

CRTs are designed to operate in one of two ways. In calligraphic CRTs (also known as *stroke displays*), the beam is directly deflected by modulating the signal applied to the deflection coils. By applying a linear ramp voltage to the coils, the locus of the beam is a straight line. The beam can be switched off, then deflected to a new screen location and switched on to generate another line. On the one hand, if the beam moves too fast, it will not excite the phosphor sufficiently to produce a bright line; but on the other hand, if it moves too slowly, it will take too long to generate an image. The major attraction of calligraphic displays is that the beam intensity can produce very clear straight lines, but the disadvantage is that the display is limited to a few thousand lines. In effect, time is lost as the beam is switched off and moved from one screen position to another and time must be allowed for the deflection coils to 'settle' after a large movement of the beam.

CRTs used in televisions are referred to as *raster displays*. With a raster display, the beam is positioned at the top left-hand corner and is then swept from left to right across the screen (deflecting the horizontal beam at a constant rate). The beam can be switched on or off as it traces a horizontal track across the screen. When the beam reaches the right edge of the screen, the beam is switched off and reset to the left edge (known as the *retrace*). At the same time, the vertical deflection coils deflect the beam to the adjacent lower row. This process is repeated line by line until the beam reaches the bottom right-hand corner of the display, whereupon both coils are reset to reposition the beam at the top left-hand corner and the process repeats. The repetition rate is known as the *display refresh rate* and typical values for commercial displays are 50–100 times per second.

The attraction of the CRT system is its simplicity. Two ramp voltages are applied to the two sets of coils, once per line for horizontal deflection and once per frame for vertical deflection. The beam intensity is modulated as the beam traverses each line. By coating the screen with phosphor dots that produce red, green and blue colours, a wide range of colours can be displayed. Moreover, the individual bits of an image stored in a computer memory can be used to modulate the beam intensity. In other words, generation of a video signal from a computer image stored in a memory is straightforward. However, the CRT does have three major drawbacks. First, the display is bulky and the amount of glass adds considerable mass. Secondly, a CRT requires a high voltage to excite the electron beam. Thirdly, and most importantly, the screen is addressed as individual pixels. While this feature simplifies the mapping between memory and screen coordinates, the screen is restricted to an integer grid. In drawing a straight line, the line can appear jagged in a way that is unrealistic. To overcome this problem, higher display resolutions can be used but this requires much higher performance amplifiers to switch the beam and deflect the coils. Alternatively, the image can be formed in a higher resolution memory and the actual pixel value is sampled to derive a best 'fit' to the pixel colour; this technique is known as *anti-aliasing* and is outlined in Section 5.2.

LCDs are based on the principle that an electrical field applied to a liquid containing liquid crystal molecules can change the optical property from transmissive to reflective (Jarrett, 2005). In the energized state, the molecules become aligned parallel to the direction of light, allowing light to pass through the liquid. In the non-energized state, the molecules (which are impregnated with a black dye) are aligned randomly and absorb the light. The application of a small potential field produces a light shutter with a very fast response. The individual crystals can be arranged as an array of rows and columns with back-projected light. The individual cells can be addressed to be switched on or off with an optical colour filter provided for each cell for the red, green and blue components of light.

The advantage of LCDs is that they can be manufactured using microelectronic fabrication methods. They are light weight and provide the same level of display resolution as CRTs. They do not require the complex analogue correction circuitry or the high voltages needed for CRTs and can be operated from a single low-power DC supply. While LCDs offer excellent colour reproduction, good contrast, high resolution and longevity, they do have several limitations. First,

Plate 1 An image from a civil flight simulator. Reproduced by permission of Thales

Plate 2 A flight simulation training device (FSTD). Reproduced by permission of Thales

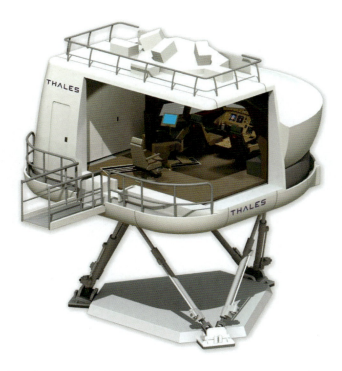

Plate 3 A civil full flight simulator. Reproduced by permission of Thales

Plate 4 A real-time instrument display

Plate 5 EFIS primary flight display

Plate 6 HUD display in the flight simulator

Plate 7 Instructor station in an Airbus A380 flight simulator. Reproduced by permission of Thales

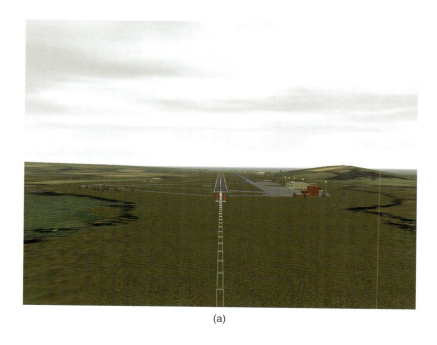

(a)

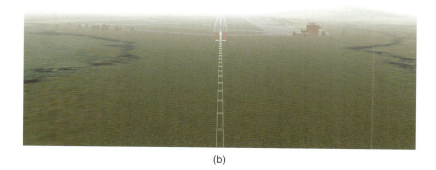

(b)

Plate 8 Bristol Lulsgate Airport approach (a) and the effect of fogging (b)

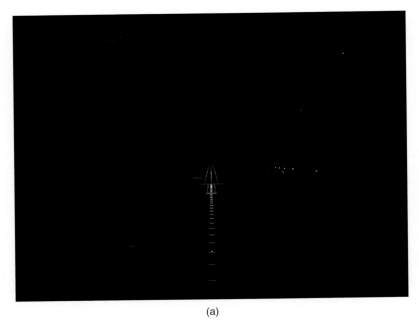

(a)

(b)

Plate 9 Bristol Airport night approach (a) and Hong Kong International Airport (Chek Lap Kok) approach (b)

Plate 10 Trees rendered as billboards

Plate 11 A wake vortex rendered as billboards

(a)

(b)

Plate 12 A Boeing 747-400 visual database model (a) wire frame version, (b) shaded and textured polygon version

Table 5.1 Monitor dimensions

Monitor size	Screen width	Screen height
12	9.6	7.2
15	12.0	9.0
17	13.6	10.2
19	15.2	11.4
20	16.0	12.0
21	16.8	12.6
24	19.2	14.4
30	24.0	18.0
32	25.6	19.2
37	29.6	22.2

progress towards large displays has been slow. Secondly, the black level is not a deep black owing to the need for fluorescent back lighting. Additional circuitry is required to adjust for ambient light levels ranging from bright sunlight to dark night. Generally, concerns over limited viewing angles (particularly in portrait mode), permanent loss of individual pixels and pixel smearing at high drawing rates have been overcome.

For the flight simulator designer, the displays need to replicate the aircraft instruments and displays not only in terms of functionality and colour, but also physical layout. Generally, CRTs are being phased out by manufacturers as the demand for CRT televisions falls. Small volume manufacture of CRTs is uneconomic and LCD and plasma displays are the only viable option for simulation displays. The display screen size is defined by the diagonal of the viewable areas. Although there is a range of aspect ratios for LCDs, Table 5.1 gives the dimensions for common 4 : 3 LCD sizes in inches.

Displays can pose a considerable limitation in flight simulators. Often, the fascia area in front of the pilot is not rectangular, sufficient space must be allowed for the pilot's leg area below the displays, CRTs require substantial depth and the display area may also contain switches and knobs. Given the size of commercial displays, the designer is constrained by the monitor size and the aspect ratio. These displays can be mounted in either landscape or portrait orientations, although care is needed with the reduced viewing angle in portrait mode and there is a potential difficulty of rotating the displayed graphics.

Modern aircraft LCDs are 8 inches square; unfortunately, these dimensions do not match the values given in Table 5.1. Conventional aircraft instruments are typically 2.5–3.5 in. diameter, but are rarely mounted in a regular rectangular pattern. In addition, the spacing of instrument panels, with their associated bezels, can be problematic (in the very compact space below the windscreen) and the use of multiple displays can introduce undesirable partitions between clusters of instruments. The selection and layout of displays will depend on the training requirement and the level of visual fidelity required. Nevertheless, the majority of aircraft cockpit geometries do not map naturally to the display sizes of commercial CRTs and LCDs.

5.2 Line Drawing

Aircraft displays form a specialized subset of computer graphics: they use few colours; there is no shading; the displays are constrained to vector graphics, alphanumeric text and aircraft-specific symbols; the display must be updated at least 20 times per second and the display functions are implemented using 2D graphics operations. Typically, the display is treated as Cartesian coordinates with the x axis horizontal and the y axis vertical. However, there are also variations in this convention. The display may be orientated in the landscape or portrait mode and the origin may

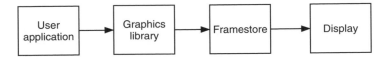

Figure 5.1 Graphics pipeline

be positioned at the bottom left-hand corner or at the top left-hand corner. These conventions are usually defined in a graphics library, allowing the user to develop graphics software independent of any hardware constraints, as shown in Figure 5.1.

The lines rendered by the graphics software are written to a memory, commonly known as the *framestore*, which is organized as conventional computer memory, with very fast access times. In addition, the memory can be accessed simultaneously by the graphics processor and the display logic generating the video signal for the display. As the display normally consists of pixels organized as rows and columns, the mapping between memory addresses and screen coordinates is organized to simplify memory access. For a display resolution 1024×768 pixels, say, the display of 786,432 pixels can be organized as 98,304 bytes (assuming 8 bits per byte) for a monochrome display, as shown in Figure 5.2.

If the memory is byte addressed, each row is stored as 128 bytes (pixels 0–1023 in the row). The pixel at (i,j) is accessed at the memory location $i*128 + j/8$ and within that byte, the pixel corresponds to the (j mod 8) bit in the byte. The computation of the memory address and the memory write cycle is performed by the graphics processor. Note however, that a line of 100 pixels will require 100 memory access cycles, and for real-time applications, the memory access speed of the framestore is a major factor in the performance of an image generator.

For colour displays, several framestores are organized as parallel memories, which can be accessed simultaneously to read or write an individual pixel, as shown in Figure 5.3.

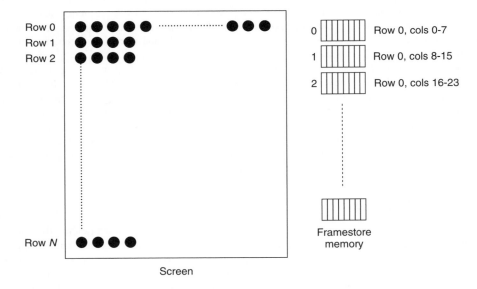

Figure 5.2 Framestore organization (monochrome display)

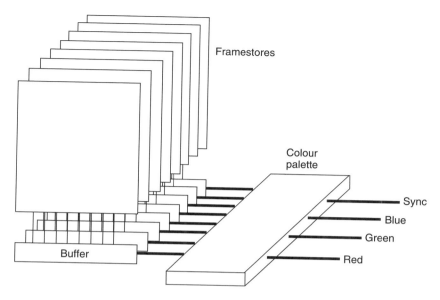

Figure 5.3 Framestore organization (colour display)

In Figure 5.3, the eight framestores can be written to simultaneously, where a single pixel is written to any selected framestores in a single write cycle. As each pixel is represented by 8 bits (in this example), there are 256 possible colour values for each pixel. During each frame, the framestores are accessed in parallel by the display logic and the 8 bits of each pixel are fed to a colour palette IC, which maps the bits from each framestore buffer byte to a video signal containing specific red, green and blue components. The colour palette IC can also be accessed independently to set the colour mappings for a specific application. The video signal generated by the colour palette IC provides a waveform that is appropriate for the display. With modern graphics cards, there may be 24 framestores, providing a potential palette of 2^{24} colours.

There is a major bottleneck with this architecture: both the graphics processor and the display logic compete for access to the framestore. One solution, which is widely used, is to provide two identical framestores. During one frame, framestore A is written to by the graphics processor and framestore B is accessed by the video output controller, which may also clear the framestore as it extracts the pixels to be displayed. In the subsequent frame, the graphics processor writes to framestore B, while the video output logic accesses framestore A to display the image generated in the previous frame. This toggling of the framestores is performed by the framestore access logic. Such a scheme, known as *double-buffering*, is used widely in real-time graphics systems and simplifies the memory organization. The benefit of double-buffering for real-time graphics far outweighs the disadvantages of the additional framestore memory.

Although line drawing is the lowest level of computer graphics, the implicit operations have a major impact on the real-time performance of a display and consequently, it is important to appreciate the basis of line drawing operations. Assuming the display has the origin at the bottom left-hand corner, with the x axis horizontal and the y axis vertical, a line drawing algorithm is used to compute the coordinates to form a straight line between (x_1, y_1) and (x_2, y_2). The display is assumed to have a fixed resolution and the pixels form an integer grid. The requirement is to find the pixels closest to the true line on the integer grid as shown in Figure 5.4.

Although this example illustrates a short line, the method of selecting the intersecting pixels, known as *Bresenham's algorithm* (Newman and Sproull, 1979), is straightforward. The line is

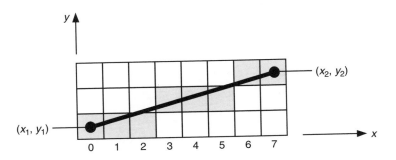

Figure 5.4 Straight line segment

incremented in the x direction for each point (the longer of the x and y components in this example). The increment in the y direction is either 0 or +1. The requirements of the line drawing algorithm are as follows:

1. The line should be as straight as possible.
2. The line should terminate at the end point.
3. The algorithm should apply correctly for line lengths of one or more pixels.
4. The line density (pixels per unit length) should be constant and independent of the line length.
5. The algorithm should be fast and simple (fast for real-time graphics and simple to minimize the complexity of the algorithm and the hardware).

Consider a line in the first octant, as shown in Figure 5.5. The algorithm increments by 1 pixel in the direction of the longer edge Δx or Δy and increments by 0 or 1 pixels in the direction of the shorter edge, as determined by the algorithm. Let the distance between the actual line and the approximated grid coordinate be ε. Consider the case where the line passes exactly through the point (0, 0), as shown in Figure 5.6.

There are two choices for the next pixel after (0, 0), either right to (1, 0) or right and up to (1, 1). The selection of the pixel therefore depends on the sign of the error term ε at each point along the line. At each iteration, the slope of the actual line $\Delta y/\Delta x$ is added to the error ε. If $\varepsilon > 0$, the exact value lies above the current y coordinate; the y coordinate is incremented and 1.0 is subtracted from ε. Otherwise the y coordinate is unchanged. The implementation of the algorithm is summarized in Example 5.1.

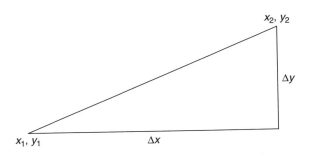

Figure 5.5 First octant

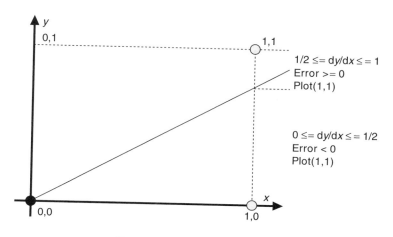

Figure 5.6 Error in line drawing

Example 5.1 Bresenham's line drawing algorithm (floating point version)

```
ε := Δy/Δx - 0.5;
FOR i = 0 TO Δx DO
  SetPixel(x, y);
  IF ε > 0.0 THEN
    y := y + 1;
    ε := ε - 1.0;
  END;
  x := x + 1;
  ε := ε + Δy/Δx;
END;
```

where SetPixel writes a pixel to the display memory at the screen coordinates (x, y).

However, there are two problems with the algorithm in this form. First, the computation $\Delta y / \Delta x$ is constant for the line and does not need to be computed in the for-loop. Secondly, the term $\Delta y / \Delta x$ cannot be represented accurately as an integer value. However, if ε is multiplied by $2\Delta x$, then the sign of ε is still valid and the division operation is removed. Admittedly, a multiplication by two is introduced, but this can be implemented as an addition ($2\Delta y = \Delta y + \Delta y$). The modified algorithm is shown in Example 5.2, where e, dx and dy are integers.

Example 5.2 Bresenham's line drawing algorithm (fixed point version)

```
e := 2dy - dx;
FOR i := 0 TO dx DO
  SetPixel(x, y);
  IF e > 0 THEN
    y := y + 1;
    e := e + (2dy - 2dx);
  ELSE
    e := e + 2dy;
  END;
  x := x + 1;
END;
```

For a given line, the common terms 2dx and 2dy are computed before the for-loop is executed.

This method of line drawing is known as *Bresenham's algorithm* and is widely used for line drawing in graphics processors for the following reasons:

- It is fast;
- It is simple, requiring two increments, one addition and one subtraction per pixel;
- It is easily extended to all eight octants.

The simplicity of the line drawing method has enabled the algorithm to be implemented in graphics hardware. However, it is important to appreciate two limitations. First, every call of SetPixel implies a memory write cycle. For example, with 50-ns memory, a 1000-pixel line will take 50 µs to write the pixels to memory. Secondly, the resultant line is drawn on a pixel grid with jagged discontinuities along the line. For example, a line from (0, 0) to (50, 1) will result in one segment from (0, 0) to (25, 0) and a second segment from (26, 1) to (50, 1), with a noticeable discontinuity from (25, 0) to (26, 1). The problem is not solved by using a higher resolution display; although the visual impact may improve, a line that consisted of two segments will contain four segments if the resolution is doubled.

The solution to the problem of jagged effects caused by pixelization is known as *anti-aliasing*. The image is generated at a higher resolution, but this image is filtered and displayed at a lower resolution (the screen resolution). Of course, there are two immediate effects. First, additional memory is needed to form the image at a higher resolution and secondly, the overall drawing rate is reduced as there are more pixels per line. Consider a line passing through three adjacent screen pixels, as shown in Figure 5.7.

The amount of the pixel P_1 covered by the line is considerably less than the amount of the pixel P_2 covered by the line. By increasing the display resolution by a factor of 4 in the x and y directions, the same line now covers more pixels, as shown in Figure 5.8.

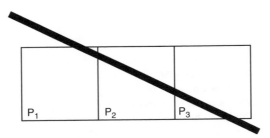

Figure 5.7 Aliasing at low resolution

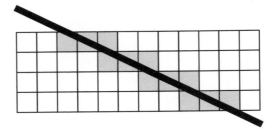

Figure 5.8 Aliasing at high resolution

Aircraft Displays

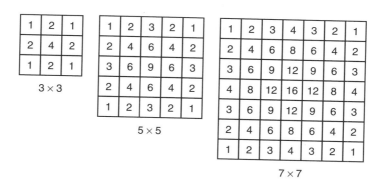

Figure 5.9 2D Spatial filters use in anti-aliasing

In each 4×4 grid, the contribution of each sub-pixel covered by the line is summed as a weighted component and a spatial filter is used to compute the overall intensity of the 16 sub-pixels, in order to adjust the intensity of the pixel at the lower resolution. Typical spatial filter values are given in Figure 5.9 for 3×3, 5×5 and 7×7 pixel filters.

Although anti-aliasing can make a dramatic improvement to image quality, it is important to appreciate the trade-off:

- Extra memory is required to form the image at a higher resolution, prior to filtering;
- Extra memory access cycles are required to accommodate the increased resolution, prior to filtering, with potential reduction in graphics throughput – for example, the 7×7 filter introduces 49 multiplications per pixel;
- Extra hardware logic is needed to provide the filtering arithmetic and to provide the video access logic – introducing a significant increase in cost.

Nevertheless, these problems have been overcome with the current generation of graphics cards and anti-aliasing is now commonplace.

5.3 Character Generation

In addition to lines, an aircraft display will also contain text for digits, alphabetic characters and symbols. Generally, only a few fonts are used for aircraft displays, with fixed font sizes. They have the same critical requirement as line drawing; they must be rendered at high speed. In some applications, the characters may be scaled or rotated although in many applications, the characters may always be upright. Three techniques are used to generate fonts for aircraft displays:

- Bitmapped characters;
- Vector-generated characters;
- Textured characters.

Bitmapped characters are probably the simplest method used to generate characters. Figure 5.10 shows a fragment of a set of bitmapped characters, showing the characters '@', 'A', 'B' and 'C', where each character is organized on a 16×8 pixel grid. In this case, each character is arranged as 16 bytes with the same memory address byte ordering and bit ordering as the framestore, simplifying the access to individual characters in the font. The 16 bytes for a specific character are then copied

Figure 5.10 A bitmapped character font

directly from an array containing the font character to the framestore. This approach has some merit:

- It is straightforward to generate and edit a font in this form;
- Rendering an individual character only requires 16 memory write operations;
- Non-standard characters are readily formed.

The main limitations with the method are that font characters are difficult to scale or rotate without introducing distortion and characters can appear jagged. Generally, the minimum size of a bitmapped character set is 8 × 5 pixels. Although higher resolution fonts are easily generated, few of the modern graphic cards support scaling or rotation of bitmapped fonts. Nevertheless, high-speed copying of bitmaps is supported by these cards; therefore, these techniques may be amenable in applications where there is no requirement to rotate the font characters.

An alternative approach is to treat a character as a set of vectors, which are rendered using a line drawing method. For example, the character 'A' is shown in Figure 5.11, also using a 16 × 8 pixel grid. The character can be rendered by five line drawing operations, as shown in Example 5.3:

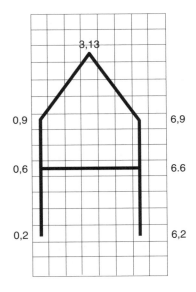

Figure 5.11 A vector-generated character

Example 5.3 Vector generation of characters

```
draw(0, 2, 0, 9)
draw(0, 9, 3, 13)
draw(3, 13, 6, 9)
draw(6, 9, 6, 2)
draw(0, 6, 6, 6)
```

where the coordinates are relative to the bottom left-hand corner of the character (0, 0). One attraction of this approach is that the line width can be set to a value that is appropriate to the size of the font. Detailed character sets can be generated that contain many line segments. The font set is then represented by the set of vectors needed for each character. One minor problem is that the number of vectors in each character may vary and an additional table must be generated to locate the start of the vectors for each character. In addition, some form of terminator is needed to separate the characters in a table of character fonts.

If line anti-aliasing is provided, scaling and rotation will not lead to distortion in contrast to bitmapped characters. Moreover, with vector-generated characters, there is no requirement to align the character segment with the pixel grid. The main disadvantage with this method is the difficulty of organizing the font table and the time taken to render characters that contain a significant number of line segments.

A third option is to capture each character of a font as an image or texture map and then render it directly as a region of texture. Modern graphics processors are able to render texture maps at high speed, minimizing the effects of distortion in rotation and without introducing performance penalties. The drawback with this approach is that each character may occupy a significant amount of storage in the texture map memory of the graphics processor and a character set typically comprises 128 characters, although only the characters needed for the aircraft display would need to be stored. A character set can be digitally scanned and stored as small graphic files in a standard image format. These images are loaded initially at the start of an application and each character is extracted from the list of texture maps and then rotated and rendered at the appropriate screen position, providing very clear and realistic characters.

There are two requirements with all three methods of character generation. First, the visual quality of the rendered character must be acceptable. Any distortion caused by rotation, aliasing or alignment to an integer pixel grid must be minimized. Secondly, each character will require several write operations to the framestore. The time for each operation must be minimized to maintain the overall real-time performance, particularly in applications where there are several hundred characters or symbols in an aircraft display. Generally, characters are produced for fixed size fonts for aircraft displays and, for most aircraft displays, there are only a few font sizes used for character generation. Consequently, the generation of characters and symbols is not too onerous.

Characters may be edited with a font editor, which allows individual pixels to be set. The characters produced by the font editor are stored in a compact form, so that they can be subsequently modified by the font editor or accessed by the application. Figure 5.12 shows font editing of the character '5' for a 24 × 10 font.

In Figure 5.12a, the bitmapped characters are stored as 24, 16-bit values. In Figure 5.12b, the vector-generated characters are stored as line segments (11 segments, in this example), to capture the curvature of the character. Note that the line width is explicit for bitmapped fonts, whereas a vector representation enables the designer to select a line width appropriate to the application. For the vector representation, if the character can be drawn in one continuous operation, each line segment can be defined by the list of the end coordinates of the segment. Generally, vector-generated characters maintain their visual appearance when rotated (if hardware anti-aliasing is provided), whereas rotation introduces distinct jaggedness for bitmapped characters. If memory capacity is not an issue and hardware texture mapping is available, then very high-quality character generation

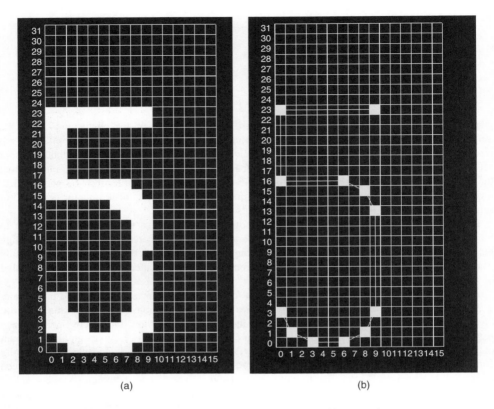

Figure 5.12 Character fonts (a) bitmapped character (b) vector character

is achieved by representing characters as texture maps and editing the character set with texture editing tools, such as Adobe Photoshop™.

5.4 2D Graphics Operations

In 2D graphics, the screen is considered as a pixel grid, typically with the origin in the bottom left-hand corner of the screen, the x axis along the bottom edge of the screen and the y axis along the left-hand edge of the screen. Objects consist of lines which are drawn on the screen. The vertices of a vector can be translated, scaled or rotated in the screen space (Foley *et al.*, 1995).

To translate the point (x, y) to (x', y')

$$x' = x + T_x \tag{5.1}$$

$$y' = y + T_y \tag{5.2}$$

where T_x and T_y are the respective translations in the x and y directions.

To scale the point (x, y) to (x', y')

$$x' = S_x x \tag{5.3}$$

$$y' = S_y y \tag{5.4}$$

where S_x and S_y are the respective scale factors in the x and y directions.

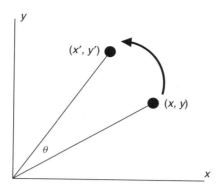

Figure 5.13 Rotation of a point

To rotate the point (x, y) to (x', y') by an angle θ about the origin, as shown in Figure 5.13.

$$x' = x \cos \theta + y \sin \theta \tag{5.5}$$

$$y' = -x \sin \theta + y \cos \theta \tag{5.6}$$

These operations are readily combined. For example, a point (x, y) can be rotated about another point (x_c, y_c) in three operations:

1. Translate the point by $(-x_c, -y_c)$ (translating the point of rotation to the origin).
2. Rotate the translated point about the origin.
3. Translate the rotated point by (x_c, y_c).

These operations can be represented by 3×3 transformation matrices:
Translation:

$$\begin{bmatrix} x' \\ y' \\ 1 \end{bmatrix} = \begin{bmatrix} 1 & 0 & T_x \\ 0 & 1 & T \\ 0 & 0 & 1 \end{bmatrix} \begin{bmatrix} x \\ y \\ 1 \end{bmatrix} \tag{5.7}$$

Rotation:

$$\begin{bmatrix} x' \\ y' \\ 1 \end{bmatrix} = \begin{bmatrix} \cos \theta & -\sin \theta & 0 \\ \sin \theta & \cos \theta & 0 \\ 0 & 0 & 1 \end{bmatrix} \begin{bmatrix} x \\ y \\ 1 \end{bmatrix} \tag{5.8}$$

Scaling:

$$\begin{bmatrix} x' \\ y' \\ 1 \end{bmatrix} = \begin{bmatrix} S_x & 0 & 0 \\ 0 & S_y & 0 \\ 0 & 0 & 1 \end{bmatrix} \begin{bmatrix} x \\ y \\ 1 \end{bmatrix} \tag{5.9}$$

Multiple transformations can be formed as a concatenation of these matrix multiplications. For example, to rotate the point (x, y) about the point $(3, 4)$ by $20°$ (anticlockwise).

$$\begin{bmatrix} x' \\ y' \\ 1 \end{bmatrix} = \begin{bmatrix} 1 & 0 & 1 \\ 0 & 1 & 0 \\ -3 & -4 & 1 \end{bmatrix} \begin{bmatrix} \cos 20 & -\sin 20 & 0 \\ \sin 20 & \cos 20 & 0 \\ 0 & 0 & 1 \end{bmatrix} \begin{bmatrix} 1 & 0 & 1 \\ 0 & 1 & 0 \\ 3 & 4 & 1 \end{bmatrix} \begin{bmatrix} x \\ y \\ 1 \end{bmatrix} \quad (5.10)$$

In practice, there is little requirement for scaling in the rendering of 2D graphics displays for aircraft instruments. Instruments are likely to be positioned at a specific (fixed) location on the screen, requiring translation. The graphical operations can be defined relative to the coordinates of an instrument, enabling pointers, symbols and characters to be rotated (for example, the markings on a compass card). Although it would seem that matrix operations for 2D operations are an unnecessary complication, as the translations and rotations can be computed explicitly in OpenGL® (Section 5.6), matrices are used for both two- and three-dimensional image rendering operations, providing a consistency in the representation of the geometric transformations while, in practice, the overhead of maintaining these transformation matrices is negligible.

5.5 Textures

During the 1980s, as the PC reduced in cost and became more widespread, the demand for new applications increased, particularly for computer games. Several vendors developed chip sets to render graphics at speed, where the memory used to render the image could be accessed at the high speeds associated with real-time graphics. These chip sets were developed to perform the 3D graphics operations (Clark, 1982) to render planar polygons, particularly transformation, clipping, hidden surface elimination, in-fill, lighting, shading and fogging. As processor speeds increased, the polygon rendering rate also increased, providing more detail for a range of graphics applications, particularly the burgeoning market for computer games.

However, these early graphics systems were limited to polygon rendering, where each polygon was a specific colour and a scene was constructed from these polygons. One consequence of these products is that it was very difficult to show objects such as a brick wall with acceptable detail. If each brick is treated as a single polygon, the rendering task can quickly exceed the real-time frame rate. Such limitations were particularly apparent in flight simulation, where a typical scene includes a runway, airfield buildings, fields, trees, aircraft and so on, and many of the early image generators produced a noticeably cartoon-like effect, despite the rapid advances in the graphics devices.

Towards the end of the 1980s, the designers of graphics chips made a very significant advance. Rather than applying a single flat colour to a polygon in the in-fill phase of the pipeline, a textured image could be mapped to the shape of the polygon. Typically, a texture is a rectangular 2D surface, whereas a 3D polygon (or more often a triangle) is distorted by the perspective operations. In other words, in 'painting' a texture on a 3D polygon, the perspective calculations must also be computed as the texture is rendered. There is also one further problem: as a texture is reduced, pixels in the texture image (known as *texels*) will no longer map directly to pixels in the 2D image, and filtering is needed to avoid the apparent break up of a detailed image.

Nevertheless, the gain of real-time texturing is considerable. A detailed surface such as a brick wall can be defined as a single texture map, possibly captured from a photograph of the wall, which is then rendered in a single operation during the polygon in-filling operation. In other words, an object containing several thousand features can be rendered as a single object rather than several thousand objects, bringing a very high degree of realism to real-time computer graphics, but without any significant reduction in processing speed.

Consider the image of the house shown in Figure 5.14a. The house can be formed as a set of polygons, where this texture of the photograph of the front wall (including the window frame and

Figure 5.14 (a) A stone wall of a building. (b) A fragment of the texture map

burglar alarm) is painted on the front face of the 3D polygons representing the house. The section of the wall marked with a dotted rectangle is shown in Figure 5.14b, where the individual texels are clearly discernible. The height and width of the 2D texture is known, and for each texel, the colour can be extracted in terms of its red, green and blue components and possibly, intensity and alpha (transparency) components. In other words, a texture map is simply a two-dimensional array where the rows and columns of the array correspond to the rows and columns of the texture image and the elements of the array are the colour components of an individual texel. Typically, the colour components are represented as 8 bits for each component, giving 2^{24} (over 16 million) possible colours.

One problem associated with this form of image rendering is the amount of storage needed for the texture map. The original image of the house was captured at a resolution of 2048×1536 pixels. If each texel is represented by 24 bits of colour (3 bytes), the required storage is 9.4 MB, for just one image. With such large images, the texture maps are likely to be static (loaded at run-time) and consequently, the resolution and number of images that can be rendered is a limiting factor, even for a modern visual system. On the other hand, the capacity of memory chips has advanced rapidly and the cost per unit of memory has fallen dramatically in 20 years. There is an additional consideration with texture memory; each textured polygon must be rendered as fast as possible and each textured pixel may be derived from the filtering of tens or hundreds of texels in the texture memory (depending on the resolution and compression of the original image). Consequently, the access speed of the texture memory must also be very fast and can impose a limit on the overall rendering rate.

The actual rendering of the texture map, which is implemented by the graphics processor, is illustrated in Figure 5.15. The perspective distortion requires a 2D mapping of texels of the original texture map to the polygon produced after the transformation, clipping and perspective operations. In order to ensure the coherence of the original texture, the pixels in the final image are derived from a set of texels in the original image, which correspond to the mapping of each transformed pixel back to the flat, rectangular 2D texture image. This filtering is executed on a per-pixel basis by the graphics processor and the quality of the filtering process depends on the number of texels accessed to determine the resultant colour of the framestore pixel.

Texture mapping is often a subjective issue in flight simulation. Poor texture mapping techniques or poor resolution texture can introduce unacceptable visual artefacts, which can become apparent during simulation: edges can become jagged, small areas of texture may appear to change shape slightly, very small areas of texture may appear to scintillate (as a result of rounding errors or poor filtering algorithms). In other words, the considerable benefits afforded by real-time texturing may be offset by irritating artefacts, which attract the attention of a pilot in a way that never occurs in flight, reducing the training effectiveness.

Figure 5.15 Perspective texture mapping

In simulation, the rendering of texture objects is performed in four steps:

1. The object and the texture of its surfaces are defined in terms of its geometry and texture – this process is performed offline, usually with a computer-aided design (CAD) tool to design and edit 3D objects.
2. The mapping of the texture to be applied during rendering is specified, so that the graphics engine will know how to apply the texture map.
3. Texture mapping is enabled – there may also be specific objects that do not require texturing.
4. The geometric coordinates of the displayed polygon and its texture map are defined to enable the graphics processor to perform the texturing operation.

Although one texture map may be applied to an individual polygon, it can also be applied to several polygons or alternatively, a texture pattern can be repeated across an object (where the edges must repeat without discontinuity, similar to wallpaper).

Fortunately, OpenGL provides an extensive range of texture operations. A texture may be produced from a photographic image or a computer-generated image (for example, using a fractal generator) and several functions are provided to convert from a wide range of texture formats (and word lengths and byte orientations) to the internal format used by OpenGL. Often, where a visual scene is represented at several levels of detail, as a texture reduces in size, the image may change in observable steps. The mipmapping capability of OpenGL enables a base texture to be stored as a correctly filtered texture (in 2D texture space). The pre-filtered mipmaps are stored as reductions of the original image in powers of two, enabling the most appropriate version of the texture to be rendered.

One of the problems of texture rendering in flight simulation applications is the wide range of viewing distance, which can occur during simulation. For example, if a 2000-m runway is represented by a 256 × 256 texture map, the texture resolution (the size of an individual texel) is 8 m × 8 m. If the aircraft is taxiing on the runway, the pilot eye point is only 4 m, say, from the runway, then the texture of the runway in the near visual field will appear very coarse. However, at distances greater than 2 nm, the runway may only occupy a few screen pixels and the resolution of a coarse texture image would be acceptable. To provide a resolution to centimetre resolution, the texture map would be over 200,000 × 200,000 (or possibly 100 repetitions of a 2048 × 2048 texture), but with no apparent benefit at distance; moreover, the time to render this texture would, in effect, be wasted. While this problem can be ameliorated by the use of mipmaps and intelligent use of level of detail in the visual database, the choice of appropriate texture resolution can have a significant influence on visual fidelity. Detailed texture is essential at near range, while the reduction to coarser texture at long range needs to be blended very carefully.

The quality of a rendered texture also depends on the filtering techniques used. After transformation, clipping and perspective operations, the pixels of the final image will not correspond to the original texels and texture filtering attempts to recover the texture. OpenGL allows the user to define the interpolation method to be used to average the texels. For 2D textures, OpenGL can select the nearest texel to the resultant pixel, or a linear average based on the 2×2 array of texels nearest to the centre of the pixel or linear averaging based on the nearest (in terms of resolution) mipmap. However, the smoothest results are also the most time consuming (on a per-pixel basis) to render and there is an inevitable trade-off between image quality and rendering rate.

The techniques summarized above provide a brief overview of the techniques used in generating textured polygons. Several vendors provide additional hardware to improve the filtering algorithms and many of the features afforded by OpenGL are not applicable to flight simulation. For example, visual artefacts are more likely to occur in rendering distant objects, but at the same time, it is likely that the visual system will provide haze (or fogging) to increase the image realism which may obscure limitations with the texture filtering. OpenGL 2D texturing methods are used in Section 5.7 to simulate mechanical aircraft instruments where the components of an instrument are defined as texture maps. In Section 8.4 texturing of 3D visual scenes is introduced, where objects such as a runway are rendered from textured images, for example, to include the effects of concrete paving or tyre skid marks.

5.6 OpenGL®

During the 1980s, as the use of the PC in the office and home increased, a number of manufacturers started to develop graphics cards to meet the increasing demand for computer games and graphics applications. Prior to this period, a few companies had manufactured specialized graphics hardware for the flight simulation industry and companies needing high-performance computer-aided design. Although a number of graphics libraries were developed to support the growing number of graphics devices, OpenGL® developed by Silicon Graphics Incorporated (SGI) for their workstation products quickly became adopted by universities and research organizations. This link was reinforced as the main graphics card vendors built in hardware support for OpenGL into their products. For the users, OpenGL provides a high degree of portability (or platform independence) and exploits the high-speed processing afforded by the graphics devices developed since the 1990s.

For the simulator designer, OpenGL is a natural choice; drivers are provided by most computer manufacturers and graphics card manufacturers and there is a high degree of compatibility between OpenGL libraries. In addition, considerable effort is invested into providing very efficient graphics architectures and low-level operations, so that there is very little incentive for users to develop alternative software. Consequently, most of the investment in software development has focused on higher level graphics applications, leaving the development of graphics cards and OpenGL drivers to the handful of equipment manufacturers.

OpenGL extends to 3D graphics, providing high-performance rendering of complex scenes. However, in simulating aircraft instruments, OpenGL affords the following practical benefits for 2D applications:

- It is supported by most graphics card vendors;
- It provides a documented straightforward interface for 2D graphics operations;
- Most graphics cards can achieve high frame rates (in excess of 50 Hz) with modest hardware;

- OpenGL supports hardware implementations of rotation, shading, in-fill, texturing, anti-aliasing and framestore management;
- Software libraries exist for most programming languages including C;
- It supports double-buffering – one framestore is displayed and cleared while the graphics is rendered using the other framestore; these operations are toggled at the simulator frame rate and are transparent to the user.

OpenGL is a very powerful graphics language and there are excellent textbooks covering programming in OpenGL and providing reference texts on OpenGL (Shreiner, 2004). This section only focuses on aspects of OpenGL that are relevant to aircraft displays. For more details of using OpenGL, the reader is advised to study these books, which cover OpenGL in much more depth than is possible in this brief section.

To illustrate the ease of using OpenGL for computer graphics, Example 5.4 is a small C program to draw a white rectangle on a black background. There are a number of points to note with this simple example, which is based on a similar example in the OpenGL Programming Guide (Shreiner et al., 2005):

- The program is relatively small;
- There are only three procedures:
  o main() – management of the program;
  o init() – initialization of the OpenGL environment;
  o display() – to draw the rectangle;
- By default, the display area is (0, 0) to (1,1);
- Library procedures are provided for low-level graphics operations.

This example is illustrated because it has the basic structure of an OpenGL program. In main, glutInitDisplayMode sets the display to a non-refresh mode (rather than double-buffering) using RGB colour components; the size and position of the window is defined and a window is created. The procedure init is called to set a background colour of black (the arguments of glClearColor) and an orthogonal projection mode is set (for 2D graphics). The procedure glutDisplayFunc passes the user-defined procedure display as its parameter, which is the procedure to be invoked by OpenGL for graphics. Finally, glutMainLoop activates the OpenGL environment and only returns when the application terminates.

Example 5.4 A simple example to draw a rectangle

```
#include <stdio.h>
#include <stdlib.h>
#include <GL/gl.h>
#include <GL/glut.h>

void display(void);
void init(void);

void display(void)
{
  glClear(GL_COLOR_BUFFER_BIT);
  glColor3f(1.0, 1.0, 1.0);
  glBegin(GL_POLYGON);
    glVertex3f(0.25, 0.25, 0.0);
    glVertex3f(0.75, 0.25, 0.0);
```

```
    glVertex3f(0.75, 0.75, 0.0);
    glVertex3f(0.25, 0.75, 0.0);
  glEnd();

  glFlush();
}

void init(void)
{
  glClearColor(0.0, 0.0, 0.0, 0.0);
  glMatrixMode(GL_PROJECTION);
  glLoadIdentity();
  glOrtho(0.0, 1.0, 0.0, 1.0, -1.0, 1.0);
}

int main(int argc, char *argv[])
{
  int t;

  glutInit(&argc, argv);
  glutInitDisplayMode(GLUT_SINGLE | GLUT_RGB);
  glutInitWindowSize(1024, 768);
  glutInitWindowPosition(0, 0);
  t = glutCreateWindow("prog 1");
  init();
  glutDisplayFunc(display);
  glutMainLoop();
  return 0;
}
```

In this example, the background is cleared, the drawing colour is set to white (all three colour components of `glColor3f` are 1.0) and the rectangle is drawn, by implication in-filled. Note the structure of the code to draw the polygon

```
glBegin(GL_POLYGON);
  glVertex3f(0.25, 0.25, 0.0);
  glVertex3f(0.75, 0.25, 0.0);
  glVertex3f(0.75, 0.75, 0.0);
  glVertex3f(0.25, 0.75, 0.0);
glEnd();
```

`glBegin` is used to initiate OpenGL graphics primitives. The code between `glBegin` and the corresponding `glEnd` is executed as a set of graphics operations, pushed into the OpenGL pipeline. In this case, `GL_POLYGON` defines a polygon of n sides where n is the number of vertices defined by calls of `glVertex3f`. Notice the naming used for `glVertex3f`; the `3f` denotes there are three floating point components. In this case, the display area is defined as a square of size 1.0, the coordinates of `glVertex3f(x,y,z)` are in the range 0–1, where the z component is redundant in 2D graphics applications.

This example typifies the low level of the graphics primitives provided by OpenGL. It is left to the user to construct higher level functions from these primitives. Ten basic geometric primitives are provided, as shown in Figure 5.16.

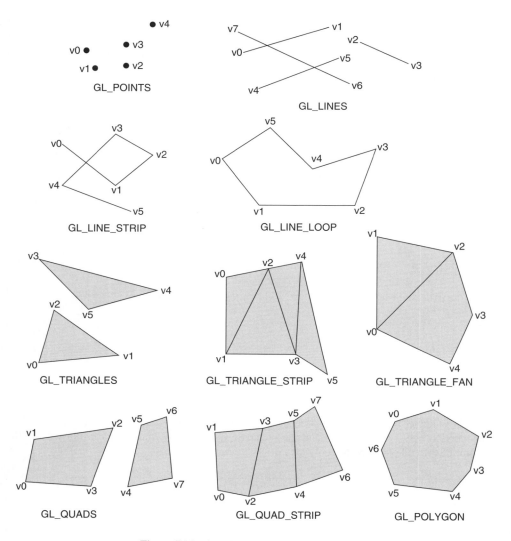

Figure 5.16 OpenGL geometric primitive types

OpenGL maintains a series of matrix transformations, which are applied to the graphics primitives as they are executed. For example, consider the following code fragment:

```
glPushMatrix();
glRotatef(dp, 0.0, 0.0, 1.0);
glTranslatef(dx, 0.0, 0.0);
glRectf(x1, y1, x2, y2);
glPopMatrix();
```

A matrix is pushed onto a run-time stack used by OpenGL and will become the transformation matrix applied to all graphics operations until it is popped. In this simple example, the code between glPushMatrix and glPopMatrix contains glRotatef, which rotates the graphics by dp degrees, in this case a rotation in the plane of the screen (for 2D operations), followed by a translation dx, which is a horizontal translation. The rectangle is defined by its bottom left-hand corner (x1,y1) and its top right-hand corner (x2,y2) and is drawn with a rotation followed by a translation. Although there are only three graphics operations shown in this example, any number of graphics primitives could be executed between glPushMatrix and glPopMmatrix, and furthermore, these matrix operations can be nested to any depth within any number of matching glPushMatrix and glPopMatrix calls. Although OpenGL appears to be very low level, in terms of its primitives, if the calls are embedded in a structured programming language, very powerful and effective graphics operations can be constructed and implemented as a direct consequence of the primitives provided by OpenGL.

OpenGL and modern graphics cards provide support for double-buffering, a feature that is essential in flight simulation. Double-buffering enables an image to be written to one framestore while a second framestore is displayed. The roles of the two framestores are reversed at the display frame rate. Once initiated, this process is autonomous – the user renders the image to the current framestore and the framestore management is handled by OpenGL. Without such a feature, it would be necessary to erase the dynamic (changing) components of an image and then redraw these components in their new positions. Care would be needed to ensure that erasing one component does not alter any of the other components. Such techniques are impractical for complex images, and the use of double-buffering for both 2D and 3D graphics is commonplace. However, for aircraft instrument displays, double-buffering requires that both the static and the dynamic parts of the image are redrawn every frame; with the rendering rates of modern graphics cards, this technique is practical. If the simulator update rate is 50 Hz, then the worst-case display time should not exceed 15 ms, say, leaving a 25% margin for future expansion. There may, of course, be a wide variation in the frame-to-frame rendering speed, so it is essential to measure the rendering times to ensure that the worst-case frame time is within a reasonable margin. Figure 5.17 shows the frame rate for a Boeing 747–400 primary flight display; in this case, the frame rate is less than 1.7 ms. It is particularly important that the frame period is not exceeded in real-time applications. In addition, the margin should allow for the time to access flight data needed in the computation of the display.

Consider a simple example of a display, with a single pointer and markings every 20°. The code given in Example 5.5 outlines the structure of a double-buffered implementation of this display, which is shown in Figure 5.18. Of course, this is not a practical display; it is intended to illustrate the method to repeatedly generate a background and a dynamic pointer. In this example the pointer motion is simulated to move backwards and forwards over a range of 300°.

The code for the procedure main is very similar to the previous non-buffered example.

Example 5.5 Initialization of a double-buffered display

```
int main(int argc, char *argv[])
{
  unsigned int t;

  glutInit(&argc, argv);
  glutInitWindowSize((GLuint) DisplayWidth, (GLuint) DisplayHeight);
  glutInitWindowPosition(0, 0);
  glutInitDisplayMode(GLUT_DOUBLE | GLUT_RGB);
  t = glutCreateWindow("test");
  glutDisplayFunc(Display);
  glutReshapeFunc(reshape);
  glutKeyboardFunc(key);
  glutIdleFunc(Update);
  glutMainLoop();
}
```

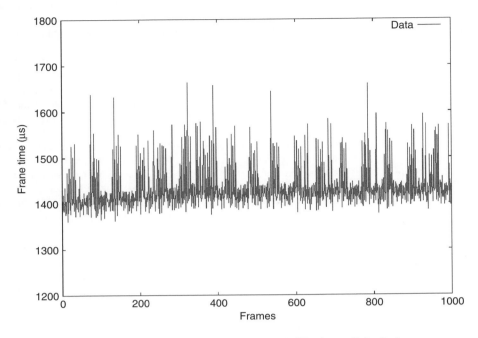

Figure 5.17 Frame rendering rate – Boeing 747-400 primary flight display

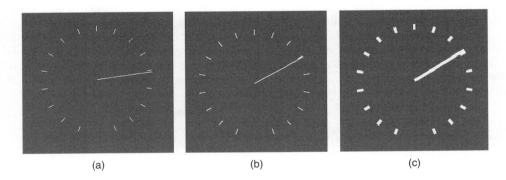

Figure 5.18 A basic instrument display

In this case, the procedure `glutInitDisplayMode` is called with an argument of `GLUT_DOUBLE| GLUT_RGB` to enable the double-buffering. A display function `Display` is still defined by `glutDisplayFunc` to implement the graphics functions in each frame. An additional procedure `glutIdleFunc` defines the procedure `Update` to be invoked by OpenGL, when no graphics is being executed. In addition, the procedure `glutKeyboardFunc` allows the application to be terminated by detecting that the escape key (ESC) has been pressed. In addition, the procedure `reshape` passed to the procedure `glRshapeFunc` enables the display to be rendered as a 2D projection using integer screen coordinates, rather than the default screen range of 0.0–1.0.

Aircraft Displays

The code for Display is given in Example 5.6.

Example 5.6 Display function for a simple instrument display

```
void Display(void)
{
  int a;

  glClear(GL_COLOR_BUFFER_BIT);

  for (a=-160; a<=160; a=a+20) {
    glPushMatrix();
    glTranslatef(200.0, 200.0, 0.0);
    glRotatef((float) a, 0.0, 0.0, 1.0);
    Draw(0.0, 90.0, 0.0, 100.0);
    glPopMatrix();
  }

  glPushMatrix();
  glTranslatef(200.0, 200.0, 0.0);
  glRotatef((float) ptr, 0.0, 0.0, 1.0);
  Draw(0.0, 0.0, 0.0, 100.0);
  glPopMatrix();

  glFlush();
  glutSwapBuffers();
}
```

In this example, the instrument is centred on the screen at (200, 200) and the radius of the instrument pointer is 100 pixels. The instrument markers are drawn as an arc from $-160°$ to $+160°$. In the for-loop, a translation by 200 pixels in the x and y directions is followed by a rotation of $a°$ for each mark. The vertical line of length 10 pixels is therefore translated and rotated by this amount. Similarly, the pointer is drawn by a similar translation followed by a rotation of $ptr°$, where ptr is the angular position of the pointer. In this simple example, the pointer is displayed as a single line.

Note that the marks and the pointer in Figure 5.15a are jagged. In Figure 5.15b, the following lines of code were added before glClear to provide smooth anti-aliased lines.

```
glEnable(GL_LINE_SMOOTH);
glEnable(GL_BLEND);
glBlendFunc(GL_SRC_ALPHA, GL_ONE_MINUS_SRC_ALPHA);
glHint(GL_LINE_SMOOTH_HINT, GL_NICEST);
```

Different forms of anti-aliasing are implemented by OpenGL (and supported by the graphic card manufacturers), providing a trade-off between line quality and rendering speed. Figure 5.15c is the same as Figure 5.15b except that a line width of four pixels is specified by the following code.

```
glLineWidth(4.0);
```

The improvement in image quality is clearly seen. The code in Update updates the instrument parameters (the angular position of the pointer in this simple example) and informs OpenGL that the display is ready to be redisplayed.

```
void Update(void)
{
  ptr = ptr + dptr;
  if ((ptr > 160.0) || (ptr < -160.0)) {
    dptr = -dptr;
  }
  glutPostRedisplay();
}
```

In this example, the only computation of the instrument parameters is to rotate the pointer by 1°, reversing the motion when it reaches ±160°. Note that OpenGL uses degrees, rather than radians. The actual line drawing is implemented using a user-defined procedure Draw, which invokes the OpenGL procedure glVertex2f, as shown below.

```
void Draw(float x1, float y1, float x2, float y2)
{
  glBegin(GL_LINES);
    glVertex2f(x1, y1);
    glVertex2f(x2, y2);
  glEnd();
}
```

There is one final point to note about this example; it is free running. As soon as the pointer is updated the graphics is invoked and, when the graphics is completed, the idle function is invoked, which updates the pointer position. Such a trivial example probably runs at over 200 Hz on a standard PC. In a real-time simulator, the display function will only be invoked once per frame and therefore, the double-buffering is slaved to the simulator frame rate.

The overall structure of this example is used in actual real-time simulator displays. Once the OpenGL environment is initialized, the displays are drawn using OpenGL primitives, as shown in the procedure Display. Similarly, the display variables are updated once per frame, as shown in the procedure Update. Note there is no OpenGL in this procedure other than an OpenGL call glutPostRedisplay(), which informs OpenGL that the display can be redrawn.

In practice, only a small set of graphics primitives is needed to generate aircraft displays and rather than use the low-level OpenGL primitives, it is sensible to provide a higher level graphics interface for user applications, so that the low-level OpenGL functions are encapsulated in a graphics library. For example, the procedure Draw simplifies the coding by providing one call rather than four for every line drawing operation. For aircraft displays, the graphics library will contain primitives for line drawing, character generation, font selection, setting colours and other general functions (for example, turning anti-aliasing on or off, or setting line drawing modes or line widths).

Although 2D graphics operations have been described, some aircraft displays are actually three dimensional; they may contain an attitude 'ball', or a slip ball or have a raised bezel. In these cases, the designer may opt to use 3D graphics with shadows and shading to provide added realism. Of course, this turns out to be a straightforward extension in OpenGL, which is strictly a 3D graphics package that can also be applied to 2D display applications.

5.7 Simulation of Aircraft Instruments

The instrument panel of a light aircraft or a transport aircraft built before 1990 will have an array of dials to display flight information. Typically these are round mechanically driven instruments with a bezel, markings on a dial and one or more pointers. The instruments are positioned 30–40 cm in front of the pilot, on a flat panel above the legs of the pilot (to provide access to the rudder pedals and brakes) and below the line of the coaming. Although there is very little standardization of instruments, the arrangement, size and format of these displays is consistent across manufactures and aircraft types. Typically the instruments are 3–4 in. in diameter, with a black face and white markings and pointers, for legibility. In practice, the primary instruments are replicated for the pilot and co-pilot, but the instrument panel will contain some 30 instruments or gauges to display flight information, the state of the engines, subsystems and radios.

Since the introduction of the Link trainer in the 1930s, instruments and gauges used in flight trainers have been fabricated to replicate the equivalent aircraft parts, with the dials and pointers driven by pneumatics or electric motors. Since the 1990s, with the availability of fast computer graphics and low-cost high-resolution computer displays, real-time computer graphics has provided an alternative method to emulate aircraft instruments, subject to the following considerations:

- The complete display must update at a sufficient rate so that there is no apparent latency or jerkiness in the display;
- The instruments should be the same size as the counterpart aircraft instruments;
- Any lettering or marks must be legible;
- The functionality of the instrument must be identical to the aircraft instrument including the rotation of dials and pointers.

In a modern PC with a graphics accelerator and OpenGL, display rates of complex images at 50 Hz are easily achievable, supporting display resolutions up 1280×1024 pixels. The hardware graphics operations supporting OpenGL include 2D rendering of lines, polygons and text and rotation and texturing of displayed entities. Two 17-in. flat-screen displays, orientated side-by-side in portrait mode, occupy an area approximately 65 cm $\times$ 38 cm. Alternatively, a single 28 in. monitor, orientated in landscape mode, occupies an area approximately 58 cm $\times$ 43 cm. In other words, there is sufficient space with off-the-shelf displays to replicate the instrumentation found in most aircraft.

While the use of computer graphics offers a major reduction in the complexity and cost of simulated aircraft instruments and a high degree of flexibility, this solution poses two problems. First, the instrument panel must be rectangular, which is not the case for all aircraft. Secondly, some instruments contain knobs, which are used for selection, for example, to set the barometric pressure on an altimeter. While it might be possible to bond a mechanical knob to a screen, a simpler alternative is to provide a touch screen, where the knob (or selector) is presented in a graphical form and is animated to respond to pilot inputs, for example, to turn a knob clockwise or anticlockwise according to the pilot interaction. Similarly, switches can be displayed in one or two positions; by touching the switch, the switch is redrawn, changing from on to off, say, or from the up position to the down position. For the majority of training applications, it is difficult to envisage a situation where a graphical form of an aircraft instrument might adversely affect the training transfer. However, considerable care is needed to ensure that there is no difference in the behaviour of the simulated instrument and the actual instrument, for example, in simulating the lag of a vertical speed indicator or the effect of aircraft acceleration on a magnetic compass.

Figure 5.19 shows a typical simulated display for a light aircraft, in this case, an aerobatic aircraft with a G-meter and an angle-of-attack meter. In addition to the primary flight instruments and a VOR instrument, the display includes engine gauges for RPM, manifold pressure and exhaust gas temperature and gauges for suction, charging, fuel quantity, fuel pressure and temperature and oil pressure and temperature.

Figure 5.19 A real-time instrument display (see Plate 4)

In this case, the display is partitioned across two 1024 × 768 pixel displays in portrait mode. Notice the rotated lettering on the horizontal situation indicator (HSI) card, the shaded bezel around each dial, the clear edges of the pointers (from anti-aliased rendering) and the reflections on the surface of the attitude indicator ball and the slip indicator glass tube and ball, which provide a three-dimensional effect.

For the simulator designer, each instrument comprises a set of vectors, characters and polygons and of course, OpenGL is intended for just such applications to render and rotate graphics objects of this form. Although it would be possible to draw each pointer as a polygon which is rotated or to draw the marks around the edge of a dial every frame, such an approach requires different fonts for the instruments and re-computation of largely static images every frame. An alternative approach, and the method adopted in this book, is to exploit the OpenGL texture operations outlined in Section 5.5. An offline drawing package is used to capture the graphics of each component of the display. During each frame, an instrument is rendered by drawing its textured components from the back layer to the front layer. Consider the compass card shown in Figure 5.19. By representing the compass card as a single textured object, the complete card is rotated and rendered by a few OpenGL operations. Whereas, a vector-generated approach would require the rotation and drawing of 48 lines and 18 digits.

A drawing package such as Adobe Photoshop™ enables each entity to be constructed in considerable detail; textures can be selected from a library of texture maps, shading can be applied to give a 2D object a 3D appearance and simple artwork tools are provided to enable the components of each instrument to be captured and edited. These drawing packages include a range of font types and font sizes, so that, in the case of aircraft instruments, a very realistic 2D image can be captured with a few hours of artwork.

Figure 5.20 shows the background of an airspeed indicator. Notice the texture of the bezel material (a burnished metal effect), the shading applied to give relief to the dial surround, the mounting screws, the shading around the instrument bezel to provide a sense of depth, the clarity

Aircraft Displays

Figure 5.20 Indicated airspeed instrument graphic background

Figure 5.21 Airspeed indicator needle texture

of the lettering and the colour bands. Using the drawing package, it is straightforward to replicate marks at regular intervals (e.g. every 5 or 10 kt) and to place the lettering accurately. Moreover, any further changes to the instrument, for example, to add a flap limit band or to change the maximum indicated speed would only require a small amount of editing. The construction of an instrument from common components can be performed by the drawing package. For example, many instruments share a common bezel and a bezel can be selected from a common library of shapes and textures. Similarly, the needle of the airspeed indicator can be drawn as a texture. The graphical rendering of this instrument reduces to first, drawing the background texture at its location on the panel and secondly, rotating the texture map of the needle and drawing it in front of the background texture. The texture map of the needle is shown in Figure 5.21, against a black background.

Packages such as Photoshop™ enable the graphic object to be written in many industry standard file formats. For the flight simulator described in this book, SGI's Image File format has been used for the following reasons:

- It is recognized by OpenGL;
- There are public domain tools to convert to and from other formats;
- The format includes run length encoding to compress the size of texture files;
- The colour components and the alpha (transparency) component of each pixel can be represented in 8 bits.

The specific detail of the conversion is given in *textures.c*, where a *.sgi* file is converted to the internal format recognized by OpenGL at run-time. In the case of the airspeed indicator, the dial and pointer texture files are converted by the function Glib_GenerateTexture in the graphics library *glib*, as follows:

```
Glib_GenerateTexture("Textures/needle1.sgi", TEX_NEEDLE_ONE, GL_CLAMP);
Glib_GenerateTexture("Textures/basic_dial_speed.sgi", TEX_SPEED, GL_CLAMP);
```

The first argument is the name of the *.sgi* file produced by Photoshop™. The second argument is the unique texture value (defined in *textureid.h*). For this application, the third argument is always GL_CLAMP, to prevent any wrapping of the texture. If the texture is successfully created, its address is entered in an array of textures *texture_objs*, managed by the graphics library *glib*. The simplicity of the function is shown in Example 5.7.

Example 5.7 Initialization of the textures of an instrument display

```
GLuint Glib_GenerateTexture(char *FileName, int textureID, GLint wrap)
{
    unsigned *sprimage;
    int iwidth, iheight, idepth;

    /* read SGI format rgb image */
    sprimage = read_texture(FileName, &iwidth, &iheight, &idepth);
    if (!sprimage)
    {
        printf("error reading texture file %s\n", FileName);
        return 0;
    }

    glGenTextures(1, &textureArray[textureID]);
    glBindTexture(GL_TEXTURE_2D, textureArray[textureID]);
    glPixelStorei(GL_UNPACK_ALIGNMENT, 1);
    glTexParameteri(GL_TEXTURE_2D, GL_TEXTURE_WRAP_S, wrap);
    glTexParameteri(GL_TEXTURE_2D, GL_TEXTURE_WRAP_T, wrap);
    glTexParameteri(GL_TEXTURE_2D, GL_TEXTURE_MAG_FILTER, GL_LINEAR);
    glTexParameteri(GL_TEXTURE_2D, GL_TEXTURE_MIN_FILTER, GL_LINEAR);
    glTexEnvi(GL_TEXTURE_ENV, GL_TEXTURE_ENV_MODE, GL_MODULATE);
    glTexImage2D(GL_TEXTURE_2D, 0, 4, iwidth, iheight, 0, GL_RGBA,
                 GL_UNSIGNED_BYTE, (GLubyte *) sprimage);

    free(sprimage);
    return textureArray[textureID];
}
```

The function read_texture is taken from SGI's library of public domain software to support OpenGL. It creates a temporary area of memory for the texture (pointed to by sprimage) or a null pointer if the file cannot be found or if a problem is encountered with the file format. glGentexures generates a single name to reference the texture, which is stored in textureArray[textureID]. glBindTexture binds this texture to a 2D texture and glPixelStorei defines a byte alignment to be used in extracting the texture. The calls to glTexParameteri define the wrapping mode (always clamped) and the filter to be 2D linear, as there is no distortion of the original texture for this

Aircraft Displays

application. `glTexEnvi` defines how the colour is modulated, based on the four *rgba* components. `glTexImage2D` specifies the 2D format, particularly the width and height of the texture stored in `sprimage`. In the case of the image used for the airspeed indicator, the texture is 256 × 256. Finally, the area of memory allocated by `read_texture` must be released by the application.

The parameters passed to the airspeed display procedure `Asi_Asi` are the indicated airspeed and the indices of the two textures (the dial and the pointer) in the list of stored textures. The indicated airspeed is used to compute the angle of rotation of the pointer. The original texture is assumed to be vertical, so that the rotation can be derived from the angle of the minimum airspeed on the dial and the angular spacing of the airspeed 'ticks'. The display is then rendered with the following code:

```
PFD_SetOrigin(PFD_AsiX, PFD_AsiY);
Glib_DrawTexQuad(0.0, 0.0, 320.0, 320.0, 0.0, Glib_ANCHOR_CENTRE, TEX_SPEED);
Glib_DrawTexQuad(0.0, 0.0, 235.0, 235.0, needle_rot,
                 Glib_ANCHOR_CENTRE, TEX_NEEDLE_ONE);
glDisable(GL_TEXTURE_2D);
```

Both the instrument texture and the pointer textures are 256 × 256 pixels. The first four parameters of `Glib_DrawTexQuad` define the scaling to be applied to the textures, in mapping to screen coordinates. The fifth parameter is the angle of rotation of the texture (in degrees). The sixth parameter defines an anchor point, either the centre of an object for an instrument or the bottom left-hand corner for a rectangular object. The final parameter is the texture to be displayed, as defined in *textureid.h*. The procedure `Glib_DrawTexQuad` is used to render textures and is given in Example 5.8.

The initial code in the procedure enables texturing, binds to the specific texture object, defines and enables the blending function (needed for anti-aliasing), translates to the anchor point and defines the colour and alpha values. A rotation is applied for a centre anchor and finally the object is rendered by defining the texture coordinates and its respective screen coordinates, in the order of the bottom left, bottom right, top right and top left corners.

The code is given to illustrate the ease of generating and rendering textures in OpenGL. The artwork of a complex display can be produced offline and saved in a standard texture format. The texture files are then initialized at the start of the simulation. During each frame, these textures are translated, rotated and rendered to form the instruments and gauges. A similar technique can also be used to display lamps, switches, knobs, radio panels and warning indicators. Figure 5.22 shows five texture files used for toggle switches, push switches, rotary knobs and selectors.

For instruments containing several elements, where one entity is in front of another, the textures are rendered as layers; a texture is drawn in front of a previously drawn texture, for example, to

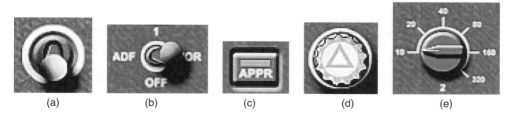

Figure 5.22 Texture files for switches and selectors

Example 5.8 Instrument texture rendering

```
void Glib_DrawTexQuad(float X, float Y, float XS, float YS, float rot,
                     int anchor, GLuint texobj)
{
    glPushMatrix();
    glEnable(GL_TEXTURE_2D);
    glBindTexture(GL_TEXTURE_2D, texobj);
    glBlendFunc(GL_SRC_ALPHA, GL_ONE_MINUS_SRC_ALPHA);
    glEnable(GL_BLEND);
    glTranslatef(X,Y,0.0f);
    glColor4ub(255,255,255,255);
    if(anchor == Glib_ANCHOR_CENTRE) {
        glRotatef(rot, 0.0f, 0.0f, 1.0f);
        glBegin(GL_QUADS);
        glTexCoord2f(0.0, 0.0);  // bottom left
        glVertex3f(-XS/2,-YS/2,0.0f);
        glTexCoord2f(1.0, 0.0);  // bottom right
        glVertex3f(XS/2,-YS/2,0.0f);
        glTexCoord2f(1.0, 1.0);  // top right
        glVertex3f(XS/2,YS/2,0.0f);
        glTexCoord2f(0.0, 1.0);  // top left
        glVertex3f(-XS/2,YS/2,0.0f);
        glEnd();
    } else {  // Glib_ANCHOR_LOWLEFT
        glBegin(GL_QUADS);
        glTexCoord2f(0.0, 0.0);  // bottom left
        glVertex3f(0.0,0.0,0.0f);
        glTexCoord2f(1.0, 0.0);  // bottom right
        glVertex3f(XS,0.0,0.0f);
        glTexCoord2f(1.0, 1.0);  // top right
        glVertex3f(XS,YS,0.0f);
        glTexCoord2f(0.0, 1.0);  // top left
        glVertex3f(0.0,YS,0.0f);
        glEnd();
    }
    glPopMatrix();
}
```

draw two pointers or if a warning flag appears behind a pointer. In the case of an altimeter, where the altitude is also displayed as a digital value, the digits can be drawn over the texture, to appear in a box within the display. The attitude indicator poses additional problems. In an aircraft, the attitude indicator is a gyroscope mounted inside two gimballed frames. The gyros are stabilized in the earth frame (although they slowly precess) to provide pitch and roll indications. If a sphere is constructed as a set of surfaces and then a texture map is applied to the surfaces of the sphere, which is rendered in 3D, a realistic attitude indicator can be rendered using standard OpenGL functions, as shown in Figure 5.19. Finally, additional static objects, such as the attitude reference symbol, are rendered over the ball.

A sphere can be generated as a set of quadrilaterals (Bourke, 1996), where the vertices of the quad strips lie on the surface of the sphere. There must be sufficient strips so that the sphere appears rounded, but if too many are generated, the rendering time could exceed the few milliseconds per frame, allocated to this graphics task. The radius of the sphere is such that it appears inside the circle of the instrument bezel, shown as a texture in Figure 5.23a. In this case, the bezel

Aircraft Displays

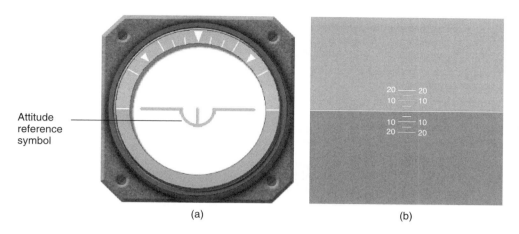

Figure 5.23 Attitude indicator textures

is drawn after the attitude is rendered so that the attitude reference symbol appears in front of the ball.

During each frame, the texture is rotated in three dimensions in pitch and roll to enable the coordinates of the texture quad strips to be computed from the quad strips of the sphere. Example 5.9 shows how the sphere is rendered as a textured 3D object. The sphere is translated and rotated, first to align with the local frame and then to accommodate the aircraft pitch and roll. The texture shown in Figure 5.23b is selected, using glBindTexture. Each quad strip is drawn by generating the surface normal for the illumination (glNormal3f) and specifying the 2D texture coordinates (glTexcoord2f) and the 3D quad strip coordinates (glVertex3f).

Example 5.9 Rendering the attitude indicator sphere

```
void Ai_DrawSphere(XYZ c, float roll, float pitch, int n, GLuint texobj)
{
    int     i,j;
    int     qstrip_pts = n*2 + 2;

    glPushMatrix();
    glTranslatef(c.x, c.y, c.z);
    glRotatef(90.0f, 0.0f, 0.0f, 1.0f);  // rotation for PFD coord system
    glRotatef(-90.0f, 0.0f, 1.0f, 0.0f); // centre texture (y axis - correct)
    glRotatef(roll , 1.0f, 0.0f, 0.0f);  // apply roll (x axis orig rotation)
    glRotatef(-pitch , 0.0f, 0.0f, 1.0f); // apply pitch (z axis - orig rotation)
    glBindTexture(GL_TEXTURE_2D, texobj);

    for ( j=0; j<n/2; j++ ) {
        glBegin(GL_QUAD_STRIP);
        for ( i=0; i<=qstrip_pts; i++) {

            glNormal3f(sphere_nor_coords[j*qstrip_pts+i].x,
                       sphere_nor_coords[j*qstrip_pts+i].y,
                       sphere_nor_coords[j*qstrip_pts+i].z);
            glTexCoord2f(sphere_tex_coords[j*qstrip_pts+i].u,
                         sphere_tex_coords[j*qstrip_pts+i].v);
```

```
            glVertex3f(sphere_pos_coords[j*qstrip_pts+i].x,
                       sphere_pos_coords[j*qstrip_pts+i].y,
                       sphere_pos_coords[j*qstrip_pts+i].z);


        }
        glEnd();
    }
    glPopMatrix();
}
```

The lighting to illuminate the sphere, in order to emphasize the three-dimensional shape of the attitude indicator ball, is initialized at the start of the simulation. The effect of the illumination is shown in Figure 5.24. Note that the apparent shadow on the left-hand side of the ball is not computed, it is part of the bezel texture in Figure 5.23a, which is overlaid on the textured sphere.

The following constants were used to generate the attitude indicator shown in Figure 5.24.

```
GLfloat m_specu[] = {1.0,1.0,1.0,1.0};
GLfloat m_diffu[] = {0.5,0.5,0.5,1.0};
GLfloat m_null[]  = {0.0,0.0,0.0,1.0};
GLfloat m_shine[] = {1.5};
GLfloat l_pos[]   = {10.0,800.0,-100000.0,1.0};
GLfloat l_wht[]   = {1.0,1.0,1.0,1.0};
GLfloat l_amb[]   = {0.2,0.2,0.2,1.0};
```

The initialization of the lighting model is shown below.

```
glShadeModel(GL_SMOOTH);
glMaterialfv(GL_FRONT, GL_AMBIENT, l_amb);
glMaterialfv(GL_FRONT, GL_DIFFUSE, m_diffu);
glMaterialfv(GL_FRONT, GL_SPECULAR, m_specu);
glMaterialfv(GL_FRONT, GL_SHININESS, m_shine);
glLightfv(GL_LIGHT0, GL_POSITION, l_pos);
glLightfv(GL_LIGHT0, GL_DIFFUSE, l_wht);
glLightfv(GL_LIGHT0, GL_SPECULAR, l_wht);
glLightModelfv(GL_LIGHT_MODEL_AMBIENT, l_amb);
glLightModeli(GL_LIGHT_MODEL_LOCAL_VIEWER, GL_FALSE);
glEnable(GL_LIGHT0);
```

The grey background of the instrument panel is initialized at the same time by `glClearColor`, saving the time needed to render the background every frame, as follows:

```
glClearColor(0.25,0.25,0.25,1.0);
```

The advantage of the use of texture to improve the appearance of the display and also to reduce the time to render the display cannot be over emphasized. Although the design of the artwork for the display is time consuming, it is straightforward to generate and for most instruments and gauges the rendering of the instrument display reduces to the rendering of a static texture and a rotated texture for a pointer or compass card. Using a basic OpenGL-compatible graphics card and a standard PC, the displays shown in Figure 5.19 were rendered in less than 2 ms per frame.

Figure 5.24 Attitude indicator.

5.8 Simulation of EFIS Displays

Since 1990, most civil aircraft have been equipped with either CRT or LCD flat panel displays (Dawson, 1992; Pallett and Coombs, 1992). Although one option in simulator design is to use original aircraft equipment, the cost can be prohibitive, particularly if the displays and associated computer systems are certified for flight. An alternative approach is to develop real-time graphics software to emulate these displays. This display must meet the requirements of display resolution and real-time rendering and must also match the functionality of the aircraft display. Many of the cards developed for aircraft are based on OpenGL and since about 1995, many of the graphics cards developed for CAD applications and computer games have also supported OpenGL, with hardware implementations of OpenGL primitives. Thus, for the simulator designer, the emulation of an EFIS display is a practical option in many simulator programs. Figure 5.25 shows an emulation of an EFIS primary flight display for a Boeing 747-400 (Koniche, 1988).

Most aircraft EFIS displays are 8 in. × 8 in., partly as a result of mounting CRTs in the panel area in front of the flight crew; partly because this size enables the aircraft manufacturer to segregate the primary flight, the navigation display, the engine display and the FMS display into separate display systems; and partly because this is the size that the manufacturers of avionics equipment have adopted for the aviation market for flat panel displays. Consequently, the progression from the classic round instruments to flat-screen computer graphics, enabled aircraft designers to redesign aircraft displays, exploiting the use of colour and graphics functions to produce more compact displays (given the limited area of an 8 in. square display) and to present information in a clear and unambiguous way. Although there has been no standardization of EFIS display formats, there is a high degree of consistency across the displays produced by the avionics companies.

Unlike the simulation of analogue instruments, where the use of textures can provide a realistic image, EFIS displays are mostly generated using 2D vector graphics and standard character generation methods. There is a requirement for the rotation of objects, which necessitates anti-aliasing for both vector generation and character generation. In addition, the size of the display should be close to the aircraft displays. Typically, if the display resolution is in the ratio 4 : 3 and the display dimensions are in the same ratio, the number of pixels per unit length will be the same in both directions. If this is not the case, then a circle will appear oval or a square will not have sides of equal length and this distortion must be corrected in the graphics.

Although most aircraft EFIS displays have various operational modes and warning states, the basic display modes will be described, covering the primary flight display (PFD), the navigation flight display (NFD) and the Engine Indicating and Crew Alerting System (EICAS) display.

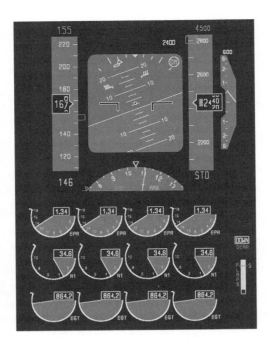

Figure 5.25 EFIS primary flight display (see Plate 5)

The PFD contains an attitude indicator, an altimeter, an airspeed indicator, a vertical speed indicator and a compass.

For these displays, lines are rotated and translated and drawn relative to the centre of the components of each display. This convention extends to characters. A graphics library, Glib was developed to provide a number of commonly used graphics primitives to simplify the rendering code, translating these primitives to their equivalent OpenGL functions. One attraction of this method is that it avoids dependency on screen orientation as the transformation can be implemented within Glib. For the EFIS displays, the screen is rotated to portrait mode. However, from the programming perspective, the bottom left-hand corner of the display is (0, 0) and the coordinates of objects are given in this axis system. The procedure PFD_SetOrigin is used to relocate the origin for each component in the display, so that drawing is relative to the centre of that component, in order to provide a transparent mapping between user coordinates and the display coordinates.

```
void PFD_SetOrigin(int x, int y)
{
  glTranslatef((float) (-y + yOrigin), (float) (x - xOrigin), 0.0);
  xOrigin = x;
  yOrigin = y;
}
```

For example, if the origin is set to the centre of the attitude indicator, the following code

```
Draw(-5, -20, 30, 40)
```

generates a line from ($Ai_x - 5$, $Ai_y - 20$) to ($Ai_x + 30$, $Ai_y + 40$), where the centre of the attitude indicator is at (Ai_x, Ai_y).

Aircraft Displays

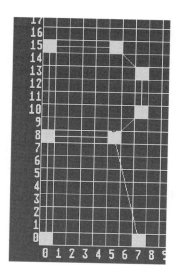

Figure 5.26 An example of the font editor

The EFIS display uses vector-based character fonts of size 8, 12, 16 and 24 (character height in pixels). These character sets were produced by a purpose-built font editor. As these fonts are vector-based, there is no distortion resulting from rotation as anti-aliasing is applied to all vector operations. Figure 5.26 shows an example where the character 'R' is captured with the font editor.

The designer defines the location of the intermediate points (shown as solid grey squares) and also the order of drawing (given by the grey connecting lines). The standard fonts provided by OpenGL Glut are not appropriate for flight simulator applications for two reasons. First, they do not resemble the fonts used in aircraft displays and secondly, they are not amenable to rotation. The font editor generates a table of offsets to access each character and a table storing the individual vectors for each character, which is accessed via the offset. Using Glib, the following procedure call

```
Glib_Chars("abcde", -20, 30);
```

renders the string 'abcde' at the position (−20, −30) relative to the centre of the current component, in the currently select font. Of course, any prevailing translations or rotations will also apply to the string of characters.

5.8.1 Attitude Indicator

The attitude indicator of the PFD is very similar to the attitude indicator provided in most aircraft in the sense that it displays the angles of pitch and roll and contains an artificial horizon separating the land and sky regions of the display, as shown in Figure 5.27.

The display shows the aircraft pitched up 6° and banked 10°. The white horizon line and the white pitch lines are aligned to the aircraft attitude, so that the 'artificial horizon' is always parallel to the actual horizon. Typically, the region below the 0° pitch line depicts ground (shown here in dark grey) and the area above this line depicts sky (shown here in light grey). The display has static markings for bank angles (in this case showing 0°, 10°, 20°, 30° and 45°) and a rotating bank angle pointer to indicate the actual angle of bank. The small rectangle below the bank angle pointer indicates sideslip and the two static bars (shown in black with a white outline) provide a

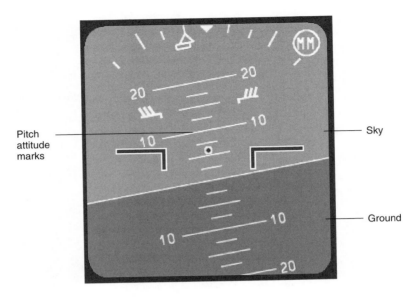

Figure 5.27 PFD attitude indicator

0° pitch 0° roll datum. In this example, the middle marker (MM) symbol is also displayed and the maximum pitch attitude marks (to protect against stalling) are shown here in light grey. Note that the pitch lines and characters are clipped to the edge of the display and the corners of the display are slightly rounded. In this example, the flight director and ILS indicators are omitted.

As the display is completely redrawn in every frame (OpenGL double-buffering), the display is formed as layers. For example, the white pitch ladder lines are in front of the blue and brown regions, but behind the black pitch reference. The first step is to draw the blue and brown regions. This is straightforward; they are rendered as in-filled rectangles where the dimension of the rectangles are chosen to exceed the boundary of the attitude indicator display, regardless of the angles of pitch and roll. Each rectangle is rotated in pitch and roll, and is clipped to the edge of the attitude indicator display, as illustrated in Example 5.10.

Example 5.10 Attitude indicator in-fill operations

```
glPushMatrix();
glRotatef(Roll_Deg, 0.0, 0.0, 1.0);
glTranslatef(-Pitch_Pix, 0.0, 0.0);
Glib_Colour(Glib_BROWN);
glRectf(0.0, (float) (-5 * Ai_aiWidth),
        (float) (5 * Ai_aiWidth), (float)(5 * Ai_aiWidth));
Glib_Colour(Glib_BLUE);
glRectf((float) (-5 * Ai_aiWidth), (float)(-5 * Ai_aiWidth),
        0.0, (float)(5 * Ai_aiWidth));
.
.
.
glPopMatrix();
```

where the rotations and translations following `glPushMatrix` apply until the corresponding `glPopMatrix` is encountered. `glRotatef` applies the rotation in roll to the objects in the display. The

Aircraft Displays

pitch attitude is converted to a translation in pixels (in this case 6.4 pixels per degree) and then the rotation and translation transformation is applied to a brown rectangle and a blue rectangle, where the width of each rectangle is five times the display width, to ensure that each rectangle fills the display area, prior to clipping.

A similar rotation and translation applies to the pitch lines and also to the pitch indication digits for the $10°$ and $20°$ pitch lines. The x and y screen coordinates (relative to the centre of the attitude indicator) of each pitch line are defined in a table for the 17 pitch lines; a similar table of coordinates is used for the pitch line digits. Other components (for example, the pitch attitude limit marker) are added in a similar way. Finally, the rounded corners are produced by pre-computing a set of lines (drawn in the background colour), which erase the blue and brown square corners.

The power of OpenGL to render real-time displays in this way is remarkable. The major part of the attitude indicator display is approximately 50 lines of code. The lines and characters are smooth and the rendering time is a few hundred microseconds, which includes complex in-filling.

5.8.2 Altimeter

The altimeter displays altitude in feet. Rather than use a circular dial with a set of pointers, EFIS displays present airspeed and altitude using vertical tape displays. In fact, the display comprises two parts: a sliding vertical scale and a digital readout of the altitude. The digital value of the altitude mimics the rolling digits of a mechanical display. This is not intended for cosmetic effect – it also provides the pilot with a useful secondary cue for the rate of change of altitude; as the climb rate increases, the digits roll faster.

A typical EFIS altitude display is shown in Figure 5.28, indicating an altitude of 5280 ft. Note that the digital readout occludes the altitude tape. In other words, the tape is drawn first and the readout box is then overlaid on the tape. The altitude tape is marked every 100 ft, with digital values

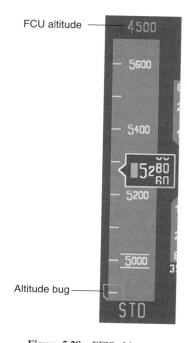

Figure 5.28 EFIS altimeter

displayed every 200 ft and additional underscore and over line marks every 1000 ft. The aircraft is at 5275 ft with tape marks from 4800 to 5600. The aircraft altitude corresponds to the datum of the altitude box. It is straightforward to compute the minimum and maximum altitude values shown and draw these marks at the relative vertical displacement from the datum, appended with digital values every 200 ft, as shown. Similarly, the altitude digits, rounded to the nearest 20 ft, are displayed in the altitude box. Leading digits for 10,000 ft and 1000 ft are shown as grey rectangles rather than leading zeros, to avoid ambiguity in the readout. The trailing digits 00, 20, 40, 60 and 80 rotate and are clipped to a small window clearly visible in the altitude box. The 100 s digit rolls above 90 ft and below 10 ft and similarly, the 1000 s digit rolls above 900 ft and below 100 ft.

Implementation of this display is straightforward. It simply requires drawing a grey rectangle, adding the altitude marks and text, drawing the black polygon (and white edge) for the altitude box and then computing the individual fields of the digital value of the altitude at the appropriate position and vertical offset in the altitude box. Finally, the barometric pressure setting (in this example, STD) is shown below the tape, the FCU altitude is shown above the tape and a marker is added at the appropriate altitude (in this case, it is 'parked' at the bottom of the tape).

5.8.3 Airspeed Indicator

The airspeed indicator is very similar to the altimeter, in terms of computer graphics. It provides a sliding tape with a digital value of airspeed in the airspeed box. A typical display is shown in Figure 5.29, indicating an airspeed of 167 kt.

The tape is marked every 10 kt, with digital values displayed every 20 kt. The airspeed is indicated as a three-digit value, with rolling digits, similar to the altimeter display. The Mach number is displayed below the tape and the FCU airspeed is shown above the tape. In this example, the 'bug' is positioned at 155 kt on the tape.

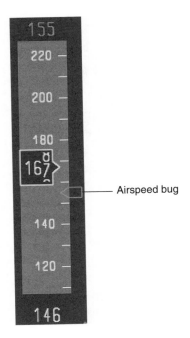

Figure 5.29 EFIS airspeed indicator

Aircraft Displays

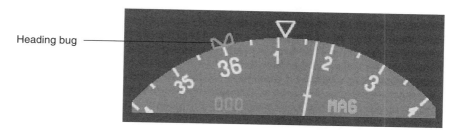

Figure 5.30 EFIS compass

5.8.4 Compass Card

The compass at the bottom of the EFIS display in Figure 5.30 shows the aircraft heading (011°). In addition, the display shows the commanded heading (a heading bug at 360°) and the aircraft track pointer (showing a track of 017°). The display can provide the aircraft heading in degrees magnetic or degrees true and this information is indicated by the MAG caption. Notice that the digits are rotated and clipped to the grey background of the display.

Only a fragment of the display (in effect, the top of the compass card) is displayed, approximately ±35° from the current heading. The clipping window is set to the left and right edges and the lower edge and top of the displayed segment. The lines and digits are then drawn by rotating them about the centre of the compass card (in this example, 411 pixels below the centre of the attitude indicator). The fragment of code in Example 5.11 illustrates the simplicity of the rendering of the compass.

Example 5.11 Rendering the EFIS compass display

```
do {
  glPushMatrix();
  t = offset * ScaleFactor;
  glRotatef(Maths_Degrees(-t), 0.0, 0.0, 1.0);
  if (h1 % 10 == 0) {
    Glib_Draw(0, radius - 10, 0, radius);
    if (h1 % 30 == 0) {
      Glib_SetFont(Glib_gFont16, 12);
      if (h1 > 90) {
        Glib_Char(h1 / 100 + '0', -10, radius - 10 - 20);
        Glib_Char(h1 / 10 % 10 + '0', 2, radius - 10 - 20);
      } else {
          Glib_Char(h1 / 10 % 10 + '0', -5, radius - 10 - 20);
      }
    } else {
      Glib_SetFont(Glib_gFont12, 8);
      if (h1 > 90) {
        Glib_Char(h1 / 100 + '0', -8, radius - 10 - 16);
        Glib_Char(h1 / 10 % 10 + '0', 2, radius - 10 - 16);
      } else {
        Glib_Char(h1 / 10 % 10 + '0', -4, radius - 10 - 16);
      }
    }
  } else {
    Glib_Draw(0, radius - 5, 0, radius);
  }
```

```
    offset = offset + DEG5;
    if (h1 >= 360) {
      h1 = 5;
    } else {
      h1 = h1 + 5;
    }
    glRotatef(t, 0.0, 0.0, 1.0);
    glPopMatrix();
  } while (!(h1 == h2));
```

where `h1` and `h2` are the limits of the display (heading ±35°), `offset` is the heading displacement from 360° and `scalefactor` is the magnification effect (1.574 in this example). The marks at 5° intervals are 5 pixels long and at 10° intervals are 10 pixels long. Similarly, the digits are drawn at 10° intervals, with larger digits at 30° intervals. For heading values above 90°, two digits are drawn rather than a single digit and the symbols are aligned accordingly. Note that all the marks and digits are computed in a vertical orientation and then rotated. The rotation by `t` degrees for each entity is performed between the OpenGL `glPushMatrix` and `glPopMatrix` procedures.

5.9 Head-up Displays

HUDs have been used in military aircraft since the 1970s. The display consists of two main components: the projection system and the combiner or lens (Spitzer, 2001). Flight information is written onto a CRT in a graphical form and is projected onto a small display directly in front of the pilot. The graphical format of the display depends on the application, with the objective to provide as much useful information as possible in the limited area of the HUD (Hall, 1993). The combiner (a semi-transparent sheet of glass) reflects the light from the CRT, but allows external light to pass through the combiner, with as little as 30% optical attenuation. A typical HUD provides a field-of-view 30° × 18°, reflecting the CRT rays so that they are collimated (parallel). The advantage of collimation is that the information appears to be focused at infinity and the pilot does not need to accommodate this information when looking at distant features. The display must also be conformal, that is to say, information displayed on the HUD must align with the outside world. Information of pitch, roll and yaw displayed on the HUD is aligned with the pilot's eye position; in simple terms, the HUD appears overlaid on the outside world with a 1 : 1 linear projection in all axes. For example, the 0° pitch line should follow the distant horizon.

The view of the outside world seen through the HUD is a combination of a slightly green forward view overlaid with bright fluorescent green graphics. Most HUDs are based on calligraphic CRT displays, where the beam is deflected to draw line segments. Consequently, the amount of information is limited to the display of primary flight data (attitude, airspeed, altitude and heading) and weapons aiming and guidance information. CRTs used in HUDs are monochrome, using a CRT coated with green phosphor. A single colour is used because the HUD optics can provide an effective narrow-band optical filter for a specific colour, and bright green stands out against the terrain and sky. The specific phosphor is also selected to have an appropriate decay time constant; the persistency ensures that each drawn pixel decays during the frame refresh period. If the decay period is too long, the moving elements of the image would appear smeared.

In recent years, there has been considerable interest in the use of HUDs in civil transport aircraft, in particular to enhance the guidance cues in landing (Sexton *et al.*, 1988). Generally, the HUD projector in a military aircraft is located below the HUD (but above the pilot's legs, to ensure a safe ejection path) and above the pilot's head in a commercial aircraft, where there is sufficient space for the projection system. In both systems the information is projected onto a small glass screen.

Aircraft Displays

To provide an equivalent facility in a flight simulator, there are three options:

- Install an aircraft HUD in the simulator and stimulate the HUD interface with appropriate inputs;
- Fabricate a HUD to project a HUD image onto a small display in front of the pilot;
- Project the HUD image onto the visual scene, so that it appears to be collimated at infinity.

The solution depends on the training application. In the first case, the cost of an aircraft HUD is likely to be high and it will probably need to be interfaced to the simulator via an aircraft databus. The second option may be attractive with the availability of low-cost projection systems, but it can prove difficult to collimate the image; if the HUD data is focused at a short distance, the pilot has to refocus between the HUD display and the visual scene, causing potential eye fatigue. The third option is arguably the simplest solution, but introduces three problems:

- The flat HUD display may appear slightly curved (or warped) if it is projected directly by the visual system projection system onto a curved projector screen;
- If a blank HUD frame is mounted in the simulator, a slight movement of the pilot's head can result in the HUD information appearing outside the frame;
- Additional time is added per frame to generate the HUD information after the visual scene is rendered.

For the simulator designer, whichever method of projection is used, it is necessary to generate a HUD, corresponding to the formats of a specific system. Either the image is fed directly to the HUD or it is overlaid as the final stage of rendering by the visual system image generator. An example of a HUD overlaid as a 2D layer on the visual scene (the centre channel) is shown in Figure 5.31.

The display is rendered in OpenGL and added to the visual scene image rendered in OpenSceneGraph. The HUD comprises a pitch ladder at 5° intervals, where the positive pitch lines (pitch up) are solid and the negative pitch lines are dashed. The heading display moves left and right to display the current heading (048°). In addition, heading information is also displayed on the 0° pitch line. The airspeed and altitude displays are similar to the EFIS strip displays and include rolling digits to display the airspeed and altitude in a digital format. A bank angle display is included to provide roll information. The fixed attitude reference enables the pilot to determine the pitch and roll attitude, and the velocity vector indicates the aircraft flight path, with respect to the terrain.

Considerable care is needed to ensure that the display is conformal. However, the display area of the HUD will be known in terms of the angular field-of-view and the pixel resolution of the visual system forward looking channel will also be known. As the display is conformal, roll angles are unaffected. If the vertical field-of-view of the display is 40°, which corresponds to 768 pixels, then the mapping is shown in Figure 5.31, where

$$\tan \theta = \frac{y}{k}, \text{ or } y = k \tan \theta$$

But, $\tan 20 = \frac{384}{k}$, giving $k = 1055.031$ (per degree) or 18.414 (per radian). In the case of a 4 : 3 aspect ratio display, the pixels can be considered to be square (so that circles appear correctly circular). However, if the aspect ratio is such that the display is not conformal, then this scaling needs to be modified for the field-of-view and display size in the x axis, as shown in Figure 5.32.

The rendering of the objects in the HUD follows the methods outlined for an EFIS display in the preceding section. The airspeed and altimeter tapes are implemented by a vertical translation of the tape segments close to the current airspeed and altitude values, respectively. Similarly, the rolling digits are generated by using a clipping window. The pitch ladder is generated at 5° intervals,

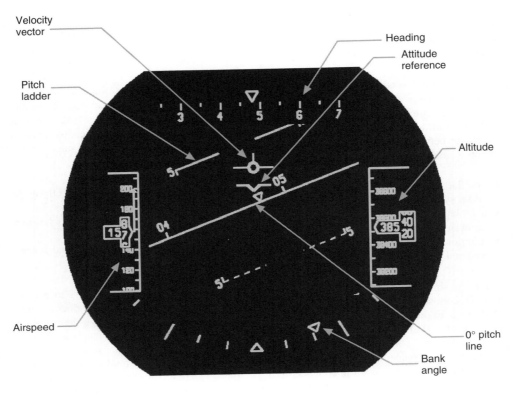

Figure 5.31 A simple HUD

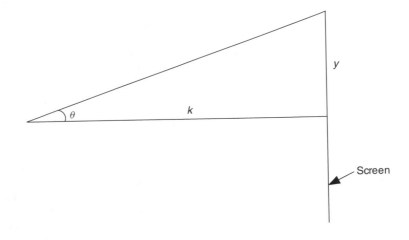

Figure 5.32 HUD projection

Figure 5.33 HUD display in the flight simulator (see Plate 6)

clipped within a rectangle to avoid drawing over the airspeed, altimeter and compass regions of the display. The flight path vector is simply an angular displacement in the flight path angle (γ) and sideslip (β). Finally, the effect of the combiner can be shown as the HUD shape (30° × 18° in this example) drawn in-filled in green with a transparency of 0.05, as shown in Figure 5.33.

Of course, the HUD is a 2D projection on a 3D projection. With rendering packages such as SGI OpenGL Performer or OpenSceneGraph, as their underlying design is based on OpenGL, it is straightforward to integrate a HUD as an overlay on the rendered image. There are some considerations to take into account: any 3D settings, such as lighting or shading models should be turned off; the projection needs to be established as an orthographic projection with the appropriate screen coordinates; the HUD image needs to be attached to the rendering process as the final phase of rendering. One further consideration is that the HUD overlay adds to the processing of the imaging pipeline and care is needed to ensure that the frame period for rendering is not exceeded.

The HUD example provided in the sample code is integrated with OpenSceneGraph (which is written in C++). The HUD is very simple and is intended to provide a template to develop the HUD formats used in military and civil aircraft. However, in most flight simulators, the HUD formats will be specified for the aircraft and the training applications and it is necessary for the simulator design to replicate the HUD functions in terms of the HUD geometry, HUD selections and the HUD functionality. One particular difficulty in implementing HUD software is that the code may need to be modified or reused for other display systems. In mapping from the HUD geometry to screen axes, care is needed to avoid the introduction of scaling factors which may be difficult to comprehend. Consequently, it is essential that, as far as possible, the code is written so that it is portable and is independent of the HUD or display coordinate systems. Moreover, the derivation of angular and linear transformations should be clearly annotated, for example, the separation of airspeed markings on an airspeed tape or the implied spacing of a rolling digit, in order to ensure the portability of the software.

The two procedures `drawHUD` and `createHUD` are defined in `cgi.cpp` to operate with the class `HUD`. `drawHUD` is simply a C++ call to the C code for the HUD functions. The `drawable` component of the class HUD ensures that any DisplayLists are turned off, the HUD construction is copied and that the procedure `drawHUD` is invoked. The HUD is created as a new node in `CreateHUD`. At this time, the lighting attributes are turned off, the bounding box is set and the rendering order is defined so that the HUD is drawn last and the `HUD` node is added to the `camera` node.

References

Bourke, P. (1996) *Parametric Equations of a Sphere and Texture Mapping*, http://local.wasp.uwa.edu.au/~pbourke/texture_colour/texturemap/index.html (accessed 9 May 2009).

Clark, J.H. (1982) The geometry engine: a VLSI geometry system for graphics. *Computer Graphics*, 16 (3), 349–355.

Dawson, W.D. (1992) C-130 glass cockpit development program. AIAA Aerospace Design Conference, Irvine, CA.

Foley, J.D., van Dam, A., Feiner, S.K. and Hughes, J.F. (1995) *Computer Graphics: Principles and Practice in C*, 2nd edn, Addison-Wesley Publishing Company, Reading, MA.

Hall, J.R. (1993) The design and development of the new RAF HUD format, in *Combat Automation for Airborne Weapon Systems: Man-Machine Interface Trends and Techniques*, AGARD CP-520.

Jarrett, D.N. (2005) *Cockpit Engineering*, Ashgate, Surrey, UK.

Koniche, M.L. (1988) 747-400 flight displays development, AIAA Aircraft Design, Systems and Operations Conference, Atlanta, GA.

Newman, W. and Sproull, R. (1979) *Principles of Interactive Computer Graphics*, McGraw-Hill, London.

Pallett, E.H.J. and Coombs, L.F.E. (1992) *Aircraft Instruments and Integrated Systems*, Pearson Education Ltd., Longman.

Sexton. G.A., Moody, L.E., Evans, J.A. and Williams, K.E. (1988) *An Evaluation of Flight Path Formats Head-Up and Head-Down*, NASA CR-4176, NASA Langley Research Centre.

Shreiner D. (ed.) (2004) *OpenGL® Reference Manual: The Official Reference Document to OpenGL Version 1.4*, Addison Wesley Publishing Company, Reading, MA.

Shreiner, D., Woo, M., Nelder, J. and Davis, T. (2005) *The OpenGL® Programming Guide: The Official Guide to Learning OpenGL Version 2.1*, Addison-Wesley Publishing Company, Reading, MA.

Spitzer C.R. (ed.) (2001) Head-up displays, *Digital Avionics Handbook*, The Blackburn Press.

6

Simulation of Aircraft Navigation Systems

6.1 Principles of Navigation

The navigation systems in an aircraft process signals from a range of sources to enable the flight crew to determine position. In a flight simulator, there are no signals or sensors and the simulator designer is faced with the 'simulate or stimulate' dilemma. The option to simulate involves fabrication of a subsystem that replicates the behaviour of the equivalent aircraft subsystem. Internally, the simulator system will replicate the functionality of the subsystem, using whatever technology is most appropriate. The simulator designer is responsible for ensuring that all operational modes are fully replicated and that there is no difference between the simulated equipment and the actual equipment. The option to stimulate allows the simulator designer to use the actual aircraft equipment but requires the equipment to be stimulated with the same signals it will receive in the aircraft. Quite often, the relative advantages of these two approaches are far from clear. For complex systems, it is very difficult to ensure absolute compatibility and any incompatibility could lead to incorrect behaviour of the aircraft systems. Similarly, for stimulated equipment, it may be difficult to generate the signals that would occur in an airborne environment. However, both approaches are used and the decision for the simulator designer is principally one of cost of the design, implementation and testing of the subsystem.

There is one very important consideration with the simulation of navigation systems. In an aircraft, the signals received by a sensor may be subject to distortion, attenuation and noise. But, in a flight simulator, the position of the aircraft and the signal source is known and navigation parameters can be computed to far higher accuracies than is achievable with aircraft equipment. The role of the simulator designer is, therefore, to provide an equivalent functionality (derived from simulator variables) and to model correctly the errors associated with specific navigation systems. This responsibility cannot be overstated. For example, if a navigation source is limited by line of sight, the signal may be lost in low-level flying. If the simulator provided this signal at low level in training, and, consequently, the pilot assumes this signal is valid at low level, such an anomaly could lead to a catastrophic outcome in an operational situation.

Navigation is based on the measurement of aircraft position, aircraft motion and time, derived from aircraft sensors. These measurements are used in flight guidance to control altitude, heading and airspeed to follow a flight plan in terms of a lateral and vertical flight path. In practice, the errors

---

Principles of Flight Simulation   D. J. Allerton
© 2009, John Wiley & Sons, Ltd

in the sensor measurements can be estimated from knowledge of the sensor or may be bounded. Aircraft navigation is also subject to constraints:

- Magnetic variation – this varies slowly with aircraft position and is easily modelled in a simulation;
- Wind – this varies with time, position and altitude and although it is difficult to measure directly in an aircraft, the direction and magnitude of wind is likely to be known in a simulator;
- Malfunctions – these can occur in sensors, displays and beacons causing erroneous measurements; in a simulator, it is essential to detect and replicate all failure modes.

A wide range of sensors have been developed for aircraft, which can be classified in the following groups:

- Direct measurement, for example, air data measurements;
- Dead reckoning, that is, extrapolating position from knowledge of an initial position and subsequent measurements of velocities, accelerations, aircraft heading and time;
- Distance from radio beacons or satellites;
- Bearing from radio beacons.

Navigation is also supported by the availability of various forms of navigation data, including published charts, databases of radio beacon frequencies, navigation procedures (e.g. approach plates) and onboard databases (e.g. flight management systems and enhanced ground proximity warning systems). Clearly, the same data must be used in both training in a simulator and in actual flight. In the case of airlines, monthly updates of navigation data are uploaded to the aircraft and simulators, and considerable care is given to assuring that the data used in the simulators is up to date.

Aircraft position is referenced to the earth frame defined by latitude (degrees), longitude (degrees) and altitude (feet above sea level), as shown in Figure 6.1. The parallels of latitude range from

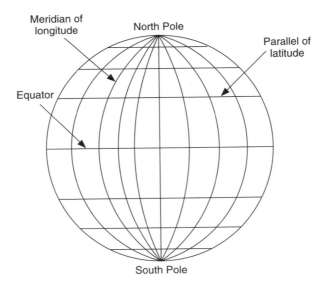

Figure 6.1 Latitude and longitude

90°S at the South Pole to 90°N at the North Pole. The meridians of longitude are aligned with reference to the prime meridian 0°, which passes through Greenwich, near London and range from 180°W to 180°E. Note that distance is derived from latitude.

$$1 \text{ nautical mile} = 1 \text{ minute of latitude}$$
$$= 1/60 * 1/360 * 2\pi R$$
$$= 6076 \text{ ft } (1852 \text{ m})$$

where R is the radius of the earth.

This unit of distance is constant over the earth, that is, it is independent of longitude or deformation of the earth. Note that distance is measured in nautical miles in aircraft navigation, rather than statute miles (1 statute mile = 1609 m)

Aircraft velocity is measured in nautical miles per hour or knots (kt) and is usually derived from measurements from an air data computer, which measures static air pressure, dynamic air pressure and air temperature to provide altitude, airspeed, Mach number and air density, with the appropriate compensations for measurement errors. Four variants of velocity are measured in an aircraft and need to be computed in a simulation:

- True air speed (TAS) – in an aircraft, this speed is measured from the dynamic air pressure and includes corrections for air density; in simulation, TAS is derived from the body frame velocity U;
- Indicated airspeed (IAS) – in an aircraft, this speed is measured directly from the dynamic pressure and is displayed by the airspeed indicator; as the measurement is not compensated by change of air density, it reduces with increasing altitude; in simulation, IAS is given by $IAS = U\sqrt{density\ ratio}$;
- Calibrated airspeed (CAS) – this speed includes corrections for location of the sensor and instrumentation; although this measurement is important in an aircraft, it is generally assumed to be zero in simulation;
- Ground speed – the speed over the ground, including effects of wind; in aircraft, ground speed is derived from INS, Doppler or GPS measurements; in simulation, ground speed is given by $\sqrt{V_N^2 + V_E^2}$, where V_N and V_E are the north and east velocities with respect to the earth.

In an aircraft, altitude is measured by an air data computer, converting a static pressure measurement to altitude. In simulation, the altitude is derived in the flight model as P_z, recalling that altitude is positive downwards. Aircraft altimeters are set to QFE, QNH or flight levels, by appropriate selection of the barometric pressure setting of the altimeters. Regional pressure setting is one of the environmental parameters that can be varied in a simulator to ensure that pilots are trained to reference altimeters to airfield altitude, regional pressure or standard pressure (1013 mb). Variation of pressure should be correctly indicated on the simulator altimeters, according to the pressure variation outlined in Section 3.2.

In an aircraft, heading is derived from a magnetic compass measurement. In simulation, true heading is given by the aircraft yaw and magnetic heading is calculated by subtracting the magnetic variation at the current aircraft location. The magnetic variation is given by isogonals (lines of common magnetic variation), which are published. These are stored in an FMS to compensate for magnetic variation and similar tables are used in simulators. The isogonals change slowly, typically less than one degree per year. However, the compass measurement is affected by magnetic fields in an aircraft and by acceleration, and in simulation, it is important to model the variation in magnetic compass readings caused by accelerating flight. Note that compass heading is always

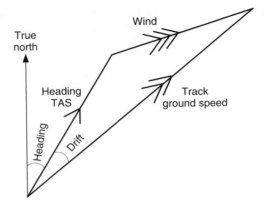

Figure 6.2 Aircraft track and ground speed

given as degrees magnetic rather than degrees true, although modern navigation displays can be set to display either true or magnetic heading. In aircraft navigation, a pilot will follow a track rather than a heading to allow for the effect of wind, as shown in Figure 6.2. The wind results in a track and ground speed, which may vary from the aircraft heading and airspeed. In simulation, the wind is known, and it is therefore straightforward to compute track and ground speed, which are used in navigation displays.

6.2 Navigation Computations

In navigation, if an aircraft at a position (λ_1, ϕ_1) needs to steer to a position (λ_2, ϕ_2), where λ and ϕ denote latitude and longitude, respectively, it is necessary to know the relative bearing of (λ_2, ϕ_2) from (λ_1, ϕ_1). In flat earth coordinates, this is a simple computation, as shown in Figure 6.3.

The relative bearing of (x_2, y_2) from (x_1, y_1) is given by α. The relative bearing of (x_1, y_1) from (x_2, y_2) is the reciprocal of α, given by $\alpha + \pi$. From simple 2D geometry,

$$\alpha = \tan^{-1}\left(\frac{x_2 - x_1}{y_2 - y_1}\right) \tag{6.1}$$

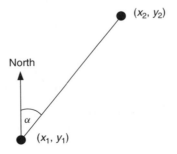

Figure 6.3 Relative bearing

and the distance D is given by

$$D = \sqrt{(x_2 - x_1)^2 + (y_2 - y_1)^2} \tag{6.2}$$

In practice, aircraft position, the location of radio beacons and the position of airfields are given by latitude and longitude, based on spherical geometry. Radio waves are assumed to follow great circle paths and computation of relative bearing to (or from) a beacon or distance to a beacon is performed in spherical coordinates.

The distance D between two points (λ_1, ϕ_1) and (λ_2, ϕ_2) on the surface of the earth, where λ denotes latitude and ϕ denotes longitude is given by

$$d = \sqrt{(\Delta \lambda)^2 \cos \lambda_1 \cos \lambda_2 (\Delta \phi)^2} \tag{6.3}$$

$$D = Rd \tag{6.4}$$

where d is the distance in radians, R is the earth radius and

$$\Delta \lambda = \lambda_2 - \lambda_1 \tag{6.5}$$

$$\Delta \phi = \phi_2 - \phi_1 \tag{6.6}$$

The relative bearing ψ from a point (λ_1, ϕ_1) to a point (λ_2, ϕ_2) is given by

$$x = \sin \lambda_2 - \sin \lambda_1 \cos d \tag{6.7}$$

$$y = \cos \lambda_2 \sin \Delta \phi \cos \lambda_1 \tag{6.8}$$

$$\psi = \tan^{-1}\left(\left|\frac{y}{x}\right|\right) \tag{6.9}$$

The absolute values of x and y are used in the computation because most arctan functions are limited to the range 0 to $\pi/2$. The actual value of ψ depends on the four possible quadrants (signs of x and y). If $x < 0$ then $\psi = \pi - \psi$. If $y < 0$ then $y = -y$. Care is also needed to cater for small values of x, which could lead to numeric overflow. In this situation, $\psi = \pi/2$ or $-\pi/2$ (depending on the signs of x and y).

Over short distances, the track between two points is a straight line. However, this path is curved for distances greater than a few hundred miles and forms part of a great circle. A great circle is a circle drawn on the surface of the earth, where the plane of the circle passes through the centre of the earth, for example, the equator and meridians of longitude are great circles. A great circle has the following properties:

- The shortest distance between two points on the surface of the earth is an arc of a great circle;
- There is only one great circle between two points.

Tracking along a great circle, from a point A to a point B, the heading relative to north changes continuously, increasing from A to B, as shown in Figure 6.4. An alternative representation is a rhumb line, which is line between two points, such that there is a constant heading from A to B. Over short distances (less than 200 miles), a great circle and a rhumb line between two points are almost identical. For oceanic routes, flown under FMS guidance, it is not unreasonable to change heading to follow a great circle route. However, for local area navigation, it is clearly preferable to fly along a straight line without requiring any heading corrections. This requirement is reinforced by the need to provide charts that are easy to follow in a cockpit and more particularly, where the track between two points is shown as a straight line.

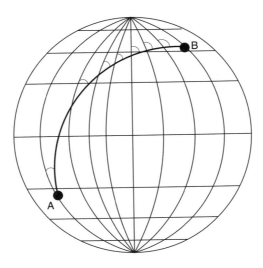

Figure 6.4 Relative bearing along a great circle track

6.3 Map Projections

Navigators have sought methods to represent a spherical earth on paper for hundreds of years. The objective in chart production is to transform between spherical coordinates (3D) and Cartesian coordinates (2D). There is a similar requirement for modern map displays in aircraft and also in flight simulation, to provide a map representation of the aircraft flight path for the instructor station. However, despite the numerous methods developed for map projection, some distortion is inevitable and most charts are designed to minimize these errors

Charting is the transformation of a point (ϕ, λ) to (x, y), where ϕ is latitude, λ is longitude and x and y are Cartesian coordinates in the axes of the chart. This transformation requires three properties:

- Equivalence – equal areas of the earth's surface are represented by equal areas on the chart;
- Equidistance – equal distances on the earth's surface must have equal length on the chart, to avoid linear distortion;
- Conformance – preservation of angles, so that the shape of a feature (e.g. a lake) seen from the air corresponds to the shape on a chart.

In addition, great circles and rhumb lines should be represented as straight lines on a chart. In summary, a map projection method has the following requirements:

- To preserve angles and bearings, that is, a conformal projection;
- To minimize distortion.

The most commonly used map projection for aircraft navigation is the Lambert conformal projection, developed by John Lambert in 1772. Conceptually, if the features of the world are drawn on a transparent sphere illuminated at its centre and a cone is placed over the sphere so that the features are projected onto the inside of the cone, then the chart is obtained by 'unwrapping' the cone, as illustrated in Figure 6.5. A point on the surface of the earth is then defined by the distance

Simulation of Aircraft Navigation Systems

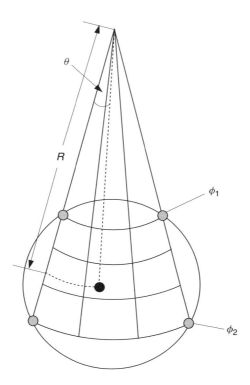

Figure 6.5 Lambert conformal projection

R from the apex of the cone and the angle θ subtended by the point. The Lambert conformal projection has four particularly useful properties for aircraft navigation:

- Lines of longitude and latitude intersect at 90°;
- Angles and bearings are preserved (i.e. shape is preserved);
- Great circle lines are straight lines;
- Rhumb lines are curved.

The projection is simplified by the use of intermediate polar coordinates:

$$(\phi, \lambda) \rightarrow (r, \theta) \rightarrow (x, y) \qquad (6.10)$$

where

$$x = r \sin \theta \qquad (6.11)$$
$$y = r \cos \theta \qquad (6.12)$$

The angle θ is proportional to the longitude, that is, $\theta = n\lambda (n < 1)$, where n is the convergence factor (given by the angle subtended by the cone). This convergence factor depends on the intersection of the cone and the sphere and, for the Lambert conformal projection, these intersections are known as the *standard parallels* ϕ_1 and ϕ_2.

Table 6.1 Map projection errors

Location	Latitude (°N)	Latitude error (%)	Longitude error (%)
Southampton	51	0.09	0.09
Cranfield	52	0.11	0.11
Nottingham	53	0.09	0.09
Edinburgh	56	0.14	0.14
Jersey	49	0.03	0.03

The mechanization of the Lambert conformal mapping (Maling, 1991) is as follows:

$$n = \frac{(\log_e \sin x_1 - \log_e \sin x_2)}{\left(\log_e \tan \frac{x_1}{2} - \log_e \tan \frac{x_2}{2}\right)} \quad (6.13)$$

$$r = \frac{\sin x_1}{n} \left(\frac{\tan \frac{x}{2}}{\tan \frac{x_1}{2}}\right)^n \quad (6.14)$$

$$\theta = n\lambda \quad (6.15)$$

where

$$x = \frac{\pi}{2} - \phi, \; x_1 = \frac{\pi}{2} - \phi_1 \text{ and } x_2 = \frac{\pi}{2} - \phi_2$$

Note that n is a constant for two given standard parallels and that at the latitudes ϕ_1 and ϕ_2, there is no distortion. Typically, standard parallels are selected near the extremities of a chart. For example, for the ICAO Southern England 1 : 500,000 chart, the standard parallels are 49°20′N and 54°40′N. Typical errors for this chart are set out in Table 6.1.

6.4 Primary Flight Information

A pilot will use the information displayed by the primary flight instruments to control and navigate the aircraft. This information includes the attitude indicator (pitch and roll), the airspeed indicator (airspeed), the altimeter (altitude), the compass (yaw), the vertical speed (rate of climb), the turn indicator (yaw rate) and the 'slip ball' (sideslip). The actual display formats are illustrated in Sections 5.7 and 5.8. In addition, military aircraft displays may include an angle of attack gauge and a g-meter.

6.4.1 Attitude Indicator

The display shows the aircraft pitch and bank angles, which are derived directly from the pitch and roll angles in the Euler (or local) frame. The display must be conformal, that is, the roll angle of the artificial horizon is parallel to the distant horizon. The pitch displacement depends on the size of the displayed attitude 'ball', which is stabilized by gyros in an aircraft, with the requirement that it rotates correctly through 360° during a loop.

6.4.2 Altimeter

The altimeter displays the aircraft altitude, typically in feet, which is computed by an air data computer. In addition, the displayed value must include the offset for barometric setting by the pilot and variation in regional pressure set by the instructor.

6.4.3 Airspeed Indicator

The airspeed indicator displays airspeed derived by an air data computer. Typically, the IAS is derived from the TAS by $IAS = V_c \sqrt{\frac{\lambda}{\lambda_0}}$ where V_c is true airspeed, λ is air density and λ_0 is air density at sea level. In addition, an airspeed indicator may also display the Mach number, which is computed by the atmospheric model.

6.4.4 Compass

The compass will display the aircraft heading, derived from the true heading given by the yaw angle (in the Euler frame) adjusted for the local magnetic variation.

6.4.5 Vertical Speed Indicator

The vertical speed is computed in an aircraft by the rate of change of the static pressure measured by the air data computer. In modern civil aircraft this is the instantaneous vertical speed but for light aircraft, this is simulated as a simple first-order lag of one to two seconds.

6.4.6 Turn Indicator

The rate of turn is displayed, typically with a simple display showing the rate one (3 °/s) turn value, enabling a pilot to establish the appropriate turn rate. The displayed value is derived from the yaw rate term in the flight model.

6.4.7 Slip Ball

The traditional slip ball is a ball bearing in a curved tube of fluid. The fluid provides damping as forces are applied to the aircraft (and therefore to the ball) and the curvature provides a small restorative force. The simplest method to implement a slip ball is a second-order model of a critically damped pendulum, where the force (in the body frame) applied to the ball is *sideforce/mass* and the restorative force, which is proportional to the displacement of the ball. The net displacement is limited to the length of the curved tube. The output of the second-order model is the ball displacement, which is modified for the slightly curved motion of the ball. Although the viscosity of the fluid is unlikely to be known, its role is to damp the motion of the ball and therefore, by choosing a critical damping value, the model will provide appropriate motion of the slip ball.

6.5 Automatic Direction Finding (ADF)

Although ADF is not approved for primary navigation, it is still widely used in remote regions of the world where there is poor navigation coverage. Marker beacons used in instrument procedure approaches are detected by an ADF receiver (Helfrick, 1994) and there are still a large number of NDBs in most countries. Most radio magnetic indicators (RMIs) on modern aircraft can take their signal source from NDB transmitters or VOR transmitters. On an aircraft, two small loop antennas are used to detect the signal. The current induced in each branch of a loop antenna is proportional

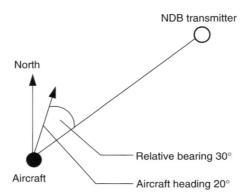

Figure 6.6 ADF relative bearing measurement

to the angle of incidence of the antenna (strictly the cosine of the angle made between the direct line to the antenna and the plane of the antenna). The use of two antennas resolves the ambiguity in the bearing measurement to the NDB. As the two loop antennas are fixed to the frame of the aircraft, the signal varies with aircraft heading, so that the antennas are aligned to provide a signal proportional to the bearing to the station, relative to the current heading.

ADF has the advantage that it can operate over long distances (up to 500 nautical miles), it is not dependent on line of sight to the transmitter and the signal can be detected by a simple antenna and receiver. However, this low frequency amplitude-modulated signal is susceptible to interference (particularly from the ionosphere) and reflections, and as the aircraft banks, errors known as *dip errors* are introduced. The signal is converted to an offset displayed on either a fixed compass card known as a *relative bearing indicator* (RBI) or a rotating compass card known as a *radio magnetic indicator* (RMI). The measurement is illustrated in Figure 6.6, where the aircraft is positioned south west of the NDB, heading 20° on a relative bearing of 30°. The RMI compass card in Figure 6.7 shows a heading of 020°M and the pointer (the bearing indicator) is 30° to the right. The display is intuitive; it displays the direction of the NDB transmitter relative to the current aircraft heading.

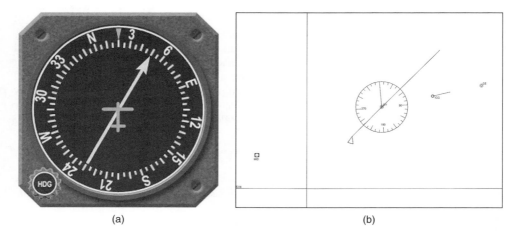

Figure 6.7 ADF display

In a simulator, the computation is straightforward. The selected ADF frequency is used to locate the NDB transmitter in the navigation database to provide the latitude and longitude of the transmitter. The pointer offset is given by $\psi - \beta$, where ψ is aircraft heading and β is the relative bearing of the transmitter from the aircraft. The main source of error for light aircraft is ADF dip, but this is readily modelled as a function of bank angle.

A computer-generated ADF display is shown in Figure 6.7a where the aircraft heading is 020°N and the relative bearing of the NDB station is 050°N. The aircraft position relative to the NDB transmitter (GY) is shown in Figure 6.7b, where the aircraft is south west of the transmitter.

6.6 VHF Omnidirectional Range (VOR)

Despite its name, VOR provides a relative bearing to a VOR transmitter (Helfrick, 1994), but does not provide any range information. Unlike ADF, the VOR signal is transmitted as a frequency-modulated signal in the VHF band, which is far less susceptible to interference; the signal is therefore line-of-sight and, consequently, is limited in range. The receiver captures two 30 Hz signals, a reference signal and a variable signal. The two signals are exactly in phase along a line due north from the transmitter. At other directions around the transmitter, the phase of the variable signal varies linearly from the reference phase, for example, by 90° due east of the transmitter, 180° due south of the transmitter and 270° due west of the transmitter. Rather than providing a relative bearing to the transmitter, the pilot can select a specific radial (with respect to the transmitter) and the signal provided by the receiver is the error (or phase offset) from the phase of the selected radial. This information is presented on a VOR as a vertical pointer, centred when the aircraft is exactly on the radial and up to 10° at maximum deflection. Consequently, this signal is independent of aircraft heading; it provides only relative bearing information. The modern VOR receiver is accurate to less than one degree and is approved for navigation procedures.

The advantages of VOR are that the receiver is simple, it provides a high degree of immunity to propagation and atmospheric effects and VOR transmitters are widespread throughout Europe and the United States. The main limitation is that the signal is line of sight, so that it can be occluded by other aircraft or hills and more importantly, the signal cannot be received if the transmitter is below the horizon. An approximation that is used for VOR range is given by

$$D \approx 1.2(\sqrt{h_T} + \sqrt{h_A}) \qquad (6.16)$$

where D is the distance in nautical miles, h_T is height of the transmitter above sea level (ft) and h_A is the height of the aircraft above sea level (ft). There is one further consideration with VOR; tracking towards a VOR station along the 35° radial, say, the receiver provides exactly the same signal as tracking outbound from the station. This ambiguity is resolved by providing a TO/FROM indication, typically in the form of an arrowhead or flag to indicate whether the radial applies to the inbound or outbound radial. However, close to the station, there is a sudden transition from TO to FROM and if an ambiguous signal is received, the indication changes to an undefined state. A typical EFIS VOR display is shown in Figure 6.8a. In Figure 6.8b, the aircraft is shown positioned 18.9 nautical miles north west of the Midhurst VOR transmitter MID with a heading of 130°. The aircraft is heading towards the VOR station (shown by the orientation of the arrowhead) and is approximately 6° to the left of the 130° radial to the station, shown by the offset of the centre bar.

A VOR receiver can be used to determine the current radial from a VOR station, by turning the omni bearing sector (OBS) until the pointer is centred. Alternatively, the VOR can be used to track towards a specific radial and to maintain a track along the radial. As a VOR displays angular information, it is necessary for a pilot to have a reasonable estimate of distance to the station, in order to judge the appropriate intersection angle to a radial.

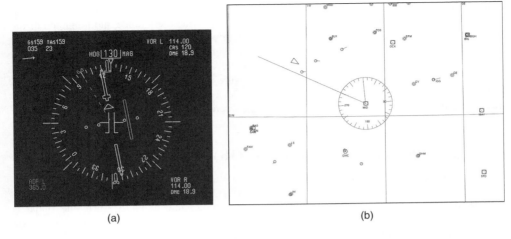

Figure 6.8 A VOR display

In terms of simulation, VOR is straightforward to model. The relative bearing of the VOR station from the aircraft can be calculated. The displayed pointer offset is simply the relative bearing less the selected radial, scaled so that full-scale deflection corresponds to a 10° error. As the receiver accuracy is of the order 1° and the OBS is selected in 1° steps, signal errors are not usually modelled. However, in situations where poor signal reception is important in training, the effect of terrain shielding and aircraft altitude, in terms of signal loss, is modelled. The TO/FROM indicator is displayed as a red flag if any of the following operational failure modes occur:

- if the VOR receiver is switched off
- if the VOR station is beyond reception range
- if the VOR frequency is incorrectly selected
- if the VOR transmitter is unserviceable.

These conditions should be provided in a flight simulator, as they can occur in operational situations.

6.7 Distance Measuring Equipment (DME)

Distance measuring equipment (DME) is used with both VOR and ILS transmitters to detect and display distance from the aircraft to a DME station (Kelly, 1984). DME transmitters are frequently co-paired with VOR and ILS transmitters, so that selection of the VOR or ILS frequency also tunes the associated DME transmitter. The combination of relative bearing and distance is often referred to as *RHO-THETA* navigation (ρ is range to the station, provided by DME and θ is the bearing to the station, provided by VOR). DME transmitters also operate in the VHF band, so the signal is line of sight, providing slant range rather than distance over the ground, for distances up to 150 nautical miles. For example, an aircraft at 6500 ft directly overhead a DME transmitter will display a distance of over 1 nautical mile. In aircraft, DME distance is measured by the transmission of a pulse that is detected and echoed by a DME ground station. These transmissions are designed so that an aircraft avoids detecting the echoed pulses intended for other aircraft. The transit time is converted to distance (1/10 nautical miles) and is displayed digitally.

In simulation, the standard distance equation given in Equation 6.2 provides the ground distance between the aircraft and the DME station. The slant distance D is derived from

$$D = \sqrt{D_g^2 + (h_A - h_D)^2} \qquad (6.17)$$

where D_g is the ground distance, h_A is the altitude of the aircraft and h_D is the altitude of the DME station. The same line-of-sight errors that apply to VOR transmissions also apply to DME and should be modelled, otherwise sensor errors of the order 0.1 nautical miles are modelled. The actual pulse-pair transmissions are Gaussian pulses, in order to minimize the interference from side bands, but then the sensor measurement is limited by taking a timing reference from a signal with a Gaussian shape, resulting in errors of the order of 500 m.

6.8 Instrument Landing Systems (ILS)

A pilot with a full instrument rating is approved to fly an ILS approach to 200 ft above ground level. At that point, if the runway is fully in view, the pilot can proceed with the landing. Otherwise, the approach is aborted and the flight crew can either divert or attempt another approach.

An ILS system comprises three elements:

- A localizer transmitter providing an azimuth (lateral) error signal relative to the centre line of the runway – in many ways this is similar to the lateral guidance signal provided by a VOR transmitter;
- A glide slope transmitter providing an elevation or glide path (vertical) error signal relative to the published descent profile for the airfield, typically 3°;
- Marker beacons at published distances from the runway threshold, comprising an outer marker at 5 nautical miles, a middle marker at 1 nautical mile and possibly an inner marker at 0.2 nautical mile.

A localizer transmitter beyond the far end of the runway transmits two signals in the VHF band with demodulated components at 90 and 150 Hz, where the depth of amplitude modulation in each lobe varies with the angular distance from the runway centre line. The receiver recovers a signal to detect the angular offset of the aircraft from the runway centre line. A similar system is used in the vertical channel. Two lobes at 90 and 150 Hz are centred on the published glide slope angle, enabling the ILS receiver to detect the angular offset of the aircraft relative to the published glide path. These two signals are fed to a cross-hair display giving an instantaneous measurement of the aircraft position relative to the ideal flight path. In older aircraft, this information was displayed using two galvanometer pointers. With modern aircraft, a vertical bar indicates the lateral flight path error and a horizontal bar indicates the vertical error. Typically, this information is displayed below and to right of the attitude indicator, respectively. It is important to appreciate that this information is independent of heading, turn rate, vertical height or vertical speed. It is simply a 'snapshot' of the aircraft position relative to the ideal approach flight path.

The marker beacons are positioned at published distances from the runway threshold. These beacons transmit a narrow vertical signal that is detected as an aircraft passes closely overhead these beacons, to provide an aural signal, which is demodulated and transmitted as predefined tones in the flight crew headsets and displayed by indicators on the flight deck. The markers are used as a cross-check during an approach, as the aircraft should pass overhead each beacon at a defined altitude.

In addition, most airfields have precise approach path indicators (PAPIs) or similar, positioned close to the touchdown point and some 50 m to the left of the runway, to provide flight crew with a visual vertical reference during the approach. These passive reflectors are set at precise angles so that they display either red or white light. The four reflectors display two red and two white lights for an approach close to the published glide slope. As the aircraft goes below the glide slope, the PAPIs change to three reds and one white and then four reds. Similarly, if the aircraft is above the glide slope, the PAPIs indicate three whites and one red and then four whites. In simulation, the PAPIs are set in the visual scene, as directional light sources set at precise angles to indicate white or red lights seen from the aircraft.

In simulating an ILS, the localizer error computation is very similar to the VOR computation, except that the runway QDM is used rather than the OBS value. Having selected an ILS frequency, the runway is located in the navigation database. The localizer error is given by the difference between the relative bearing to the runway and the runway QDM (degrees magnetic). A similar procedure is used to compute the glide slope error. The aircraft elevation angle is computed from the ground distance and the aircraft altitude above the runway. The glide slope error is the difference between the aircraft elevation angle and the published glide slope for the runway. Typically, full-scale localizer deflection is $2°$ (rather than $10°$ for VOR) and the full-scale glide slope deflection is $0.7°$. Although ILS is susceptible to ground reflections from the transmitter, ILS transmitters are designed to minimize these effects. False glide slope signals can also occur at multiples of the glide slope angle, but to avoid this situation, it is standard practice to acquire the glide slope from below the published glide slope. An ILS also has a failure flag to indicate loss of the transmitter signal, or receiver failure or selection of an invalid ILS frequency.

The major consideration in simulating an ILS is to ensure that its operational characteristics match airfield ILS systems. For example, the signal must only be valid within a given distance from the transmitter; the angular offset for the localizer and glide slope must be accurately modelled; the display should become more sensitive close to the transmitter and all operational failure modes should be fully implemented. A typical EFIS ILS display is shown in Figure 6.9, the central part of a primary flight display. The glide slope indicator is shown as a white diamond symbol on the right-hand side of the display in this example, indicating that the aircraft is marginally low on the glide path. The localizer indicator is shown as a white diamond symbol at the bottom of the display, where the aircraft is slightly to the left of the localizer.

6.9 The Flight Director

The VOR and ILS provide a pilot with guidance information. The VOR provides a lateral error from the selected VOR localizer. The ILS provides a lateral error from the runway QDM and a vertical error from the runway glide slope. However, for both systems, the information provided on the respective displays must be interpreted by a pilot, to determine the flight path needed to reduce these errors to zero. Both sensors provide guidance information that must be interpreted by the pilot to derive the flight path to acquire the localizer or glide slope.

The pilot selects a heading to acquire the localizer and a rate of climb/descent to acquire the glide slope, which requires the pilot to select a bank angle appropriate to the turn rate to acquire the desired heading and a pitch attitude appropriate to the rate of descent, respectively. During an ILS approach, the pilot must interpret the 'raw data' provided by the display to take account of distance, offset and wind in capturing the localizer and glide slope.

The flight director performs these functions, providing a guidance display that can be followed directly to track either a VOR radial or an ILS localizer and glide slope. There is often some confusion between an ILS and a flight director as both can be implemented as a cross-hair display whereas an ILS is just a snapshot of the aircraft position relative to the localizer and glide slope, independent of the aircraft heading or vertical speed, a flight director computes the elevator and aileron inputs needed to acquire and track an ILS and displays the difference between the current pilot inputs and the desired inputs. If the cross-hair is centred, the elevator and aileron inputs are correct. If the vertical bar moves to the left, the pilot should move the aileron input to the left, until the bar is re-centred. Similarly, if the horizontal bar moves up, the pilot pulls the elevator control backwards (to raise the aircraft nose) until the display is centred. In following these commands, the aircraft should follow the ideal flight path to minimize localizer and glide path deviations during an approach. The flight director provides an optimal trajectory to acquire the localizer and glide slope.

In a modern aircraft, the ILS indicators are displayed to the right of and directly below the attitude indicator, typically as small white diamond-shaped symbols, as shown in Figure 6.9. The flight director is displayed here as dark-grey cross-hairs superimposed on the attitude indicator display for modern EFIS displays. For example, in Figure 6.9 the flight director is commanding a bank right and pitch down action. By increasing the bank angle the vertical bar can be centred and similarly, by pitching down, the horizontal bar will become centred. The flight director guidance is

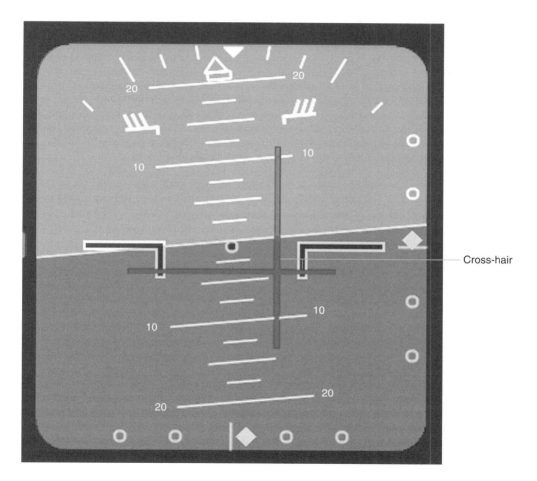

Figure 6.9 ILS and flight director display

very similar to the autoland functions covered in Section 4.7. For an autoland system, the localizer offset is used to determine the intercept heading; the turn rate is computed to achieve this desired heading; finally aileron input is applied to achieve the desired turn rate. The difference with a flight director is that the aileron input is displayed to the pilot rather than being driven directly by the flight control system. For the flight director, this final actuation is made by the pilot who is given a left/right aileron command and a forwards/backwards elevator command.

However, there is one important difference between an autoland system and a flight director. The autopilot system is likely to update at a typical rate of 50 Hz, which is well beyond the bandwidth of the human pilot. The dynamics of the flight director must therefore take into account the dynamics of the aircraft and the response time of a human pilot. One solution is to take the autopilot equations and display the commanded aileron and elevator inputs, which are filtered with an appropriate time constant. If this time constant is too large, the pilot will not be able to correct errors owing to the lag in the display. Alternatively, if the time constant is too small, the display is likely to move abruptly at rates that cannot be followed by a human pilot. In practice, considerable effort is needed to match the overall system response to pilot inputs. Weir, *et al*. (1971) and Kudlinski and Ragsdale (1999) outline methods to design flight director algorithms with acceptable responsiveness. Adams (1983) describes the implementation of a flight director in a simulator study. The overall guidance loop is shown in Figure 6.10. The flight director computes the position of the flight director guidance bars from the state variables of the aircraft dynamics (altitude, airspeed, heading, localizer error, etc.). The summation point shown in Figure 6.10 includes the time for visual acquisition of the flight director guidance and the cognitive processes of the pilot to adjust the current pilot input applied to the aircraft. Consequently, the response of the flight director system must be matched to the dynamic response of the pilot and the aircraft, so that the overall system response enables the pilot to command the aircraft effectively. In other words, the flight director system must be designed so that the poles of the system are appropriate to the frequency response of a human pilot.

An aircraft is more responsive in pitch (to changes in flight path angle) than it is in roll to changes in aircraft track. Consequently, different time constants may be used for vertical and lateral guidance. A time constant of 2 s was found to be appropriate for the Boeing 747-200 flight model. The autoland code is invoked to compute the desired elevator and aileron inputs. The differences between these commanded values and the actual values are used to drive the horizontal and vertical command bars, as shown in the fragment of code for the localizer in Example 6.1.

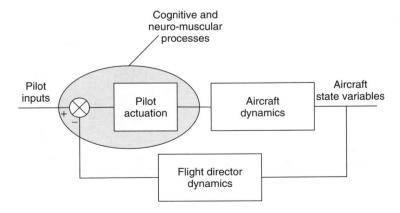

Figure 6.10 Flight director guidance

Example 6.1 Flight director command law (lateral channel)

```
Autoland(&de, &da);
dh = (Aileron - da) * 140.0;
Limit(dh, -140.0, 140.0);
Integrate(fdhx, 0.5 * (dh - fdhx));
```

Where `de` and `da` are the commanded elevator and aileron values, `dh` is the offset of the vertical bar in pixels (limited to ±140) and `fdhx` is the filtered position of the vertical command bar. The integration term (with a constant of 0.5) provides a first-order lag with a 2-s time constant.

6.10 Inertial Navigation Systems

An INS is commonplace on both military and commercial aircraft. The aircraft attitude, velocity and position are computed by sensing linear and angular accelerations on the airframe. Although inertial navigations systems prior to the 1970s were based on mechanically stabilized platforms, with complex gimbal arrangements and torque motors to continuously realign the platform, the availability of microprocessors and ring laser gyros transformed inertial navigation with the strap-down INS. The microprocessor afforded sufficient speed of computation and accuracy of the navigation equations, particularly the trigonometric operations needed for axes transformations. The ring laser gyro was (effectively) a solid-state device providing accurate and reliable measurements of the vehicle rotation rates, in all three axes.

Prior to the availability of GPS, inertial navigation provided reasonable accuracy for transoceanic navigation and found use as a passive position sensor for military aircraft. Inertial navigation systems are still widely used in both military and civil transport aircraft, providing measurements of acceleration, velocity and position for both linear and angular motion in all axes. The introduction of GPS has actually strengthened the case for the use of inertial sensors. Although GPS provides good long-term measurements of position, it can introduce discontinuities, resulting from changes in the satellite constellation, whereas, an INS has good short-term accuracy but accumulates drift errors as a consequence of the method of attitude and position determination. By integrating the INS and GPS sensors, the GPS can realign the INS, bounding the drift error and the INS can smooth GPS errors resulting from position jumps caused by changes to the constellation of tracked satellites.

Consequently, simulation of an INS is often required in applications where navigation system integrity is relevant to the training role. Moreover, as an actual INS is subject to physical accelerations, which cannot be reproduced in a flight simulator, the possibility of using aircraft equipment is not a viable option and the INS must be modelled. Strictly, a perfect INS would derive motion from the same set of accelerations used in the equations of motion, modelling both the navigation equations used in an INS and also the errors associated with an INS. A specific INS will have a known maximum drift rate and the simplest form of error model can be implemented by adding a drift offset to the true aircraft position, where the drift is proportional to the time since the last alignment. For military aircraft, a pilot may realign the INS by flying over a known feature such as a peak or a village church, where the latitude and longitude coordinates are known and can be entered to realign the INS. In applications where a GPS sensor is simulated, if the maximum bound of GPS error is also known, then the INS can be realigned whenever it exceeds this bound.

The alternative, and more general approach to INS simulation, is to model the navigation equations used in INS computations. In fact, an INS will solve many of the equations inherent in the equations of motion described in Section 3.7. From raw accelerometer and rate gyro measurements, an INS will compute latitude, longitude and altitude outputs, which are used by the other aircraft systems. If required, an INS can also provide linear acceleration and velocity measurements and angular position, velocity and acceleration, typically transmitted over the aircraft data bus. For example, the fire control computer of a military aircraft, which is responsible for the release of weapons, is provided with an estimate of aircraft motion, which is essential for accurate weapon deployment. Similarly, a head-up display may include flight path angle information or speed trend information, derived from INS data.

The following section contains a brief overview of the modelling of an INS. For a more detailed explanation of the equations used in an INS, the reader is directed to the standard textbooks on the subject, particularly Titterton and Weston (2004), Farrell (1976), Siouris (1993) and Biezad (1999).

6.10.1 Axes

As with the equations of motion developed for long range navigation in Section 3.11, three reference frames are used:

- The *e-frame*, an inertial ECEF non-rotating frame with its origin at the centre of the earth;
- The *n-frame*, a local or navigation frame, with axes continuously realigned to point north, east and down, with the origin directly below the aircraft, on the surface of the earth;
- The *b-frame*, or body frame, to represent the aerodynamic forces and propulsive forces on the aircraft, with axes pointing forwards, along the right wing and downwards, with the origin at the aircraft cg.

Three transformations are used:

- To transform the accelerations measured by the accelerometers in the body frame to the navigation frame to enable the aircraft velocities (V_N, V_E and V_D) to be computed, which are needed to compute latitude, longitude and altitude (λ, φ, h);
- To transform the earth rotation rate (once per 23 hours and 56 minutes) to the navigation frame;
- To transform the navigation frame rate to the body frame in order to compute the quaternion rates.

Unfortunately, there is no standardization on the notation used to represent navigation axes systems and the use of subscripts and superscripts varies widely, even within the handful of textbooks on the subject. The convention adopted in this book applies to rotation vectors. A vector ω_{ab}^c defines the rotation vector from frame a to frame b, defined in frame c. Where appropriate, the true inertial frame is also used, indicated by the subscript i.

6.10.2 INS Equations

Figure 6.11 shows an INS model. The model is computed in the n-frame, with inputs from the raw accelerometer measurements a_X, a_Y and a_Z and the raw gyro measurements P, Q and R, measured in the body frame. The outputs are the three Euler angles θ, ϕ and ψ pitch, roll and yaw, respectively) and the aircraft position λ, φ and h (latitude, longitude and altitude, respectively).

An INS of this form is often referred to as a north, east, down (NED) INS as it is valid in a north, east and down navigation frame. Although the equations include terms for aircraft accelerations, body rates, earth rate and Coriolis accelerations, the equations are not valid for navigation near the poles and in such cases, the equations must be extended to include wander-axes computations (Titterton and Weston, 2004).

The earth rate Ω in the earth frame is given by

$$\omega_{ie}^e = \begin{bmatrix} 0 \\ 0 \\ \Omega \end{bmatrix} \qquad (6.18)$$

where Ω is the earth rotation rate $= 7.292115 \times 10^{-5}$ rad/s.

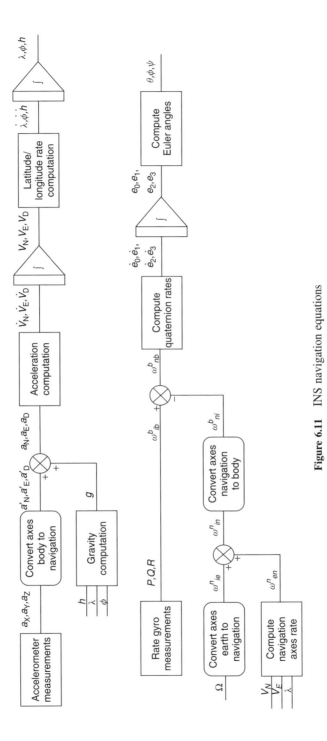

Figure 6.11 INS navigation equations

The earth rotation vector is transformed to the navigation frame as follows:

$$\omega_{ie}^n = \begin{bmatrix} \Omega \cos \lambda \\ 0 \\ -\Omega \sin \lambda \end{bmatrix} \quad (6.19)$$

The aircraft transport rate in the navigation frame is given by

$$\omega_{en}^n = \begin{bmatrix} \dfrac{V_E}{R+h} \\ \dfrac{-V_N}{R+h} \\ \dfrac{-V_E \tan \lambda}{R+h} \end{bmatrix} \quad (6.20)$$

where R is the nominal earth radius $= 6378137.0$ m, h is the aircraft altitude (above the surface of the earth), λ is the aircraft latitude and V_N and V_E are the earth north and east velocities, respectively.

The earth rate and the aircraft transport rate are combined to give the following rotation vector in the navigation frame:

$$\omega_{in}^n = \omega_{ie}^n + \omega_{en}^n \quad (6.21)$$

which is then transformed to the body frame, as follows:

$$\omega_{ni}^b = C_n^b \omega_{in}^n \quad (6.22)$$

The transformation matrix C_n^b is the transpose of the DCM used to transform body velocities to Euler axis velocities in Section 3.7. This vector is subtracted from the body rates sensed by the rate gyros to give the net rotation vector in the body frame ω_{nb}^b, where

$$\omega_{nb}^b = \omega_{ib}^b - \omega_{ni}^b \quad (6.23)$$

These body rate terms are used in the computation of the system quaternions as described in Section 3.6, which enables the aircraft attitude (the Euler angles θ, ϕ, ψ) to be determined.

The raw accelerometer measurements are transformed to the navigation frame using the standard body to Euler DCM. The gravity component g is computed as a function of altitude, latitude and longitude. The approximation used in WGS 84 (National Imagery and Mapping Agency, 2000) for the magnitude of the normal gravity (g_0) on the surface of the reference ellipsoid is given by

$$g_0 = \gamma_e \frac{1 + k \sin^2 \lambda}{\sqrt{1 - e^2 \sin^2 \lambda}} \quad (6.24)$$

where

γ_e is the gravity at the equator $= 9.7803253359$ m/s^2
k is the gravity formula constant $= 0.00193185265241$
e is the first eccentricity of the earth, where $e^2 = 0.00669437999014$

An alternative commonly used approximation is the Airy formula (Biezad, 1999)

$$g_0 = g_{eq}(1 + 0.0052884 \sin^2 \lambda - 0.0000059 \sin^2 2\lambda) \tag{6.25}$$

where g_{eq} is the gravity at the equator and λ is the geocentric latitude.

The gravity term is modified to take account of altitude, given by

$$g = g_0 \frac{R^2}{(R+h)^2} \tag{6.26}$$

The gravity component, computed in Equation 6.26, is added to the accelerations measured in the body frame $[a_x, a_y, a_z]^T$, after they are transformed to the navigation frame.

$$\begin{bmatrix} a_N \\ a_E \\ a_d \end{bmatrix} = C_b^n \begin{bmatrix} a_x \\ a_y \\ a_z \end{bmatrix} + \begin{bmatrix} 0 \\ 0 \\ g \end{bmatrix} \tag{6.27}$$

In addition to the acceleration terms from airframe accelerations and the gravity component, the accelerations from the earth rotation and transport rate are computed in the navigation frame.

$$\dot{v}_N = a_N - 2v_E \Omega \sin \lambda + \frac{v_N v_D - v_E^2 \tan \lambda}{R+h} \tag{6.28}$$

$$\dot{v}_E = a_E + 2v_N \Omega \sin \lambda + 2v_D \Omega \cos \lambda + \frac{v_N v_E \tan \lambda + v_E v_D}{R+h} \tag{6.29}$$

$$\dot{v}_d = a_d - 2v_E \Omega \cos \lambda - \frac{v_E^2 + v_N^2}{R+h} \tag{6.30}$$

The aircraft position (latitude, longitude and altitude) is computed from the navigation frame accelerations as follows:

$$v_N = \int \dot{v}_N dt \tag{6.31}$$

$$v_E = \int \dot{v}_E dt \tag{6.32}$$

$$v_d = \int \dot{v}_d dt \tag{6.33}$$

$$\dot{\lambda} = \frac{v_N}{R+h} \tag{6.34}$$

$$\dot{\varphi} = \frac{v_E}{\cos \lambda (R+h)} \tag{6.35}$$

$$\lambda = \int \dot{\lambda} dt \tag{6.36}$$

$$\varphi = \int \dot{\varphi} dt \tag{6.37}$$

$$h = \int V_d dt \tag{6.38}$$

Although these equations provide the basis of a model of an NED INS, the terms involving tan λ result in a singularity as latitude approaches $\pm 90°$ and the model is therefore inappropriate for simulated flight near the poles. To simulate an INS for worldwide navigation, the equations must be extended and the reader is advised to study wander-axes implementations covered in the appropriate textbooks on navigation systems.

6.10.3 INS Error Model

The equations covered in the previous section assume perfect measurements of body frame accelerations and body frame rates. In the case of a flight simulator, the INS model should match the kinematics of the equations of motion. In practice, error terms in these measurements greatly influence the accuracy of INS measurements. A set of INS sensor measurements will contain errors inherent in the physical properties of the sensors (Collinson, 1996). Moreover, integration of accelerations and body rates will also integrate (or accumulate) any noise in these measurements, which is the main limitation of an INS. These errors can include

- Initial alignment of accelerometers or gyros – in effect, the platform should be north pointing and horizontally aligned;
- Accelerometer bias errors – mechanical forces within an accelerometer that cannot be distinguished from external forces;
- Gyro drift rate – the gyro does not represent the correct aircraft attitude – resulting in errors in transforming between axes;
- Computation errors – for example, rounding or truncation errors in navigation algorithms and in the resolution of the sensor measurements (typically truncation errors in converting from analogue accelerometer measurements to digital values);
- Inaccuracies in the gravity vector model – gravity varies over the surface of the earth and needs to be modelled accurately.

Ideally $x = \iint \ddot{x} dt dt$ where $\ddot{x}$ is known as the *specific force*. In practice, a bias error B is also integrated so that the distance error Δd increases with t^2.

$$\Delta d = \iint B dt dt = \frac{Bt^2}{2} \tag{6.39}$$

Similarly, a gyro drift rate W will introduce a tilt error $\Delta \theta = Wt$, introducing an error gWt in computations involving the gravity vector, which will contribute to the distance error.

$$\Delta d = \iint gWt dt dt = \frac{gWt^3}{6} \tag{6.40}$$

While B and W can be reduced by manufacturing tolerances, the accumulation of these errors cannot be eliminated. If the INS is realigned (to reset the position) by using measurements from other sensors, or from a radio fix or from a visual fix, the effect of drift can be reduced. Although these errors cannot be eliminated, they are to a large extent bounded. An aircraft flying on a trajectory parallel to a tangent to the earth surface will not maintain this parallel path. A stabilized mechanical INS platform must be continuously realigned so that it is always level and an equivalent mechanization is implemented in software for strap-down INS platforms. This is shown in Figure 6.12, where the aircraft platform at A must be tilted through $\theta°$ to remain level at B (with respect to the navigation frame).

Simulation of Aircraft Navigation Systems

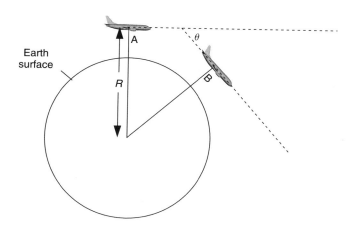

Figure 6.12 INS platform alignment

The transport rate is given by

$$v = R\dot{\theta} \qquad (6.41)$$

where v is the aircraft velocity, R is the radius from the centre of the earth and $\dot{\theta}$ is the rate of change of angle from the horizontal. In terms of the angular acceleration of the platform,

$$\dot{\theta} = \frac{v}{R} \qquad (6.42)$$

$$\ddot{\theta} = \frac{a}{R} \qquad (6.43)$$

where a is the acceleration sensed by the accelerometers. This equation is similar to the equation of a simple pendulum of length L, where the pendulum period is

$$t_p = 2\pi\sqrt{\frac{L}{g}} \qquad (6.44)$$

where $a = g$ and $L = R$. For the earth radius of 6378137 m, the period of the pendulum is 84.4 minutes. In other words, continuously realigning the platform excites (the equivalent of) a pendulum of length R, producing a sinusoidal response with a period of 84 minutes. The accelerometer measurements plus the accelerations from the earth rotation rate and the vehicle transport rate are used to derive V_E and V_N. The vehicle rate terms $\frac{V_E}{R}$ and $\frac{V_N}{R}$ are then fed back to torque the north and east gyros, respectively, so that the platform follows the local vertical.

The net accelerometer output is

$$a - g\Delta\theta + B \qquad (6.45)$$

where B is the accelerometer bias

Rotation of the platform over a period is time is given by

$$\theta_p = \iint \frac{a}{R} \, dt dt \qquad (6.46)$$

resulting in a platform tilt

$$\Delta\theta = \theta_p - \iint \frac{a}{R} dt dt \qquad (6.47)$$

The rotation rate of the platform is therefore defined in terms of the measured acceleration, the accelerometer bias and the gyro bias as follows:

$$\dot{\theta}_p = \int \frac{a - g\Delta\theta + B}{R} dt + W \qquad (6.48)$$

where W is the gyro drift rate. Rearranging the equation in terms of the platform rotation,

$$\theta_p = \iint \frac{a - g\Delta\theta + B}{R} dt dt + \int W dt \qquad (6.49)$$

$$\Delta\theta = \iint \frac{a - g\Delta\theta + B}{R} dt dt + \int W dt - \iint \frac{a}{R} dt dt \qquad (6.50)$$

This is a second-order differential equation given by

$$s^2 \Delta\theta = \frac{a}{R} - \frac{g\Delta\theta}{R} + \frac{B}{R} + sW - \frac{a}{R} \qquad (6.51)$$

$$\Delta\theta \left(s^2 + \frac{g}{R}\right) = \frac{B}{R} + sW \qquad (6.52)$$

where s is the Laplace operator.

To consider the error introduced by an accelerometer bias B, let $W = 0$ and assume B is a constant. The initial conditions (given by solving for $s = 0$) are

$$\frac{g\Delta\theta}{R} = \frac{B}{R} \qquad (6.53)$$

$$\Delta\theta_0 = \frac{B}{g} \qquad (6.54)$$

$$\Delta\dot{\theta}_0 = 0 \qquad (6.55)$$

The solution is a second-order equation given by

$$\Delta\theta = \frac{B}{g} \cos \omega_0 t \qquad (6.56)$$

where $\omega_0 = \sqrt{\frac{g}{R}}$ (the Schuler loop natural frequency of 84.4 minutes)
The acceleration error is

$$g\Delta\theta = B \cos \omega_0 t \qquad (6.57)$$

The velocity error is

$$\int B \cos \omega_0 t \, dt = \frac{B}{\omega_0} \sin \omega_0 t \qquad (6.58)$$

The position error is

$$\iint B \cos \omega_0 \, dt \, dt = \frac{B}{\omega_0^2}(1 - \cos \omega_0 t) \tag{6.59}$$

In other words, an accelerometer bias will result in an oscillation of the position error with a period of 84.4 minutes and an amplitude proportional to B.

A similar approach can be used to assess the effect of the gyro bias, by setting $B = 0$ and assuming that W is a constant. The initial conditions are given by

$$\Delta \theta_0 = 0 \tag{6.60}$$

$$\Delta \dot{\theta}_0 = W \tag{6.61}$$

with a solution

$$\Delta \theta = \frac{W}{\omega_0} \sin \omega_0 t \tag{6.62}$$

The acceleration error is

$$g \Delta \theta = \frac{gW}{W_0} \sin \omega_0 t \tag{6.63}$$

The velocity error is

$$\int \frac{gW}{W_0} \sin \omega_0 t = WR(1 - \cos \omega_0 t) \tag{6.64}$$

The position error is

$$\int WR(1 - \cos \omega_0 t) \, dt = WR \left(t - \frac{1}{\omega_0} \sin \omega_0 t \right) \tag{6.65}$$

In other words, the position error increases linearly with time, with amplitude dependent on W. The effect of INS errors is summarized in Table 6.2 where ω_s is the Schuler frequency, R_0 is the earth radius and t is the time since the previous alignment. The errors are bounded by the Schuler tuning with the exception of the gyro bias error, which increases linearly with respect to time.

By selecting appropriate bias values for the accelerometers and gyros, the characteristics of a specific INS can be verified by simulation. With the increasing utilization of uninhabited air vehicles (UAVs) and the availability of low-cost inertial sensors (with relatively poor performance),

Table 6.2 Summary of INS errors

Error	Effect
Initial position δx_0	δx_0
Initial velocity δv_0	$\delta v_0 \frac{\sin \omega_s t}{\omega_s}$
Initial attitude $\delta \theta_0$	$\delta \theta_0 R_0 (1 - \cos \omega_s t)$
Accelerometer bias δf_{xb}	$\delta f_{xb} \left(\frac{1 - \cos \omega_s t}{\omega_s^2} \right)$
Gyro bias $\delta \omega_{xb}$	$\delta \omega_{yb} R_0 \left(t - \frac{\sin \omega_s t}{\omega_s} \right)$

there is need for off-line simulation of both the UAV and its sensor platforms, to establish mission limitations in terms of navigation accuracy or to confirm the effectiveness of sensor fusion algorithms. Similarly, in evaluation of mission systems using integrated sensors and sensor fusion, the characteristics of INS errors must be modelled accurately so that the effect of these errors on overall system performance can be assessed.

6.10.4 Validation of the INS Model

Having developed an INS model, it is essential to validate the model. Of course, it would not be appropriate to test the model with a non-rotating earth model for the aircraft equations of motion. In addition, it is necessary to assign error values to the INS model so that the characteristics are matched to the accuracy of the INS. Using the full equations of motion given in Section 3.11, the simulator aircraft could follow a flight plan over several hours where the true position and attitude is compared with the INS derived position and attitude terms. An alternative method, and one that is commonly used, is to position the simulated INS at rest and then introduce velocity initialization errors, or a gyro bias or an accelerometer bias. In this situation, the errors will start to build up and attitude errors and position errors (relative to the initial conditions of the stationery INS) can be plotted over several hours.

Example 6.2 shows a fragment of C code implementing an NED INS, where the north gyro is given a drift rate error of 0.002 °/hour. Note that the P and R gyros equations include a term for the earth rate, as they are stationary and will therefore sense the earth rate. Note also that the vertical velocity Vd is set to zero in this test. The INS is positioned at 45 °N 0 °E pointing north. The simulation has an iteration rate of 100 Hz and runs for 30 hours. Although a first-order forward Euler integration method is used, it does not introduce significant errors in the simulation. The INS outputs (latitude, longitude, pitch, roll and yaw) are written to a file every second of simulated time.

The results of the simulation are shown in Figure 6.13. In Figure 6.13a, there is an initial velocity error where $V_n = 1.0$ m/s. In Figure 6.13b, the north gyro has a drift rate error of 0.002 °/s. The

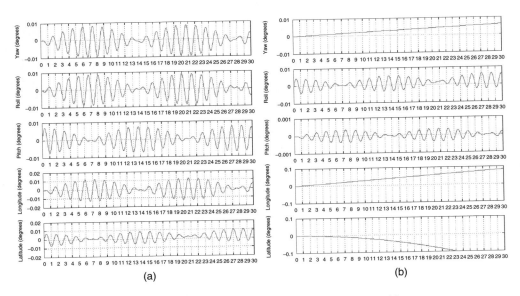

Figure 6.13 Simulation of an NED inertial navigation system with errors

84.4-minute Schuler loop is clearly seen in both plots. In Figure 6.13a, the increasing drift in latitude, which is not fully bounded by the Schuler loop, is also evident.

Example 6.2 The primary equations of an NED inertial navigation system

```
for (i = 0; i <= (30 * 60 * 60 * 100); i += 1)
{
    SetDCM();
    Pitch = asin(-A31);
    Roll = atan2(A32, A33);
    Yaw = atan2(A21, A11);

    Ax = 0.0;
    Ay = 0.0;
    Az = -start_g;

    /* note the deliberate gyro drift error term on the next line */
    Pgyro = EarthRate * cos(latitude) + Radians((double) (0.002 / 60.0 / 60.0));
    Qgyro = 0.0;
    Rgyro = -EarthRate * sin(latitude);

    Wn1 = EarthRate * cos(latitude) + Ve / (Re + h);
    Wn2 = -Vn / (Re + h);
    Wn3 = -EarthRate * sin(latitude) - (Ve * tan(latitude)) / (Re + h);
    nav2body(&Wb1, &Wb2, &Wb3, Wn1, Wn2, Wn3);

    P = Pgyro - Wb1;
    Q = Qgyro - Wb2;
    R = Rgyro - Wb3;

    Quaternions(P, Q, R);
    SetDCM();

    body2nav(&An, &Ae, &Ad, Ax, Ay, Az);
    Gravity(&g, latitude, h);

    VnDot = An - 2.0 * Ve * EarthRate * sin(latitude) +
            (Vn * Vd - Ve * Ve * tan(latitude)) / (Re + h);
    VeDot = Ae + 2.0 * Vn * EarthRate * sin(latitude) +
              2.0 * Vd * EarthRate * cos(latitude) +
            (Vn * Ve * tan(latitude) + Ve * Vd) / (Re + h);
    VdDot = Ad - 2.0 * Ve * EarthRate * cos(latitude) -
            (Ve * Ve + Vn * Vn) / (Re + h) + g;
    Integrate(&Vn, VnDot);
    Integrate(&Ve, VeDot);
    Integrate(&Vd, VdDot);
    Vd = 0.0;

    latitudeDot = Vn / (Re + h);
    longitudeDot = Ve / ((Re + h) * cos(latitude));
    hDot = -Vd;

    Integrate(&latitude, latitudeDot);
    Integrate(&longitude, longitudeDot);
```

```
    Integrate(&h, hDot);

    if ((i % (60 * 100)) == 0)
    {
        printf("%f %f %f %f %f %f\n", (float) i / (60.0 * 60.0 * 100.0),
               Degrees(latitude - start_lat), Degrees(longitude - start_long),
               Degrees(Pitch), Degrees(Roll), Degrees(Yaw));
    }
}
```

6.11 Global Positioning Systems

GPS has been in widespread use in aircraft navigation for several years, providing guidance previously available from LORAN, DECCA, VOR and ADF systems. Although their use in navigation procedures is still very restricted, GPS receivers are used in both civil and military aircraft and need to be modelled.

For the simulator designer, the difficulty in modelling GPS is to choose the appropriate level of modelling. At one extreme, the satellite orbits can be modelled, together with the signal generation by the satellites, the signal decoding and computation of position by the aircraft receiver and the sources of error including propagation delay through the ionosphere, multi-path reflections and receiver noise. At the other extreme, GPS (for civil users) is typically accurate to less than 30 m and knowing the aircraft position accurately in the simulation, some form of noise can be added to the values to give a GPS measurement of position with error properties similar to those associated with actual receivers. In addition to position, GPS provides a measurement of velocity (including vertical speed) and an estimate of signal quality, known as the *dilution of precision*.

Although the position error with GPS changes slowly, it also varies with the satellites in view, which depends on the continually changing GPS constellation, the aircraft position in the world and, to some degree, the aircraft attitude, which may detect different satellites, particularly as the aircraft bank angle changes. The satellites have orbits of just less than 12 hours. At elevations below 15°, the signal is very poor owing to refraction through the ionosphere at low incident angles. Accordingly, GPS is prone to sudden jumps as satellites appear above or drop below the horizon. Values for availability, continuity and accuracy of service are mostly based on 95% figures, so considerable care is needed to produce a statistical model of GPS errors that has a high degree of equivalence to the actual system.

GPS orbit information (or ephemeris data) is provided by the National Geodetic Survey (NGS), which is part of the National Oceanic and Atmospheric Administration agency in the United States. Although this data is 8–10 days old, for most simulation purposes, it provides an accurate set of satellite orbits. For non-military users, the coarse acquisition (C/A) code is subject to refraction through the ionosphere. The approximation used in GPS is based on a method by Klobuchar (1987), which uses a simple cosine, with zero phase at 14:00 hours local time, where the amplitude and period vary with time and geomagnetic latitude (third-order polynomial). The night-time model is simply a 5-ns fixed delay. The model contains eight coefficients that are broadcast in the GPS message for two polynomials. The polynomial predicts the time delay of the signal from the satellite to the aircraft and takes the form

$$\Delta t = F \left(5 \times 10^{-9} + \sum_{i=0}^{3} \alpha_i \phi_m^i \left(1 - \frac{x^2}{2} + \frac{x^4}{24} \right) \right) \qquad (6.66)$$

where

$$F = 1 + 16(0.53 - E)^3 \qquad (6.67)$$

E is the elevation of the satellite, ϕ_m is the geomagnetic latitude and

$$x = \frac{2\pi(t - 50{,}400)}{\sum_{i=0}^{3} \beta_i} \qquad (6.68)$$

The coefficients α and β are passed in the NAV message for each satellite and are available in the published ephemeris data; t is the time in seconds since the last GPS epoch.

The signal is also delayed in the transit through the troposphere. The model developed by Smith and Weintraub (1953) is used to predict tropospheric delay:

$$N = 77.6 \times \frac{P}{T} + 3.73 \times 10^5 \times \frac{e}{T^2} \qquad (6.69)$$

where P is the total atmospheric pressure (millibars), T is the absolute temperature (°K) and e is the partial water vapour pressure. The term $77.6 \times P/T$ is known as the *dry component* and accounts for 90% of the error (P and T are known for an aircraft). The angle of elevation also effects these terms, as given in Table 6.3.

Knowing the position of the satellites in the WGS-84 coordinate frame and the position of the aircraft, and by modelling the delays through the ionosphere and troposphere, the signal transit time can be computed to give the pseudo-range from the aircraft to each satellite in view. Based on the signals from four satellites, which are more than 15° above the horizon and spaced to provide a good intersection, the aircraft position is computed (typically at a rate of 1 Hz) by applying the following simultaneous equations (Hofman-Wellenhof et al. 1992):

$$R_1 = \sqrt{(x - x_1)^2 + (y - y_1)^2 (z - z_1)^2} + ct \qquad (6.70)$$

$$R_2 = \sqrt{(x - x_2)^2 + (y - y_2)^2 (z - z_2)^2} + ct \qquad (6.71)$$

$$R_3 = \sqrt{(x - x_3)^2 + (y - y_3)^2 (z - z_3)^2} + ct \qquad (6.72)$$

$$R_4 = \sqrt{(x - x_4)^2 + (y - y_4)^2 (z - z_4)^2} + ct \qquad (6.73)$$

where
(x, y, z) is the aircraft position (WGS-84 coordinates)
t is the (unknown) aircraft clock bias
c is the speed of light
(x_i, y_i, z_i) is the position of the ith satellite
R_i is the pseudo-range measurement from the ith satellite

Table 6.3 Effect of satellite elevation on the tropospheric delay

Elevation	Dry error (m)
90	2.3
30	4.6
10	13.0
5	26.0

R_i is derived from the actual aircraft to satellite distance plus an allowance for delay through the ionosphere and troposphere (which will also be corrected by the GPS receiver). The satellite position (x_i, y_i, z_i) is extrapolated from the satellite ephemeris. The aircraft clock bias t is unknown but can be eliminated as there are four equations in four unknown variables. Note that the equations are non-linear and an implicit solution is needed to solve for the aircraft position (x, y, z).

These approximations for GPS errors do not include selected availability (currently switched off), multi-path reflections or receiver noise, which occurs in decoding the gold code signals and in tracking the Doppler shift of the carrier, as a satellite may be moving at high speed with respect to the aircraft. Moreover, as the aircraft banks, different satellites may come into view leading to a discontinuity in position measurements during the manoeuvre. In addition, a different model is needed for military GPS as the delay through the ionosphere is eliminated and the receiver resolution is approximately 10 times the resolution found in the civilian receivers. Plots of GPS errors for stationary receivers are widely available and provide a means of validating a GPS model. The errors can be modelled as a simple Markov chain with occasional abrupt changes to represent constellation changes, where GPS position is given by the true aircraft position plus errors modelled in the simulation. However, such a model is very simple and the reader needing to implement a more exact model of GPS is advised to study the numerous text books that describe the sources of error in GPS measurements (Forsell, 1991).

As with all forms of simulation, the simulation of GPS sensors should be based on the requirements of the simulation. If GPS navigation procedures are to be practised, a rudimentary GPS model may be acceptable. However, if the GPS simulation is part of a detailed system evaluation, a detailed dynamic model of GPS and its errors may be necessary. In the former case, it may sufficient to acquire GPS data using a receiver at a static location to record position errors, which can added to the true position of a simulated aircraft, so that the errors exhibited in the simulator are similar to those measured with an actual GPS receiver. As GPS is a complex dynamic system, the use of pre-recorded error data provides a viable method of simulating GPS characteristics. However, a large amount of data may be required to avoid repetition of the data and it may be difficult to introduce failure modes in a realistic way.

The other option, and the one used in the book, is to develop a dynamic model of GPS. However, there are several sources of error with GPS and the detail or fidelity required of the model should be established. In the model, the only errors considered are the delay through the ionosphere, the delay through the troposphere and the local clock bias. Other errors, including satellite clock drift, orbital corrections, multi-path reflections and noise in the receiver are omitted. Details of more exact orbital models and error models are given in Farrell and Barth (1999) and Kaplan (1996). These errors also contribute to the measured pseudo-range and could be added to the pseudo-range term in the software model to provide a more complex model.

Each satellite is in a circular orbit, circling the earth every 11 hours and 58 minutes. There are six separate orbital planes, each inclined at 55° to the equator. These six planes are spaced at 60° intervals so that each satellite in a plane intersects the equatorial plane at a specific longitude. Within each orbital plane, the three or four satellites are spaced with a known phase relative to each other (they are not spaced equally). Fortunately, the data defining the constellation is known. The constellation configuration at midnight on 1 July 1993 is taken as the GPS reference time given in Brown and Hwang (1997). More recent ephemeris data could be used from data downloaded from GPS, but in this example, the original data is used. A 1-s update is simulated, so that each satellite moves by 1/43, 082 °/s in its orbit. A fixed GPS receiver is assumed to be positioned at the Amy Johnson Building at the University of Sheffield at a latitude N53 22 53.55 N and longitude W01 28 41.39. In an actual simulation, the true position of the simulated aircraft will be known and the GPS position would normally be computed with respect to the moving aircraft.

Simulation of Aircraft Navigation Systems

Although GPS operates in ECEF coordinates, the transformation to and from geodetic coordinates is straightforward. Given a geodetic position (λ, ϕ, h), the ECEF position (x, y, z) is given by

$$x = (n + h) \cos \phi \cos \lambda \tag{6.74}$$

$$y = (n + h) \sin \phi \cos \lambda \tag{6.75}$$

$$z = \left(\frac{R_p^2}{R_e^2} n + h \right) \sin \lambda \tag{6.76}$$

where

λ = latitude
ϕ = longitude
h = altitude
R_p = earth polar radius (6356752.3142 m) and
R_e = earth equatorial radius (6378137 m)

and

$$n = \frac{R_e}{\sqrt{1 - e^2 \sin^2 \lambda}} \tag{6.77}$$

where e is the eccentricity of the earth, given by

$$e^2 = \frac{R_e^2 - R_p^2}{R_e^2} \tag{6.78}$$

Similarly the ECEF position (x, y, z) can be transformed to geodetic position (λ, ϕ, h) as follows:

$$\phi = a\tan 2 \left(\frac{y}{x} \right) \tag{6.79}$$

$$\lambda = \tan^{-1} \left(\frac{z + e'^2 R_p \sin^3 \theta}{p - e^2 R_e \cos^3 \theta} \right) \tag{6.80}$$

$$h = \frac{p}{\cos \lambda} - n \tag{6.81}$$

where

$$p = \sqrt{x^2 + y^2} \tag{6.82}$$

$$e'^2 = \frac{R_e^2 - R_p^2}{R_p^2} \tag{6.83}$$

$$\theta = \tan^{-1} \left(\frac{zR_e}{pR_p} \right) \tag{6.84}$$

The position of each satellite is computed in the ECEF frame, as a function of the time since midnight and the satellite positions at midnight. For each satellite,

$$\theta = \theta_0 + t \frac{360}{48{,}082} \tag{6.85}$$

$$\Omega = \Omega_0 - t \frac{360}{86{,}164} \tag{6.86}$$

$$S_x = R_s(\cos\theta \cos\Omega - \sin\theta \sin\Omega \cos 55°) \tag{6.87}$$

$$S_y = R_s(\cos\theta \sin\Omega + \sin\theta \cos\Omega \cos 55°) \tag{6.88}$$

$$S_z = R_s(\sin\theta \sin 55°) \tag{6.89}$$

where θ is the phase of the satellite within its orbital plane, θ_0 is the phase of the satellite at the previous midnight, Ω is the rising longitude of the orbital plane to take account of the rotating earth, Ω_0 is the rising longitude of the orbital plane at the previous midnight and the inclination of the six orbital planes is 55°.

Only the satellites in view can be used in the navigation computation, that is, satellites above the horizon. In practice, satellites below an elevation of 5° are ignored. The satellite position is transformed from the ECEF frame to a tangent plane at the receiver and this is achieved by the following transformation:

$$dx = S_x - R_x \tag{6.90}$$

$$dy = S_y - R_y \tag{6.91}$$

$$dz = S_z - R_z \tag{6.92}$$

$$p_n = -dx \sin\lambda_R \cos\phi_R - dy \sin\lambda_R \sin\phi_R + dz \cos\lambda_R \tag{6.93}$$

$$p_e = -dx \sin\phi_R + dy \cos\phi_R \tag{6.94}$$

$$p_d = dx \cos\lambda_R \cos\phi_R + dy \cos\lambda_R \sin\phi_R + dz \sin\lambda_R \tag{6.95}$$

where the receiver ECEF position is (R_x, R_y, R_z) and the receiver geodetic latitude and longitude are λ_R and ϕ_R, respectively. The position of the satellite in the tangent plane at the receiver is given in NED coordinates p_n, p_e and p_d, respectively.

Typically eight or more satellites may be visible above the horizon. However, some selections of these satellites may produce a poor fix and ideally, the selection of the most appropriate four satellites will be used in the navigation equation. Assume there are 10 satellites above 5°. Then, the permutations of possible satellites are given by the set of 10-bit numbers in the range 0–1023, which contain 4 bits. It is likely that an estimation of the user position is available and, moreover, the relative motion of the satellite will not change significantly in one second. The DCM (which is also used in the navigation equation) can be computed for each potential set of four satellites to search for the optimal position dilution of precision (PDOP) as follows:

$$A = \begin{bmatrix} \alpha_{11} & \alpha_{12} & \alpha_{13} & 1 \\ \alpha_{21} & \alpha_{22} & \alpha_{23} & 1 \\ \alpha_{31} & \alpha_{32} & \alpha_{33} & 1 \\ \alpha_{41} & \alpha_{42} & \alpha_{43} & 1 \end{bmatrix} \tag{6.96}$$

where

$$\alpha_{i1} = (\hat{R}_x - S_{ix})/\hat{R}_i \qquad (6.97)$$

$$\alpha_{i2} = (\hat{R}_y - S_{iy})/\hat{R}_i \qquad (6.98)$$

$$\alpha_{i3} = (\hat{R}_z - S_{iz})/\hat{R}_i \qquad (6.99)$$

$$\alpha_{i4} = 1 \qquad (6.100)$$

where $(\hat{R}_x, \hat{R}_y, \hat{R}_z)$ is the estimated receiver position in the ECEF frame, (S_{ix}, S_{iy}, S_{iz}) is the position of each of the ith satellite ($i = 1, \ldots, 4$) in the ECEF frame. GDOP is given by

$$GDOP = \sqrt{trace[(A^T A)^{-1}]} \qquad (6.101)$$

where

$$trace(A) = \sum_{i=1}^{4} A_{ii} \qquad (6.102)$$

For $i = 1, \ldots, 4$, the dilution of precision is known as the geometric dilution of precision and includes the local clock bias covariance. For $i = 1, \ldots, 3$, the DOP is known as the position dilution of precision (PDOP). For 10 satellites, there are 210 distinct bit patterns containing four 1s. Each of these patterns is used to select 4 satellites from 10 to compute the PDOP of this configuration. By searching through these groups of four satellites, the configuration with the minimum PDOP is found and this configuration is used in the subsequent computation of the navigation equation.

Recall that four equations are needed to solve for the receiver position (R_x, R_y, R_z) and the local clock bias b, given by the following equation

$$R_i = \sqrt{(R_x - S_{xi})^2 + (R_y - S_{yi})^2 + (R_z - S_{zi})^2} + cb \qquad (6.103)$$

where R_i is the measured pseudo-range, (S_{xi}, S_{yi}, S_{zi}) is the position of the ith satellite (computed in simulation but extracted from the ephemeris data in aircraft navigation) in the ECEF frame and c is the speed of light. The clock bias b is common to all four pseudo-range measurements. The equations can be linearized, using initial estimates of the receiver position, iterating the solution until it converges to an acceptable value. The linearized solution in matrix notation is as follows:

$$x = A^{-1} r \qquad (6.104)$$

where A is the DCM (defined in Section 3.6), x is the relative position error vector $[\Delta x, \Delta y, \Delta z, c\Delta t]^T$ and r is the relative pseudo-range error vector $[\Delta R_1, \Delta R_2, \Delta R_3, \Delta R_4]^T$ where

$$x = \hat{x} + \Delta x_i \qquad (6.105)$$

$$y = \hat{y} + \Delta y \qquad (6.106)$$

$$z = \hat{z} + \Delta z \qquad (6.107)$$

$$t = \hat{t} + \Delta t \qquad (6.108)$$

$$R_i = \hat{R}_i + \Delta R_i \qquad (6.109)$$

A Gauss–Jordan method can used to compute the solution of Equations 6.70–6.73, partly because of the simplicity of the method and also because it converges with reasonable stability.

The solution of these equations produces a better approximation for (x, y, z, t) and these new values are then used as new estimates; this process is repeated until the set of computed values and the set of estimated values converge, typically to an accuracy of a few metres (and usually within less than 10 iterations). In fact, an estimate of an initial user position at the centre of the earth (0, 0, 0, 0) can be used. Note that although the algorithm converges to centimetre accuracy, this does not imply that the navigation equations are solved to that accuracy.

The actual value of the measured pseudo-range is readily computed in a simulation from the distance between the aircraft position and the satellite position. At that point, the error terms in the pseudo-range measurement can be added, where the errors are based on the noise properties of GPS measurements, for example, the added delay in passing through the ionosphere and troposphere, or multi-path path reflections or the receiver clock bias.

Having formulated the GPS equations, the GPS sensor is simulated by rotating each satellite by one second in its orbit and solving the navigation equations every second, using a suitable GPS error model. Figure 6.14 shows the number of satellites visible (about 5° elevation) during a 24-hour period, based on satellite positions given in Brown and Hwang (1997). Figure 6.15 shows the variation in PDOP during the same period, which exhibits the sensor characteristics shown in actual GPS recordings as given in Farrell and Barth (1999). Figure 6.16 shows the position error over the same 24-hour period. The major error is caused by the ionosphere and the delay (increased pseudo-range) is based on the model in the NOAA GPS toolkit (Harris and Mach, 2007). This model is based on Klobuchar's (1987) prediction of GPS errors and includes computation of

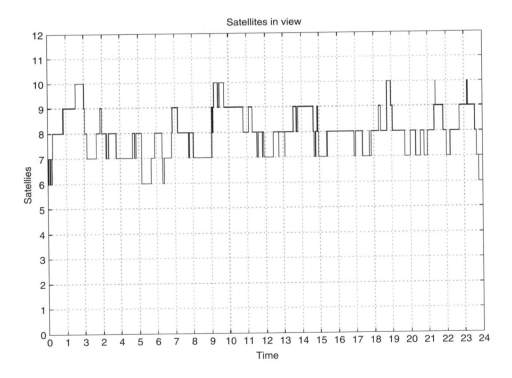

Figure 6.14 GPS satellite coverage

Simulation of Aircraft Navigation Systems

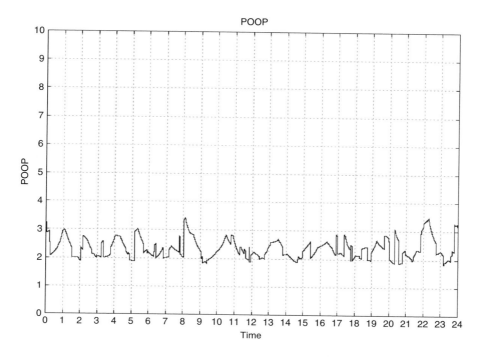

Figure 6.15 GDOP figures over a 24-hour period

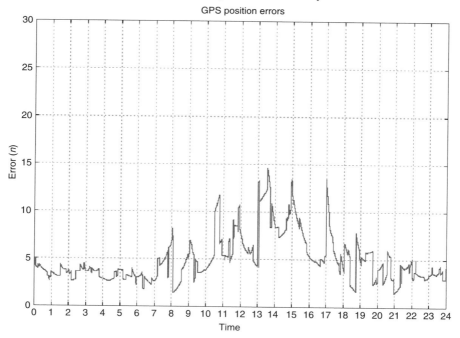

Figure 6.16 Position errors (static location) over a 24-hour period

the refraction caused by the satellite elevation angles and also the phase and amplitude of the delay, which varies over the period of a day. The increased delay around 14:00 hours local time is evident, with a more constant delay during the hours of night-time. The model also includes terms for the delay through the troposphere, as a function of each satellite's angle of elevation. The error associated with the clock bias in the GPS is eliminated in the solution of the navigation equation. In this example, a clock bias of 1 ms was assumed.

This model of a GPS receiver is only used to illustrate that sensor models can be developed from physical principles. To study GPS characteristics in more depth, a more detailed model would be needed, which would include more exact trajectories of satellite motion and more detailed models of the error terms. However, in flight simulation, a model of the form described provides representative sensor characteristics and enables sensor errors (for example, loss of satellite signals, receiver noise or ephemeris errors) to be simulated.

References

Adams, J.J. (1983) Simulator Study of a Flight Director Display System, NASA TM-84581, Langley Research Centre, Hampton.

Biezad, D.J. (1999) *Integrated Navigation and Guidance Systems*, American Institute of Aeronautics & Astronautics, Reston, VA.

Brown, R.G. and Hwang, P.Y.C. (1997) *Introduction to Random Signals and Applied Kalman Filtering*, John Wiley and Sons, New York.

Collinson, R.P.G. (1996) *Introduction to Avionics*, Chapman and Hall, London.

Farrell, J.A. (1976) *Integrated Aircraft Navigation*, Elsevier Science & Technology Academic Press Inc, San Diego, CA.

Farrell, J.A. and Barth, M. (1999) *Global Positioning System and Inertial Navigation*, McGraw-Hill.

Forsell, B. (1991) *Radio Navigation Systems*, Prentice Hall, Hemel Hempstead.

Harris, B.J. and Mach, R.G. (2007) The GPSTK: an open source GPS toolkit. *GPS Solutions*, **11**(2), 145–150.

Helfrick, A.D. (1994) *Modern Aviation Electronics*, Prentice Hall, Eaglewood Cliffs, NJ.

Hofman-Wellenhof, B., Lichtenegger, H. and Collins, J. (1992) *GPS Theory and Practice*, Springer Verlag, New York.

Kaplan, D. (1996) *Understanding GPS: Principles and Applications*, Artech House Inc, Boston, MA.

Kelly, R.J. (1984) System considerations for the new DME/P international standard. *IEEE Transactions on Aerospace and Electronic Systems*, **AES-20**(1), 2–24.

Klobuchar, J.A. (1987) Ionospheric time-delay algorithm for single frequency GSP users. *IEEE Transactions on Aerospace and Electronic Systems*, **AES-23**(3), 325–331.

Kudlinski, K.E. and Ragsdale, W.A. (1999) Design and Development of Lateral Flight Director, NASA TM-1999-208957, Langley Research Centre, Hampton.

Maling, D.H. (1991) *Coordinate Systems and Map Projections*, Elsevier Science and Technology Butterworth-Heinemann Ltd, London.

National Imagery and Mapping Agency (2000) Department of defense world geodetic systems 1984: its definition and relationships with local geodetic systems. Technical Report No. 8350.2, 3rd edn, National Imagery and Mapping Agency, Washington, DC.

Siouris, G.M. (1993) *Aerospace Avionics Systems*, Academic Press, San Diego, CA.

Smith, E.K. and Weintraub, S. (1953), The constants in the equation for atmospheric refraction index at radio frequencies, *Proceedings of IRE*, **41**, 1035–1037.

Titterton, D. and Weston, J. (2004) *Strapdown Inertial Technology*, Institution of Engineering and Technology, Stevenage.

Weir, D.H., Klein, R.H. and McRuer, D.T. (1971) Principles for the Design of Advanced Flight Director Systems Based on the Theory of Manual Control Systems, NASA CR-1748, NASA, Washington, DC.

Further Reading

Thomas, P.D. (1965), Mathematical models for navigation systems. Technical Report, No. 182, US Naval Oceanographics Office, Washington, DC.

7
Model Validation

7.1 Simulator Qualification and Approval

With the proliferation of civil airline flight simulators in the 1980s, there were obvious benefits in ensuring a high degree of compatibility between simulators developed by different manufacturers and operated by different airlines. In addition to variations in simulator technology, there was also some variability in the methods of acceptance of simulators. Several attempts were made by manufacturers to standardize on simulator Approval Test Guides (ATGs). In situations where one simulator might have to meet the requirements of several ATGs, in order to be accepted by different countries, the simulator might need to be taken out of service several times a year to enable the relevant authorities to conduct initial and recurrent checks to ensure that a simulator met its training requirements.

In 1992, a common set of criteria for the evaluation and qualification of full flight simulators was agreed following a series of working party meetings and international conferences. The 'International Standards for the Qualification of Airplane Flight Simulators' set down the requirements for an 'International Qualification Test Guide' (IQTG). In 1995, ICAO published the 'Manual of Criteria for the Qualification of Flight Simulators' (ICAO, 1995) to provide guidance to regulators and operators for the installation and recurrent checking of flight simulators (the second edition was produced in 2003). These international standards provide objective testing and guidelines for subjective testing of simulators, but more importantly, simulators qualified to this standard are then accepted by all countries represented in ICAO.

The FAA has also set out guidelines in their CFR (Code of Federal Regulations) FAR (Federal Aviation Regulation) Part 60 (FAA, 2008), which defines the requirements for the evaluation, qualification and also maintenance of FSTDs. The document includes six appendices covering requirements, objective tests for full flight simulators, subjective evaluation, sample documents, wind shear training and FSTD directives. In addition, the FAA has published Advisory Circulars for Airplane Simulator Qualification (FAA, 1991), Aeroplane Flight Training Device Qualification (FAA, 1992) and Helicopter Simulator Qualification (FAA, 1994). Data provided by the manufacturer of an FSTD must comply with the IATA Flight Simulator Design and Performance Data Requirements (IATA, 2002).

The situation is somewhat complicated as both the FAA and the JAA (Joint Aviation Authorities) have produced equivalent documents covering the qualification of flight training devices. The JAA has published four sets of documents for fixed wing aircraft JAR-STD 1A-4A, covering aeroplane flight simulators (JAA, 1998), flight training devices (JAA, 1999), flight and navigation procedure trainers and basic instrument training devices, respectively. JAR-STD 1A is equivalent to FAA AC 120–40B and covers full flight simulators from Level A to Level D, where Level D applies to zero

---

Principles of Flight Simulation D. J. Allerton
© 2009, John Wiley & Sons, Ltd

flight time approval for experienced flight crews. JAR STD-2A applies to flight training devices, for which there are seven levels, with some overlap of the four levels covered by FAA AC 120–45A. Despite the apparent duplication and confusion, the attempt has been to categorize levels of flight simulators and to standardize the processes to qualify these devices for flight training.

In addition to the published standards, a guideline to the testing and evaluation, which is undertaken by engineers, pilots and regulatory authorities, known as the *Evaluation Handbook*, was published under the auspices of the Royal Aeronautical Society and is currently in its third edition (RAeS, 2005). The regulatory authorities are responsible for answering the question 'Is this an acceptable simulation of the aircraft'? The Evaluation Handbook sets out clearly the sequence of tests to be conducted to ensure that this evaluation is conducted objectively and analytically. There is one further important aspect of simulator evaluation that is also addressed by the Evaluation Handbook. If the simulator is altered in any way, possibly with a software update or the replacement of a faulty part, then it is necessary to ensure that these changes have not altered the characteristics or behaviour of the simulator. Therefore, repeatability of tests and the comparison of sets of data with baseline data are fundamental to evaluation.

To a lesser degree, the same problems are faced by the simulator designer. Having implemented the flight model and the real-time simulation software and having constructed the simulator, its displays, visual systems and possibly motion platform, it is then necessary to check out the simulator. With so many interacting systems, a fault may be difficult to isolate and the subjective views of pilots, tasked with validation of the simulator, may include a wide degree of variance. Generally, for full flight simulators, the data provided by a manufacturer contains a large amount of flight test data, obtained from actual aircraft trials, which enable the simulator designer to compare the simulator performance and response with the actual aircraft.

Consider an example; having developed the simulator, a pilot is asked to assess the simulator in the take-off phase and concludes that the simulator is sluggish to become airborne. There are many possible reasons for this problem:

- The airspeed indicator may be incorrectly calibrated – the pilot is possibly attempting to rotate at a significantly low airspeed;
- The effective gearing of column displacement to elevator angle is incorrect;
- The trimming mechanism used in the simulator differs from the aircraft, causing a different trim point to be set in the simulator;
- The dynamic response of the landing gear model is too stiff, inhibiting the pitching rate;
- The flight model may contain errors, for example, the elevator is less effective than the aircraft or the damping in pitch is higher than the aircraft;
- The loading on the control column is higher than the aircraft, giving the sense that the aircraft is less responsive to the pull that would normally be used for the aircraft;
- The atmospheric model may be incorrect, for example, the ground temperature is too high, reducing take-off power or air density is too low, reducing lift;
- The runway may have been set to a slight uphill slope by the instructor;
- The aircraft may have a flight control law in the pitch channel, which may not have been correctly implemented in the simulator;
- The projector geometry may be incorrect, giving a false impression of height and rotation rate;
- The effect of flaps is incorrectly modelled in terms of the effect on lift, drag or pitching moment;
- There may be latency in the system, so that there is a significant delay between the pilot input and the visual system response;
- The visual system may have a restricted field of view, affecting the visual cues close to the ground;
- The pilot eye point in the visual system may not be correct, causing a false impression of rotation;
- The simulator may have a tail wind, set from a previous exercise, which has not been cancelled.

These faults are not necessarily the only causes of the problem and, of course, a pilot's subjective assessment may be influenced by a lack of experience of that particular aircraft type. Furthermore, this is only one very specific example of simulator validation; there will be several hundred similar test points where the simulator must be carefully checked against the aircraft performance and response.

As a result of this specific test, a pilot might be adamant that the problem is caused by the elevator effectiveness and it would be tempting to increase the value of the associated variables in the flight model, repeating this exercise, until the pilot is satisfied with the response in the simulator. Later, the control loading system might be modified and another pilot might undertake a similar test, leading to a different assessment, perhaps making another change to elevator effectiveness. In such situations, the Evaluation Handbook (or the test methodology it summarizes) is essential. It sets out, in a logical order, a series of tests and measurements to ensure standardization in simulator evaluation. In addition, it summarizes the information that should be obtained from the simulator, so that objective tests can be repeated to allow data from a test to be checked against a baseline. In these cases, the regulator works with the airline test pilots and the simulator engineers to ensure that the results obtained from the simulator are as objective as possible.

A detailed series of tests are set out in the Evaluation Handbook to minimize any subjective assessment. For example, the column control loading plots will be carefully checked against similar tests on the aircraft; the aircraft acceleration will be recorded and compared with the aircraft data; the landing gear response will be checked for static loads and pitching rotation; the visual system geometry will be carefully checked in terms of the pilot eye position and so on. Consequently, when any variation is detected, it is assessed against as much flight data as possible to gain an understanding of the problem before attempting to rectify the problem.

Although it might seem reasonable to obtain data from as many test exercises as possible conducted in the aircraft and then compare these results against the data from equivalent tests in the simulator, there is likely to be some variability between an aircraft and its simulator:

- The resolution of the visual system in a flight simulator is far less than the visual acuity of a pilot;
- The field of view in a simulator may be reduced;
- The motion system is not able to provide the actual forces experienced in an aircraft;
- The model may be an approximation of the actual aircraft dynamics;
- The simulator introduces latency, which is not experienced in the real world;
- A pilot is aware that they are flying a simulator, reducing levels of fear or anxiety, which may change their behaviour.

Consequently, the evaluation of a simulator is not simply an exercise in data collection and data comparison. The test pilot is an essential element of simulator evaluation. Once approved, the simulator will be used for pilot training and although any variations between the simulator and its aircraft need to be minimized, the simulator models may need to be adapted to provide effective training.

Data obtained from airborne exercises is likely to include pilot inputs. Nowadays, data acquired from flight tests is recorded at high data rates and resolution. Although it is difficult to produce identical sets of inputs in a simulator using manual input (even for pulse or doublet inputs), it is possible to 'drive' the simulator with recorded pilot inputs in order to replicate actual pilot inputs.

Finally, although a large part of simulator evaluation focuses on the flight model and engine model, in terms of performance and response, in a modern aircraft with flight control systems, head-up displays, flight management systems, EFIS and EICAS displays, GPWSs, TCASs and so on, it is also necessary to validate these systems in situations where they may be used in training. To give some idea of the problem, a GPWS takes inputs from the flight model and warnings must be integrated with the visual system (including the fog model) and navigation systems so that, in a training exercise, alerts occur in situations where the flight crew can appreciate the severity of a situation. If the simulator and GPWS use different coordinate frames or different units or data is incorrectly transferred between systems, the alerts may occur at incorrect positions, giving the flight crew a false sense of security in potentially hazardous situations. One further possibility is that the terrain database in the GPWS system is different from the database used in the visual system. Consequently, the evaluation of the GPWS in the simulator requires extensive testing to ensure that it does not generate false or misleading alerts or, more importantly, does not miss critical alerts.

7.2 Model Validation Methods

Development of aircraft flight models comprises three stages:

1. derivation of a mathematical model;
2. implementation of the model in code and
3. validation of the model.

Anyone who has developed complex flight models, only to watch the model plummet to the ground or pitch up to the vertical at rest on the runway within two seconds, will appreciate the importance of the model testing or validation stage. Often, validation can take over 80% of the project development time and software tests must be designed with as much rigour as the model itself. One further observation is that the quality of the model validation depends, to a high degree, on the quality and quantity of data provided for comparison. It is not uncommon for students to take models from a textbook, implement the model in a modelling language and then be surprised at the system response. Without any test data, it is possible to assume a level of fidelity that is not warranted by the formulation of the model.

Validation consists of two forms of model testing, although strictly both are interdependent. Performance tests are, for the most part, based on steady-state measurements, for example, maximum level speed, engine-out climb rate or time to reach V_1 from brakes release or sideslip angles for rudder input. The other form of tests involve the system dynamics to ensure that the response of the system to inputs is correct, for example, the pitching moment change caused by a change of engine thrust, response in yaw and roll resulting from an engine failure or pitch rate response with a flight control law.

One of the most difficult aspects of model validation is the isolation of variables. For example, the ground run during take-off is affected by engine thrust, aircraft drag, the landing gear dynamics (to follow the runway centre line) and weather conditions set by the instructor. Any variation or error in the associated terms could influence the ground run distance or time measurements. For example, consider Figure 7.1 that shows the ground roll for a Boeing 747-100 aircraft (260,000 kg) with Pratt and Whitney JT9D engines, at sea level.

Full power is applied at time $t = 0$ and the aircraft becomes airborne at 32 s at approximately 155 kt. There are a number of observations from this data:

- The aircraft is at rest when full power is available – in practice, as power is applied the aircraft will start to roll forwards, so that full power occurs at about 10 kt;
- The landing gear takes several seconds to settle down;
- The aircraft is airborne once the main gear oleo length is positive (after 32 s);
- A constant elevator input of $-8°$ is applied in this test throughout the ground run – in practice, this value would change as the pilot rotates the aircraft to the appropriate climb attitude – the performance will also change slightly for different values of elevator input;
- There are clear non-linearities in the aircraft response, which might introduce significant differences if simplified to a linear model;
- There is an abrupt change in the lift and drag forces as the aircraft transitions from the ground roll to climbing flight.

Model Validation

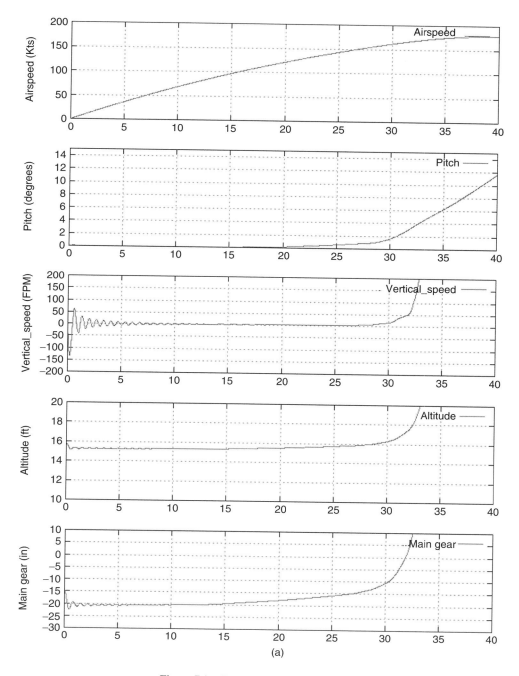

Figure 7.1 Ground roll: (a) performance

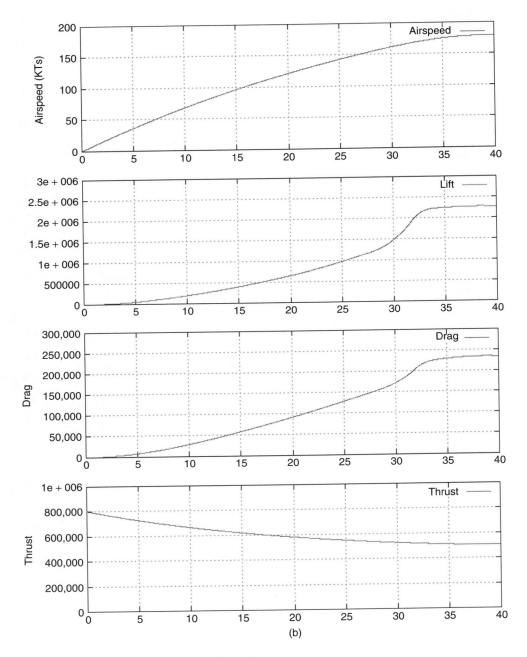

Figure 7.1 (b) forces

7.2.1 Cockpit Geometry

Aircraft cockpits and flight decks are designed to accommodate a wide range of size of pilot, providing adjustment for seat position, seat height and also rudder pedal position for some aircraft. The location of the flight controls, the position of panels containing switches and knobs and the pilot eye height should correspond very closely to the aircraft. These are simple measurements but they are critical. For example, a side-stick will have a specific spring force, but it will also be located where it can be reached by the pilot, possibly with a vertical offset in mounting. The resultant moment, in terms of the pilot's hand and arm, needs to be same as the aircraft; if the side-stick is in a different position, the pilot will need to apply one force in the simulator and another in the aircraft for the same manoeuvre. Similarly, the pilot eye position needs to be adjusted in terms of height and distance from the windscreen. The graphics generated by the visual system will be based on an exact and specific point in the cockpit. Small variations in the eye point can give rise to some degree of distortion for the pilot, seeing one geometric image in the simulator and another in the aircraft. The control column will have artificial loading, but the dimensions of the control column (or centre stick for a military aircraft) are also critical, as the forward/backward and left/right displacements and moment arm must be the same as the aircraft.

7.2.2 Static Tests

From knowledge of the aircraft data, there are fixed data points where the simulator should produce the same values as those produced by the aircraft. The weather model is a good example, where air density, pressure ratio and density ratio are defined at specific altitudes. Similarly, indicated airspeed and Mach number can be checked at a specified altitude, ambient air temperature and true airspeed.

The engine data should provide details of thrust, fuel flows, engine temperatures and so on at known flight conditions. The simulator does not necessarily need to be flown in these conditions; specific test values can be applied to the engine model, recording the engine variable values at these fixed points. The thrust provided by the engines is critical to the validation of the aerodynamic model, so considerable care is given to ensuring that the engine model performance is valid at a large number of test points covering the aircraft flight envelope (altitude versus airspeed). Jet engine performance may also vary with the angle of attack and the angle of sideslip, which can occur in asymmetric flight conditions (e.g. one engine out), where engine performance is critical to the safe operation of the aircraft, but with much reduced operational margins for speed. Prior to testing the aircraft response to engine failure conditions, the engine performance is rigorously checked to ensure that the thrust delivered is very close to the published data for the engine.

The pilot inceptors are critical to the fidelity of the simulator in terms of their size, location, tactile feel and forces. The forces applied are not necessarily linearly proportional to the control displacement and some controls may exhibit mechanical backlash. Plots are recorded for force versus position for the primary flight controls and also for toe brakes, engine levers, mechanical trim wheels and selectors.

For aircraft dynamics, steady-state conditions occur in level, climbing and turning flight enabling checks to be made against pitch attitude, true airspeed, rate of climb or bank angle. These tests should also include flap and gear settings. Similar checks are made for ground handling, for example, minimum turn radius tests at specific ground speeds for both rudder pedal inputs and also the nose wheel steering controller (if applicable). Figure 7.2a shows the steady-state aircraft performance at 3000 ft, 160 kt, flaps 30, gear down. Figure 7.2b shows the steady-state aircraft performance at 15,000 ft, 320 kt, flaps up, gear up. Although these are 'straight line' plots, they show that initial conditions can be computed for steady-state flight and they provide steady-state values for specific flight conditions.

Note the different values for pitch attitude, elevator setting and engine lever setting for the different flight conditions. Note also the additional check that $\alpha = \theta$ for level flight. These plots

292   Principles of Flight Simulation

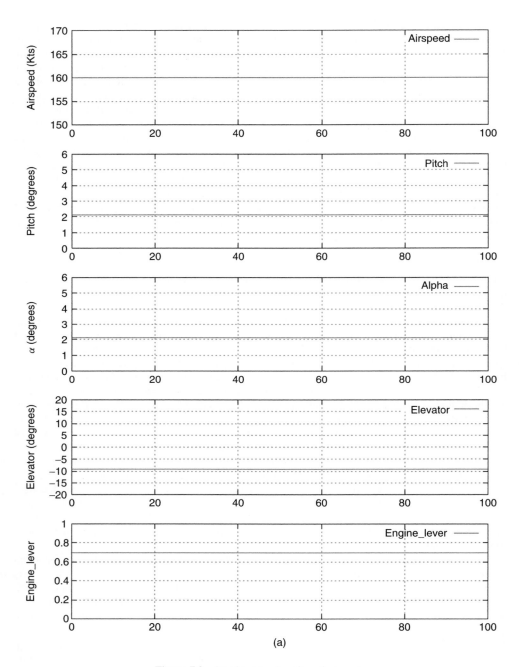

Figure 7.2   Steady-state aircraft performance

Model Validation

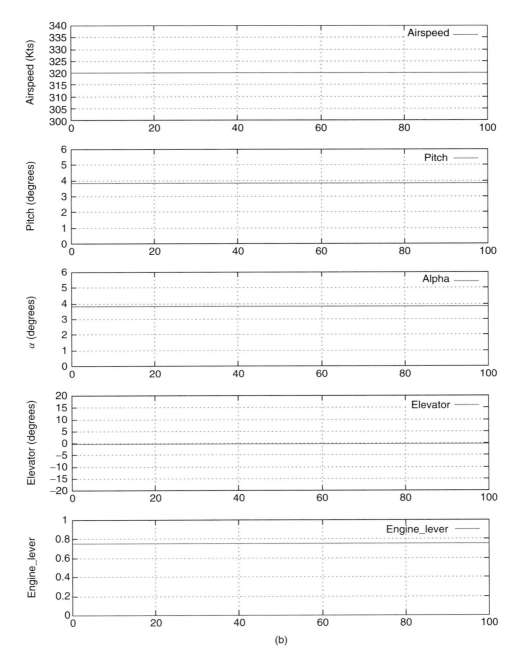

Figure 7.2 (*continued*)

were produced using the auto-trim function (described in Section 4.4). For these examples, the aircraft steady-state condition is valid for 100 s, giving further confidence in the aerodynamic and engine models and also providing values that can be recorded in both graphical and tabular form. Alternatively, values can be recorded from trimmed flight conditions, flown by a test pilot.

7.2.3 Open-loop Tests

In applying a step input, a pulse input or a doublet input to the aircraft or engines, the dynamic modes of the models are excited. The response of the model to these inputs provides useful insight into the system characteristics. This is particularly valid where the models are non-linear, or an input results in changes over a relatively large operating range, where the assumption that a variable might be constant is not valid. It is also possible to generate pilot inputs during an aircraft flight test to obtain airborne data based on specific inputs. Although, as we shall see in the following sections, exciting the modes of motion of the aircraft provides a set of responses that characterize the aircraft dynamics, primary inputs also provide a useful check of system gains or damping factors to ensure that the simulation has correctly captured the aircraft behaviour. In these open-loop tests, the controls are fixed. Figure 7.3a shows the aircraft response to a rudder step input of 5° at 3000 ft, 180 kt, flaps 20, gear down (with the yaw damper disengaged). The aircraft yaws through 50° in 40 s, inducing a rate of descent of 1400 ft/min. Note the spiral mode, with slowly increasing yawing rate. Although this is not a standard aircraft manoeuvre, in a flight test the pilot trims the aircraft and then applies a required rudder input for a fixed time. Figure 7.3b shows the aircraft trimmed at 180 kt at 3000 ft with an engine lever input of 20% of the lever range. The subsequent increase in thrust causes a pitch-up motion from 4° to just over 8° in 40 s, accompanied by an increase in altitude of 400 ft and a 7 kt variation in airspeed. Although the onset of phugoid motion is apparent from this plot, it does provide an objective (and repeatable) measurement of engine performance for a specific trimmed state.

These tests should also include flap and gear retractions and extensions and operation of the speed brake. The effect of flap and gear in the landing and take-off is critical and changes in airspeed, pitch attitudes and vertical speeds need to be checked closely against the aircraft flight manual. It is often argued that the landing phase is the most critical aspect of flight simulation and errors in both performance and handling are most evident in this phase of flight.

The 'perfect' step inputs shown in Figure 7.3 are simply waveforms and they could easily be replaced with less exact pilot inputs obtained from flight trials. For certification, the number of data plots to be acquired and the accuracy (or resolution) of the data must be established. For example, the open-loop rudder input exercise could be undertaken for airspeeds from 150 to 450 kt in steps of 20 kt (15 tests) over a range of altitude from sea level to 40,000 ft in steps of 1000 ft, requiring a total of 600 tests and producing 600 sets of data for comparison, just for this single variable. On the one hand, too many tests generate far too much information, but on the other hand, insufficient testing could miss a problem introduced by a change to the simulator.

A very important point, which is clear from the previous example, is the need to document test results very carefully. In addition to the information provided in the plots, it is also essential to record essential data that is required to repeat the exercise, for example, the version numbers of any software, the time and date of the recording, any initial conditions (e.g. ground temperature, winds, and flight control modes), the location of the recorded files and the names of the investigators who conducted the exercise.

7.2.4 Closed-loop Tests

In closed-loop tests, the pilot is asked to execute specific manoeuvres, for example, to turn to a new heading or climb to a new altitude. It is assumed that the pilot would operate the aircraft using

Model Validation

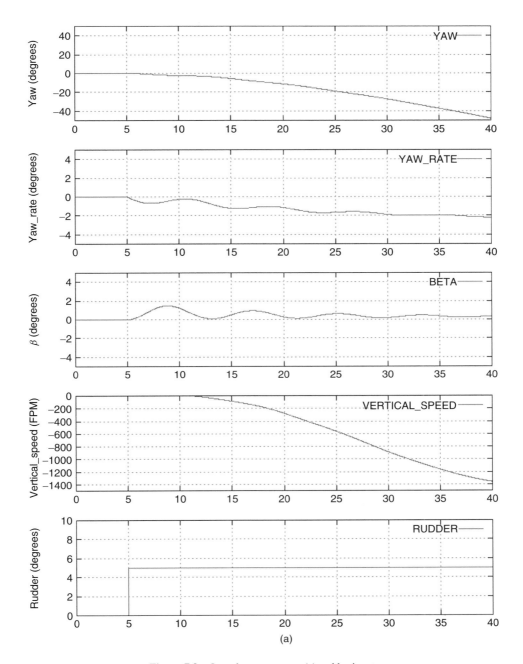

Figure 7.3 Open-loop response: (a) rudder input

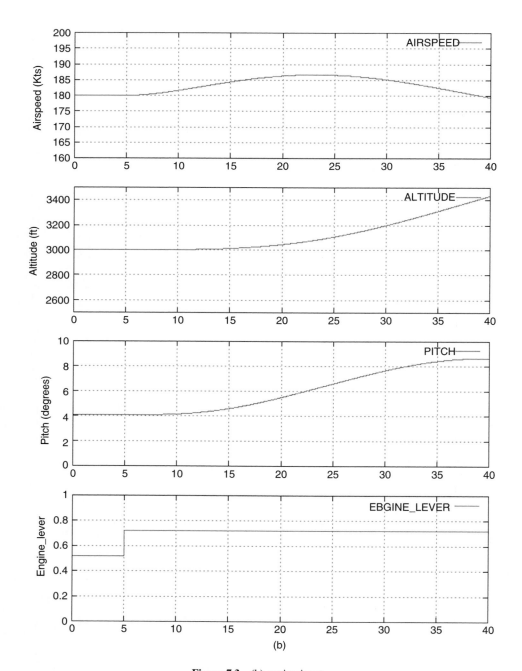

Figure 7.3 (b) engine input

normal flying techniques, for example, limiting the bank angle in turns or the vertical speed in climbing manoeuvres. Such tests include steady-state sideslip, crosswind landing, speed changes, altitude changes, heading changes, stall characteristics, engine failures and so on. Although there is some variability between the flying techniques of different pilots, for simulator evaluations in airlines, the tests are conducted by senior captains, experienced on the aircraft type and in many cases holding test pilot qualifications.

A large part of these tests include system failures. For example, a pilot requires sufficient control authority to manoeuvre the aircraft safely if an engine fails in the reverse thrust mode, following touchdown. In such tests, the engine is failed at different speeds, and the pilot inputs to maintain the runway centre line in the presence of the failed engine are monitored. Such events must be demonstrated by pilots in their recurrent simulator checks, so the tests are conducted on the basis that the pilot's inputs are representative of less experienced flight crew.

Figure 7.4 shows recorded pilot input, turning from a heading of 20 °M to 50 °M at 3000 ft at 180 kt. Aileron input, bank angle, turn rate and heading are recorded during the manoeuvre. Note the range of aileron input to maintain a steady turn rate at this airspeed. Repeating this manoeuvre should require similar inputs, in terms of aileron input in the range ±8° to achieve a 30° heading change in 25 s.

Figure 7.5 shows the pilot rudder input response to a reverse thrust failure of the outboard port engine of a Boeing 747-100 at 100 kt (35.5 s), where the runway QDM is 77°. Note that the pilot is able to contain the failure with a heading change of less than 2° and a maximum rudder input of 21° following the failure. If the rudder control loading system was replaced or modified at a later stage or software updates were made to the undercarriage model, this plot provides a baseline for piloted checks that the modifications have not introduced any anomalies.

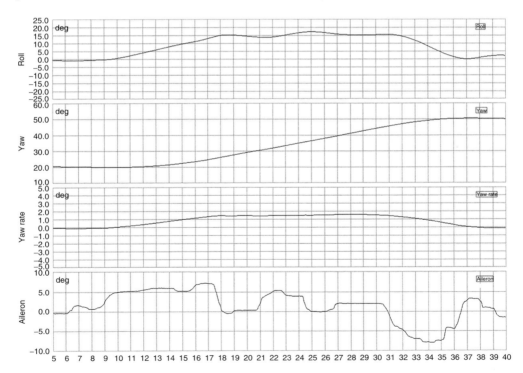

Figure 7.4 Pilot in-the-loop 50° heading change

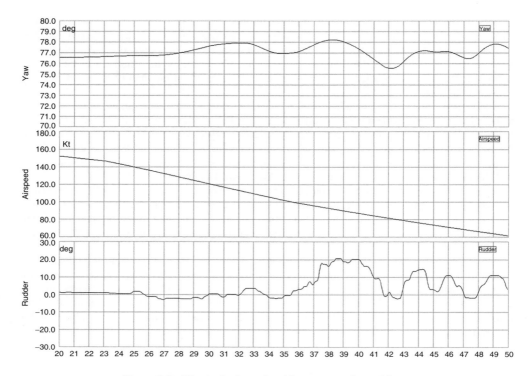

Figure 7.5 Pilot in-the-loop closed-loop reverse thrust failure

In addition to pilot inputs, automatic flight modes must also be checked extensively. For example, an autopilot height hold system needs to be checked over a range of airspeeds, acquiring an altitude from both above and below the selected altitude and in the presence of turbulence. Such tests extend to checking the operational modes, for example, accidentally switching off both autopilots or disengaging selections. Autoland functions are particularly important because flight crews are trained to monitor autoland operations and to disengage the autopilot in response to indicators and warnings. The fidelity of the autopilot functions is critical; any variation between the simulated system and the aircraft system would not only lead to negative transfer but the differences could also endanger the aircraft operation. The equipment manufacturers will themselves use flight simulation to verify their designs and should have flight test data from flight trials that are used by the simulator designer to validate the implementation of automatic functions.

7.3 Latency

Flight simulators are designed using real-time computer systems. This means that the equations are solved at a fixed frame rate, which is sufficiently fast that there are no apparent discontinuities in the simulation of the aircraft, its engines and subsystems, as perceived by the pilot. From Shannon's sampling theorem, the frame rate must be at least twice the fastest time constant of the modelled system (Herman and Grant, 1974). In other words, with a frame rate of 50 Hz (20 ms frame time), the aircraft dynamics should not have any time constants faster than 40 ms (25 Hz). However, as McLean (1990) points out, with uniform sampling, the sampling period depends on the Nyquist frequency (also known as the *folding frequency*), $0 < \omega_n \leq \pi/T$, where ω_n is the natural frequency of the system, T is the sampling period and π/T is the folding frequency. A rule of thumb for many control systems is that T must be less than one-tenth of π/ω_n. Powell and Katz (1975) report that values up to one-third of π/ω_n are sufficiently conservative for flight control applications.

The modern flight simulator is likely to be constructed from a number of computers connected by network cables, with analogue channels sampling inputs at a predefined rate. Consequently, inevitable delays occur in real-time simulation, where an input occurring in one computer is not accessed or processed by another computer for a significant amount of time. These delays are referred to as *lags* or *latency* and if they are sufficiently large, they can introduce major problems in real-time simulation. This effect is particularly difficult to analyse for non-linear systems. A simplified diagram of the primary sources of latency in flight simulation is shown in Figure 7.6.

Latency L_1 arises from the pilot's response, that is, the time to acquire information, process the information (cognition) and then activate the muscles to implement the desired control function. Typical pilot delays are of the order 0.3–0.5 s. L_2 occurs because there is a finite time to acquire the inputs needed by the flight model and engine model, which will include analogue to digital conversion sampling times. The delay L_3 is the time needed to compute the aircraft dynamics before the updated aircraft position and attitude can be passed to the visual system. Finally, the visual system needs to complete its current frame, render the image for the next frame and then project the image for the pilot. L_4 can vary from one to four frames, depending on the rendering methods used by the image generator. With a network architecture, if packets of 1000 bytes are sent between computers at 10 MB/s, delays of the order 1 ms will be encountered for each transfer, simply from the time to transmit over 8000 bits over the network bus.

Latency can introduce delays that alter the overall system behaviour to the point where a stable system is actually unstable in the presence of latency. In such cases, it is important to analyse the overall system transfer function (including latency) to determine any instability, particularly in the case where latency cannot be reduced. The simplest way to model latency is a first-order lag, given by

$$\frac{1}{1+\lambda s} \tag{7.1}$$

where λ is the latency. However, this does not strictly generate a lag, as shown in Figure 7.7, where the upper plot shows an input pulse of 1 s. The lower plot shows a true latency of 0.2 s, whereas the middle plot models the latency as a first-order lag. The attraction of using a first-order lag is that the overall transfer function is still a polynomial in s and analysis of the characteristic equation will yield the system stability. The lower output is very easy to simulate; it is simply a queue where data is removed from the queue 0.2 s after it is entered into the queue. Although this

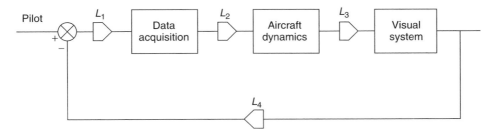

Figure 7.6 Latency in flight simulation

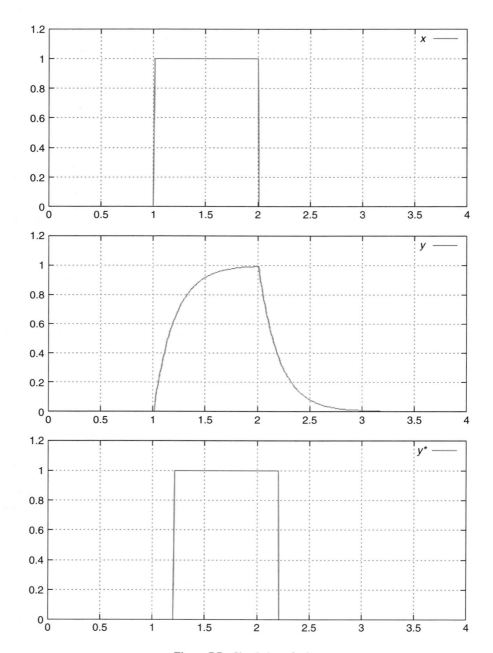

Figure 7.7 Simulation of a lag

Model Validation

method allows a lag to be simulated, it provides no insight into the effect of the lag on system stability.

A more exact method used to simulate and analyse the effect of a lag is Padé's approximant (Franklin et al. 1980). The Laplace transform of a pure time delay $f(t - \lambda)$ is given by $F(s)e^{-s\lambda}$. Assume that e^{-s} could be represented to an accuracy ε by a transfer function of the form

$$\frac{b_0 s + b_1}{a_0 s + 1} \tag{7.2}$$

then

$$e^{-s} - \frac{b_0 s + b_1}{a_0 s + 1} = \varepsilon \tag{7.3}$$

e^{-s} can be expanded using a Maclauren series, given by

$$e^{-s} = 1 - s + \frac{s^2}{2!} - \frac{s^3}{3!} + \frac{s^4}{4!} - \cdots \tag{7.4}$$

Similarly, the series expansion of $\dfrac{b_0 s + b_1}{a_0 s + 1}$ is given by Dwight (1961)

$$\frac{b_0 s + b_1}{a_0 s + a_1} = b_1 + (b_0 - a_0 b_1)s - a_0(b_0 - a_0 b_1)s^2 + a_0^2(b_0 - a_0 b_1)s^3 + \cdots \tag{7.5}$$

Equating the terms of Equations 7.4 and 7.5

$$b_1 = 1 \tag{7.6}$$

$$(b_0 - a_0 b_1) = -1 \tag{7.7}$$

$$-a_0(b_0 - a_0 b_1) = \frac{1}{2} \tag{7.8}$$

$$a_0^2(b_0 - a_0 b_1) = -\frac{1}{6} \tag{7.9}$$

With four equations in three unknowns, it is straightforward to solve for a_0, b_0 and b_1, by matching the first three coefficients. Substituting λs for s

$$e^{-\lambda s} \cong \left(\frac{1 - \dfrac{\lambda s}{2}}{1 + \dfrac{\lambda s}{2}} \right) \tag{7.10}$$

Extending this approach to a second-order equivalent form gives

$$e^{-\lambda s} \cong \left(\frac{1 - \dfrac{\lambda s}{2} + \dfrac{(\lambda s)^2}{12}}{1 + \dfrac{\lambda s}{2} + \dfrac{(\lambda s)^2}{12}} \right) \tag{7.11}$$

Figure 7.8 shows the response to a pulse unit. Bearing in mind that such a discontinuity is the worst-case input, the higher order model of a lag is much closer to the actual lag than the simple first-order model given in Figure 7.7.

Assuming that the latency is known, the technique based on Padé's approximant can provide insight into changes in the time response; by replacing the lag with a second-order transfer function, the system stability can be analysed in terms of the poles, or root locus to establish the degree to which latency might contribute to any system instability.

In a flight simulator, most sources of latency can be either measured or estimated. For example, the sampling period for an analogue to digital converter can be determined from a data sheet. However, the time for the operating system to respond to the interrupt from an A/D conversion or to run specific processes may vary during simulation, introducing additional delay during a frame. These problems are compounded with real-time networks, where packets are transmitted between computers to fulfil a protocol (avoiding contention for the network). Generally, the order of packet transmission and the size of each packet are known, enabling the transmission delay to be estimated.

There is also some variability associated with visual systems. If data is received by the visual system at the start of a frame, the image generated may be based on values of aircraft pitch and roll that were computed several milliseconds ago, during the previous frame. The image is then formed in a frame store, which may be completed in a single frame but, with some visual systems, may require

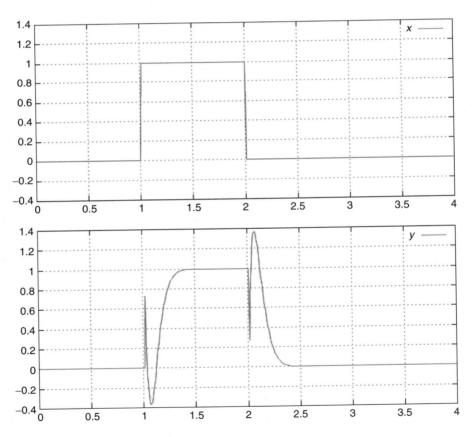

Figure 7.8 Simulation of lag using the Padé approximant

an additional frame to post-process the image, for example, to overlay HUD imagery. The image is projected in the next frame, but it is conjectural as to when this information is actually processed by a pilot. If the frame rate is 20 ms (50 Hz), then it would seem reasonable to assume that, on average, a pilot acquires the information 10 ms into the frame, adding a further element of latency.

An additional source of latency is the human pilot. Although the latency of a pilot is likely to be the same in a simulator as it is in an aircraft, it may be affected by the visual feedback cues, motion feedback cues and tactile cues provided in the simulator. In tests in a simulator with an Airbus side-stick, a pilot was asked to use the side-stick to control a cursor on a display. A small circle was drawn at a new position on the display with a small random variation in the repetition rate; the pilot was asked to respond to the change, where the side-stick provided direct position control of the cursor, by placing the cursor inside the circle. Figure 7.9 shows the results for five inputs of longitudinal side-stick movement and an expanded view of the fourth input.

Note the consistent lag of 0.3 s for longitudinal movement of the stick. Although an Airbus side-stick requires a considerable force, this does not affect the response time. Even for small changes (rather than the coarse stimuli used in this experiment), a pilot still needs time for visual acquisition, cognitive processes and activation of the hand and arm muscles. In simulator validation, these latencies must be included in pilot studies attempting to identify or assess simulator latency (300 ms is equivalent to six frames at a 50-Hz frame rate). Of course, in designing active flight controls, the control system must compensate for pilot latency, in order to avoid pilot-induced oscillation.

A further consideration is the actual measurement of latency. In some cases, the regulatory authorities may require a demonstration of the system latency to ensure that it falls below a defined value. Often, latency is an accumulation of a number of latencies so that it is necessary to measure several sources of latency to arrive at an overall figure of latency. Generally, latency is the time between a pilot input and its perceived output in the projected image or the reactive motion of the platform. One way to characterize this delay is to detect the system response to a step input from steady-state conditions. However, the difficulty is compounded; first, because a step input at one place in the simulator may result in a slow response where it is difficult to determine the onset of the response and secondly, because there may be a wide variation in latency, which is often referred to as jitter. It may be more important to know the worst-case latency rather than the average latency. As simulators are designed to meet a fixed frame rate, a margin is usually designed into the system to allow for small variations in latency. However, while latency values may accumulate, encroaching into the frame margin, in some cases if the frame period is exceeded, the latency may jump by an integer value of the frame rate. These events are particularly true for visual systems, where the system has to start processing to make use of the full frame time, with the consequence that the frame rate may drop from 50 frames per second to 25 for a short time. Although such events may not be discernible to a pilot, many of the simulator algorithms are designed for a fixed frame rate and, if this violated, unexpected behaviour can ensue, for example, because the integration step length has, in effect, been doubled.

In the case of a simulator based on a distributed system of computers, it is necessary to record events on several computers in order to track sources of latency, which implies that all measurements are aligned with respect to a common clock, in turn necessitating specialist hardware for the measurements. The alternative option is to measure events with local processor clocks, but these can drift by 10^{-6} s/s (almost one-tenth of a second per day) and the complexity of synchronizing clocks can be considerable. Often, packets sent from one computer to another are buffered, so it is important to trace the packet to ascertain whether the source of the latency is caused by the system hardware, the application software or even the operating system software. Diagnostic software can be used to detect an input and then record the reaction of the motion system using accelerometers or the reaction of the visual system using an image-detection system. However, specific diagnostic software can miss latency introduced by the application software. For example, display rendering times can vary with aircraft attitude, or a helicopter rotor model may require the implicit solutions

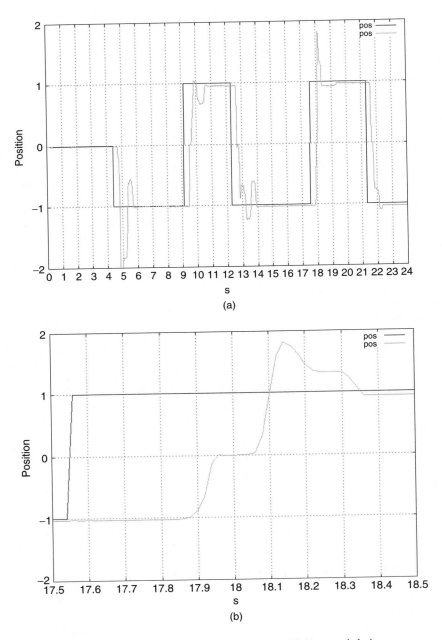

Figure 7.9 Pilot response time: (a) five stimuli and (b) expanded view

of equations that may take a varying time to converge. Consequently, measurement of latency in flight simulators is often overlooked or measured crudely or estimated from knowledge of the system architecture and interfaces. For civil airline flight simulators, figures of 100 ms are often quoted as the worst-case acceptable latency. For military systems, where the aircraft is much more responsive, latency figures of 20–40 ms are more typical. However, it is important to appreciate two limitations: first, one to two frames of latency are inevitable with the distributed architecture of a modern flight simulator. Secondly, if the latency is unacceptable, the only viable option is to try to increase the frame rate, but in practice this may prove impractical for the motion platform or visual system, without a major redesign of the simulator.

Finally, a test pilot may be asked to validate the model, undertaking a series of handling tests, to ensure that the latency is acceptable. Problems caused by latency are very difficult to detect with subjective tests of this form. By way of illustration, consider the traces shown in Figure 7.10, which show the response of a Boeing 747-100 to a sinusoidal input, applied at 5000 ft, 250 kt, flaps up and gear up. Figure 7.10a shows the normal aircraft response to the input. Three faults were introduced independently into the simulator and are shown in Figures 7.10b–d. In Figure 7.10b, a dead band is introduced to the elevator control, so that no movement of the control occurs within $\pm 10°$ of the full elevator movement ($\pm 23°$), albeit a relatively large dead band. The limited input of the elevator control is shown in the lower trace. There is a noticeable reduction in the peak of the pitch response and a discernible discontinuity in the pitch rate response. In Figure 7.10c, the damping in pitch (C_{mq}) was increased by 50% from the baseline model. Note the slight reduction in the magnitude of both the pitch and pitch rate response but with no apparent change in phase. In Figure 7.10d, a 0.1 s lag was added to the pitch rate variable q. Although a small phase shift is apparent in the pitch response, the system appears to be much more damped by the introduction of latency (and in this artificial example, just for one variable).

The important point to draw from these simple examples is the difficulty of discriminating between simulator problems and how difficult latency is to detect, even though it may introduce subtle changes to the system. In particular, latency can give the impression of a change in the system dynamics. Much more detailed analysis is needed to isolate problems of this form, but it is very easy to draw the wrong conclusions, where the simulator validation is only based on pilot-in-the-loop tests.

One solution to reducing the effects of latency is to provide delay compensation (Guo et al., 2005). In Figure 7.11, the difference between the path represented by the sampled inputs and the average value of the variable over a number of frames is evident. Using the current value, plus recently sampled values, gives a better estimate of the variable during the frame. The most common method is to derive a polynomial passing through the last n points and then apply the polynomial to extrapolate to the adjacent point. For example, a simple first-order extrapolation algorithm is given by

$$y_i^* = y_i + 3(y_i - y_{i-1})/2 \qquad (7.12)$$

where y_i denotes a sampled value and y_i^* denotes the predicted value.

7.4 Performance Analysis

In addition to validation of the flight model dynamics, the engine model dynamics, the landing gear dynamics and the flight control systems, it is also essential that the simulator meets its real-time criteria. Most real-time flight simulators have a fixed frame period, say 20 ms, with a small margin

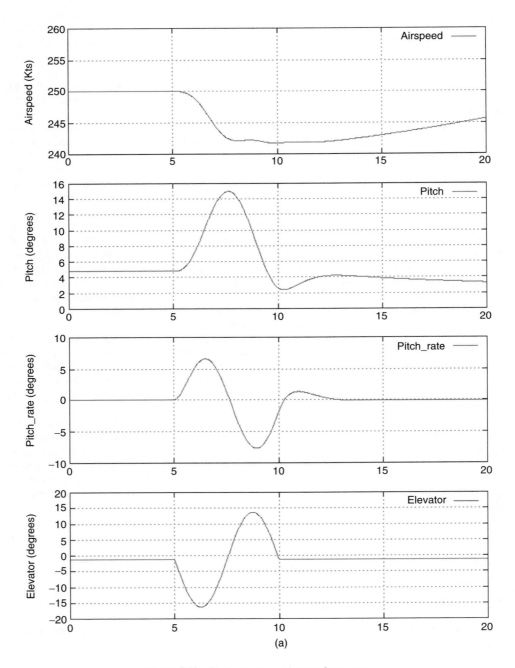

Figure 7.10 Pitch response: (a) normal response

Model Validation

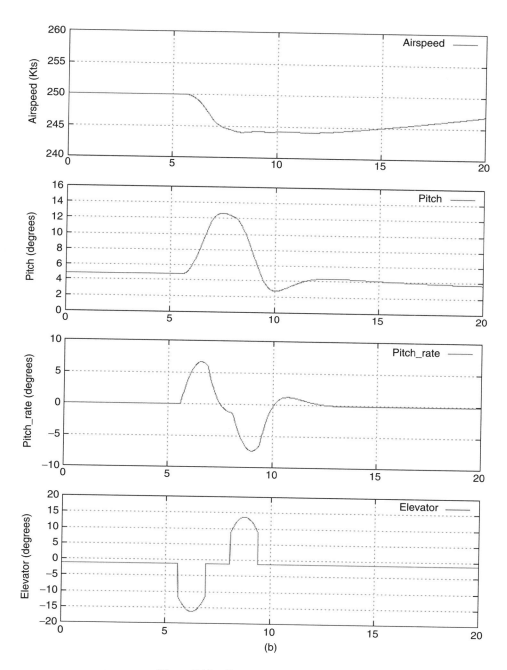

Figure 7.10 (b) elevator dead band 5°

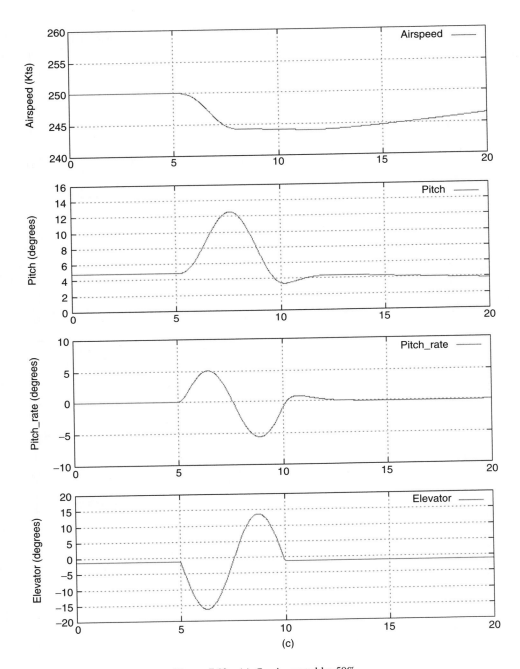

Figure 7.10 (c) C_{mq} increased by 50%

Model Validation

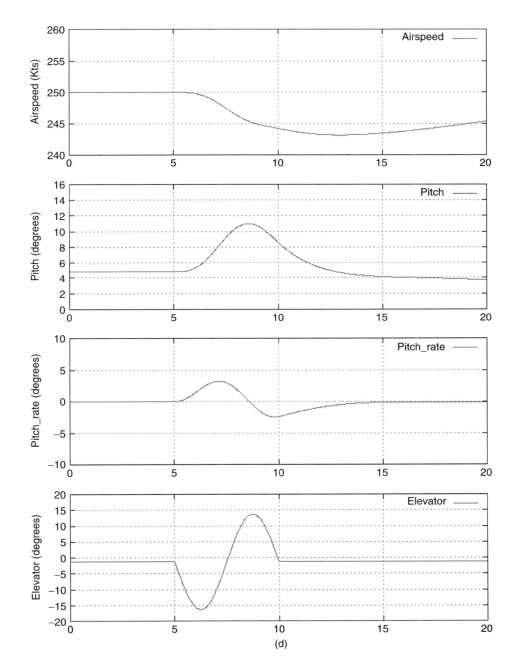

Figure 7.10 (d) latency 0.1 s (q)

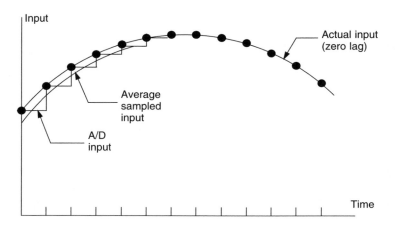

Figure 7.11 Delay compensation

to allow for some variability in timing, say 2 ms. This constraint implies that the computations will never exceed the frame period at any time. However, there may be considerable variation in the amount of computation per frame. For example, if the yaw damper is engaged, the code for the yaw damper is executed once per frame, whereas if it is disengaged, this code is omitted from the computations in each frame. In addition, if extra code is needed in the simulator, it is necessary to know if there is a sufficient margin to cater for this increased load without violating the timing constraint and also to confirm the performance following the installation of the software.

One of the difficulties in taking timing measurements of a real-time system is the processing time needed to execute the timing measurements. This problem is often referred to as the *probe effect*, where timing measurement software may even alter the real-time performance of the system it is attempting to analyse. One very important consideration to bear in mind is that most modern operating systems perform input and output, which includes writing to a screen or writing to a file. If any diagnostic software writes to the screen or writes to a file, the overhead of these transfers may distort any timing information. For example, writing 20 characters to the screen may involve several hundred accesses to the screen display hardware including scrolling, which involves rewriting all the screen pixels. Such operations can take several milliseconds. Similarly, while writing to a disk file, the characters may be buffered in memory but after 1024 bytes, say, are written, the buffer will be written to the disk. Single disk transfers can take several milliseconds, particularly if the disk writing head moves from one track to another. Introducing additional monitoring tasks that take several milliseconds each frame may very well distort the data acquired by the monitoring software. Consequently, timing or diagnostic information is usually written to memory during the data capture phase and then written to the screen or disk once the data acquisition task is completed.

The simplest method is to analyse the compiled code. Each machine instruction will have a specific addressing mode and the execution times (in terms of machine cycles) can be estimated from the data sheets for the processor. These figures also depend on memory access times to fetch instructions and data, but they can provide a good estimate of execution times. The method does have several disadvantages:

- There may be a lot of instructions to analyse in a large segment of code;
- If the source code is changed, the machine instructions can vary considerably;

Model Validation

- Modern processors optimize the instruction pipeline of a processor and make efficient use of caches for both code and data, which can distort estimates of processor performance.

Moreover, if the code contains conditional branches, it is necessary to analyse the code to determine the worst-case (in terms of timing) paths through the code. The timing is also dependent on the algorithms in the code. For example in searching, where a loop is proportional to some variable n, which is determined at run-time and where there may be inner loops also dependent on n, it is very difficult to estimate the timing and worst-case values of n should be assumed.

Most processors provide a system clock that can read by an operating system call. However, the clock resolution may not be sufficiently accurate to time a few instructions. An alternative method is to time a small section of code a large number of times, 10,000 say, and then compute the average time to execute the section. Although this method can be used to time small fragments of code execution, it provides no insight into worst-case timing. In addition, the time needed to read the system clock 10,000 times should be deducted from the measurement.

Further insight can be gained by tracing the flow of software. On entry to a procedure, a software counter is incremented. Control flow analysis of these counters can show which sections of code are executed most frequently, particularly for inner loops where small timing increments can have a significant impact on the overall performance. A related technique is to execute a background monitoring process, which regularly monitors the code of the process under test, to identify critical code segments. Of course, the background monitoring process also consumes processor cycles with the possible effect of distorting any timing measurements, even to the point where the processor can become overloaded.

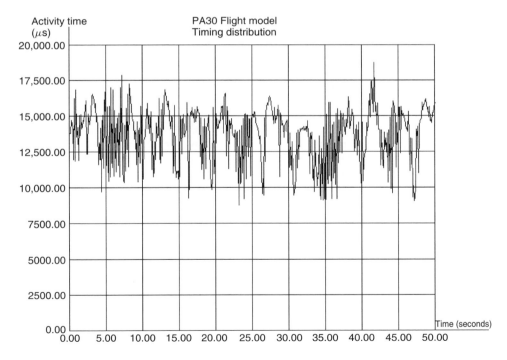

Figure 7.12 Timing analysis

Another commonly used method is to provide a set of hardware counters that can be started or stopped, but under program control. A counter is started on entry to a procedure and stopped on exit from the procedure. After eliminating the time needed to access the counter, the amount of time spent in specific procedures or sections of code can be identified. A variation of this technique is to generate a digital-to-analogue output, connected to an oscilloscope or logic analyser. Setting a high output on entry to a procedure and a low output on exit from the procedure produces a waveform that can be captured and analysed.

The importance of performance analysis cannot be overstressed. It is very easy to write code that executes more times than anticipated or where an algorithm fails to converge at the desired rate. In these situations, processing delays are hidden in the code and possibly not observable unless frame times are violated or the iteration rate drops to an unacceptable value. It is not uncommon for an algorithm to be data dependent, so that for certain values of data, the execution period increases unexpectedly. Figure 7.12 illustrates system timing figures for a real-time system with a 20 ms frame rate. Although the average frame time is approximately 13 ms, several frames exceeded 17 ms during the 50 s period of measurement.

7.5 Longitudinal Dynamics

Aircraft are designed to meet specific performance criteria. For civil aircraft, the main handling requirement is that the aircraft can be manoeuvred safely and accurately by an airline pilot within the range of aircraft operations without placing major demands on the pilot (Cook, 1997). On the one hand, the aircraft needs sufficient stability to be insensitive to external disturbances caused by turbulence or wind, while at the same time, the pilot needs a sufficiently responsive aircraft for evasive manoeuvres or to make heading or flight path adjustments during an approach. In the longitudinal axis, a pilot controls the aircraft pitch and airspeed by applying the elevator and engine controls. Application of the elevator will change the aircraft pitch attitude and the velocity vector, leading to a transfer between kinetic energy and potential energy. The relationships between the elevator and engine inputs and the resultant outputs in terms of pitch, pitch rate, angle of attack, altitude, airspeed and vertical speed can be analysed, in order to predict the aircraft handling. The techniques that are used by the aircraft designer to establish the aircraft response to inputs can also be used by the simulator designer to validate the dynamics of the simulated aircraft.

Generally, the longitudinal dynamics can be separated from the lateral dynamics in order to analyse the longitudinal dynamics. This simplification is valid for an aircraft where the aircraft geometry is symmetric about the $x-z$ plane of the body axes and where the normal engine thrust can be assumed to contribute negligible yawing or rolling moments. In this case, with centred aileron and rudder inputs, the rolling and yawing moments are zero and the lateral forces are zero. Although this simplification is useful in the analysis of fixed wing aircraft dynamics, for rotary wing aircraft, the effects of the main rotor and tail throughout the flight envelope make it difficult to analyse the longitudinal dynamics in isolation.

In analysing the modes of motion, it is common practice to linearize the equations of motion and to use dimensional stability derivatives (Stevens and Lewis, 2003). Considerable care is needed with the conversion of non-dimensional derivatives to dimensional derivatives. The two main notations are referred to by Babister (1962) as the American and British units, and he clarifies their usage. Of course, dimensional derivatives can also be defined in SI units, which is the practice adopted in this book. The scaling factors between the non-dimensional derivatives and the dimensional derivatives, for the longitudinal terms, are given in Table 7.1.

In state-space notation, the system of equations to represent longitudinal motion is

$$\dot{x} = Ax + Bu \qquad (7.13)$$

Model Validation

Table 7.1 Longitudinal dimensional derivatives

Non-dimensional derivative	Multiplier	Dimensional derivative
C_{x_u}	$\rho V S / 2m$	X_u
C_{x_w}	$\rho V S / 2m$	X_w
$C_{x_{\dot{w}}}$	$\rho S \bar{c} / 4m$	$X_{\dot{w}}$
C_{z_u}	$\rho V S / 2m$	Z_u
C_{z_w}	$\rho V S / 2m$	Z_w
$C_{z_{\dot{w}}}$	$\rho S \bar{c} / 4m$	$Z_{\dot{w}}$
C_{z_q}	$\rho V S \bar{c} / 4m$	Z_q
$C_{z_{\delta e}}$	$\rho V^2 S / 2m$	$Z_{\delta e}$
C_{m_u}	$\rho V S \bar{c} / 2I_y$	M_u
C_{m_w}	$\rho V S \bar{c} / 2I_y$	M_w
$C_{m_{\dot{w}}}$	$\rho V S \bar{c}^2 / 4I_y$	$M_{\dot{w}}$
C_{m_q}	$\rho V S \bar{c}^2 / 4I_y$	M_q
$C_{m_{\delta e}}$	$\rho V^2 S \bar{c} / 2I_y$	$M_{\delta e}$

where x is the state vector, u is the vector of control inputs and the matrices A and B contain the aircraft dimensional derivatives, as follows:

$$\begin{bmatrix} \dot{u} \\ \dot{w} \\ \dot{q} \\ \dot{\theta} \end{bmatrix} = \begin{bmatrix} X_u & X_w & 0 & -g \\ Z_u & Z_w & V & 0 \\ M_u + M_{\dot{w}} Z_u & M_w + M_{\dot{w}} Z_w & M_q + M_{\dot{w}} V & 0 \\ 0 & 0 & 1 & 0 \end{bmatrix} \begin{bmatrix} u \\ w \\ q \\ \theta \end{bmatrix}$$

$$+ \begin{bmatrix} X_{\delta e} & X_{\delta t} \\ Z_{\delta e} & Z_{\delta t} \\ M_{\delta e} + M_{\dot{w}} Z_{\delta e} & M_{\delta t} + M_{\dot{w}} V \\ 0 & 0 \end{bmatrix} \begin{bmatrix} \delta_e \\ \delta_t \end{bmatrix} \qquad (7.14)$$

where V is the true airspeed, derived from the aircraft body frame velocities (u, v, w) given by $\sqrt{u^2 + v^2 + w^2}$ and g is the gravitational acceleration.

Note that two simplifications can be applied. First, it is assumed that the aircraft is in the trimmed state; therefore, terms including $M_{\dot{w}}$ can be ignored. Secondly, for a jet aircraft, the thrust is assumed to be constant; $X_{\delta e}$ is also assumed to be zero, as the elevator produces a negligible force in the aircraft body frame X axis, giving the simplified form of the longitudinal equations of motion:

$$\begin{bmatrix} \dot{u} \\ \dot{w} \\ \dot{q} \\ \dot{\theta} \end{bmatrix} = \begin{bmatrix} X_u & X_w & 0 & -g \\ Z_u & Z_w & V & 0 \\ M_u & M_w & M_q & 0 \\ 0 & 0 & 1 & 0 \end{bmatrix} \begin{bmatrix} u \\ w \\ q \\ \theta \end{bmatrix} + \begin{bmatrix} X_{\delta e} & X_{\delta t} \\ Z_{\delta e} & Z_{\delta t} \\ M_{\delta e} & M_{\delta t} \\ 0 & 0 \end{bmatrix} \begin{bmatrix} \delta_e \\ \delta_t \end{bmatrix} \qquad (7.15)$$

The derivatives $C_{x_u}, C_{x_u}, C_{z_u}$ and C_{z_w} are not usually provided in a simulator data package and the following approximations are normally made for jet engine aircraft:

$$C_{x_u} = -2C_d \qquad (7.16)$$

$$C_{x_w} = C_{l_d} \qquad (7.17)$$

$$C_{z_u} = -2C_l \qquad (7.18)$$

$$C_{z_w} = -C_d \qquad (7.19)$$

The remaining dimensional derivatives are derived from the corresponding non-dimensional derivatives in the longitudinal equations of motion. The dynamic stability of the longitudinal motion of the aircraft is therefore given by the eigenvalues of the A matrix, which are found by solving the linear equation:

$$|\lambda I - A| = 0 \tag{7.20}$$

where I is the identity matrix.

Expanding the determinant produces a polynomial in λ, given by

$$\lambda^4 + a_1 \lambda^3 + a_2 \lambda^2 + a_3 \lambda + a_4 = 0 \tag{7.21}$$

For a stable aircraft, the roots of this polynomial (the eigenvalues) should be negative, typically with a pair of roots close to the imaginary axis and a pair of roots some distance from the imaginary axis. In other words, the motion is characterized by two second-order equations of the form

$$(\lambda^2 + 2\zeta_{lp}\omega_{lp} + \omega_{lp}^2)(\lambda^2 + 2\zeta_{sp}\omega_{sp} + \omega_{sp}^2) \tag{7.22}$$

where the subscript lp denotes the long-period phugoid and the subscript sp denotes the short-period phugoid.

The term *phugoid* was introduced by Lanchester (1908), who analysed the oscillatory longitudinal motion of light aircraft, and is based on the Greek word for flight (but unfortunately choosing the meaning of fleeing rather than flying). Nevertheless, the term has stuck and pilots using elevator input to produce the desired short-term response can induce a longer term response, causing a change in altitude and airspeed. The short-period mode gives the aircraft its responsiveness, while the longer mode is readily damped by the aircraft dynamics and basic flying skills.

Consider the example of a Boeing 747-100 in the landing configuration, with full flaps and gear down. Assuming the aircraft is close to sea level at 155 kt, this is a common manoeuvring speed where the aircraft may be flown manually and, in this situation, the pilot needs to make changes to control the aircraft flight path. In the trimmed state, the aircraft pitch attitude is 2.15°, the elevator is at $-22.9°$ and the power lever is at 68% of travel.

The following terms were computed for this trimmed state:

U_0	79.788094
X_u	-0.031890
X_w	0.112561
Z_u	-0.244705
Z_w	-0.571591
M_u	0.000238
M_w	-0.007013
M_q	-0.395607
$X_{\delta e}$	0.0
$M_{\delta e}$	-0.503491
$Z_{\delta e}$	-3.006194

For the system equations

$$\dot{x} = Ax + Bu$$
$$y = Cx$$

Model Validation

$$A = \begin{bmatrix} -0.031890 & 0.112561 & 0 & -9.81 \\ -0.244705 & -0.571591 & 79.788094 & 0 \\ 0.000238 & -0.007013 & -0.395607 & 0 \\ 0 & 0 & 1 & 0 \end{bmatrix}$$

$$B = \begin{bmatrix} 0 \\ -3.006194 \\ -0.503491 \\ 0 \end{bmatrix}$$

$$C = \begin{bmatrix} 0 & 0 & 0 & 1 \end{bmatrix}$$

giving the following eigenvalues for A:

$$-0.490858 \pm 0.750656$$

$$-0.008686 \pm 0.150039$$

corresponding to a natural frequency $\omega_{lp} = 0.150$ rad/s and a damping ratio $\zeta_{lp} = 0.058$ for the long-period phugoid mode and a natural frequency $\omega_{sp} = 0.897$ rad/s and a damping ratio $\zeta_{sp} = 0.547$ for the short-period phugoid mode.

The Octave (Eaton *et al*. 2008) program Example 7.1 enters the state-space matrices A, B and C, obtains the eigenvalues of A, produces the transfer function and generates the pitch response shown in Figure 7.13 (in radians) for a step input. The pitch response transfer function produced by the Octave function `ss2tf` is

$$\frac{\theta(s)}{\delta e(s)} = \frac{-0.802297s^2 - 0.450577s - 0.035780}{s^4 + 0.999088s^3 + 0.844067s^2 + 0.036149s + 0.018170} \tag{7.23}$$

Example 7.1 Octave pitch response program

```
A = [-0.031890   0.112561    0.0        -9.81;
     -0.244705  -0.571591   79.788094    0.0;
      0.000238   0.007013   -0.395607    0.0;
      0.0        0.0         1.0         0.0];

B = [0.0; -3.006194; -0.503491; 0.0];

C = [0.0 0.0 0.0 1.0];

D = 0;

Lambda = eig(A);

[num, den] = ss2tf(A, B, C, D);

ptf = tf(num, den);

step(dtf, 1, 100, 100);
```

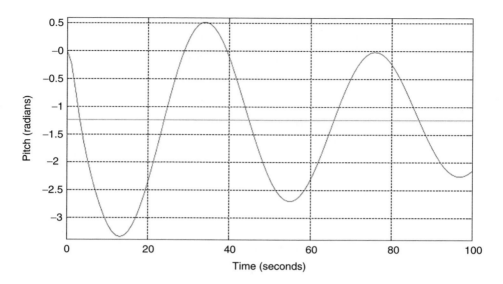

Figure 7.13 Pitch response – linearized model

The responses $u(s)/\delta e(s)$, $w(s)/\delta e(s)$ and $q(s)/\delta e(s)$ can be obtained by setting the appropriate element of the matrix C to 1. By comparison, the response obtained from the flight simulator, for the same initial conditions, is shown in Figure 7.14. Note that both traces have a period of approximately 43 s and a similar damping ratio. Clearly, the short-period response, which occurs in the first 20 s of the non-linear model, is not evident in the linear model.

Figure 7.15a shows the long-period phugoid response for a Boeing 747-100. The aircraft is positioned 3000 ft at 200 kt with flaps up and gear up. An elevator backward input of 10° was applied for 5 s. Figure 7.15b shows the response for the same initial conditions, but with the auto-throttle engaged (200 kt). Notice the considerable damping resulting from the effect of the auto-throttle. Similar effects are noticeable with variable-pitch constant-speed propellers, where the propeller attempts to 'constant speed' during the phugoid.

Figure 7.16a shows the short-period phugoid response for the same initial conditions. The pitch response and the damped pitch rate response are clearly illustrated in the top two traces. A doublet input (stick backwards for n seconds, then stick forwards for n seconds) is shown in Figure 7.16b. The doublet manoeuvre shows the aircraft pitch rate response to both positive and negative inputs, providing valuable insight into the short-period response.

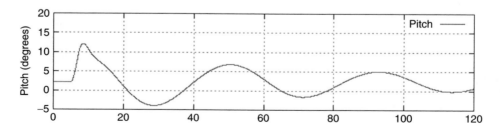

Figure 7.14 Pitch response – non-linear model

Model Validation

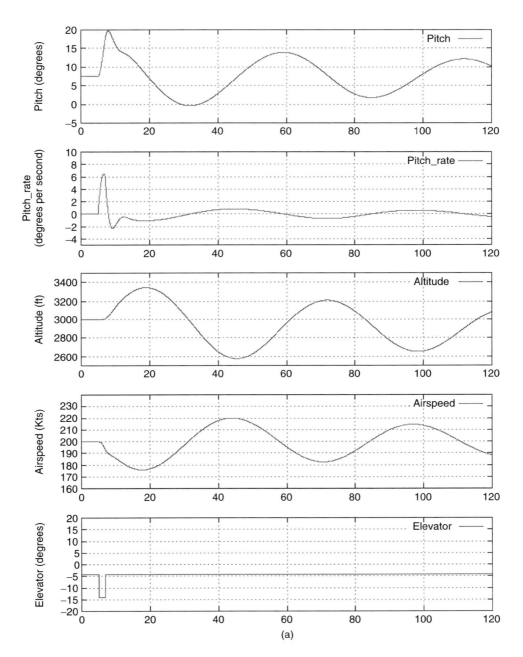

Figure 7.15 Long-period phugoid: (a) normal

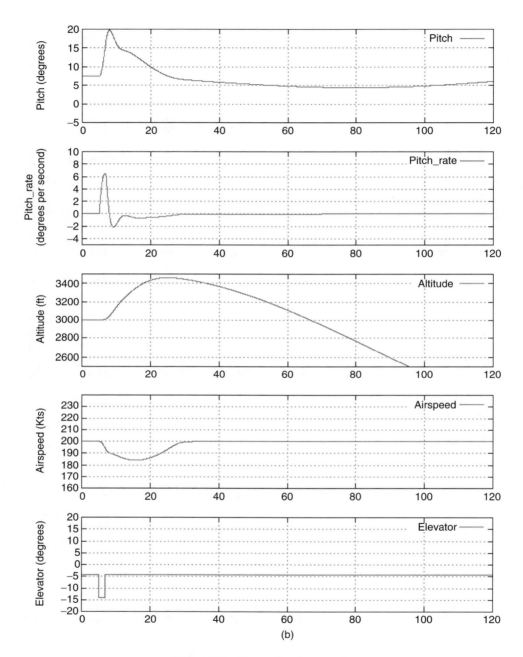

Figure 7.15 (b) auto-throttle engaged

Model Validation

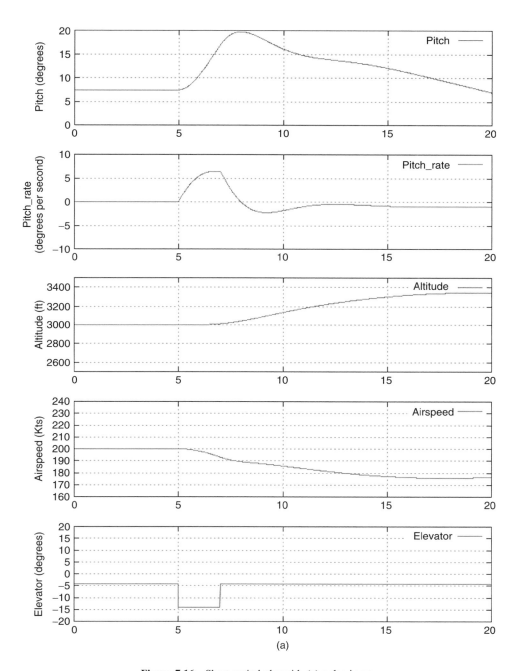

Figure 7.16 Short-period phugoid: (a) pulse input

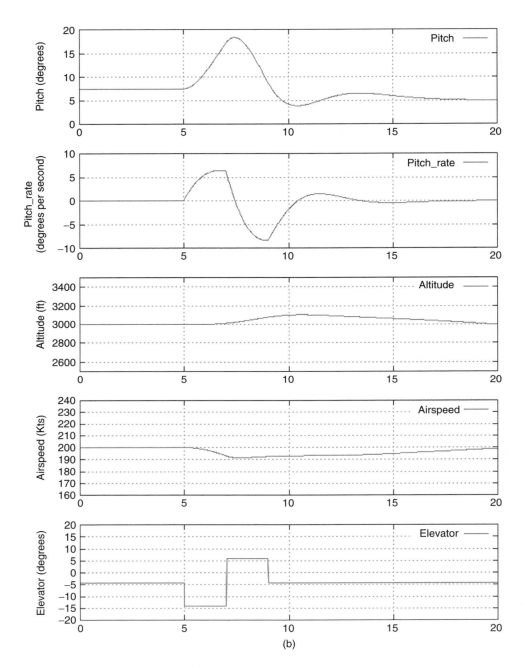

Figure 7.16 (b) doublet input

Model Validation

For comparison with trials data, the damped natural frequency ω_d and the damping ratio ζ can be measured from the dynamic responses. ω_d is defined as

$$\omega_d = \frac{2\pi N}{t_2 - t_1} \qquad (7.24)$$

where N is the number of peaks (not including the first peak), t_1 is the time of the first peak and t_2 is the time of the Nth peak. Consider the pitch response in Figure 7.17, taken from Figure 7.15.

Taking $t_1 = 32$ s, $t_2 = 85$ s and $N = 1$, $\omega_d = 0.119$ rad/s (a period of approximately 53 s). The damping ratio is given by

$$\zeta = \frac{-\ln\left(\frac{A_n}{A_1}\right)}{\sqrt{4\pi^2 N^2 + \ln\left(\frac{A_n}{A_1}\right)}} \qquad (7.25)$$

where A_n is the amplitude at time t_N and A_1 is the amplitude at time t_1. From the plot, $A_1 = 7.5$ and $A_2 = 5.5$, giving $\zeta = 0.0496$. The natural frequency is given by

$$\omega_n = \frac{\omega_d}{\sqrt{1 - \zeta^2}} \qquad (7.26)$$

giving $\omega_n = 0.119$ rad/s.

Two further simplifications are possible to estimate the long-period and short-period characteristics. For the long-period mode, the changes in pitch and airspeed are dominant, so that the terms for w and q could be omitted. The simplified state-space equation is given by

$$\begin{bmatrix} \dot{u} \\ \dot{\theta} \end{bmatrix} = \begin{bmatrix} X_u & -g \\ \frac{-Z_u}{V} & 0 \end{bmatrix} \begin{bmatrix} u \\ \theta \end{bmatrix} \qquad (7.27)$$

The eigenvalues from the determinant of this matrix are given by

$$\lambda^2 + X_u \lambda - \frac{Z_u}{g} = 0 \qquad (7.28)$$

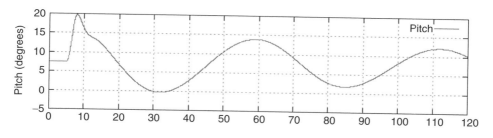

Figure 7.17 Pitch response data

The natural frequency and damping ratio of this second-order equation are then

$$\omega_n = \sqrt{\frac{-Z_u g}{V}} \tag{7.29}$$

$$\zeta = \frac{-X_u}{\omega_n} \tag{7.30}$$

Assuming $Z_u = -2C_L$ and the average coefficient of lift is approximately 1, then

$$\omega_n \approx \sqrt{2}\frac{g}{V} \tag{7.31}$$

$$\zeta = \frac{1}{\sqrt{2}\frac{L}{D}} \tag{7.32}$$

In other words, the period depends on the entry speed to the phugoid (independent of the aircraft mass, shape or dynamics) and the damping ratio depends on the lift/drag curve, which is normally designed for endurance or range for civil transport aircraft.

A similar analysis is used to estimate the characteristics of the short-period phugoid. The dominant terms are the pitch rate and angle of attack response, so that the terms for u and pitch can be ignored. The simplified state-space equations are

$$\begin{bmatrix} \dot{w} \\ \dot{q} \end{bmatrix} = \begin{bmatrix} Z_w & V \\ M_w + M_{\dot{w}} Z_w & M_q + M_{\dot{w}} V \end{bmatrix} \begin{bmatrix} w \\ q \end{bmatrix} \tag{7.33}$$

Given that

$$Z_\alpha = V Z_w \tag{7.34}$$

$$M_\alpha = V M_w \tag{7.35}$$

Then

$$\begin{bmatrix} \dot{\alpha} \\ \dot{q} \end{bmatrix} = \begin{bmatrix} \frac{Z_w}{V} & 1 \\ M_\alpha + M_{\dot{\alpha}} \frac{Z_\alpha}{V} & M_q + M_{\dot{\alpha}} \end{bmatrix} \begin{bmatrix} \alpha \\ q \end{bmatrix} \tag{7.36}$$

Solving for the eigenvalues

$$\lambda^2 - \left(M_q + M_{\dot{\alpha}} + \frac{Z_\alpha}{V}\right) + M_q \frac{Z_\alpha}{V} - M_\alpha = 0 \tag{7.37}$$

gives the following estimated values for the natural frequency and the damping ratio of the short-period phugoid

$$\omega_n = \sqrt{M_q \frac{Z_\alpha}{V} - M_\alpha} \tag{7.38}$$

$$\zeta = \frac{-\left(M_q + M_{\dot{\alpha}} + \frac{Z_\alpha}{V}\right)}{2\omega_n} \tag{7.39}$$

The main benefit of estimating the natural frequency and damping of the longitudinal modes of motion, in terms of modelling the aircraft dynamics, is that it provides a degree of confidence in the early stages of development of a model. With the different conventions used for units, it is very easy to introduce major errors (particularly factors of two) in the data, which can have adverse effects on the dynamics of the model and are often difficult to locate. An early test of any model is to initialize the aircraft with reasonable values for altitude and airspeed and then start the dynamic model. Some evidence of a phugoid response should be noticeable, whereas incorrect signs, wrong units or simple coding errors can produce a model with no aerodynamic characteristics whatsoever. Later on in the development, the availability of a representative off-line linear model can provide a useful tool for the development of control laws and can also help to identify problems in the model.

7.6 Lateral Dynamics

In analysing the lateral dynamics of an aircraft, it is assumed that they are independent of the longitudinal dynamics of the aircraft. This simplification is useful in eliminating the longitudinal dynamics from the analysis. However, it is likely that lateral motion of the aircraft, particularly turning manoeuvres, will cause a change in airspeed and altitude. Again, in linearizing the equations of motion, it is assumed that the change of state of many variables will be small and can be neglected over a small range. However, the lateral dynamics introduce a further problem; rolling motion can induce yawing motion and vice versa, so there is a coupling between angular rotations about the roll and yaw axes that must be included in the analysis. Generally, the effects of engine thrust are also ignored, although this may not be valid for propeller wash or for helicopter rotor contributions or for engine failure conditions.

Three forms of instability are observable in the lateral dynamics of fixed wing aircraft:

- Roll subsidence – as the aircraft banks, there is a resistance to the rolling motion, which is usually highly damped;
- Spiral divergence – with poor lateral stability, if the aircraft starts to turn, this motion can gradually lead to the turn becoming steeper and tighter (often resulting in a spiral dive);
- Dutch roll oscillation – a yawing motion will induce a small rolling motion and vice versa; although this motion damps out, it is undesirable for passengers and most civil aircraft are fitted with a yaw damper to suppress this motion.

The state-space form of the lateral equations of motion is given by

$$\begin{bmatrix} \dot{\beta} \\ \dot{p} \\ \dot{r} \\ \dot{\phi} \end{bmatrix} = \begin{bmatrix} Y_\beta & 0 & -1 & \frac{g}{V} \\ L_\beta & L_p & L_r & 0 \\ N_\beta & N_p & N_r & 0 \\ 0 & 1 & 0 & 0 \end{bmatrix} \begin{bmatrix} \beta \\ p \\ r \\ \phi \end{bmatrix} + \begin{bmatrix} 0 & Y_{\delta r} \\ L_{\delta a} & L_{\delta r} \\ N_{\delta a} & N_{\delta r} \\ 0 & 0 \end{bmatrix} \begin{bmatrix} \delta a \\ \delta r \end{bmatrix} \tag{7.40}$$

As in the linear longitudinal model, the derivatives defined in the A and B matrices are dimensional. The state vector comprises the angle of sideslip β, the roll rate p, the yaw rate r and the angle of bank ϕ. The subscripts δa and δr refer to the aileron and rudder inputs, respectively. The dimensional derivatives are obtained directly from the non-dimensional derivatives, using the scaling factors given in Table 7.2.

The linearized small perturbation model is fourth order (for this representation of the state variables), giving four transfer functions for aileron input: $\beta(s)/\delta a(s)$, $p(s)/\delta a(s)$, $r(s)/\delta a(s)$ and $\phi(s)/\delta a(s)$ and similarly, four transfer functions for rudder input: $\beta(s)/\delta r(s)$, $p(s)/\delta r(s)$, $r(s)/\delta r(s)$ and $\phi(s)/\delta r(s)$. The characteristic equation is a fourth-order polynomial of the form:

$$\left(s + \frac{1}{T_s}\right)\left(s + \frac{1}{T_r}\right)(s^2 + 2\zeta_d \omega_d s + \omega_n^2) \tag{7.41}$$

Table 7.2 Lateral dimensional derivatives

Non-dimensional derivative	Multiplier	Dimensional derivative
$C_{y\beta}$	$\rho V^2 S / 2m$	Y_β
$C_{L\beta}$	$\rho V^2 S / 2I_x$	L_β
C_{L_p}	$\rho V S b^2 / 4I_x$	L_p
C_{L_r}	$\rho V S b^2 / 4I_x$	L_r
$C_{N\beta}$	$\rho V^2 S b / 2I_z$	N_β
C_{N_p}	$\rho V S b^2 / 4I_z$	N_p
C_{N_r}	$\rho V S b^2 / 4I_z$	N_r
$C_{y\delta r}$	$\rho V^2 S / 2m$	$Y_{\delta r}$
$C_{L\delta a}$	$\rho V^2 S b / 2I_x$	$L_{\delta a}$
$C_{L\delta r}$	$\rho V^2 S b / 2I_x$	$L_{\delta r}$
$C_{N\delta a}$	$\rho V^2 S b / 2I_z$	$N_{\delta a}$
$C_{N\delta r}$	$\rho V^2 S b / 2I_z$	$N_{\delta r}$

where T_s is the time constant of the spiral mode, T_r is the time constant of the roll subsidence mode and ζ_d and ω_d are the damping ratio and natural frequency of the Dutch roll mode, respectively. Note that the roots at $-1/T_s$ and $-1/T_r$ are real – there is no oscillatory motion associated with these terms, although the damping of the spiral mode can be positive, in which case, the spiral mode slowly diverges.

Figure 7.18 shows the sideslip response to a rudder step input for the linear model. The 747-100 aircraft is trimmed at 3000 ft, 180 kt, flaps 20°, gear up and a 10° rudder input is applied. The following terms were computed for this trimmed state:

V	92.827721
Y_v	-0.099593
L_β	-1.700982
L_p	-1.184647
L_r	0.223908
N_β	0.407420
N_p	-0.056276
N_r	-0.188010
$Y_{\delta r}$	0.740361
$L_{\delta a}$	0.531304
$L_{\delta r}$	0.049766
$N_{\delta a}$	0.005685
$N_{\delta r}$	-0.106592

For the systems equations

$$\dot{x} = Ax + Bu$$
$$y = Cx$$

where

$$A = \begin{bmatrix} -0.099593 & 0 & -1 & 0.1056796 \\ -1.700982 & -1.184647 & 0.223908 & 0 \\ 0.407420 & -0.056276 & -0.188010 & 0 \\ 0 & 1 & 0 & 0 \end{bmatrix}$$

Model Validation

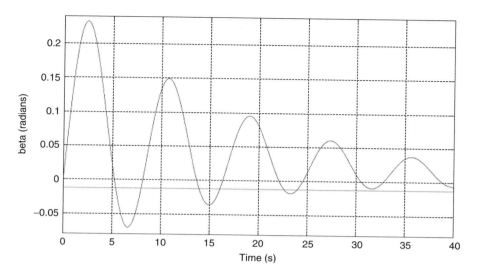

Figure 7.18 Rudder response – linearized model

$$B = \begin{bmatrix} 0.740361 \\ 0.049766 \\ -0.106592 \\ 0 \end{bmatrix}$$

$$C = \begin{bmatrix} 1 & 0 & 0 & 0 \end{bmatrix}$$

giving the following eigenvalues for A:

-0.0318

-1.31984

-0.06027 ± 0.075553

The pole at -0.0318 is the spiral mode and the pole at -1.3198 is the roll subsidence mode. The two other poles result from the Dutch roll characteristics, where

$$\omega_d = 1.32 \text{ rad/s}$$

$$\zeta_d = 0.079523$$

for the natural frequency and damping ratio of the Dutch roll mode, respectively. The transfer function for the sideslip response to rudder input, produced by the Octave function `ss2tf`, is

$$\frac{\beta(s)}{\delta r(s)} = \frac{0.14807s^3 + 0.22457s^2 + 0.061712s - 0.00030669}{s^4 + 1.472250s^3 + 0.779453s^2 + 0.781569s + 0.024156} \qquad (7.42)$$

The rudder response, together with the roll rate and yaw rate responses, is shown in Figure 7.19a. The period of approximately 8.5 s and the damping ratio correspond with the values obtained from the linear response.

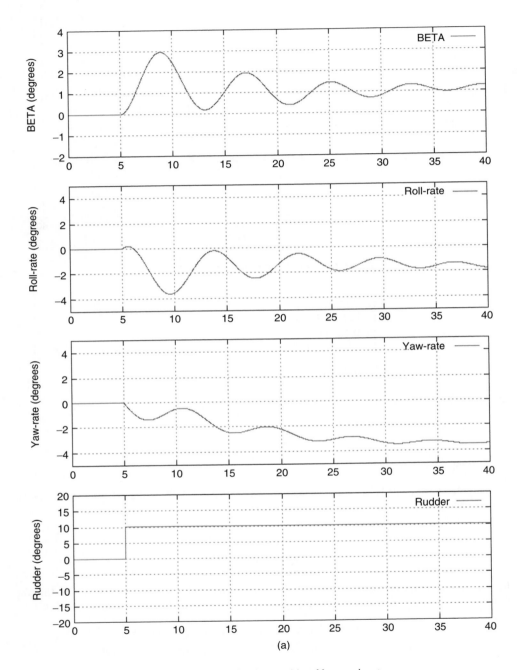

Figure 7.19 Lateral response: (a) rudder step input

Model Validation

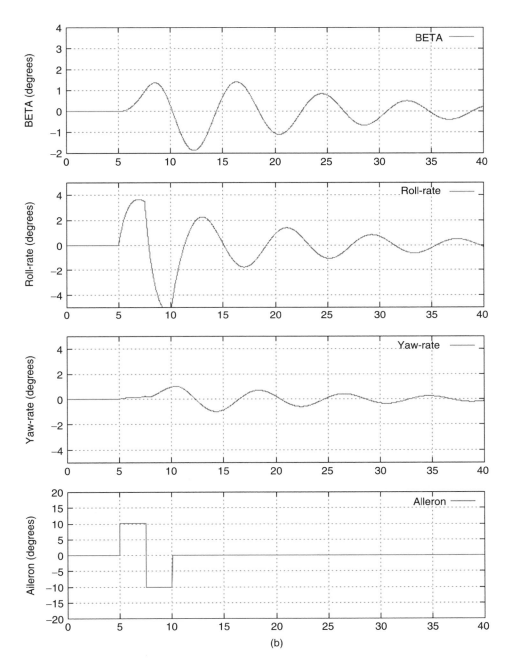

Figure 7.19 (b) aileron doublet input

Notice that the yaw rate is damped out and there is a discernible spiral mode with the yaw rate increasing over 40 s. Figure 7.19b shows the sideslip response to aileron, in this case an aileron doublet, but plotted on the same scales as the rudder input, to indicate the significance of the cross-coupled terms. These non-linear plots were obtained with the yaw damper disengaged to give the basic response of the aircraft to rudder and aileron inputs.

Remarks: Although the representation of system dynamics by an equivalent linear system is a very powerful technique for the design of control systems, for the flight simulator designer, a degree of caution is advisable with such simplifications. The approximations and estimates needed to linearize the equations and the limited range, over which certain variables are assumed to be constant, may result in a model with very different characteristics from the actual aircraft dynamics. A control system designed for optimal performance with a linear model may not perform as expected with the actual non-linear system. Often, if the data for an aircraft or a non-linear model is available, the preferred method to extract a linear model is to record the system response to a step input and then use system identification techniques to match a linear model (of an appropriate order) to the system response. Even so, the validity of this linear model is likely to be restricted to a relatively small region of the flight envelope of the aircraft. To apply the model outside this region is courting disaster.

7.7 Model Validation in Perspective

The importance of validation of the aircraft dynamics cannot be overstated. Having been asked, on several occasions, to review the shortcomings of specific simulators, where pilots had expressed concerns with the fidelity of the simulator, it was surprising to find faults that could have been identified during validation of the model. Strictly, criticisms such as this do not apply to airline flight simulators, where international regulations, published standards and adherence to procedures set out in the Evaluation Handbook have ensured that these simulators do meet the highest standards. However, it is also worth noting that the data package for a simulator costing some $15M is likely to cost over $1M and contain the equivalent of several hundred pages of graphs, tables and data.

Pilots fly by numbers; that is to say, in a specific phase of flight, a pilot will know the attitude and engine settings to achieve the desired performance. It is therefore essential to match the aircraft performance very closely to avoid negative transfer of training, where one set of performance data applies to the simulator and another applies to the aircraft. Similarly, the response and handling of the simulator, albeit with a limited motion system and visual system, must match the aircraft. A pilot transferring from a Boeing 757 to a Boeing 767 will need to appreciate the subtle handling differences and variation in performance between the two aircraft and should experience the change in aircraft handling in a flight simulator. The fidelity of the simulator is critical if the maximum level of positive training transfer is to be achieved in the simulator.

Similarly, in a military simulator, the pilot will fly close to the margins of the flight envelope of the aircraft, potentially in the non-linear regions associated with high angles of attack. In training military pilots, a small difference in manoeuvrability or performance in a combat situation may be critical and the fidelity of the model must be assured. Increasingly, combat tactics are developed and practised in simulators and errors in modelling the aircraft response or performance could be hazardous if techniques learnt in a poor fidelity simulator are used in actual combat.

In Sections 7.4 and 7.5, it was clear that every aircraft has a unique signature in terms of its longitudinal and lateral handling and that methodical and exhaustive validation of the flight model is needed to ensure that the model has correctly captured the aircraft behaviour throughout the flight envelope, from engine start up and taxiing through to stalls, collision avoidance and major system failures. However, the task of validation nowadays also extends to the integration of complex avionics systems, including displays, engine management systems, databuses and datalinks. The techniques covered in this chapter provide some insight into the validation of flight models, but the reader should appreciate that the validation of a full flight simulator takes several months, involving pilots and engineers from the manufacturers, the operator and the regulator. Inevitably, validation is a combination of analytic data analysis and pilot flight tests. The important point for any simulator designer is that formal and rigorous validation of the simulator software is the critical component of validation and is based on many of the techniques outlined in this chapter.

References

Babister, A. W. (1962) *Aircraft Stability and Control*, Elsevier Science and Technology.
Cook, M. V. (1997) *Flight Dynamics Principles*, Arnold.
Dwight, H. B. (1961) *Table of Integrals and Other Mathematical Data*, Macmillan, New York.
Eaton, J. W., Bateman, D. and Hauberg, S. (2008) *GNU Octave Manual Version 3*, Network Theory Limited.
FAA (1991) Advisory circular 120-40B, *Aeroplane Simulator Qualification*, Federal Aviation Administration.
FAA (1992) Advisory Circular 120-45A, *Aeroplane Flight Training Device Qualification*, Federal Aviation Administration.
FAA (1994) Advisory Circular 120-40B, *Helicopter Simulator Qualification*, Federal Aviation Administration.
FAA (2008) 14 CFR FAR Part 60, *Requirements for the Evaluation, Qualification and Maintenance of Flight Simulation Training Devices*, Federal Aviation Administration.
Franklin, G. F., Powell, J. D. and Emami-Naeini, A. (1980) *Feedback Control of Dynamic Systems*, Addison-Wesley, Reading, MA.
Guo, L., Cardullo, F. M., Houck. J. A., Kelly, L. C. and Wolters, T. E. (2005). A *comprehensive study of three delay compensation algorithms for flight simulators*. AIAA Modelling and Simulation Technologies Conference, San Francisco.
Herman, H. and Grant, R. (1974) Design principles for digital autopilot synthesis. *Journal of Aircraft*, 11 (7), 414–422.
IATA (2002) *Flight Simulator Design and Performance Data Requirements*, 6th edn, International Air Transport Association.
ICAO (1995) Doc. 9625-AN/938, *Manual of Criteria for the Qualification of Flight Simulators*, 1st edn, International Civil Aviation Organisation.
JAA (1998) JAR-STD 1A, Aeroplane Flight Simulators. Joint Aviation Authority.
JAA (1999) JAR-STD 2A, Flight Training Devices. Joint Aviation Authority.
Lanchester, F. W. (1908) *Aerodonetics*, A. Constable and Co. Ltd., London.
McLean, D. (1990) *Automatic Flight Control Systems*, Prentice Hall.
Powell, P. and Katz, P. (1975) Sample rate selection for aircraft digital control. *AIAA Journal*, 13 (8), 975–979.
RAeS (2005) *Flight Simulator Evaluation Handbook*, vol. 1, 3rd edn, Royal Aeronautical Society, London.
Stevens, B. L. and Lewis, F. L. (2003) *Aircraft Control and Simulation*, John Wiley & Sons, Ltd.

8

Visual Systems

8.1 Background

Taxiing out from the apron to the runway in a large transport aircraft, the pilot will see airport buildings, taxiways, other aircraft, grassed areas, marker boards, edge lights, runway holding position markings and so on, in the aircraft windscreen. The area occupied by these objects in the windscreen will depend on the distance from the objects and the position of the objects in the windscreen will vary as the aircraft heading changes. Similarly, during an approach, the pilots will see areas of forest and woodland, rivers, lakes and roads and perceive the shape of the terrain formed by hills and valleys. In addition to the detail of the terrain, lighting from the sun, which varies with the time of day and the level of cloud will illuminate the terrain and any mist or fog will restrict visibility.

A pilot in a flight simulator expects to see similar detail and identifiable terrain features. In training airline flight crews, the simulator must replicate specific airport approaches. The scene rendered by the visual system must provide sufficient detail and realism that the image seen in the simulator replicates an actual airfield approach very closely. This requirement is very demanding. To provide an image of the approach into an international airport may require in excess of a million objects in the scene for the image to be acceptable to an airline (in terms of realism, or visual fidelity) in a flight simulator. These requirements pose two problems for the simulator designer: first, a visual scene must be constructed to represent the required level of detail and secondly, the image must be rendered so that the pilot sees a continuously changing scene through the windscreens of the flight simulator.

The first requirement, the construction of a visual model, introduces several pragmatic problems. The visual system designer must acquire the physical dimensions of all the entities in the scene, for example, the position and colour of lights around the airfield, the size and shape of airport buildings, the layout of road systems and so on. Collection of this data, often obtained from surveying but, more recently, derived from the merging of satellite data, photographs, maps, charts and survey data is a daunting task. It is also introduces the classification of data. The accuracy of some features is critical, for example, taxiway lighting, whereas other data can be representative, for example, an area of forest, where the specific type of vegetation or even the shape of forested areas will not contribute significantly to the training of airline pilots.

The second requirement is to project the image in a form where it is seen by the flight crew in a simulator, with the same perspective, depth, resolution and possibly stereo effect seen in real life. The method of projection must avoid introducing any artefacts that might affect the flight training. In particular, the geometry of the landscape and its objects perceived by the human brain

---

Principles of Flight Simulation D. J. Allerton
© 2009, John Wiley & Sons, Ltd

in the real world must be faithfully replicated in the flight simulator. At the same time, the level of detail (e.g. bricks in the wall of a building) and the resolution of the image must meet levels where they sustain a sense of reality in the flight simulator. For example, a runway is not simply a grey rectangle of concrete; it will contain cracks, runway markings, skid marks, oil stains and undulations and exhibit the texture of worn concrete. Although some of these features will be useful in taxiing, in practice, they would not be visible during an approach. Consequently, these features may be displayed at different levels of details for different phases of flight.

The actual image, or images, will be displayed nowadays by means of computer graphics. However, this poses a constraint, in the sense that image resolutions are limited to the order of 1000–1500 pixels, in the x and y directions of a display. Any projection system will provide optical magnification of screen pixels, so there is a lower bound for resolution, beyond which individual pixels are discernible. There is also an upper bound, which is subject to two constraints. First, the video amplifier must produce a video signal for the projector. As the display resolution increases, so does the bandwidth of the video amplifier, typically constraining displays to significantly less than 2000×2000 pixels. Secondly, the projector technology must be capable of producing sufficient optical power at the display resolution and switch the projection beam as each pixel is drawn. Consequently, projector technology also limits the display resolution, which is currently an order of magnitude below visual acuity.

In addition to requiring a display with sufficient detail to provide acceptable visual cues, these images must be rendered at a sufficient rate so that the scene changes smoothly and continuously; for the real-time displays used in flight simulation, the image must be both redrawn and rendered at least 50 times per second. A display of 1000×1000 pixels, say, rendered at 50 Hz will generate 50,000,000 pixels per second. Put another way, only 20 nanoseconds are available to generate each pixel. The computations inherent in rendering a scene in real-time (Akenine-Moller *et al.* 2008) include retrieving the objects to be drawn, performing the geometric computations to determine the coordinates of each object in the view port and rendering the image as a set of pixels. Although there have been major advances in the performance of image generators, considerable attention must also be given to minimizing delays in the path from locating objects in the database through to their subsequent rendering and projection.

The visual system is integrated with other simulator systems, so that it corresponds to the aircraft motion and position; other entities in the simulation (e.g. threats and targets) are very closely aligned to the dynamics of the motion platform. The visual system is a critical element in many training tasks. Its primary function is to sustain the illusion of the real world in a simulated cockpit, where the levels of acceptance of this synthetic environment depend on the visual fidelity of the visual system. However, the requirement of a visual system to achieve total fidelity is impractical. On the one hand, the cost of the visual system increases dramatically with the scene content and image resolution and on the other hand, the visual system must be designed to meet the training task. Notwithstanding these constraints, for some training tasks, it may possible to reduce the visual fidelity without compromising the training effectiveness.

8.2 The Visual System Pipeline

The visual system is an integral component of a flight simulator. Its basic function is shown in Figure 8.1. The image generator (Schachter, 1981) is the heart of any visual system and is typically a set of graphics cards that perform the 3D graphical operations (Foley *et al.*, 1995), which are needed to transform 3D objects defined in the scene database of objects to a 2D image stored in the framestore. The image generator produces an image seen from the position of the pilot eye point (PEP), which is derived from the aircraft position and attitude, computed by the flight model. In addition, this image will also depend on the geometric dimensions which define the position of the pilot relative to the aircraft cg, the field of view of the projection system and the display

Visual Systems

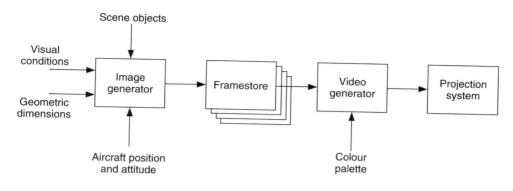

Figure 8.1 Visual system

resolution. The instructor can also change the ambient conditions, such as time of day, or cloud base or visibility and this information is also used in the computations to render the image, which is written as pixels to the framestore.

The framestore is simply a set of memory planes (typically 8–24). Each framestore corresponds to the resolution of the display, so that the (x, y) pixel of the display is located at the same memory address in each framestore. The colour of a pixel is encoded in the bits extracted (in parallel) for each pixel from the n framestores, giving a possible range of 2^n colours. These pixels are extracted at the frame rate of the display system, typically 50–60 Hz, by the video generator, which produces a video signal for a projector. Note that the framestore is written to by the image generator and is also read by the video generator. The framestore is often the bottleneck in any visual system, so delays accessing the framestore must be minimized. One way to avoid contention to the framestore memory is to provide two framestores, where one is written to by the image generator while the other one is read by the video generator. These roles are toggled every frame – a technique that is referred to as *dual buffering*.

The actual colour component of each pixel is obtained from the parallel word extracted from the set of framestores. A colour palette provides a mapping between the binary patterns encoded in the framestores and the video signal format recognized by the projectors.

This transformation, from a geometric description of each object in the visual database to an optical projected image is often referred to as the *visual pipeline*. As each object is extracted from the visual database, it is formed as pixels in the framestore which are subsequently converted to a video signal for a projector. During each frame, this sequential process of extraction of an object, rendering the object to form a 2D image from a 3D description, is a processing pipeline which involves a large number of computations and must be sustained at maximum computation speed. Any delays in this pipeline will reduce the time available for the graphic processing and therefore reduce the number of objects rendered per frame.

Within the image generator, there is a further pipeline, which is known as the *graphics engine* or *graphics processor* and is illustrated in Figure 8.2. Objects defined in the visual database as 3D objects are transformed from coordinates in the external 'world' frame to coordinates in the 3D pilot view port. This is very similar to the transformation used to transform body frame accelerations or

Figure 8.2 Graphics processor

velocities to the navigation frame. There is a slight complication because the pilot is not usually positioned at the aircraft cg but the transformation between coordinate frames covered in Chapter 2 also applies to coordinates in a visual system. Each graphical object is defined as a set of faces, or planar polygons (typically triangles); the vertices of these polygons are transformed from the 3D coordinates in the world frame to 3D coordinates in the pilot eye frame, typically where x is to the right of the pilot's line of sight, y is upwards and z gives the forward or depth component. Note that straight lines in the world frame are straight lines in the pilot's view port. In other words, the polygon shape is retained in the transformation.

Of course, not all objects are in view; some objects may be behind the aircraft; some objects may be outside the field of view and some objects may be too distant to be seen. To optimize the graphics pipeline, objects outside the pilot's view port can be discarded, avoiding any further computations. Some objects will be completely within the pilot's view port and they will need to be converted to a 2D image. There is however, one further category, that is, an object which is partially inside the pilot's view port and partially outside the view port. In this case, the parts of the object outside the view port can either be eliminated or adjusted to the edge of the view port – a process known as *clipping*. The clipping process, which follows the transformation process, is based on a viewing frustum extending from the pilot's eye through the corners of the display screen, as shown in Figure 8.3.

The view port is a pyramid with the apex at the pilot's eye. The shape of the pyramid depends on the distance of the pilot's eye from the aircraft windscreen and also the shape of the windscreen. The object at A is outside the viewing frustum and can be discarded following the transformation. The object at B is completely inside the viewing frustum and can be rendered as a 2D object. However, the object at C is partially inside the frustum and must be clipped to the frustum. At this stage, the objects are still represented as 3D objects, possibly with clipped edges.

The vertices of a polygon seen in the pilot's view port intersect the windscreen at a point where the straight line from a vertex of the polygon to the pilot's eye passes through the screen, as shown in Figure 8.4. This transformation from 3D to 2D provides a perspective effect, similar to a painting,

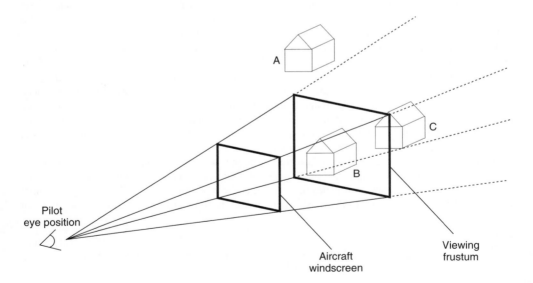

Figure 8.3 The viewing frustum

Visual Systems

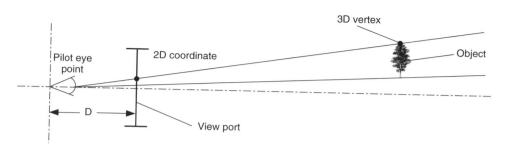

Figure 8.4 Perspective projection

and is referred to as the *perspective projection*, where distant objects appear smaller. In effect, a 3D vertex coordinate is transformed to a 2D coordinate in the screen space.

After the projection stage, each polygon is defined by the 2D coordinates of its vertices. In the rendering phase, each pixel within a polygon is written to the frame store. This process, which is often referred to as *polygon in-fill*, takes 'wire frame' polygons, with edges joining the vertices, to form filled polygons. The rendering process requires repeated access to the frame store to write the pixels within the polygon edges; the memory of a graphics processor is optimized for this application, to maximize the rendering rate. However, it is also necessary to locate the intermediate pixels along the edges of the polygon to determine the in-fill boundary, which consumes further processing.

Simply writing pixels of a uniform colour for each polygon produces a somewhat bland and unrealistic display. There are three potential improvements during the rendering phase:

- As the screen coordinates of each pixel are computed, the z distance of the pixel (from the eye point to the object) is also written to the framestore. Subsequently, if a pixel is written to the frame with a smaller z value, it must be nearer the eye point, whereas a pixel with a larger z value must be beyond the current pixel and the writing of this new pixel is inhibited. This simple comparison of the z value, known as z *buffering*, provides hidden surface elimination, where one object appears correctly in front of another object, independent of the order of drawing;
- Any illumination can be modelled across the surface of a polygon to provide shading. Although such effects in flight simulation have minimal benefit at the polygon level, illumination models can provide lighting levels from midday to dusk and night conditions, which is essential in training simulators;
- Rather than simply filling a polygon with uniform colour, it is possible to render a texture pattern over the polygon. This is undoubtedly the major advance incorporated in graphics cards developed since the mid-1990s. However, it is important to appreciate that the rendering of texture of a 3D surface on a 2D polygon requires a spatial transformation, which is applied at the time of rendering.

One final effect that can be added in a flight simulator is to model the visibility of each polygon to simulate the effects of haze and fog. If fogging is applied, as each pixel is written to the framestore, its z value indicates its distance from the pilot eye position and this value is accessed by a fogging algorithm to modify the colour of the pixel to simulate the effects of visibility.

The resultant image, formed in the framestores, is then accessed pixel by pixel by the video access hardware of the video generator. The framestore memory is organized as rows and columns corresponding to the display axes to simplify the framestore access. Further speed is achieved by accessing up to 256 pixels in a single memory read operation. The video generator produces video signals, which comply with the standards used in projection systems, allowing for the display resolution, video bandwidth and any other parameters needed for the projection system.

8.3 3D Graphics Operations

To put the requirement of the graphics operations of a geometry engine (Clark, 1982) into perspective, if a scene contains 2000 polygons, rendered at a 50 Hz frame rate, then these polygons must be rendered at an average rate of 10 μs per polygon. Some of these computations require floating point multiplications and divisions, in addition to the memory accesses to write pixels to the framestore. Consequently, considerable effort is given to minimizing the underlying arithmetic and graphics engines use custom hardware to achieve this high throughput.

The first stage is the transformation of vertices from a world frame to the pilot eye frame. By world frame, some form of reference frame used by the visual system is implied. It may well be a coordinate frame used to capture the visual database, but in practice it is likely to be similar to an Euler frame, with a north, east and down axis system, where the angular orientation of the pilot eye frame relative to this frame is defined by the Euler angles. The problem reduces to transformations between orthogonal frames, as described in Section 3.5. There is a further axis transformation because aircraft motion is defined with respect to the aircraft cg, whereas the PEP is located at some fixed distance from the cg. As the airframe is rigid, and the offset to the pilot eye station is a constant, this translation is trivial.

Each vertex of an object in the world frame is transformed by three rotations in yaw, pitch and roll to the body frame, and then to the pilot eye frame. These transformations are the inverse of the rotations provided by the DCM to transform body velocities u, v and w to the navigation frame velocities V_N, V_E and V_d, respectively. However, a point defined in the world frame is both translated *and* rotated in the pilot eye frame and consequently, homogeneous coordinates are used to allow these operations to be combined, where the matrix operations for 3D geometric transformation are based on 4 × 4 matrices.

The translation of a vector is given by

$$\begin{bmatrix} x' \\ y' \\ z' \\ 1 \end{bmatrix} = \begin{bmatrix} 1 & 0 & 0 & T_x \\ 0 & 1 & 0 & T_y \\ 0 & 0 & 1 & T_z \\ 0 & 0 & 0 & 1 \end{bmatrix} \begin{bmatrix} x \\ y \\ z \\ 1 \end{bmatrix} \qquad (8.1)$$

$$x' = x + T_x \qquad (8.2)$$

$$y' = y + T_y \qquad (8.3)$$

$$z' = z + T_z \qquad (8.4)$$

where T_x, T_y and T_z are the translations in the x, y and z axes respectively.

Rotation in yaw by an angle ψ is given by

$$\begin{bmatrix} x' \\ y' \\ z' \\ 1 \end{bmatrix} = \begin{bmatrix} \cos\psi & -\sin\psi & 0 & 0 \\ \sin\psi & \cos\psi & 0 & 0 \\ 0 & 0 & 1 & 0 \\ 0 & 0 & 0 & 1 \end{bmatrix} \begin{bmatrix} x \\ y \\ z \\ 1 \end{bmatrix} \qquad (8.5)$$

Rotation in roll by an angle ϕ is given by

$$\begin{bmatrix} x' \\ y' \\ z' \\ 1 \end{bmatrix} = \begin{bmatrix} \cos\phi & 0 & -\sin\phi & 0 \\ 0 & 1 & 0 & 0 \\ \sin\phi & 0 & \cos\phi & 0 \\ 0 & 0 & 0 & 1 \end{bmatrix} \begin{bmatrix} x \\ y \\ z \\ 1 \end{bmatrix} \qquad (8.6)$$

Rotation in pitch by an angle θ is given by

$$\begin{bmatrix} x' \\ y' \\ z' \\ 1 \end{bmatrix} = \begin{bmatrix} 1 & 0 & 0 & 0 \\ 0 & \cos\theta & \sin\theta & 0 \\ 0 & -\sin\theta & \cos\theta & 0 \\ 0 & 0 & 0 & 1 \end{bmatrix} \begin{bmatrix} x \\ y \\ z \\ 1 \end{bmatrix} \quad (8.7)$$

During each frame, every vertex is translated from the world frame to the pilot eye frame and then rotated, using the same set of transformations. In practice, the translation is straightforward and the resultant vector is then transformed by rotations in yaw, roll and pitch (and in that order), given by the concatenation of the three rotation matrices above.

$$\begin{bmatrix} x' \\ y' \\ z' \\ 1 \end{bmatrix} = \begin{bmatrix} \cos\psi\cos\theta & \sin\psi\cos\theta & -\sin\theta & 0 \\ \cos\psi\sin\theta\sin\phi - \sin\psi\cos\phi & \cos\psi\cos\phi + \sin\psi\sin\theta\sin\phi & \cos\theta\sin\phi & 0 \\ \cos\psi\sin\theta\cos\phi + \sin\psi\sin\phi & \sin\psi\sin\theta\cos\phi - \cos\psi\sin\phi & \cos\theta\cos\phi & 0 \\ 0 & 0 & 0 & 1 \end{bmatrix} \begin{bmatrix} x \\ y \\ z \\ 1 \end{bmatrix}$$
$$(8.8)$$

In other words, the DCM is computed once per frame (based on this transformation derived from the Euler angles) and is then used to transform each vertex $[x, y, z]^T$ in the world frame to $[x', y', z']^T$ in the pilot eye frame.

Following transformation of the vertices of an object, clipping is used to discard objects completely outside the viewing frustum and to modify the edges of objects partially inside the viewing frustum, as shown in Figure 8.5. It is tempting to assume that clipping can be applied to 2D images after the perspective transformation. However, clipping needs to be performed in 3D. In the case where part of a polygon is in front of the eye point and part is behind the eye point, 2D clipping will not generate the correct 2D coordinates. Note also in Figure 8.5 that polygons A, E and F have become disconnected as a result of clipping (edges have been discarded) and new edges are formed by the edge of the viewing frustum.

The initial clipping test is simple. If both vertices of an edge are to the left, right, below or above the viewing frustum, the edge can be discarded. In addition, if both vertices are behind the eye point, the edge is not in view and can be discarded. If both vertices of an edge are within the

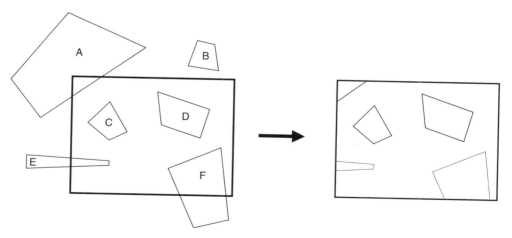

Figure 8.5 Clipping objects

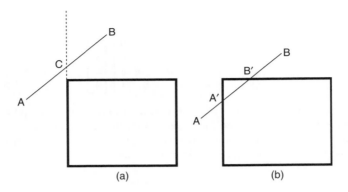

Figure 8.6 Clipping edges: (a) discarded edge (b) clipped edge

boundaries of the viewing frustum, the edge will be seen and no clipping is required. Both tests only require a comparison of the vertex coordinates with the coordinates of the viewing frustum, which are derived from the z distance of the vertex. At this stage, an edge can also be to the left of the viewing frustum and also above the viewing frustum, as shown in Figure 8.6.

In Figure 8.6a, the edge AB is clipped to the left edge of the viewing frustum at C. The edge AC is discarded because it is to the left of the viewing frustum and the edge BC is discarded because it is above the viewing frustum, effectively discarding the edge AB in two operations. In Figure 8.6b, the edge AB is clipped to the left edge of the viewing frustum at A′ and then the edge A′B is clipped to the top edge of the viewing frustum giving the clipped edge A′B′.

For edges which are partially inside the viewing frustum, the intersection of the edge and the viewing frustum is simple trigonometry, as the sides of the viewing frustum are horizontal or vertical, as shown in Figure 8.7, where

$$\frac{x_3 - x_1}{x_2 - x_1} = \frac{y_3 - y_2}{y_1 - y_2} \tag{8.9}$$

giving

$$y_3 = y_2 + \frac{(y_1 - y_2)(x_3 - x_1)}{x_2 - x_1} \tag{8.10}$$

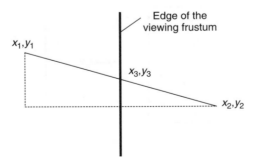

Figure 8.7 Clipping an edge

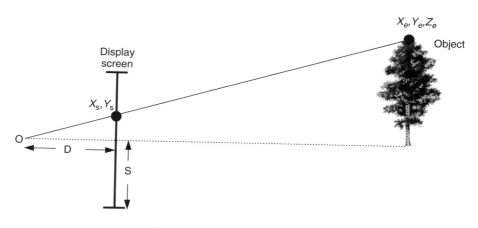

Figure 8.8 Perspective computation

where x_3 is the left edge (in this example) of the viewing frustum. Similar computations are used to clip to the right, upper and lower edges of the viewing frustum.

The final phase of the geometric transformation is the perspective computation, as shown in Figure 8.8, to compute the screen y coordinate.

Assuming the origin O is at (0, 0, 0), then

$$X_s = \left(\frac{D}{S}\right)\frac{X_e}{Z_e} \tag{8.11}$$

$$Y_s = \left(\frac{D}{S}\right)\frac{Y_e}{Z_e} \tag{8.12}$$

where D is the distance of the pilot eye position from the display (the windscreen) in pixels and S is half the display width in pixels. Note that D/S is a constant for a specific display.

By this stage of the visual system pipeline, the vertices of a 3D object have been transformed to 2D vertices in screen coordinates. For each vertex, the transformation requires nine multiplications and six additions, clipping can require eight additions, four multiplications and two divisions and the perspective projection requires two multiplications and two divisions. These are worst-case figures, because no clipping may be required for many edges. However, clipping a vertex can take up to 14 additions, 15 multiplications and 4 divisions, before any pixels are written to the framestore. The processor used in a graphics card is therefore a major constraint on the overall drawing rate. With care, these equations can be implemented in integer arithmetic if considerable care is given to scaling to avoid overflow or underflow. Nowadays, the geometry engines used in graphics cards have particularly powerful floating point processors. Even so, a scene containing only 1000 polygons, with four vertices, say, updating at 50 Hz can require of the order $1000 \times 4 \times 50 \times 33$ floating point operations per second or over 6 million floating point operations per second, just allocated to the trigonometric computations.

A major consideration in any visual system is the elimination of redundant computations as early as possible in the graphics pipeline (Watt, 1993). One particular example is back-face elimination. For a simple 3D object such as a rectangular building, at least one of the sides will not be visible. Consequently, this surface will not be rendered and time spent on transformation, clipping and perspective computations would be wasted. If the object was, for example, an aircraft consisting of a 1000 polygons, several hundred of these polygons could be eliminated by detecting a back face. The criterion is straightforward: if the angle between the normal of the surface and the line of sight is greater than 180°, then the surface cannot be seen. The normal vector of each surface can be computed when an object is entered into the visual database (avoiding a computation at run-time), where the angle between two vectors is derived from the vector dot product.

Clearly, wire frame images rendered by simply drawing straight lines between vertices, produce very unrealistic images. A first step to improving the realism in an image is to fill in surfaces. One in-fill method is to locate a pixel inside a polygon and then examine neighbouring pixels in four directions until the edge of the polygon is encountered. However, this method works on a pixel-by-pixel basis, needing access to the framestore to test each pixel and also to write each pixel. An alternative method, illustrated in Figure 8.9, is to detect the highest point (the largest of the y coordinates of the vertices) and then, using Bresenham's algorithm (Newman and Sproull, 1979), detect the x coordinates of the left edge and right edge of the polygon, descending row-by-row, until the lowest vertex is reached. Knowing the y ordinate of the scan line and the left and right ordinates of the end points of the scan line, the pixels can be written. In practice, they can be written as 16-bit, 32-bit or even 256-bit words in a single write operation, reducing significantly the number of framestore accesses. The time to render the polygon depends on the number of scan lines and the number of write operations, plus the time to compute the left edge and right edge ordinates. But, as described in Section 5.2, Bresenham's algorithm (Newman and Sproull, 1979) uses integer arithmetic, avoiding multiplication or division.

There is one further problem with polygon in-fill. The scan line method will not work if an internal angle of the polygon is obtuse, in which case the polygon is sometimes referred to as *re-entrant*. Such a polygon is shown in Figure 8.10a, where the left edge ABCDE changes its vertical direction at B, contravening the conditions of the algorithm.

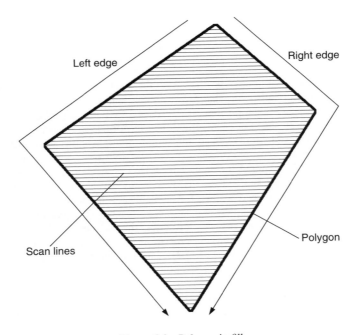

Figure 8.9 Polygon in-fill

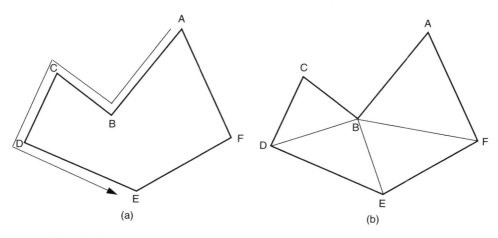

Figure 8.10 Polygon re-entrancy: (a) a re-entrant polygon (b) a non-re-entrant polygon

One solution is to ensure that polygons defined in the world coordinate are non-re-entrant. As shape is preserved, no 2D shapes should be re-entrant. However, rounding in the graphics computations can introduce re-entrant shapes. The accepted practice, which avoids re-entrant polygons occurring at the in-fill stage, is to construct planar surfaces from triangles. Whether rounding occurs or not, a triangle cannot contain an obtuse internal angle, as illustrated in Figure 8.10b.

One further improvement, in terms of realism, is to provide hidden surface elimination. This concept is illustrated in Figure 8.11. The objects on the right-hand side exhibit more solid visual characteristics as a result of removal of part of the surfaces occluded by the nearer objects. Although the clipping algorithm could be modified to detect edge intersections of this form, it would be impractical to test for the intersections of thousands of edges in a scene.

Early visual systems used the painter's algorithm, where the objects were sorted in order of depth and the more distant objects were drawn before the nearer objects. The drawback with this method is the time needed to sort the objects. In addition, parts of a surface can be both in front of and also behind parts of another surface. If the test is simply applied to the centroid of a surface, visual anomalies can occur, depending on the eye position and orientation of the surfaces. Nowadays,

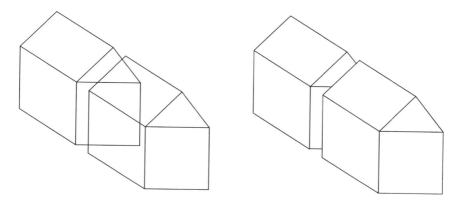

Figure 8.11 Hidden surface elimination

with the availability of cheap memory, Z buffering is used, where a pixel is only written if its z component is less than the z component of the corresponding framestore pixel. Although extra pixels are written, there is no need to sort the polygons or to compute edge intersections. However, it is important to bear in mind that a Z buffer has a fixed depth. If, say, the Z buffer is 16 bits, then rounding errors in computing the z distance as pixels are written can cause flickering where one object in the distance appears to move behind, and then in front of, another object, as the aircraft manoeuvres, simply as a consequence of the resolution of the Z buffer depth.

One method of increasing the visual fidelity is to increase the number of polygons rendered, but this clearly places a heavy load on the graphics processor, to the point where it may not sustain the frame rate. Consider a brick wall of a terminal building. It may contain several thousand bricks and it would be absurd to allocate a polygon to each brick. Since the mid-1990s, the lower cost graphics cards have supported texture rendering. An image is acquired from a drawing package or alternatively, as a photographic image. This 2D image can be applied to a surface in very much the same way that in-filling is used to add solidity to visual objects. For example, a single photograph could be taken of the wall of the terminal building, which is then applied to the image in the rendering phase.

However, such effects are difficult to implement in real-time. First, the original texture image will have one resolution, typically based on a rectangular shape, which needs to be mapped to a shape that contains a different number of pixels in the image and which is distorted in terms of the perspective computations. In other words, the texture image must be filtered. Although the original image might be 512×512 pixels, the shape of the object may become trapezoidal (as a result of the perspective transformation), covering a region of the screen 30×20 pixels, say. First, the texture processor has to filter the original image to determine the resultant pixels for the small shape and secondly, the texture mapping of a 3D image occurs on a 2D display, so that the perspective projection which applies to the vertices of the shape also applies to the texture.

Fortunately, real-time texturing is feasible, even with low-cost graphics cards. However, there are still limitations with such an approach:

- Memory is required to store the original textures. A texture of 1024×1024 pixels using 16-bit colours requires 2 MB of memory, just for that image. With increasing memory capacity and reducing costs of memory, the amount of texture memory is not a significant limitation for most visual systems;
- The filtering of the texture image needs to be based on the 3D footprint of the texture, requiring a large number of per-pixel computations; poor filtering can introduce jagged artefacts reducing the visual fidelity;
- The resolution of the original textured image restricts the point at which the image in the visual system starts to lose fidelity. For example, a clump of forest may look very detailed at 2000 m but at a distance of 20 m, 1 pixel of the original image may correspond to 10 pixels of the displayed image, resulting in a blurred image, say. On the other hand, if the clump of trees is digitized so that it is detailed at 20 m, the region of texture memory may only occupy a few pixels at 2000 m.

Nevertheless, textured images do afford considerable saving (in terms of the polygon count) and modern graphics processors are capable of generating very high quality real-time images, exploiting texture rendering for runways, areas of grass, buildings, clouds, aircraft markings and so on.

Finally, although attention has been given to the rendering performance required in geometric transformation, clipping, perspective and texturing operations, it is also important to stress the significance of the visual system database management methods. If redundant polygons are not detected in the polygon processing pipeline, a considerable amount of processing is, in effect, wasted. However, if only potential 'active' objects are extracted from the visual database, then processing is allocated to polygons which are probably visible in the scene. These issues are discussed further in Section 8.6.

8.4 Real-time Image Generation

In this section, a simple real-time image generation (IG) system is developed using OpenGL,[1] to illustrate some of the ideas covered in this chapter. As with the other software in the book, it is compiled under gcc for Linux (and MinGW).

8.4.1 A Rudimentary Real-time Wire Frame IG System

Although wire frame drawings are used in illustrations and for cut-away diagrams, they are clearly impractical in flight simulation applications as they lack realism. Nevertheless, the operations involved in wire frame rendering constitute the first three stages of an IG pipeline and are performed by all graphics engines. A very simple visual system is developed to show the operations of axis transformation, clipping and perspective computation. A rudimentary wire frame IG can be written in a few hundred lines of C code.

To illustrate these concepts, a simple visual database is used comprising approximately 50 polygons, representing a set of large rectangular fields and an airfield with a runway and a small hangar, as shown in Figure 8.12.

The database is north-east aligned, with the runway positioned on the y axis and 50 m below the x axis. A large polygon (not shown) is defined as a bounding box for the other polygons. One purpose of this bounding box is that it provides a distant horizon; it needs to be sufficiently distant that it is effectively the visible horizon, but not so distant that it could cause numeric overflow in the geometric computations.

Simple structures are defined for 3D vertices and 2D vertices, as follows:

```
typedef struct {
    float x;
    float y;
    float z;
} vertex3;

typedef struct{
    float x;
    float y;
} vertex2;
```

A polygon is then defined as four vertices using the following structure:

```
typedef struct {
    vertex3 p0;
    vertex3 p1;
    vertex3 p2;
    vertex3 p3;
    polycol col;
} polygon;
```

where `polycol` defines the colour of a polygon, with the red, green and blue components defined as follows:

---

[1] Although OpenGL is used in the implementation, it is only used to establish the double-buffered environment and to draw the 2D vectors.

```
typedef struct {
    GLfloat r;
    GLfloat g;
    GLfloat b;
} polycol;
```

In this simple database, x is north, y is east and z is negative up. Although the database of an IG is likely to be read from a file, as this database contains only 50 or so polygons, the polygons are defined in a C header file *database.h*. For example, the runway is 1650 m long by 50 m wide, defined as follows:

```
{ {  -50.0,   -25.0, 0.0}, // main runway
  { 1600.0,   -25.0, 0.0},
  { 1600.0,    25.0, 0.0},
  {  -50.0,    25.0, 0.0}
}
```

Having established an OpenGL rendering environment, the main part of the IG is to render each polygon of the database, as given in the following C procedure:

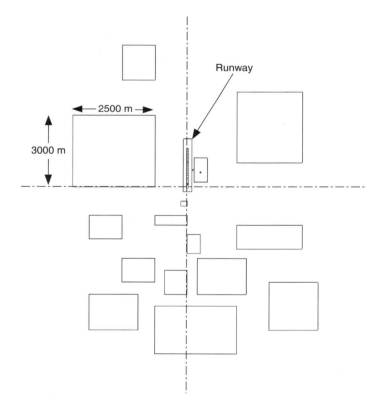

Figure 8.12 A basic visual database

```
void cgi()
{
  unsigned int i;

  transformation();
  for (i=0; i<51; i+=1) {
    poly(polylist[i]);
  }
}
```

The procedure `transformation` computes the DCM from the world coordinate frame to the pilot eye frame. This 3 : 3 matrix, which must be recomputed every frame, is used in the rendering of each polygon by the procedure `poly`, where the argument of the procedure is a pointer to a polygon. It is assumed that each polygon contains four connected sides. In the procedure `poly`, each vertex is transformed from world coordinates to pilot eye coordinates and then the four transformed edges (e0, e1, e2 and e3) are clipped (in 3D) and drawn.

```
void poly(polygon p)
{
  vertex3 e0, e1, e2, e3;

  transform(p.p0, &e0);
  transform(p.p1, &e1);
  transform(p.p2, &e2);
  transform(p.p3, &e3);

  DrawClipped(e0, e1);
  DrawClipped(e1, e2);
  DrawClipped(e2, e3);
  DrawClipped(e3, e0);
}
```

The procedure `transform` translates and rotates a vertex in the world frame to a vertex in the pilot eye frame, using the DCM (a_{11}, a_{12}, a_{13} ... a_{33}) generated in the procedure `transformation`. In line with flight dynamics conventions, the z axis is negative up, so care is needed in the procedure `transform` to transform each vertex to the pilot eye frame, where the z axis is assumed to be positive up. In the subsequent perspective computations, this convention accords with the 2D axes, where the y axis is positive up.

```
void transform(vertex3 v1, vertex3 *v2)
{
  float x = v1.x - Px;
  float y = v1.y - Py;
  float z = -v1.z + Pz; /* Pz -ve up */

  (*v2).x = x * a11 + y * a12 + z * a13;
  (*v2).y = x * a21 + y * a22 + z * a23;
  (*v2).z = x * a31 + y * a32 + z * a33;
}
```

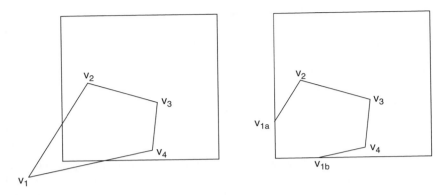

Figure 8.13 3D clipping

The procedure `DrawClipped` makes a local copy of the vertices of the edge, which are clipped against the 3D viewing frustum (defined by the field of view of the display screen) by the procedure `clip3d`, which returns false if the clipped edge is completely outside the viewing frustum. Otherwise, some part of the edge is visible and the perspective computation is applied to the 3D vertices of the edge (`a1` and `a2`) and the resultant 2D vector is drawn. The reason for taking a local copy of the edge is important as two edges may be clipped, as shown in Figure 8.13; the vertex v_1 should not be modified as it is used in clipping the edges $v_1 - v_2$ and $v_1 - v_4$.

```
void DrawClipped(vertex3 p1, vertex3 p2)
{
  vertex3 a1;
  vertex3 a2;
  vertex2 l1;
  vertex2 l2;

  a1.x = p1.x;
  a1.y = p1.y;
  a1.z = p1.z;
  a2.x = p2.x;
  a2.y = p2.y;
  a2.z = p2.z;

  if (clip3d(&a1, &a2)) {
    perspective(a1, &l1);
    perspective(a2, &l2);
    Draw(l1.x, l1.y, l2.x, l2.y);
  }
}
```

The procedure `clip3d` is based on the method given in Newman and Sproull (1979), with the appropriate change of coordinate conventions on entry to the procedure and on exit. If an edge is returned by `clip3d`, it can be guaranteed to be visible in the viewing frustum, so that the end points

Visual Systems

of the edge can be projected into screen coordinates by the procedure `perspective`. The code for the perspective computation is given below.

```
void perspective(vertex3 v1, vertex2 *v2)
{
  (*v2).x = v1.y * SCREENWIDTH / v1.x + SCREENWIDTH / 2.0;
  (*v2).y = v1.z * SCREENDEPTH / v1.x + SCREENDEPTH / 2.0;
}
```

The constants SCREENWIDTH and SCREENDEPTH define the display size in pixels. The display origin (0, 0) is at the bottom left-hand corner of the screen; the offsets are added as it is assumed that the pilot looks along the centre of the screen.

Two sample images from this visual system are shown in Figure 8.14, where the left and right images are taken from an approach at 10 km and 1 km, respectively.

Clearly, these images would not be acceptable in a modern flight simulator; the lines are jagged, they provide poor visual cues, the image lacks detail, there is no colour or texture and so on. However, it does illustrate that the basis of a real-time image generator is built on a small set of equations to implement the geometric transformations. The rendering code in *wire.c* is approximately 50 lines and *trans.c* (which includes the 3D clipping algorithm) is less than 200 lines of code.

A word of caution is needed for the enthusiast developing a custom IG: considerable care is needed with coordinate frames because both left-hand and right-hand frames may be used. In particular, aerodynamic models usually have the z axis negative upwards, whereas computer scientists tend to refer to this axis as the y axis, which is normally positive upwards.

8.4.2 An OpenGL Real-time IG System

In the previous section, graphic specific 2D and 3D graphics primitives were developed to provide transformations, clipping and perspective operation to render simple polygons. OpenGL was only used to establish a rendering environment and to draw the resultant 2D lines. However, OpenGL contains a library of graphics operations to support 3D rendering (Shreiner, 2004) and the software developed in the previous section can be modified to use OpenGL for all rendering processes. As

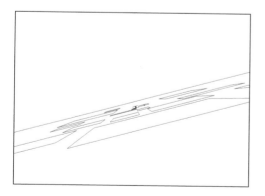 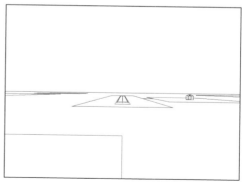

Figure 8.14 A wire frame visual system

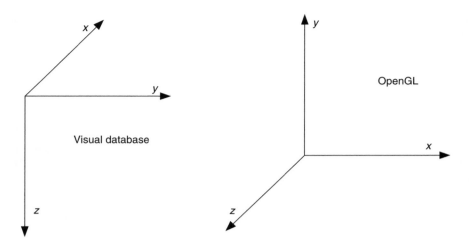

Figure 8.15 Reference frames

most modern processors include hardware to support OpenGL, a real-time IG based on OpenGL affords the following advantages:

- It is portable – software developed on one platform should run on other platforms, independent of the graphics hardware;
- The IG code is considerably simplified, in comparison with custom geometric and rendering operations;
- Considerable design effort has gone into OpenGL to optimize its run-time performance for real-time applications.

Using the visual database developed in Section 8.4.1, it is necessary to perform any transformations with respect to the appropriate axes conventions. Figure 8.15 shows the axes used for the visual database and by OpenGL. Care is needed to extract the appropriate coordinate (e.g. the y coordinate in OpenGL corresponds to the negative of the z coordinate in the visual database frame) and with the sign conventions used in rotations. In practice, these conventions are straightforward and only occur at the time coordinates are accessed in the visual database or when rotations are computed.

In the first instance, OpenGL will be used to replicate the functions developed in Section 8.4.1. The procedure display is established as the display function by glutdisplayfunction during the initialization and is invoked every frame.

```
void display(void)
{
    glClear (GL_COLOR_BUFFER_BIT);

    glPushMatrix();
        glRotatef(roll, 0.0, 0.0, 1.0);
        glRotatef(-pitch, 1.0, 0.0, 0.0);
        glRotatef(yaw, 0.0, 1.0, 0.0);
        glTranslatef (Py, Pz, Px);
        cgi();
    glPopMatrix();
```

```
    glutSwapBuffers();
    glFlush ();
}
```

Note the negation of the pitch angle and the mapping of the x, y and z components. Within the calls to `glPushMatrix()` and `glPopMatrix()`, the transformation is established in terms of the eye-point position (`Px`, `Py`, `Pz`) and the aircraft attitude in `pitch`, `roll` and `yaw` and the scene is rendered by a call to `cgi()`. The procedure `cgi` simply accesses the polygons in the database, as before:

```
void cgi()
{
  unsigned int i;

  for (i=0; i<51; i+=1) {
    poly(polylist[i]);
  }
}
```

However, the code needed to render an individual polygon is considerably simplified:

```
void poly(polygon p)
{
    glColor3f(p.col.r, p.col.g, p.col.b);
    glBegin(GL_POLYGON);
      glVertex3f(p.p0.y, p.p0.z, -p.p0.x);
      glVertex3f(p.p1.y, p.p1.z, -p.p1.x);
      glVertex3f(p.p2.y, p.p2.z, -p.p2.x);
      glVertex3f(p.p3.y, p.p3.z, -p.p3.x);
    glEnd();
}
```

In this case, the colour of the current polygon is set from the database object and the four vertices are passed to `glVertex3f`. In addition, rather than drawing the polygons as wire frame polygons, they are rendered as in-filled polygons (`GL_POLYGON`). One other addition is to clear the background to the sky colour at the start of each frame, so that the sky is, in effect, the background colour.

```
glClearColor(0.0f, 0.5f, 0.5f, 1.0f)
```

Two sample images from this visual system are shown in Figure 8.16, where the left and right images are taken from an approach at 10 km and 1 km, respectively. The images were taken using the same visual database and at the same position as the image generated in Figure 8.14.

Again, it is clear that such an IG would not be acceptable in a modern flight simulator. Nevertheless, the code for this implementation is less than 120 lines (excluding the visual database), which includes the code to establish the OpenGL environment. The database can be formed as a simple list of planar polygons, which are rendered sequentially, subject to the performance of the graphics processor, avoiding the need to implement any geometric processing. Although an IG of this quality would have been state of the art in the 1970s, nowadays such systems would only be appropriate for simple visualization tasks. The main criticism is the lack of texture in the polygons;

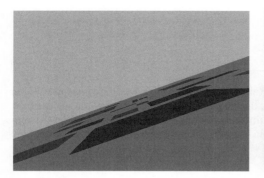 

Figure 8.16 A OpenGL in-filled visual system

during an approach, the polygons gradually increase in size but there is a distinct lack of motion provided by such images, whereas motion cues (referred to as *vection*) are evident in textured images.

8.4.3 An OpenGL Real-time Textured IG System

In Section 5.7, aircraft displays were rendered using texture maps for instrument dials and pointers. Exactly the same approach can be extended to 3D graphics. A texture file is generated for a specific object, for example, a field of crops or a runway. During each frame, the texture map of the object is mapped to the surface shape following transformation, clipping and perspective operations. The texture map is subject to exactly the same geometric transformation as the polygons and is filtered to minimize the effects of visual artefacts caused by the transformations.

For each textured polygon, a texture map is generated using digitized photographs, CAD tools or, in the case of large areas of terrain, from a supplier of satellite imagery. Generally, the texture map dimensions must be a factor of 2. Figure 8.17 shows imagery for a section of a runway, with worn tarmac and repeated slabs. The texture file is loaded into texture memory using the method described in Section 5.7 for aircraft instruments. OpenGL provides a set of primitives (Shreiner *et al.*, 2005) to define the type of texture, to bind it to a set of polygons (so that the memory area allocated for the texture map is associated with an index, to simplify subsequent access to the texture map) and to specify the unpacking alignment, the texture filter method to be applied and also whether the texture is clamped or repeated (c.f. wallpaper).

During the rendering of the polygons, texturing is enabled and the polygon is rendered by the procedure DrawTexPoly as follows:

```
void DrawTexPoly(polygon p, GLuint texobj)
{
    glEnable(GL_TEXTURE_2D);
    glBindTexture(GL_TEXTURE_2D, texobj);
    glBlendFunc(GL_SRC_ALPHA, GL_ONE_MINUS_SRC_ALPHA);
    glEnable(GL_BLEND);
    glColor4ub(255,255,255,255);


    glBegin(GL_QUADS);
      glTexCoord2f(0.0, 0.0);  // bottom left
```

```
        glVertex3f(p.p0.y, p.p0.z, -p.p0.x);
        glTexCoord2f(1.0, 0.0); // bottom right
        glVertex3f(p.p3.y, p.p3.z, -p.p3.x);
        glTexCoord2f(8.25, 8.25); // top right
        glVertex3f(p.p2.y, p.p2.z, -p.p2.x);
        glTexCoord2f(0.0, 8.25); // top left
        glVertex3f(p.p1.y, p.p1.z, -p.p1.x);
    glEnd();
}
```

In this example, the runway is 1650 m long and 50 m wide and the runway texture map is 512 pixels wide by 128 pixels high. If the texture extends across the full width of the runway, 1.0 in texture coordinates is equivalent to 50 m. By mapping the texture over an area 33.0 × 1.0, the texture map (1.0 × 1.0 in texture coordinates) is repeated 33 times. The resultant texture is shown in Figure 8.18, where the runway detail given by the texture provides the motion vection cues that are missing with IG rendering using in-filled polygons. The white edge markings and centre line are drawn after texturing, as in-filled polygons.

Clearly, it would be possible to associate a texture map with every polygon and then apply this technique to the complete scene. However, this simplistic approach is not without its problems:

- A detailed texture map (e.g. a runway) may use a lot of texture space which is only noticeable for a relatively small proportion of the time, for example, the tyre skid marks on a runway image are typically visible below 200 ft on an approach – at all other times, this detail merges into the overall runway scenery and similarly, the runway merges with the airfield scenery at distances beyond 5 nautical miles;
- If the scene contains 1000 polygons say, then 1000 texture maps may be needed, exceeding the available texture storage – in practice, the same texture maps can be applied to similar objects, for example, two fields, several miles apart may have a different shape but can use the same texture, or the walls of several buildings could use the same brick texture or window textures;
- Care is needed with the rounding that can occur in texture filtering, to avoid the introduction of small gaps between adjacent textures. To some degree this problem can be ameliorated by having a background colour that blends naturally with the majority of the terrain textures;
- Textures may be generated for different levels of detail for the same object, requiring additional software to select the appropriate texture for an object.

What has been demonstrated in this section is the ease with which texture can be applied to surfaces to increase the visual fidelity of a scene. However, in a scene containing up to one million polygons, say, it would be very tedious to generate the correct geometry and texture for each polygon. As a consequence, a number of products have been developed to aid both the production and rendering of real-time scenes.

Figure 8.17 Runway texture

Figure 8.18 A textured runway

8.4.4 An OpenSceneGraph IG System

CAD packages have been used in mechanical drawing for both 2D and 3D applications. With these packages, shapes can be formed and edited where the objects of a drawing can be stretched, scaled, rotated, translated and replicated to reduce the amount of time needed to produce a detailed drawing. Similar tools have been developed to support the creation of visual databases. The designer can enter a 3D object, move it to a specific location, and stretch or rotate the object in three dimensions. Objects can also be grouped together and then scaled, rotated or replicated, providing a hierarchy of scene detail. In addition, each surface of an object can be given a colour, level of transparency or texture. From the designer's perspective, there is no need to have detailed knowledge of every vertex; libraries of objects and textures can be selected so that very detailed landscapes can be generated in a few hundred hours of editing. Moreover, where exact detail is needed, measured coordinates can be entered, or a photographic image can be applied to texture a single polygon, or alternatively, a package can provide tools to apply textures, in effect, 'painting' a set of polygons in 3D.

This approach to visual database production can produce very detailed and realistic images for scenes for terrain and objects, such as aircraft and land vehicles; packages such as Creator™ (http://www.presagis.com/) and Blender™ (www.blender.org/) and are widely used in flight simulation. These modelling tools not only assist in the development of visual databases, but they provide four additional functions that are essential in real-time rendering:

- The scene can be defined as a hierarchy, for example, an airfield may contain buildings such as a control tower and hangars and each hangar may contain windows, which in turn are defined by a specific texture;
- The scene is automatically reduced to a set of wire frame triangles, needed for high-speed in-fill and texture operations, for example, Figure 8.19 shows an airport scene, where the topology of the terrain, runway markings and the airport buildings are clearly shown;
- The level of detail in a scene can be defined, in order to optimize the scene rendering and to avoid rendering objects where the detail is irrelevant (e.g. at long range) reducing the number of objects rendered per frame.

- The scene can be stored in the various formats that are used in the flight simulation industry. One of the most commonly used formats is OpenFlight® and loaders for this file format are widely available.

The OpenFlight format is published and real-time rendering packages can read OpenFlight databases directly, simplifying access to the visual database.

Four real-time rendering packages are widely used in the flight simulation industry. SGI's OpenGL Performer (http://www.sgi.com) was developed by Silicon Graphics Incorporated (SGI) as a commercial package for Unix platforms and subsequently ported to the PC, including Linux platforms (Eckel and Jones, 2004). As its name implies, it is based on OpenGL, providing a library of high-level graphics primitives which can be linked to an application. An alternative approach was adopted for the Vega Prime™ (http://www.presagis.com) renderer, where many of the user functions are defined using an application programming interface (API), simplifying the interface between a user application and the run-time IG software. OpenSceneGraph (Kuene and Martz, 2007) is an open source graphics toolkit, written in C++, providing a library of functions for real-time scene rendering, including the loading of visual databases, real-time rendering and scene effects including fogging, lighting and shading. DirectX (Gray, 2003) was developed for Microsoft platforms and, like OpenGL, supports many of the hardware rendering functions provided by modern graphics cards. DirectX is a low-level library of graphics functions, which has been adopted widely by the games industry.

OpenSceneGraph (OSG) has been used extensively with the flight simulation software covered in this book and will be used in the remainder of this section. The advantages of C++ are particularly beneficial in real-time rendering; an object-orientated approach can be applied to the hierarchy of objects, a suite of independent modules are dedicated to specific rendering tasks and the language is supported on most platforms. There is one further advantage: C code can be embedded in C++ code.

For reasons of brevity, it is not possible to cover all aspects of OSG and the reader is directed to the detailed documentation. However, the basic concepts of OSG are important and they provide the basis for real-time rendering. The advantage of a toolkit such as OSG is that it removes much of the responsibility for low-level detailed rendering operations from the user, allowing the user to focus on the application. The overall structure of the IG system covered in this book is illustrated in Figure 8.20.

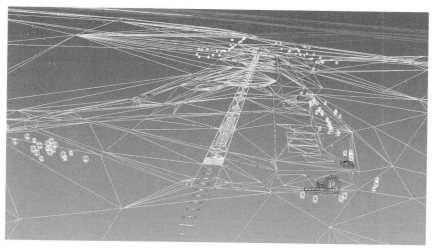

Figure 8.19 Wire frame triangles of an airport scene

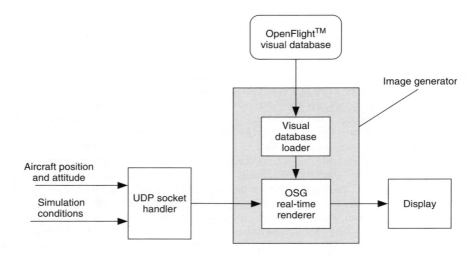

Figure 8.20 IG system structure

An OpenFlight database is loaded at the start of the simulation by OSG and organized as a tree structure of linked nodes. During each frame, the aircraft position and attitude, computed by the flight model, are passed to the IG. In addition, other simulation and flight conditions (possibly set by the instructor), such as lighting conditions, cloud base or visibility are also updated. In the simulator covered in this book, data is transferred to the IG as broadcast Ethernet UDP packets. In other flight simulators, the communication is often implemented as a shared memory. Note that the direction of the transfers is one way. However, if events such as terrain collisions are needed or if height-above-terrain measurements are used, the IG system may also transfer data to other simulator subsystems.

OSG is underpinned by a tree structure, where the nodes of the tree are created by the user at run-time but the linkage between the nodes of the trees is managed by OSG. These nodes may refer to objects in a scene but they can also combine geometric operations. For example, the world scene may contain a town and a number of military tanks. Each tank has a position and attitude in the world (which is a translation) and for each tank, its turret can be rotated and raised and lowered, as shown in Figure 8.21. An operation inserted at a high level in the tree is inherited by all the child nodes at a lower level in the tree.

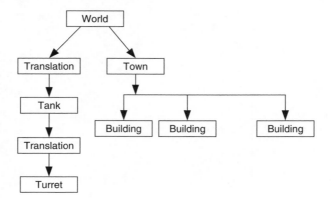

Figure 8.21 A scene graph

Visual Systems

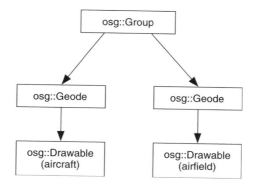

Figure 8.22 A scene graph for an airfield and an aircraft

In practice, in a flight simulator IG, the terrain is mostly static and is therefore subject to a single translation and rotation by the PEP and dynamic vehicles operate independent of the terrain, simplifying the structure of the visual database. There may be a few dynamic objects in a terrain database, such as an explosion which may alter the shape of objects in the terrain or lighting where the colour of a light may change, for example PAPI lighting or 'running rabbit' lights on a runway approach.

A node in OSG is defined by the `osg::Node` class with three commonly used subclasses, `osg::Geode`, `osg::Group` and `osg::PositionAttitudeTransform`. A node can only be attached to an instance of an `osg::Group` (or one of its subclasses), whereas an object that can rendered (referred to as *drawable*) is attached to an instance of `osg::Geode`, as shown in Figure 8.22 for a scene containing an airfield and an aircraft.

Of course, if the aircraft is a dynamic object in the scene, it is likely to translate with respect to the scene; this is achieved by adding an `osg::PositionAttitudeTransform` node (Pat) above the aircraft node, as shown in Figure 8.23. Although these structures are boundless, in practice, a scene in a flight simulator is likely to contain relatively few Geode nodes and moreover, the structure defined in an OpenFlight file will be translated to this form automatically by the OpenFlight database creation routines in OSG.

One other aspect of OSG is the use of classes to simplify resource management, particularly the allocation and de-allocation of memory. In a general form, a class enables the allocation and de-allocation as the class is used. Consider a simple example of a wrapper for an item:

```
Class ItemWrapper
{
    public:
        ItemWrapper() {
            myhandle = allocatetheitem();
        }
        ~ItemWrapper() {
            deallocatetheitem(myhandle);
        }
        myhandletype & get() {
            Return myhandle;
        }
    Private:
        Myhandletype myhandle;
}
```

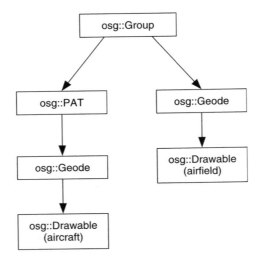

Figure 8.23 A scene graph including an aircraft with translation

An item can now be associated with this class and the procedure to use an item is as follows:

```
ItemWrapper myitem
usetheitem(myitem.get());
```

The item is allocated when it is instantiated and released when it is no longer in scope, freeing the responsibility for memory management from the user.

For example, a child node can be added to a group as follows:

```
osg::ref_ptr<osg::Group> group (new osg::Group());
osg::ref_ptr<osg::Geode> childnode (new osg::Geode());
group->addChild(childnode.get());
```

The specific code for an OSG real-time renderer is given in *cgi.cpp*. The detail of the code is beyond the scope of this book, but several features of this implementation are outlined where they are common to real-time rendering packages. The variables to hold the current aircraft (the camera position), a target aircraft (for conflicting traffic or formation flying) and OSG scene pointers are initialized as data definitions, to define the fog, the source of sunlight, a target node, the position and attitude transform to be used and the root node of the scene, as follows:

```
osg::Vec3f vecPosAircraft;
osg::Vec3f vecAttAircraft;
osg::Vec3f vecPosTarget;
osg::Vec3f vecAttTarget;

// Global scene graph objects/nodes
osg::ref_ptr<osg::Fog> fog = new osg::Fog();
osg::ref_ptr<osg::LightSource> sunLight;
osg::ref_ptr<osg::Node> TargetNode;
```

```
osg::ref_ptr<osg::PositionAttitudeTransform> TargetXForm new
             osg::PositionAttitudeTransform();
osg::ref_ptr<osg::Group> SceneRoot = new osg::Group();
```

One action of the procedure `main` is to load the visual database. This is passed in the command argument list and checked against a table of known databases. The database is loaded in a single OSG call.

```
osg::ref_ptr<osg::Node> TerrainNode = osgDB::readNodeFile(filename);
```

which returns zero if an error was encountered during the loading of the visual database. Typically, a single database is used which includes the terrain, the airfield, trees, buildings and so on. Alternatively, these subgroups of objects could be loaded separately and then linked to the root node. In fact, several target models are used with this simulator, where a new target (a separate OpenFlight file) is loaded (at run-time) as follows:

```
if (TargetLoaded)
{
    sunLight->removeChild(TargetXForm.get());
    TargetXForm->removeChild(TargetNode.get());
}

TargetNode = osgDB::readNodeFile(Filename);
if (!TargetNode)
{
    std::cerr << ''Failed to load target geometry!\n'';
    exit (1);
}

TargetXForm->addChild(TargetNode.get());
sunLight->addChild(TargetXForm.get());

TargetLoaded = 1;
```

where `filename` is the name of the OpenFlight file for the target aircraft. The existing target node is removed (together with its link to the `sunLight` node). If the database file is successfully loaded, it is added with a link to the `sunLight` node. Although the main terrain database is loaded before the scene rendering starts, the visual database for a target is loaded during rendering. This can lead to a delay of several seconds for a large file and normally, all the visual databases would be loaded prior to the start of the simulation.

The sky model is then loaded. This is also an OpenFlight file, but it differs from other objects in the sense that it is a surrounding set of textured surfaces which does not translate with respect to the aircraft. In other words, the sky model is attached to the point on the ground directly below the current aircraft position with a constant attitude. A linear fogging model is enabled using the fogging functions provided in OSG. The effectiveness of a good fogging model is very important in most flight simulation applications. It can reduce the visual artefacts which occur with distant objects and provide realistic haze as well as actual fog. Depending on the fogging techniques used, objects obscured by the fog can be omitted from the rendering pipeline to reduce the frame rendering time.

The sub-nodes for the terrain, sunlight, sky and possibly a HUD are attached to the scene as follows:

```
SceneRoot->setStateSet(fogStateSet.get());

sunLight = createSunLight();
osg::ref_ptr<osg::Group> LitObjects = new osg::Group();
LitObjects->addChild(sunLight.get());
sunLight->addChild(TerrainNode.get());

SceneRoot->addChild(SkyBoxXForm.get());
SceneRoot->addChild(backdrop.get());
SceneRoot->addChild(SkyNode.get());
SceneRoot->addChild(LitObjects.get());
SceneRoot->addChild(createHUD());
```

The final phase of initialization is to set the camera field-of-view. In the case of multiple projectors, the IGs are likely to use the same database and the identical OSG code, with the exception that the camera viewing angle of each IG is aligned with its respective projector direction. With multiple projectors, the projectors are aligned to overlap by a few degrees to avoid a visible gap between adjacent projectors. For example, the curved projection screen used in the flight simulator at the University of Sheffield is approximately $160° \times 40°$. The horizontal field of view of each projector is $63.1°$ and the left and right projector images are offset by $43.45°$ from the centre line. This alignment occurs in both the IG software and the physical mounting of the projectors. Considerable care is needed to ensure that the field of view and offset value do not introduce any distortion or misalignment. The setting in OSG is given by

```
viewer.getCameraConfig()->getCamera(0)->
    setLensPerspective(63.1, 43.0, 1.0, 200000.0);
```

where the near and far clipping fields are set to 1 m and 200,000 m, respectively.

Most visual systems are captured with a local coordinate system. An airline may have over 10 visual databases for the airports covered by its route structure for a specific simulator. Alternatively, an engineering simulator may only have a single generic airport as the simulator is mostly used in system development. In the latter case, this single representative airport database can be used to provide an airport image for all the airports defined in a navigation database. A visual database will have been developed with a specific coordinate frame and it is necessary to transform the pilot-eye position and attitude to the visual database frame. In addition, visual databases are defined in feet or metres using a Cartesian frame rather than latitude and longitude. A typical visual database coordinate frame is shown in Figure 8.24. The aircraft position is defined by its latitude, longitude (a_{lat}, a_{long}) and altitude. In addition, the latitude and longitude of the airport (defined by the start of the runway in this example) will also be known (r_{lat}, r_{long}). The runway position will also be defined in the visual database frame (x_{run}, y_{run}), where the origin of the database is also known but is not necessarily aligned with the runway. There is a further complication if a single generic airfield is replicated, the specific QDM (magnetic heading) and altitude above sea level of each airfield must be included in the transformation. In other words, the eye-point position must be referenced to the Cartesian coordinate frame used by the IG, which is likely to be based on the visual database frame.

Let

$q = $ the true runway QDM (radians)

$d = $ the distance from the aircraft to the runway (metres)

Visual Systems

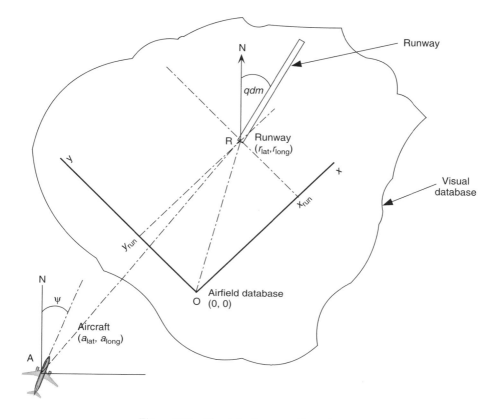

Figure 8.24 Visual database coordinate frame

b = the bearing of the runway from the aircraft (radians)
r_x = the runway x coordinate in the database frame (metres)
r_y = the runway y coordinate in the database frame (metres)
r_o = the runway offset from true north in the database frame (radians)
e_x = the eye position x offset from the aircraft cg in the Euler frame
e_y = the eye position y offset from the aircraft cg in the Euler frame
e_z = the eye position z offset from the aircraft cg in the Euler frame

An additional rotation of the pilot eye view port about the cg is essential as the IG view port is positioned at the pilot station.

Computing the aircraft position (x', y', z') relative to the runway datum

$$x' = -d \sin(b) + e_y \qquad (8.13)$$
$$y' = -d \cos(b) + e_x \qquad (8.14)$$
$$z' = -P_z + e_z \qquad (8.15)$$

The resultant position (x, y, z) in the coordinate frame of the visual system is given by

$$a = 2\pi - q - r_o \tag{8.16}$$
$$x = x' \cos(a) + y' \sin(a) + r_x \tag{8.17}$$
$$y = y' \cos(a) - x' \sin(a) + r_y \tag{8.18}$$
$$z = z' + r_z \tag{8.19}$$

The resultant attitude (p, h, r) in the coordinate frame of the visual system is given by

$$p = \theta \tag{8.20}$$
$$h = r_o - \psi + q \tag{8.21}$$
$$r = \phi \tag{8.22}$$

where θ, ϕ and ψ are the aircraft Euler angles.
The view port position and attitude is then set in OSG as follows:

```
vecPosAircraft.set(px, py, pz);
vecAttAircraft.set(h, p, r);
```

where the attitude angles h, p and r are in degrees (in OSG) and P_z is the aircraft altitude in the Euler frame.

The pilot eye position offset (e_x, e_y, e_z) is computed in the flight model, from the pilot station offset (ps_x, ps_y, ps_z) defined in the body frame. For the Boeing 747-100, the pilot station offset is given by

$$ps_x = 25.19$$
$$ps_y = 0.51$$
$$ps_z = -3.98$$

where $CG_Z = -4.78$

The eye position offset is computed with the following body frame to Euler frame transformation, where $a_{11}, a_{12}, a_{13} \ldots$ are the elements of the DCM.

$$e_x = a_{11} e_x + a_{12} e_y + a_{13} e_z \tag{8.23}$$
$$e_y = a_{21} e_x + a_{22} e_y + a_{23} e_z \tag{8.24}$$
$$e_z = a_{31} e_x + a_{32} e_y + a_{33} e_z \tag{8.25}$$

Figure 8.25 shows two images[2] captured during an approach into Bristol Lulsgate Airport. Note the image quality provided by texturing and the realistic fog shown in the lower image. Figure 8.26 shows a night approach into Bristol Airport, by reducing the illumination level. The lower image shows an approach in Hong Kong International Airport (Chek Lap Kok), showing the effects of sea texture, sky texture and hidden surface elimination provided by Z buffering in rendering the islands.

---

[2] These images were produced by OpenSceneGraph using an nVidia FX-100 graphics card.

Visual Systems

(a)

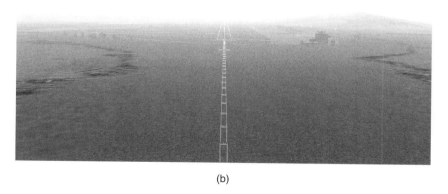

(b)

Figure 8.25 Bristol Lulsgate Airport approach (a) and the effect of fogging (b) (see Plate 8)

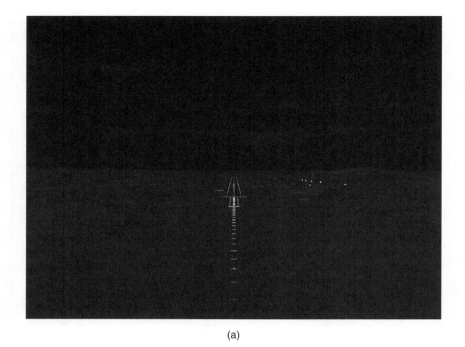

(a)

(b)

Figure 8.26 Bristol Airport night approach (a) and Hong Kong International Airport (Chek Lap Kok) approach (b) (see Plate 9)

Figure 8.27 Trees rendered as billboards (see Plate 10)

For real-time visual systems, the designer often has to compromise between image quality (or realism) and rendering rate. A significant aspect of the implementation of real-time visual systems is the identification of graphic activities which can reduce the rendering activities per frame. A very good example where excellent visual effects are achieved for minimal effect on the rendering rate is the use of billboards. A billboard is simply a flat textured object placed in a scene but with the property that it is always oriented facing towards the viewer. This is readily achieved by a rendering package because the location of the viewer and the object are known, allowing the object to be rotated so that the front face is seen at all times by the viewer. In addition, the texture can be transparent so that other objects behind a billboard can still be seen, providing very powerful depth cues.

One example of the application of billboards is the rendering of trees. By capturing the texture image of a tree, the tree can be 'planted' in the scene. As trees are representative objects, the viewer is unlikely to notice that the identical view is provided in all directions from the tree and moreover, at distances greater than a few metres, the rate of rotation of the textured image is not discernible. For example, a small copse of 50 trees can be placed on a hillside, adding valuable height and depth cues (as well as increased realism) to the scene. Figure 8.27 shows a typical scene containing trees, where the computational cost of rendering each tree reduces to the rendering of one or two textured rectangles.

One other area where billboards can contribute to visual fidelity is the production of smoke, or trails. In the flight simulator at the University of Sheffield, as part of a research programme on the real-time simulation of wake vortices (Spence *et al.*, 2007), it was necessary to display the wake vortices generated. As the contour of a vortex (its centreline) and the extent (area along the trail) was known, it was straightforward to represent the vortex as a series of billboards, spaced at intervals along the trail, where each billboard was given a texture similar to smoke, as shown in Figure 8.28. The computational cost of rendering a few hundred textured billboards, located along

Figure 8.28 A wake vortex rendered as billboards (see Plate 11)

the vortex is relatively small for a very significant benefit in visual fidelity. Of course, the visual fidelity of an object which is asymmetric will vary with viewing position. However, as is evident in Figure 8.28, the level of realism viewed from behind the aircraft generating the vortex would be acceptable in most training simulators.

8.5 Visual Database Management

A visual database will contain all the graphics entities seen in the simulator including airfields, runways, buildings, trees, road, lakes, rivers, forests, fields, lights and vehicles. Some of these features may be generic, for example, a small village of 50 houses or a copse of trees, whereas it may be necessary to replicate other features to a very high degree of accuracy, for example, the approach lights to an airfield. Many of these features will be static; once constructed they cannot change, whereas the attributes of other objects may be dynamic, for example, the undercarriage of an aircraft may be raised or lowered or a tree may have leaves in summer but not in winter conditions or the reflectivity of a runway may change in rainy conditions.

All these characteristics must be embedded in the visual database and then extracted at run-time so that each object can be rendered in the scene. The database provides three primary functions:

- It defines the geometry of objects in the database;
- It enables structure to be associated with the objects;
- It allows the visual system to access the objects.

The objects are commonly entered in the visual database using a three-dimensional CAD drawing package, such as Creator™, Photoshop™ or AutoCad™. These packages enable the database designer to enter simple objects such as lines, rectangles and circles, which can be combined to form more complex objects that can be scaled, rotated, reflected and stretched. These graphic entities can also be replicated, often in combination with other functions to reduce the time needed to design complex objects. For example, a wheel may have 30 spokes. One spoke can be designed in detail and then the 30 spokes can be replicated at regular angular interval around the hub of the wheel. Each entity is defined in specific measurement units and placed in a specified geometric frame, which can be a local or global frame, where the axes orientation can also be defined. Each surface can also be defined in terms of reflectivity, transparency, colour and texture. For example, a brick wall can be defined in terms of its coordinates and a specific brick pattern can be pasted onto the wall from a selection of brick patterns. Many of these operations take only a few mouse clicks and provide considerable savings in terms of design effort.

Of course, the actual measurements and location of objects must be known. For large objects, this information is derived from maps, or survey data or from satellite tomography. For buildings, plans are often available and mechanical drawings are also available for aircraft and other vehicles. In some cases, photographs can be used to derive measurements. The capture process requires transferring these measurements to the objects in the visual database. In fact, database capture is very time consuming and is therefore very expensive and the CAD tools used in database design are designed to minimize the effort needed to construct these objects. Even so, the capture of a database of an airfield may take several man months of effort, depending on the level of detail required. For military operators, the owners of facilities may not necessarily cooperate in the capture of the data and, in such cases, the visual data is merged from a combination of satellite imagery and photographs. To some degree, this task has been simplified with the availability of digital cameras and GPS to determine the camera position. Military users may also need to define the features in terms of optical, radar or infrared properties for use with a range of aircraft sensors and this information must also be captured and entered into the database.

The capture of this data may generate a large list of objects, each defined by its position, dimensions, colour, texture and so on. CAD tools are used to reduce the objects to triangles, typically using Delaunay triangulation (Lee and Schachter, 1980; Schneider and Eberley, 2002) and also to organize the objects in a format that is recognized by image generators, for example, OpenFlight (Anon. 2007) and 3ds Max (http://usa.autodesk.com/). Figure 8.29 shows the wire frame mesh of a geometric model of a Boeing 747-400 aircraft comprising several hundred Delaunay triangles and the corresponding model with shaded and textured polygons.

However, accessing all the objects in the database at run-time may overload the image generator. Rather than reducing the number of objects in the database, by structuring the information stored in the database, it may be possible to access only the visible or 'active' objects in the database. The database design tools also allow the designer to describe objects in terms of hierarchies, partly to facilitate access to objects in the database, but also to allow levels of detail (Luebke et al., 2002) to be embedded in the database, as shown in Figure 8.30.

At 10 nautical miles, the airport may be visible but individual buildings or the runway will probably not be distinguishable. At 5 nautical miles, large buildings may be discernible but the runway centre line or edge lines will merge into the airfield. However, at 50 ft, the centre line and skid marks will be seen in detail or taxiing back to the terminal, a company name and a 'no-unauthorized access' notice may be visible on a hangar door. By organizing the visual database in this hierarchical form, access to irrelevant polygons can be suppressed, with considerable saving in rendering time and potentially increasing the fidelity.

In Section 8.4, techniques for back face removal and hidden surface removal were outlined. However, these techniques cannot be implemented until each surface has been transformed to the pilot eye frame. Nevertheless, with knowledge of the aircraft position and attitude, objects can be excluded from the image or accessed more efficiently. For example, several levels of detail can be embedded in an aircraft visual model. Taxiing past a parked aircraft, the detail is visible, but groups of back-facing surfaces can be eliminated. At 5 nautical miles, the same aircraft,

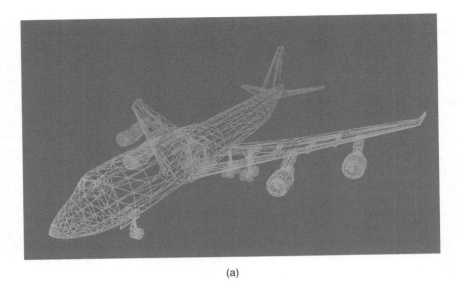

(a)

(b)

Figure 8.29 A Boeing 747-400 visual database model (a) wire frame version, (b) shaded and textured polygon version (see Plate 12)

taking off from the runway, can be represented in much less detail, without any noticeable loss of fidelity. Another method to detect redundant objects is to project the viewing frustum onto the earth surface, as illustrated in Figure 8.31, where the grey area denotes the extent of the visual database. The actual footprint $ABCD$ is expanded to the rectangle $PQRS$ and only objects within this rectangle need to be extracted from the visual database. In practice, the footprint may change rapidly with changes in aircraft attitude and also (but more slowly) with changes in aircraft position. However, it is not difficult to predict the probable footprint a few seconds in advance

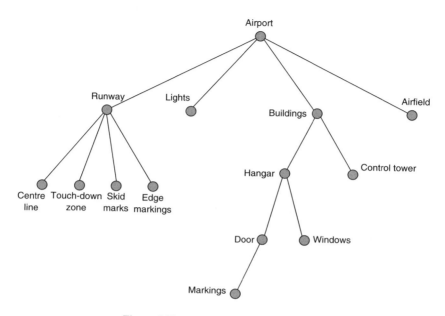

Figure 8.30 Visual database organization

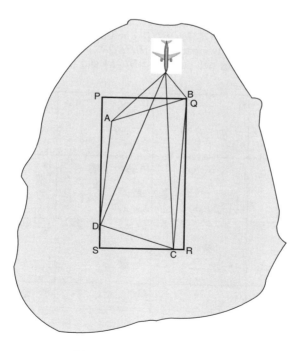

Figure 8.31 Frustum footprint

and maintain a cache of potentially active objects, where the cache is continually refreshed as the aircraft manoeuvres.

An alternative technique is to monitor the image generator loading. As the level approaches an upper limit, the level of detail threshold is reduced for objects towards the extremities of the image, or where there are significant reductions in processing at this lower level of detail. Similarly, if the image generator is less loaded, the level of detail can be increased. There are, however, two problems associated with these techniques. First, the blending between levels of detail must be smooth to avoid discrete and discernible 'jumps' in detail which would be noticeable and secondly, it is very difficult to ensure visual fidelity if the scene content is continuously changing – there is a high likelihood of introducing sudden changes to the visual scene which are apparent to the flight crew. Such artefacts are not just irritating; they can be a distraction in a flight simulation, reducing the training effectiveness.

It is possible to organize the visual database as a tree structure. To illustrate this concept, Figure 8.32 shows a database containing 30 objects. At the top level, the terrain is represented as four quadrants (north-west, north-east, south-west and south-east) given by quadrants *ABEF*, *CDGH*, *IJMN* and *KLOP*. In turn, each of these quadrants is reduced, for example, the north-east quadrant contains quadrants *C*, *G*, *D* and *H*, where *C* and *D* are empty. The process repeats until a quadrant contains only a few (or zero) elements. For example, quadrant *A* contains sub-quadrants *W*, *X*, *Y* and *Z*. The tree is shown in Figure 8.33, where only quadrant *A* is shown expanded. If a node of a tree is empty or not in view, all the objects given by that node (i.e. its sub-trees) can be discarded at that point, reducing the amount of processing. The actual objects are the leaf nodes of the tree. Note that this is a recursive data structure – a tree contains sub-trees which contain sub-trees and so on. Moreover, the complete database is defined as a single pointer to a tree structure.

One other tree structure that is commonly used is the binary-spaced partition (BSP) tree. Each node of the tree has two sub-nodes where each sub-node contains half the objects of its parent node, as shown in Figure 8.34, for the database given in Figure 8.32. The first cut partitions the space into the set of nodes {*ABCDEFGLMNOPTUV*} and {*HIJKQRSWXYZabcd*}, each containing 15 nodes. The second cut generates two partitions {*HIJKQRS*} and {*WXYZabcd*} of 7 and 8 nodes, respectively. The third cut produces two partitions {*HIQR*} and {*JKS*} and so on, until a partition

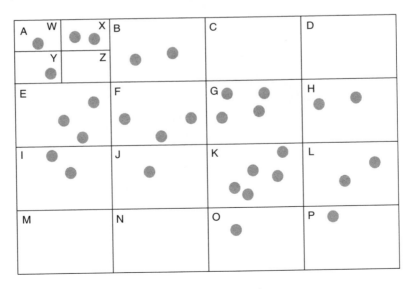

Figure 8.32 A terrain quad-tree

Visual Systems

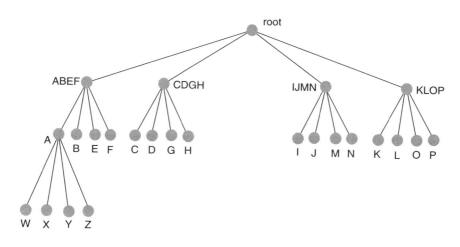

Figure 8.33 Quad-tree tree structure

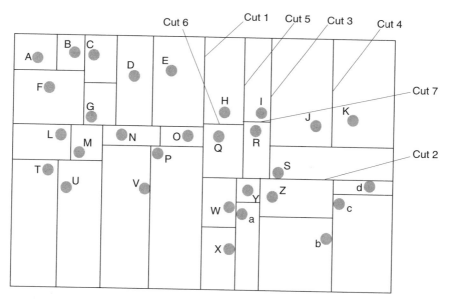

Figure 8.34 Partitioning a BSP tree

contains only one node. The resultant BSP tree is shown in Figure 8.35. The criterion for partitioning is usually based on balancing the area occupied by each of the sub-partitions. The leaf nodes of the BSP tree are the objects of the database. For both quad-tree and BSP tree representations of visual databases, the tree structure used to search or access objects in the database provides a significant speed improvement as the search time is proportional to $\log_4 n$ or $\log_2 n$ for quad-trees or BSP trees, respectively, where n is the number of objects.

The file generated by these visual database construction tools is typically organized as an Open-Flight structure (Anon. 2007), which is recognized by real-time rendering packages such as SGI

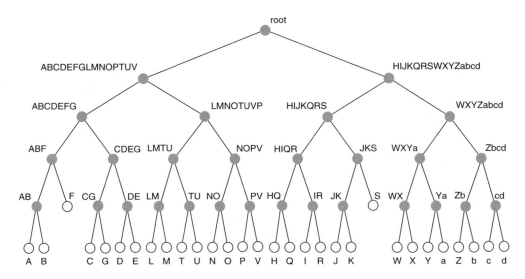

Figure 8.35 A terrain BSP tree

Performer and OpenSceneGraph. OpenFlight files contain descriptions of the geometry of objects and also the structure of the database to facilitate fast access to the objects and to implement culling algorithms to sustain real-time performance. One particularly useful property of efficient organization of a visual database is that an object can be defined once and then invoked by the renderer to instantiate several copies of the object in a model. For example, an aircraft undercarriage may contain 20 identical wheels. With an OpenFlight database, there would be one definition, with each wheel instantiated at its appropriate place in the undercarriage assembly.

An important point to appreciate with visual database management systems is that they can have a significant impact on the overall performance of a visual system. By implementing methods for the efficient extraction of active polygons, combined with a database that embeds a hierarchical structure to organize the objects, the visual system pipeline is continuously loaded with active polygons, very few of which need to be discarded. The alternative, a flat structure with no coherence between objects, means that polygons are discarded after considerable processing has been expended and the overall performance of the visual system is reduced.

8.6 Projection Systems

An image generator produces an image stored in memory. These bit patterns are retrieved by the video processor every frame to form a video signal for a monitor or projector. Simply placing a monitor in front of a pilot has several limitations as a projection system in a flight simulator. First, the human eye focuses at infinity beyond approximately 5 m. To provide a monitor covering a lateral field of view of 60° at 5 m would require a monitor over 5.7 m wide. At this distance, a 28-in. monitor occupies only 9°. Abutting monitors, as used in outdoor displays is not practical in a flight simulator. Similarly, to provide a 60° field of view with a 28-in. monitor requires the monitor to be placed some 24 in. in front of the pilot, at a distance where the pilot needs to focus on the screen detail at this short range. Although CRT displays were used in the early flight simulators, their obvious limitations have restricted their application to amusement arcades.

Consequently, consideration has been given to the optical amplification of images generated on a monitor. Initially, this was achieved by collimation (Spooner, 1976) but more recently, has been

Visual Systems

Figure 8.36 A collimated projection system. Courtesy Rockwell Collins, Inc

implemented as a result of advances in classroom projector systems. A typical collimated projection system is shown in Figure 8.36. In this system, the monitor is mounted above the display, with the screen pointing downwards. The pilot looks through a windscreen angled at 45°, which is a semi-silvered mirror, known as a *beam-splitter*. Light from the monitor is reflected off the back of the beam-splitter, to a concave mirror, known as the *collimating mirror*, with its focal point at the pilot eye position, as shown in Figure 8.37. Light from the collimating mirror passes through the beam-splitter; the rays seen at the pilot's eye position appear to be parallel and therefore to originate from a distant source. The complete unit of a monitor, beam-splitter and a collimating mirror is encased in a light-tight box.

The advantage of this system is the quality of the projected image. A high-resolution CRT is an off-the-shelf product and the cost of a beam-splitter and a collimating mirror can provide a display 60° wide by 40° high at a modest cost, where the beam-splitter is located at the position of the aircraft windscreen. Indeed, for civil aircraft with rectangular windscreens, several collimators can be placed on the simulator flight deck to provide a wide field of view. However, there are significant disadvantages with collimated projectors:

- The collimation is focused at the PEP; movement of the pilot's eye position can result in considerable distortion;
- The cost of grinding the collimating mirror accurately can be high;
- The mass of the glass mirrors and the projectors, particularly with the off-axis positioning of the projectors, can add significant loading to a motion platform;

- Shock mounts are needed to avoid damage to the mirrors caused by motion system impacts;
- It is difficult to abut these systems to provide a continuous wrap-around display;
- For a two crew operation, pilots can only see the image in their own collimator; the image in the other pilot's collimator is distorted beyond recognition.

Advances in classroom projection systems in the 1990s led to the availability of low-cost projectors, either from direct projection onto a screen in front of the cockpit, with the projector mounted above the cockpit or by back projection. One limitation of the early systems is that the projector is positioned above the pilot eye axis, causing keystone effects in projection, where a rectangular image appears trapezoidal in shape. However, this distortion is easily corrected by circuitry in the video amplifier of the projector. To increase the field of view, additional flat screens can be placed in front of the cockpit, with one projector per screen. However, there is some angular distortion towards the extremity of a flat screen; for a projector at 5 m, 1° at the centre of the projector screen is 87.3 mm, whereas 1° corresponds to 115.2 mm at the edge of a 60° screen. Of more significance, at the vertical joint between screens, there is sufficient distortion that the effect is irritating. An alternative method is to construct a screen which is continuously curved laterally and vertically, to form a segment of a sphere, centred at the PEP with three (or more) projectors located above the cockpit to project onto this curved surface. However, there are three problems with such configurations:

- The cost of constructing a large screen, curved in two dimensions, is high;
- The edges of each projected image must align very accurately;
- The image is generated for a flat screen but is projected onto a curved image.

The first problem is readily overcome by manufacturing segments which can be moulded and then assembled to form a curved screen. To avoid gaps at the edge of the projected images, each projector can overlap its neighbouring projector by a few degrees. If the projectors are correctly aligned the overlap is not noticeable. The third problem has been overcome with the availability of digital projectors. The scanning waveform of the raster can be adjusted so that the image is correctly distorted to map from a flat image to a curved image. In fact, this is usually achieved by

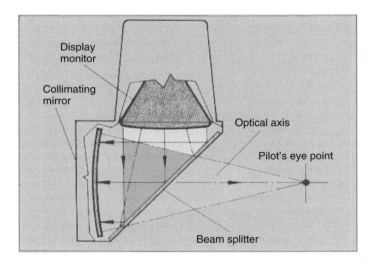

Figure 8.37 Principle of a collimated projection system. Courtesy Rockwell Collins, Inc

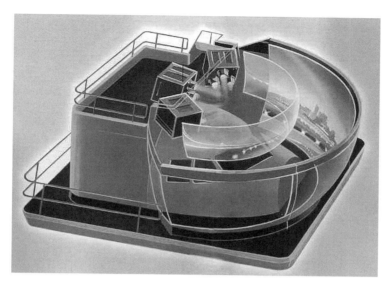

Figure 8.38 Curved mirror projection system. Courtesy Rockwell Collins, Inc

projecting a set of grids and grid points onto the screen and then adjusting the distortion on the screen until the grid appears orthogonal and evenly distributed over the complete screen. Modern projectors affording this facility usually provide calibration software to generate patterns to align the projectors. In addition, the drift with digital projectors is negligible and the correction values used to calibrate the projectors can be stored and reset to previous values.

The concept of a continuously curved screen has been extended further for both airline and military flight simulators (Blackman, 1995). If the images of three projectors, say, are projected onto a curved mirror rather than a plain white screen, the optical attenuation is reduced and moreover, if the geometry of the screen is set correctly, the rays seen by the pilot are parallel (or collimated) giving the effect of depth seen with collimated projection systems. This concept is illustrated in Figure 8.38.

The image is projected onto a small screen and the spherical mirror is designed so that this image fills the mirror when seen at the PEP, as shown in Figure 8.39. The projectors are placed above the cockpit and typically, each projector covers an area approximately 60° laterally by 40° vertically, providing a total lateral field of view up to 220° with four projectors. Although the load of a wide curved glass mirror positioned above the simulator cockpit would exceed the performance limits of a motion platform, a system developed in the 1990s has reduced this mass by a significant factor and has become the industry standard for projection systems. Rather than using a solid curved mirror, a light flexible sheet (typically the thickness of aluminium foil) is coated with a reflective surface. This continuous sheet is then sucked into a curved shape by a vacuum pump on the external side of the mirror. The resultant shape is close to spherical and variations in the pressure of the sealed vacuum unit are minimized. In addition, motion and vibration of the platform is isolated from the screen.

Curved mirror systems are relatively expensive but they have brought a major advance to flight simulation. First, they reduce the cabin load on the motion system allowing hydraulic actuation to be replaced with electrical actuation for motion platforms. Secondly, the visual system is continuous over the complete area of projection. Thirdly, and most importantly for airline training, both pilots see the same projected image, that is to say, cross-cockpit viewing is very similar to an actual aircraft. Strictly, when the aircraft is on the runway, the pilots will see a view that is only a few metres away and in this case, each pilot will see a slightly different image. However, this limitation is negligible for most training applications.

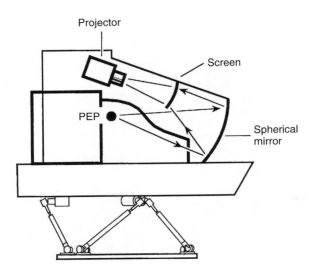

Figure 8.39 Curved mirror projection system geometry. Courtesy Rockwell Collins, Inc

In military applications, the requirements of a projection system are very different. In particular, pilots need a very wide field of view and in target acquisition, need a visual target where the image resolution exceeds standard projection systems. Dome projection systems are now in widespread use by many military organizations, most of whom have dispensed with motion platforms in favour of G-seats. Within a semi-spherical dome, a series of projectors are used to provide a near 360° view in azimuth. In addition, laser projectors provide very detailed high-resolution images of target aircraft.

One other form of projection system has also been used successfully in military flight simulation. helmet-mounted display (HMD) technology, developed for military aircraft (particularly for night vision applications) has been adapted for flight simulation. The images are generated in a bundle of fibres and this fibre optic cable is attached to the pilot's helmet so that the image is projected directly onto the pilot's eyes via a small optical lens system attached to the helmet or goggles. Both the pilot's eye position and head position are tracked and an image is then generated for each eye, based on where the pilot is instantaneously looking. With two separate visual channels, stereoscopic vision is achievable, if appropriate. In applications where the pilot is training with night vision goggles, the transition to an HMD is minimal. In other applications, the disadvantage of the additional weight on the pilot's helmet, has to be considered against the benefits of a very wide angle of view. Tracking the pilot head position and eye position is far from straightforward. The pilot eye position is detected by shining an infrared beam to track the pilot's iris while both inertial and optical techniques have been developed to measure the pilot's head movement. The tracking needs sufficient angular resolution (in three dimensions) and minimal latency to avoid nausea brought on by the pilot seeing inconsistent images. In addition, minor slippage of the helmet or small changes in the characteristics of the pilot's eye during a training mission can also introduce unacceptable errors.

8.7 Problems in Visual Systems

Invariably, a visual system is a compromise between price and performance. It is actually difficult to rate or classify visual systems. The performance may be defined by the number of triangles rendered by second, but this value also depends on the size of the triangles (in pixels) as it measures both the geometry processing and the framestore accesses. In addition, with good quality texturing, efficient texturing may produce far better visual results than a polygon rendering visual system. Much of the impetus for graphics hardware development has been driven by the games industry, although many of the effects provided are of little use in flight simulation.

Although screen resolution is an obvious limitation, apart from problems with distant light points, the resolution of the current generation of image generators (typically 1600×1200 pixels) is only a factor of 4 below good human acuity. However, scene content is far more important. Consider a human face: with only 10 triangles, the resultant face is unlikely to be recognizable. Even with a few hundred triangles the image is at best cartoon-like. Nevertheless, there is some point where the computer generated image will start approaching photographic quality. However, if a tree requires one million triangles for acceptable realism and there are 10,000 trees in a scene say, it is clearly impractical to render such scenes in real-time. Although there has been tremendous progress with visual systems, daylight images still lack content to provide realistic images and consequently, attention has focused on organizing the scene content to maximize the visual fidelity. In fact, techniques such as fogging can help, reducing the scene content in the distance, enabling scene detail to be concentrated in the near field view.

Most real-time visual systems are slaved to a fixed frame rate of 50 or 60 Hz. Even at this update rate, at high aircraft speeds, objects near the edge of a display can appear to move in a slightly jerky manner. Reduction of the frame rate, caused by overloading of the visual system, may be discernible with high-rate manoeuvres in agile aircraft. If load balancing is used, the scene content can change in a subtle way with objects rendered with less detail. These changes must be small and undetectable but major changes such as omitting objects from the rendering process can produce distinct changes that are noticeable. The avoidance of undesirable effects depends, to a large degree, on the performance of the level of detail algorithm and the organization of the visual database.

Anti-aliasing methods are now widely adopted by graphics cards manufacturers, but the quality of the filtering algorithms can vary widely, largely as a function of the amount of memory used to represent the unfiltered image and the size of the filter footprint. The effects of aliasing may still be discerned at angles very close to the vertical or horizontal, particularly where the line width is near to 1 pixel. In such cases, a continuous line can break up in an unnatural way, to form what appears to be a dashed line.

Latency can also add undesirable effects to a simulator, including additional frames introduced by the visual systems (particularly image post-processing). In most simulators, it is the variation in latency between the flight model, the motion platform and the visual system that causes problems, in severe cases leading to 'simulator sickness'. Even so, the visual channel provides very important cues and, for high-rate manoeuvres, latency in the visual system may introduce the apparent effect of altering the flight model dynamics. Although delay compensation methods have been applied to visual systems (McFarland, 1988; Guo et al., 2005), considerable care is needed to ensure that they do not introduce undesirable effects, which are perceived as false visual cues.

Temporal effects occur when there is an interaction between the display refresh rate and vertical movement of an object, giving the effect of blurring the edges of the object, for example, if the refresh rate is 50 Hz and the motion of the object on the display corresponds to 49 lines per second. Generally, this effect is eliminated at high refresh rates. Rounding errors in the geometric computations can lead to a small change in the size of an object over a few frames. In effect, the object appears to slowly creep, even though it is a static object. Similar effects can also cause small objects (the height or width of one pixel) to disappear and then reappear, causing them to scintillate.

To a large degree, OpenGL has become an accepted standard in flight simulation. However, there are different revision levels of OpenGL, with earlier versions not supporting later extensions. Of more concern, drivers for OpenGL are produced by the graphics card manufacturers which may not be fully compliant with the standard. Problems encountered may be caused by the OpenGL driver, the graphics hardware, the rendering package or the application software and it may be difficult to isolate these problems, particularly as hardware vendors produce new cards in remarkably short time frames. Certain anomalies may be data-dependent and therefore difficult to replicate and extensive testing is needed to isolate problems of this form. Nevertheless, the integrity of the visual system is critical to effective flight training. If artefacts are noticed, they are distracting and can affect the training effectiveness. With graphics cards used in flight simulators, based on games technologies developed for this very fast moving industry, the integrity of hardware developed for gaming may be considerable lower than technology developed for the simulation industry.

References

Akenine-Moller, T., Haines, E. and Hoffman, N. (2008) *Real-time Rendering*, A K Peters.

Anonymous (2007) *OpenFlight® Scene Description Database Specification*, Version 16.3, MultiGen-Paradigm, Inc.

Blackman, G.H. (1995) A review of display systems for flight simulation. RAeS Conference Flight Simulation Technology, Capability and Benefits, London.

Clark, J.H. (1982) The geometry engine: a VLSI geometry system for graphics. *Computer Graphics*, 16(3), 349–355.

Eckel, G and Jones, K. (2004) *OpenGL Performer™ Programmer's Guide*, 007-1680-100, Silicon Graphics Inc.

Foley, J.D., van Dam, A., Feiner, S.K. and Hughes, J.F. (1995) *Computer Graphics: Principles and Practice in C*, 2nd edn, Addison-Wesley Publishing Company, Reading.

Gray, K. (2003) *Microsoft DirectX 9 Programmable Graphics Pipeline*, Microsoft Press.

Guo, L., Cardullo, F.M., Houck, J.A. *et al.* (2005) A comprehensive study of three delay compensation algorithms for flight simulators. AIAA Modelling and Simulation Technologies Conference, San Francisco.

Kuene, R. and Martz, P. (eds) (2007) *OpenSceneGraph Reference Manual v2.2*, Skew Matrix Software and Blue Newt.

Lee, D.T. and Schachter, B.J. (1980) Two algorithms for constructing a Delaunay triangulation. *International Journal of Computer and Information Sciences*, 9(3), 219–242.

Luebke, D.P., Reddy, M., Cohen, J.D. *et al.* (2002) *Level of Detail of 3D Graphics*, Elsevier Science & Technology.

McFarland, R.E. (1988) Transport Delay Compensation for Computer Generated Imagery Systems, NASA TM-10084.

Newman, W. and Sproull, R. (1979) *Principles of Interactive Computer Graphics*, McGraw-Hill.

Schachter, B.J. (1981) Computer image generation for flight simulation. *IEEE Computer Graphics and Applications*, 1(4), 29–68.

Schneider, P.J. and Eberley, P.H. (2002) *Geometric Tools for Computer Graphics*, Elsevier Science & Technology.

Shreiner, D. (ed.) (2004) *OpenGL® Reference Manual: The Official Reference Document to OpenGL Version 1.4*, Addison Wesley Publishing Company.

Shreiner, D., Woo, M., Nelder, J. and Davis, T. (2005) *The OpenGL® Programming Guide: The Official Guide to Learning OpenGL Version 2.1*, Addison-Wesley Publishing Company.

Spence, G.T., Le Moigne, A., Allerton, D.J. and Qin, N. (2007) Wake vortex model for real-time flight simulation based on Large Eddy Simulation. *AIAA Journal of Aircraft*, **44**(2), 467–475.

Spooner, A.M. (1976) Collimated displays for flight simulation. *Optical Engineering*, 15(3), 215–219.

Watt, A.H. (1993) *3D Computer Graphics*, Pearson Education (US) Addison-Wesley Educational Publishers Inc.

9

The Instructor Station

9.1 Education, Training and Instruction

The instructor station is sometimes referred to as the Cinderella of flight simulation. While considerable effort is given to ensuring the fidelity of flight models, image generation, projection and the motion platform dynamics, the instructor station can be an afterthought – a console simply providing the instructor with a set of buttons and knobs to control the simulator. Certainly, there are very few papers in the literature providing guidelines for the development of instructor stations or the user interface (Goode and Evans, 1992; Ahn, 1997). Instructor stations may be developed by engineers where the instructional facilities are designed without reference to training requirements or perhaps worse, where scant regard is given to the human factors issues of a computer system that is operated on a daily basis by an instructor. It is important to recognize that the combination of an instructor and an instructor station is a critical aspect of the flight training provided by a flight simulator and also that effective design of an instructor station can improve the quality of training. The role of the instructor in a flight simulator is very different from airborne training, and the simulator designer needs to give considerable thought to the design of the instructor station in order to make training as effective as possible (Raskin, 2000).

The very nature of airborne flight training poses severe problems in training:

- Stopping to think is not feasible in airborne training situations;
- The safety of the student and the instructor is paramount;
- The learning is conducted in a hostile environment (in comparison with many other forms of training) with many unavoidable distractions.

Most flight training organizations are based on a model of training that has been in place since the 1930s, with basic flying training comprising three distinct activities: ground school, dual training and solo training. The purpose of ground school training is to provide the theoretical background and understanding needed to operate an aircraft. For example, a pilot needs to understand how lift and drag vary to appreciate the onset of a stall. In dual airborne training, an instructor can demonstrate manoeuvres or provide feedback on student performance, where the emphasis is on perfecting technique. Once the instructor is confident that the pilot is safe to practise an exercise, solo flying consolidates the training.

For many years, the division between ground school training and airborne training was quite distinct. Classroom lessons were provided using books, blackboards and simple training aids to meet the requirements of a syllabus, enabling students to pass the ground school examinations. Although many of the ideas of teaching and learning in traditional education also apply to flight training,

---

Principles of Flight Simulation   D. J. Allerton
© 2009 John Wiley & Sons, Ltd

there are aspects of flight training which differentiate it from other forms of training. Arguably, it is surprising that learning to fly is referred to as *training* rather than education. Education implies the provision of knowledge and an understanding to be able to apply that knowledge to new situations, whereas training implies practising a skill to execute tasks effectively and (in the case of aviation) safely. For example, a civil engineer's understanding of mechanical strain can be applied to designing a cantilever bridge or a suspension bridge. However, the designer will be trained to use a specific design package or to fulfil engineering regulations, where it is assumed that the designer has a sound understanding of the underlying principles. The same situation applies in flight training. For example, a pilot has to understand the principles of ADF but is trained in the techniques to fly an NDB hold pattern, to allow for the effects of wind during a holding manoeuvre, say. Admittedly, the distinction between education and training is often blurred but the aims are different and moreover, the facilities to fulfil these aims are also different.

When flying training was limited to ground school and airborne training, the boundaries of these regimes of training were clearly defined. However, with the availability of synthetic training devices, the flight simulator overarches these two activities. For example, a simulator could be used in an educational role to enable a student to observe the angle of attack or the coefficient of drag during the onset of a stall. Alternatively, in a training role, stalling could be practised in a simulator during preflight briefing, to enable a student to experience level flight with a high pitch attitude. Both simulator sessions involve stalling, but with distinctly different objectives.

On the one hand, the flight simulator is a useful training tool, enabling a student to gain a better understanding of a concept or to attain a skill in a shorter time than conventional training. On the other hand, the flight simulator actually replicates an aircraft and its systems and affords an opportunity to reduce the cost of training. In this training environment, the instructor fulfils two roles: as an educator, the instructor uses the simulator to enhance training, devising ways to use the simulation equipment effectively in terms of the training syllabus and as a trainer operates the complex flight equipment in training sessions where it facilitates productive training. The instructor is provided with an instructor station to operate the simulator effectively and to manage a training session. The instructor can initiate a training exercise to give a student better understanding of a concept or set a task to improve a skill by a selection of the functions provided by the simulator but activated by the instructor at the instructor station.

9.2 Part-task Training and Computer-based Training

Although many tasks may need to be learnt as a single activity, certain activities can be broken down into a sequence of part-tasks, which can be practised separately (Caro, 1972); this approach to training is often referred to as *part-task training*. The training objectives for each part-task are established and the student is coached to perform a specific task to a satisfactory standard. These sub-tasks are then combined so that the complete task can be performed to a satisfactory standard. This model of training applies to basic flight training, where the sub-tasks consist of take-off, climbing flight, turning flight, straight-and-level flight and stalling etc. Similarly, climbing flight, say, can be further decomposed into sub-tasks such as effects of attitude and power, lookout checks and transition to or from level flight.

The introduction of flight simulation has provided opportunities to exploit the benefits of part-task training successfully demonstrated in other industries (Hayes and Singer, 1989). For example, the aid of a freeze button in a simulator can enable a student to practise specific sub-tasks to achieve a desired standard. As an illustration, the stall comprises the following actions:

- Pre-stall checks;
- Reduction of power;
- Maintaining straight-and-level flight with reducing airspeed;

- Directional control by rudder input;
- Detection of the stall;
- Recovery from the stall.

Similarly, recovering from the stall involves lowering the nose of the aircraft, applying power, removing the carburettor heat and recovery to level flight, without an excessive increase in airspeed or loss of altitude. In airborne flight, these tasks quickly merge and are difficult to demonstrate in isolation, whereas in a flight simulator, specific part-tasks can be initiated and repeated. Moreover, the performance of the task can be monitored, allowing training to progress as tasks are accomplished to an appropriate standard.

Up to the mid-1990s, the cost of flight training devices limited their use in basic flying training to instrument training. Since that time, several training organizations have exploited CBT, where animated diagrams, video clips and interactive graphics enable course material to be presented in a form that increases understanding, while at the same time allowing students to progress at their own pace. The drawback (and arguably, the lack of progress) with CBT has been the effort needed to generate course material, often requiring over 50 hours to produce 1 hour of material.

With the increase in the speed of PC technology, one area where CBT has made significant inroads into flight training has been the re-hosting of software from a full flight simulator on a laptop or workstation. For example, in airline training, FMS training can be conducted on a laptop, away from the training organization, using software that is also used in the simulator FMS. Nevertheless, the take up of CBT in flight training has been surprisingly slow and many flight training organizations continue to conduct training in full flight simulators, rather than exploiting the benefits of CBT.

9.3 The Role of the Instructor

In civil flight training, the instructor is seated on the flight deck. A large part of training and checking involves flight crew training rather than pilot training, so the instructor is positioned behind the pilot seats, either in the 'jump' seat, so that the instructor can see the displays and the outside view, or to one side, where a limited side view of the flight situation is provided. The instructor station for an Airbus A380 flight simulator is shown in Figure 9.1. The instructor's seat provides an adequate view of the outside scene and the flight instruments. Two screens are provided for menu selections and to display the flight situation.

In military flight training, the instructor is usually located away from the simulator; seated in a separate room. Cameras are provided to enable the instructor to monitor the pilot and communication is via headsets. This is a sensible solution; in the actual aircraft the instructor is often positioned behind the student pilot, so the only form of interaction is from the instructor monitoring the displays and the outside view, communicating via a headset. There is limited benefit in replicating this environment in a simulator, and from the perspective of the student pilot, there is no difference in the training.

In basic training, the role of the instructor depends on the training application. In some cases, it may be best for the instructor to be in the right-hand seat in order to demonstrate manoeuvres and to monitor the student pilot closely; in other cases, the instructor may simply initiate events and monitor flight conditions.

One further role of the instructor in simulation is to provide various forms of external environments. This might include ATC radio communication, setting turbulence or wind conditions and, in military training, introducing and controlling external forces in the simulation, for example, enemy aircraft or ground threats.

In engineering flight simulation, the acquisition and analysis of flight data are paramount, and the instructor (or the flight test engineer) has the additional responsibility of recording flight data.

Figure 9.1 Instructor station in an Airbus A380 flight simulator (see Plate 7) Reproduced by permission of Thales

The instructor sets up specific flight conditions and specifies which flight variables are acquired and recorded. Although detailed analysis of the flight data is normally performed off-line, it may be necessary to scan through the data to check if all the data has been acquired or to mark specific events for further processing. The data is time-stamped and the storage of large amounts of flight data in an organized format, to facilitate recovery of the data, is very important. Quite often, several megabytes of data are recorded per minute, necessitating that data is written to disk storage without affecting the real-time constraints of the simulation.

In all these applications, the instructor station is designed as part of the flight simulator in very much the same way that the flight model software is designed to meet very specific requirements of the simulator. The role of the instructor is a critical aspect of the training effectiveness of a simulator, and the simulator designer needs to ensure that this role is fully provided. In many simulators, the instructor station is designed to meet the training requirements and to provide the instructor with an effective user interface to monitor a training session and interact with the student pilot to provide useful training. However, there are certainly many examples of early simulators where the instructor station was an afterthought and consequently the quality of training was reduced, or the interface made it difficult to manage a training session, or the instructor's work load increased dramatically as a consequence of a poor user interface.

9.4 Designing the User Interface

The instructor station can be viewed as a software package where the user is an instructor and the application is the management of training sessions. In this sense, the simulator designer must provide a user interface that is optimized to enable instructors to perform their functions as effectively as possible (Schneiderman, 1992). An instructor fulfils many functions in a flight simulator, which include the following:

- Training and mentoring trainee pilots, including flying skills, airmanship, navigation and communication;
- Establishing flight conditions, for example setting the time of day, visibility or take-off weight;
- Monitoring the aircraft performance, for example climb settings and speeds;
- Monitoring flight paths, for example ILS glide path or NDB holding patterns;
- Injecting failures, for example an engine fire, tyre burst on take-off or instrument faults;
- Recording information for debriefing;
- Establishing a series of situations by initiating specific events in the training session;
- Role play, for example to replicate ATC communications.

In all these activities, the instructor is interacting with the flight simulation software. For example, if the instructor sets a cloud base of 3000 ft, visibility in the visual system is set relative to this cloud base setting. If the instructor introduces a fuel leak, this information is needed by the engine model and the fuel management module, in order that the fault and its effects are correctly simulated. There is a further requirement in many training applications that the instructor is unobtrusive. For example, if the instructor is required to reach up to select a switch to fail an engine after take-off, or if a switch has a noticeable mechanical click, the flight crew may be forewarned of the event; such indications will not occur with an equivalent failure in an aircraft, possibly introducing negative transfer of training.

An early design decision is the choice of instructor input. Generally, keyboards are considered to be too cumbersome and inflexible for an instructor station. There are three viable options: mouse input, tracker ball (also known as a *trackball*) or touch screen. In fact, mouse input and tracker ball input are very similar; mouse input allows more flexible motion of the mouse cursor, but requires a large flat area for the mouse movement, whereas a tracker ball is mounted on a flat surface and the cursor movement is provided by rotating the ball. With mouse and tracker ball input, the user can make detailed and accurate selections, for example, to move a slider scale or to select one option from a list of 20 options.

A touch screen affords many of the facilities provided by mouse input, but at reduced resolution. Consequently, a touch screen is ideal for selection of a few options, but the coarse method of selection requires special forms of input to achieve the equivalent of mouse input. For example, to set wind speed in the range 0–80 kt, a slider scale could provide straightforward input using a mouse. With a touch screen, the user might be presented with a keypad to enter the wind speed, which then needs further checking of the maximum value entered and should allow for correction or cancellation of an invalid entry.

Consider a simple example: setting the fuel load at the start of a session. The instructor will need to enter the amount of fuel (for each tank), the units of fuel and, in addition, check that the maximum fuel load has not been exceeded. The instructor will need to locate the fuel menu, select the tank, select the units and then enter an amount of fuel, which does not exceed an upper limit. By recording the number of mouse and menu operations needed for this simple function, methods can be investigated to simplify these operations, without restricting the settings offered or requiring the operator to have detailed knowledge of the instructor station. For example, the training analysis may show that only fuel quantity percentage settings of 0, 25, 50, 75 and 100 will be required, simplifying the options presented to the instructor.

Arguably, there is no right answer to specific user interface designs and some users may have a preference for one form of input over another. However, given the specific tasks of an instructor station, one form of input may be more reliable (less prone to error) or more efficient (fewer key inputs to achieve the desired outcome). For the simulator designer, it is important to classify the types of input entry to clearly define the tasks to be performed and to assess the efficiency of a particular interface. One method that is commonly used to assess a user interface is to give a group of potential users a set of tasks and monitor their performance with candidate user interfaces to compare the characteristics of these interfaces. Such assessments are also combined with a questionnaire to solicit the user's opinion of each interface in order to identify aspects of a user interface, which were useful, and to highlight areas where the user interface is deficient.

In classifying the basic functions of an instructor station, similarities with other software packages may be identified; good practice, developed for other packages, could then be incorporated in the instructor station. For example, zooming functions may be needed to display a chart in the instructor station and these functions are similar to mechanical drawing packages where the user pans across a drawing, zooms in, zooms out and aligns the drawing with respect to an origin. In some packages, zooming is achieved by highlighting a rectangle that is displayed on the drawing to indicate the new area to display. Alternatively, zooming could be provided by a factor of 2 or 10, say; in the former case, to zoom in by a factor of 8 requires three zooming actions, whereas it is not possible to zoom in by a factor of 12, requiring an additional function to be provided to zoom by factors other than a power of two. To zoom by a factor of 12 would then require a mouse selection followed by keypad entry of the scale factor digits.

Another common function is the setting of variables that take a limited set of values. For example, to set the direction of the wind (in the range 1–360°), the user could be prompted to enter a numeric value for wind direction. But, it would be necessary to check for invalid entries, for example, numbers greater than 360, negative numbers and non-integer values. By providing the user with a rotatable arrow, for example, the user could readily select the wind direction from the direction of the arrow, reducing the actions required for this selection. A similar situation arises with file entry, where data is read from a file. By allowing the user to enter the file name, the entry needs to be validated and some retry mechanism needs to be provided for invalid entries. Alternatively, if certain file extensions are allocated to file types and specific files are associated with a user directory, then the search is constrained and the user is presented with the selection of a file from a list, eliminating the possibility of incorrect entry of a file name.

9.4.1 Human Factors

The instructor station will be operated by instructors, who may have limited experience of conventional computer user interfaces. Moreover, they may not have the opportunity to spend a long time learning the intricacies of the commands and options afforded by the instructor station. The instructor may also spend a considerable amount of their working day seated at the instructor station. Consideration must therefore be given to simplifying the user interface, but without compromising the interaction and control provided for the instructor.

It is essential that basic human factors issues are addressed. For example, the instructor station may be located in a dark area where the glare of a display can be distracting. In this case, consideration needs to be given to the choice of colours and light levels, with the provision of a simple means of reducing the display brightness. The instructor is likely to be positioned about 30 cm from a screen. Therefore, the size of character fonts must be chosen so that the textual information is legible at this range. Colour in displays can provide important cues for the classification and prioritization of information. Typically, useful information is given in white, green or blue colours, whereas warnings are displayed in yellow or red. Flashing components can also be used to attract the user's attention, but care is needed that such features are restricted to essential items and are neither irritating nor unnecessarily distracting.

Some instructor stations contain several hundreds of pages of information and the choice of menu pop-ups and the selection of options in such systems can be overwhelming. However, an instructor station should be intuitive and index pages and footers can be provided to 'navigate' around the pages of an instructor station. These pages must be set out logically, so that the functions are easily recognized and selection is quick and unambiguous. For example, buttons must be sufficiently large to be hit first time and an indication should be provided that the correct selection has been made. Inevitably, an instructor will make an incorrect selection and the ability to retrace or undo previous steps in a straightforward manner is essential.

There are numerous pitfalls to avoid in terms of the human factors of the user interface, and the simulator designer is strongly advised to read standard texts on this subject (Galitz, 2007) before developing a user interface for an instructor station.

- Lack of consultation with users – so that the designer is unaware of the needs of the user or the way the user will operate the system;
- Clutter – where useful information is difficult to locate or is hidden by less important information;
- Ambiguity – where the user is unsure of the validity of information provided or how it can be used;
- Lack of clarity – presenting information in a format where useful information is lost, for example, displaying a variable in a digital form where its rate of change is important but can only be discerned by the rate at which the digits appear to change;
- Poor use of colour – where contrast between colours is difficult to detect or where colour is unrelated to the information;
- Lack of structure – where the user is able to become 'lost' in the interface or is unable to return to a known state.

This is not an exclusive list of design problems, but if the user ever reaches the point of saying 'why is it doing that?', then the blame is more likely to be attributable to the interface designer than the user. Newman and Sproull (1979) give an excellent example of the effect of a badly designed user interface to highlight the difference between the expectations of the designer and the user.

9.4.2 Classification of User Operations

Prior to starting the design of a user interface, it is important to categorize the classes of operations needed to support the package. For example, in a word processor, the user may open a file, set the page width, select a font, set the number of columns and so on. For an instructor station, the following actions define the generic set of operations:

- Access an item, typically where the information is organized in a file;
- Select an item from a group of items, often to access a directory of files;
- Enter numeric settings;
- Select one option from several options offered;
- Enable or disable certain options;
- Confirm an action or selection;
- Display information appropriate to the application.

These top-level options correspond to the primitives provided by most graphical user interfaces (GUIs). However, in selecting a GUI, there are also essential non-functional requirements that need to be considered.

- The GUI should allow the user to define how the menu functions and dialogue boxes can be organized to meet the requirements of the instructor station;
- The GUI may need to be integrated with graphical displays of charts and flight data;

- The GUI actions should not contravene the real-time constraints of the simulator;
- The GUI should not be specific to a platform or operating system.

There may be additional non-functional requirements that influence the selection, including programming language compatibility, time to learn the package and even the 'look and feel' of the GUI.

A wide range of GUI packages are available for Windows and Linux, which meet these criteria, including the Fox toolkit (http://www.fox-toolkit.org/), Motif (Fountain and Ferguson, 2000), QT (Alexandria, Human Resources Research Org and Dalheiner, 2002) and GTK (Krause, 2007). For the flight simulator covered in this book, several GUIs were evaluated and an initial implementation of GTK was developed, offering the following advantages:

- It is supported on a wide range of Linux and Windows platforms;
- It supports C, with wrappers for other languages;
- It provides the range of functions appropriate to the application;
- Integration with OpenGL is straightforward.

In practice, it is possible to design a user interface where the GUI software components are contained in a few modules, simplifying the effort to change to an alternative GUI. However, after the initial development, the GTK user interface was abandoned in favour of an OpenGL customized version for the following reasons:

- Installation of GTK requires a large number of packages;
- Establishing real-time interaction is not straightforward – the integration of a GUI event handler, which is based on callbacks in GTK, with real-time loops is difficult; particularly, the management of independent sockets;
- GTK is not directly amenable to touch screen applications;
- OpenGL is portable and available for most platforms.

The design of a user interface for any software package is a balance of the implementation issues and the user requirements. In the case of an instructor station for a flight simulator, the requirements are simplified; there is only one application and all users will use a common set of instructor functions provided by the GUI. Although one specific approach has been adopted in this book in favour of a general purpose GUI, many of the concepts are common to all GUIs and it is possible to minimize the amount of GUI-specific software if the instructor station software is designed in a modular form, which simplifies reimplementation with an alternative GUI.

9.4.3 Structure of the User Interface

Considerable effort needs to be invested in the design of a user interface and the starting point is the structure of the interface (Galitz, 2007). Nowadays, most software packages allocate a few lines at the top of the screen for a menu of user functions and selections, leaving the remainder of the screen to display information specific to the application. The lines at the top of the screen define the major grouping of user commands and options. If one of these commands is selected, a dialogue is opened and according to the user selection, further levels of dialogue are invoked. In practice, most applications using a GUI are limited to two or three levels of interaction; with too many levels it is easy to become 'lost' and wrong selections can result in repetitive unwinding.

At this topmost level, the main operations performed by the instructor are as follows:

- Overall management of the instructor operating station (IOS);
- Repositioning the aircraft to establish initial flight conditions;

- Displaying charts, so that the instructor can monitor the aircraft flight path;
- Selection of weather conditions;
- Initiation of failures and malfunctions;
- Control of other traffic in the simulation;
- Selection of display modes, for example to display a chart, an approach or flight data;
- Data recording, particularly in engineering simulators;
- Replay functions to be able to review previous situations and pilot actions in a training role.

These operations are not an exclusive list and are only outlined to illustrate one possible classification of the type of functions provided for an instructor station. The importance of these initial selections would result from discussions with instructors, analysis of the training requirements and an understanding of common instructor actions. Of course, these functions also depend on the application of the simulator. For example, in a civil flight simulator, it is necessary to set the simulator conditions for a specific airport, with weather conditions appropriate to the training session. Alternatively, in a military simulator, intelligent agents can be introduced to simulate hostile aircraft where the selection of the performance and manoeuvrability of these aircraft is critical to the training effectiveness.

In the initial stage of design, the GUI is designed on paper to establish the feel of the interface and to show how the range of functions would be implemented, using simple templates to illustrate the user dialogue. Figure 9.2 shows the top-level structure of a provisional interface.

An early consideration is the persistency of this upper-level dialogue. In some applications, the user may be issuing different commands frequently. In such applications, it is sensible to display these top-level commands at all times. However, with this convention, a static region of the display is dedicated to these top-level selections. Alternatively, the importance of the display region (for the instructor station, typically a navigation chart) may override the need to retain these menus. In the case of a touch screen, by touching the display at any time, the top-level menu options could be invoked but remain hidden at other times. The penalty for reclaiming menu space for the display area is that an additional screen press (or mouse click) is added to every command. Figure 9.3 shows the format of a typical top-level display for a conventional menu interface and a touch screen alternative. The remainder of the display region is allocated to the map display, to show the aircraft track, airfields, beacons and so on.

In the flight simulator covered in this book, the touch screen form of GUI is used exclusively, enabling the GUI to be developed using OpenGL graphics and avoiding dependence on a specific GUI. Some initial constraints are apparent for the conventional GUI shown in Figure 9.3a. The permanent list of primary commands must fit on the top line of the display and the 'hit' area to select a command is much smaller than its touch screen counterpart.

For the touch screen GUI in Figure 9.3b, these top-level commands provide access to the specific command functions by selecting the appropriate button on the display. These different levels of GUI displays provide the basis of a hierarchical structure of menus, as shown in Figure 9.4, which illustrates the map options, at two levels of the tree below the root.

In this case, seven commands that would be used to select commands specific to the map display were identified, which in turn result in quite different actions. For example, to change the map scale, a numeric value is entered, whereas to centre the map, the cursor may be positioned to re-centre

Figure 9.2 Top-level structure of instructor operations

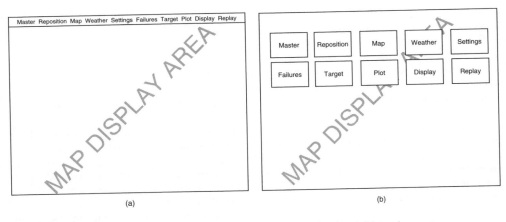

Figure 9.3 Top-level design of the GUI: (a) conventional and (b) touch screen

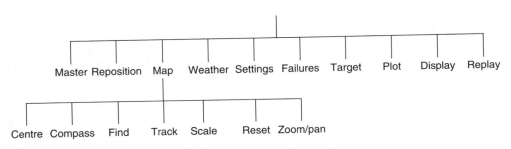

Figure 9.4 The map command settings

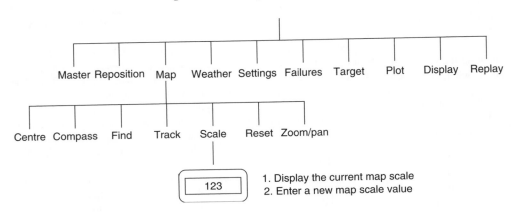

Figure 9.5 Changing the map scale value

the map at a specific location. Figure 9.5 illustrates the menu selection actions to change the map scale. First, the map command is selected, then the scale function is selected, displaying the current map scale value and allowing the user to enter a new numeric value. The specific methods used to select these functions, to enter a new value and to confirm the selections are left to a later stage in the design. At this stage, the purpose of these provisional design options is to confirm the structure

The Instructor Station

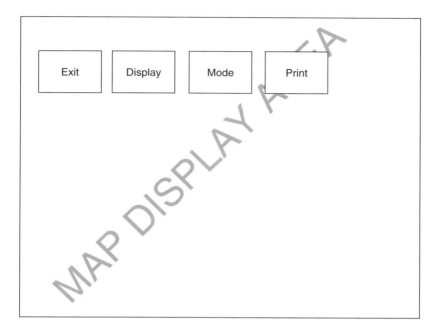

Figure 9.6 Master command options

of the user interface and, in consultation with users, ensure that they provide an effective means of performing essential instructor functions.

When the map command is invoked, the map display could be created as a separate process or display. However, it is not clear if the provision of multiple map displays provides any specific benefits. An early design decision was taken to provide a single display area, which can be used to display the map, an approach or flight data. Clearly, this decision depends on the requirements of the training analysis, whereas for an engineering flight simulator application, a single display, which can be set to one of several display modes, is acceptable.

Selecting one of the main commands activates the corresponding lower level of selections, according to the structure of the tree, to provide the options for that command. For example, consider the *Master* command, shown in Figure 9.6. The selection of this command offers four options:

- The *Exit* command terminates the simulator session;
- The *Display* commands determines the type of information displayed;
- The *Mode* command selects the running state of the simulator (for example to freeze the simulator during a lesson);
- The *Print* command allows the instructor to capture the current display screen as a graphics file.

Selecting the *Exit* option invokes an *Exit* dialogue box, as shown in Figure 9.7. In this simple case, the only choice offered is to exit from the instructor station software or to cancel the option (in the case where it may have been selected in error). Note that the actual wording and the geometry and layout of the menus and boxes can be finalized at a later stage. In this initial design phase, it is simply the functionality that needs to be agreed.

9.4.4 User Input Selections

Having selected a command, the user is presented with the options to execute the command. We have already seen in Figure 9.7 a simple example where the user is offered the choice of accepting or declining an option, in this case, to exit the instructor station program. From inspection of the commands needed for the instructor application, clearly there are only a few groups of commands and, within these groupings, the input methods are very similar. These user commands can be grouped as follows:

- Accept/decline;
- Select one of several options;
- Enter a name to specify an object (typically a file name);
- Select an object from a list of objects;
- Enter a single numeric value;
- Enter a set of numeric values.

As it is possible to select an unintended command, it is common practice to provide a cancel button, before any selection is made. The accept/decline option is used to confirm a selection. For example, in erasing a set of captured aircraft tracks displayed on the map, it is sensible to confirm that this is really what the user intended to do, because the tracks may have been generated during a 2-hour training session and would be difficult to recreate.

There are very few file operations needed for an instructor station. The flight data from a simulation exercise may be written to a file and the user could be prompted for the file name. Normally, the file would be written to the user's directory and only the file name is needed. An appropriate file extension can be added to the file name depending on the context of the command. Alternatively, the file name can be generated explicitly by the instructor station software. For example, a flight data recording file can be allocated a name based on the time and date of creation, which may be useful in organizing the files if many data files are generated. If an existing file needs to be accessed, a list of files can be presented to the user, requiring only the selection of the appropriate file, avoiding the need to enter a specific file name. In the case of a touch screen, the

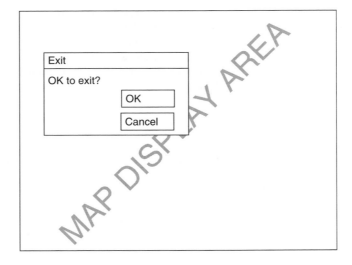

Figure 9.7 Exit command options

The Instructor Station

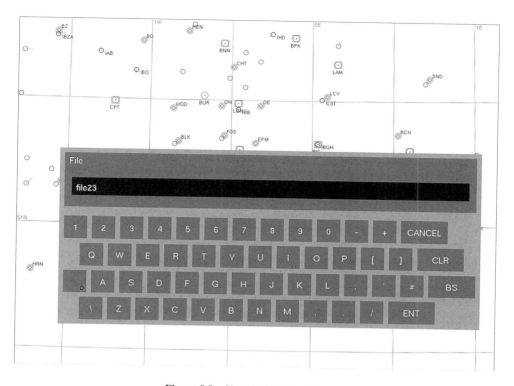

Figure 9.8 Simulated keyboard entry

user can enter a specific file name using a simulated keyboard, as shown in Figure 9.8, where the user has entered the file name *file23*.

The instructor may wish to recall a specific incident, saving the simulator state to a file or, alternatively, generate a set of files to establish different flight conditions, for example to reposition an aircraft at various distances on the approach to an airfield. The instructor may also need to take a snapshot of the instructor screen, perhaps to discuss an approach during debriefing. In these cases, the instructor needs to name the file explicitly to simplify recalling the file name at a later stage. For example, *Heathrow10.sav* might be a saved file for 10 nautical miles from Heathrow.

In other situations, to avoid requiring the instructor to enter a specific file name, the name can be generated automatically. For example, screen snapshots could be stored with names of the form *snapxxxx.png*, where *xxxx* denotes a four digit number and the extension *.png* implies that the file is stored in an image format (*p*ortable *n*etwork *g*raphics). The local directory is initially searched for files beginning with *snap* and with an extension *.png* to locate the file name with the highest numeric value. Each time a screenshot is captured, this value is incremented to generate a new file name, for example *snap123.png*, *snap124.png*, *snap125.png* and so on. A similar method is used to name flight data files in the simulator, by using the date and time of creation of the file in the file name. For example, the file *0312081452.dat* was created on the 3 December 2008 at 2.52 p.m. In this case, the extension *.dat* is used to indicate that this file contains flight data.

For many activities, it is necessary to set specific parameters, entered as numeric values, for example, to set the visibility. In this case, a small keypad can be provided for numeric entry. Figure 9.9 shows an example for a touch screen IOS, where the user has entered a value of 3500 (m). The keypad also offers a clear function (CLR) to reset the entry, a back space function (BS)

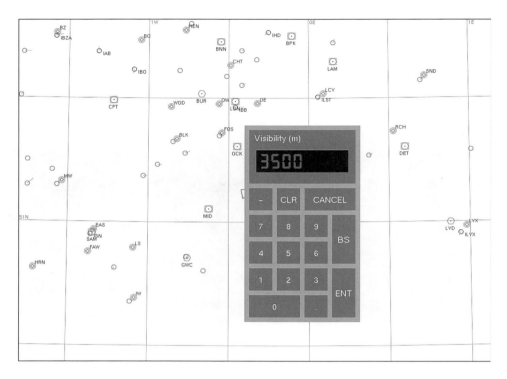

Figure 9.9 Numeric entry of visibility

to delete the previously entered character, a cancel function (CANCEL) to abandon the command and an enter function (ENT) to confirm the selection.

The GUI should also detect erroneous entries and provide the user with a simple method to revise the entry. The value can either be validated when it is entered (pressing the ENT key) or during entry, as each digit is entered. Figure 9.10 shows an example where the user has entered an inappropriate visibility value; the instructor is required to acknowledge the error (pressing the red OK button) before proceeding to correct the error.

An alternative form of numeric entry is to provide a slider scale, as shown in Figure 9.11. The range of the scale defines the range offered to the user and the currently selected value is displayed. The main problem with this scheme is that the resolution of user input may be too coarse to select specific values or alternatively, that preferred values are difficult to enter, for example, to set altitude in steps of 100 ft. These problems are particularly common with touch screen input. For example, in Figure 9.11, there are 400 steps of 100 ft between 0 and 40,000 ft. If the range of the scale on the display is less than 400 pixels, it would be difficult to select specific altitude values to the nearest 100 ft using a touch screen.

Some simulator variables are constrained to specific states. For example, an engine failure is either activated or cancelled or a variable is plotted in a specific set of units. In these cases, the user selects one of a set of options. For example, in recording simulator data using the *Plot* command, the *Record* option enables the recording to be started, stopped or continued, and the user is able to select one option from three. An example of this form of selection, which is sometimes referred to as *radio buttons*, is shown in Figure 9.12. In this case, the current state of the *Record* option is indicated by highlighting the associated button. The selection can be altered by selecting one of

The Instructor Station

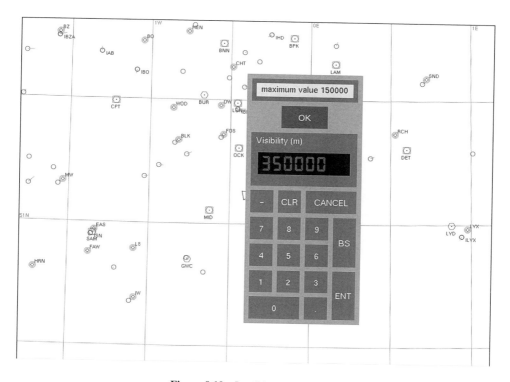

Figure 9.10 Invalid numeric entry

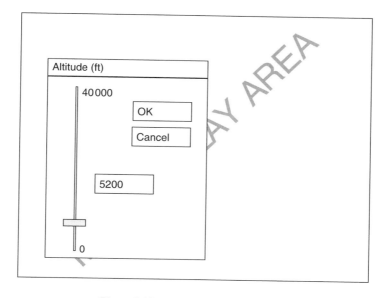

Figure 9.11 Slider scale numeric input

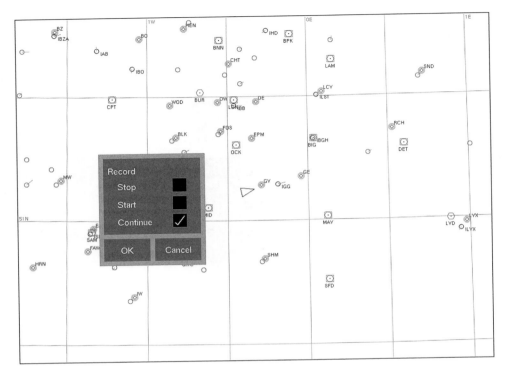

Figure 9.12 Radio button selection

the other options (*Stop* or *Start*). As with the other commands, the *OK* button is used to confirm the selection and the *Cancel* option provides a means of cancelling the command.

There may also be a few commands that can be selected directly at a high level, without requiring a dialogue box. For example, to position the aircraft in the centre of the display, no parameters are required and a *Find* command could be provided on the *Map* command menu. Simply selecting the *Find* command realigns the map centred at the current aircraft position. The effect of inadvertently selecting this command is minor and there is no case to provide an additional layer of dialogue to confirm the operation. Similarly, there are commands where the cursor is used to set a location or enter a track, for example, to centre the display at the current cursor position, or to draw a track to annotate the map display. In Figure 9.13, a compass rose has been positioned over the VORs at Midhurst (MID) and Detling (DET) and a track is being drawn from Midhurst to Detling. The latitude and longitude of the end point of the track and the distance and bearing of the track are displayed as the cursor moves.

Some commands may need to be guarded against inadvertent selection, such as the *Exit* command or in clearing information displayed on the map. In these cases, it is essential to provide a final check to provide the user with a 'last chance' to cancel the selection; alternatively, an additional *undo* command can be provided, requiring a log of previous commands to be maintained, so that previous commands can be cancelled correctly.

These commands are given only to illustrate the basis of preliminary design of the GUI. Two points are significant. First, there is a clear hierarchy to the commands and secondly, there are only a few classes of commands that cover the functions likely to be encountered in the design of an instructor station for a flight simulator.

The Instructor Station

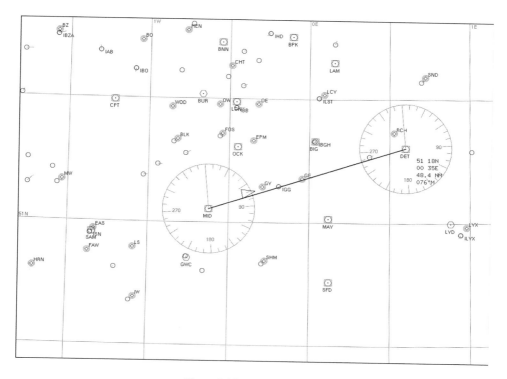

Figure 9.13 Track selection

Rather than constrain a GUI to a specific command structure or format, it is preferable to allow the user to define the commands and options and then organize the structure of the commands according to the application. The instructor activates specific commands that are broadcast in a packet containing the command and its arguments. These commands are linked by a table of constants defined in the instructor station header file *iosdefn.h* and the corresponding menu definition file *menu.txt*. In *iosdefn.h*, the commands are identified by an integer value, for example IosDefn_MapScale is 66; the same value is defined for the *Scale* command in *menu.txt*. A fragment of a typical menu definition file is given below.

```
Exit\q              31  <OK to EXIT?>
Display\o           36  <Map Approach Flightdata Rawdata>
Mode\o              124 <Run Hold Freeze>
Print\f             32  <File>

Restore\l           34
Save\f              35  <File>
Position\c          61
Altitude\n          121 <Aircraft Altitude (ft)> 0 50000 3000
Heading\n           122 <Aircraft Heading> 1 360 360
Airspeed\n          123 <Aircraft Airspeed> 0 400 150

Centre\c            62
```

```
Compass\c              63
Find\x                 64
Track\c                65
Scale\n                66  <Map Scale> 5 5000 250
Reset\q                67  <OK to reset map?>
Zoom/Pan\x             68

Turbulence\n           91  <Turbulence> 0 1 0
Wind Speed\n           92  <Wind Speed (kt)> 0 80 0
Wind Dir\n             93  <Wind Direction> 1 360 360
QNH\n                  94  <Area QNH> 995 1035 1013
Magnetic Var.\n        95  <Magnetic Variation> -20 20 -5
```

The file *menu.txt* is read at the start of the simulation to define the menu structure, the naming of commands and options and the type of the arguments of each command. The first field is the name of the command, which will be indicated in the displayed dialogue box. Each command has a qualifier \q, \f, \l, \o, \c, \x, \n or \d, which defines the argument types given in Table 9.1.

For each command, a numeric identifier is provided followed by the prompt to be given in the dialogue box. For numeric arguments, the minimum value, the maximum value and default value are also given. For options, the list of possible options is defined, where the default option is the first item in the list. For example, command 91 is used to set the turbulence level in the range 0–100%, with a default value of 0. These structures are established in *menu.c* and then used by the user interface procedures to display the appropriate menus and dialogues.

9.4.5 Instructor Commands

The actual structure and scope of the instructor commands are derived from the requirements analysis of the instructor station, consultation with potential users and from studying the training application. The instructor station covered in this book is based on a three-level tree of commands, which were designed for use in an engineering flight simulator rather than a training simulator. The overall command structure is shown in Figure 9.14. The type of each command is given by the qualifier following the '\' symbol, as defined in Table 9.1. The complete list of commands, the type of each command and the associated parameters are defined in the file *menu.txt*. A summary of the commands is given in Table 9.2.

Although these specific commands were developed for an engineering flight simulator in a university, the structure of the commands, the type of arguments and the option and numeric ranges can be easily modified for other applications, simply by modifying the text file *menu.txt*. Of course, a general aviation flight trainer, an airline flight trainer and a military trainer will have very different requirements. Nevertheless, the grouping of commands and the type of commands needed for the instructor station are likely to be equivalent to the set of commands outlined in this

Table 9.1 Menu command qualifiers

q	a *q*uery – the user is prompted to confirm the selection
f	a *f*ile name – the user is prompted for a file name
l	a *l*ist of file names – the user selects a file name from a list of file names
o	an *o*ption – the user selects one option from a list of options
c	a latitude/longitude *c*oordinate is displayed and the option is selected directly
x	an *e*xecutive command – no argument is given and no dialogue is needed
n	a *n*umeric value – the user is prompted for a numeric value
d	a flight *d*ata recording value – the user is prompted for data recording settings

The Instructor Station

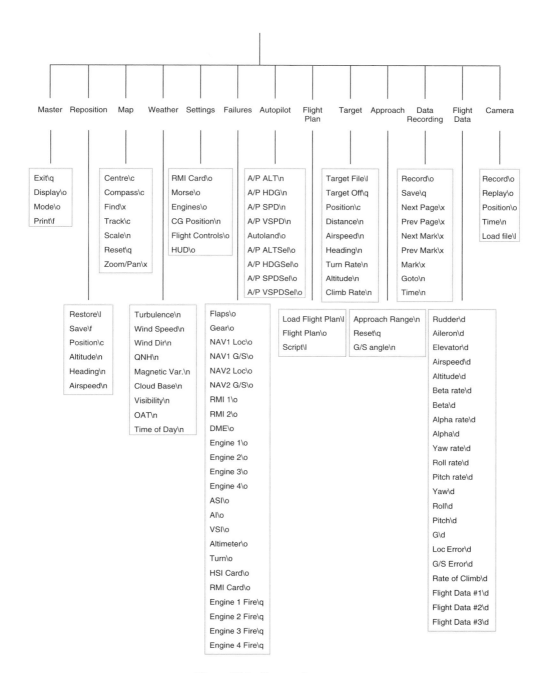

Figure 9.14 Command structure

Table 9.2 Summary of instructor station commands

Main Group	Command	Type	Details
Master	Exit	Confirm	Terminate the instructor station session
	Display	Option	Select the information displayed
	Mode	Option	Control the simulator state (run, hold, freeze)
	Print	File	Print the screen to a file
Reposition	Restore	File list	Restore a saved file
	Save	File	Save the simulator state to a file
	Position	Cursor	Move the cursor to position the aircraft
	Altitude	Value	Set the aircraft altitude
	Heading	Value	Set the aircraft heading
	Airspeed	Value	Set the aircraft airspeed
Map	Centre	Cursor	Centre the map display at the cursor position
	Compass	Cursor	Draw a compass at the cursor position
	Find	Command	Centre the display at the aircraft position
	Track	Cursor	Draw a track line of the map using the cursor
	Scale	Value	Set the map scale
	Reset	Confirm	Reset the map display to the initial state
	Zoom/pan	Command	Use the cursor to zoom and pan
Weather	Turbulence	Value	Set the turbulence conditions
	Wind speed	Value	Set the wind speed
	Wind dir	Value	Set the wind direction
	QNH	Value	Set the regional QNH
	Mag. Var.	Value	Set the magnetic variation
	Cloud base	Value	Set the cloud base
	Visibility	Value	Set the visibility
	OAT	Value	Set the outside air temperature
	Time	Value	Set the time of day
Settings	RMI card	Option	Set the RMI to a moving card or fixed card
	Morse	Option	Enable or disable Morse idents
	Engines	Option	Use one lever for the engines
	CG pos.	Value	Set the CG position
	Controls	Option	Select the centre stick or side stick
	HUD	Option	Enable or disable the HUD
Failures	#1	Option	Fail a selection of aircraft systems
Autopilot	A/P ALT	Value	Set the autopilot altitude
	A/P HDG	Value	Set the autopilot heading
	A/P SPD	Value	Set the autopilot speed
	A/P VSPD	Value	Set the autopilot vertical speed
	A/P ALT	Option	Enable/disable autopilot altitude hold
	A/P HDG	Option	Enable/disable autopilot heading hold
	A/P SPD	Option	Enable/disable autopilot speed hold
	A/P VSPD	Option	Enable/disable autopilot vertical speed hold
	Autoland	Option	Enable/disable autopilot autoland
Flight Plan	Load FP	File	Load a flight plan
	FP	Option	Engage/disengage a flight plan
	Script	File	Activate a script
Target	Target	File	Load a target
	Off	Option	Enable/disable the target
	Script	File	Activate a target script

(continued overleaf)

Table 9.2 (*continued*)

Main Group	Command	Type	Details
Approach	Range	Value	Set the range for plotting
	Reset	Option	Reset the approach plot
	Angle	Value	Set the glide path angle
Data record	Record	Option	Start, stop or continue recording
	Save	Confirm	Set the recorded data to a file
	Next page	Command	Move to the next page
	Prev. page	Command	Move to the previous page
	Next mark	Command	Move to the next mark
	Prev. mark	Command	Move to the previous mark
	Mark	Command	Mark the current position
	Goto	Value	Start plotting from a specific time
	Time	Value	Set the time axis
Flight Data	#2	Value	Set the plotting range for recorded variables
Camera	Record	Option	Start, stop or continue recording
	Replay	Option	Start, stop or continue replaying
	Position	Option	Set the camera position
	Time	Value	Set the recording time
	Load file	File	Load a data file to replay

Table 9.3 Arguments for failures and flight data recording

#1	Flaps, Gear, NAV1 Loc, NAV1 G/S, NAV2 Loc, NAV2 G/S, RMI 1, RMI 2, DME, Engine 1, Engine 2, Engine 3, Engine 4, ASI, AI, VSI, Altimeter, Turn, HSI Card, RMI Card, Engine 1 Fire, Engine 2 Fire, Engine 3 Fire, Engine 4 Fire
#2	Rudder, Aileron, Elevator, Airspeed, Altitude, Beta rate, Beta, Alpha rate, Alpha, Yaw rate, Roll rate, Pitch rate, Yaw, Roll, Pitch, G, Loc Error, G/S Error, Rate of climb, Flight Data #1, Flight Data #2, Flight Data #3

chapter. In this sense, the design of the user interface covered in this chapter provides a template to develop an instructor station.

The options for the arguments of the failures and data recording commands are given in Table 9.3. For the failures, each failure can be activated or cancelled. For the recorded variables, the units, range and plotting intervals can be selected.

Several examples of menu selection are shown in Figures 9.14–9.19 to illustrate the structure of the touch screen commands, the sub-options and parameter settings. In Figure 9.15, the top level of the 14 main commands is shown. Normally, this display is activated by touching the screen when no menus are active. In Figure 9.16, the user has selected the *Master* command in the main menu and the four sub-options *Exit*, *Display*, *Mode* and *Print* are shown. The options to display aircraft flight data are shown in Figure 9.17. The user is prompted to set the altitude scale from 3000 to 8000 ft at intervals of 1000 ft to set the units to feet and to enable plotting, as shown in Figure 9.18.

Although the autopilot functions can be selected from the FCU panel, these may also be selected from the instructor station. By selecting an *Autopilot* command in the main menu, a selection of autopilot settings is provided to set specific autopilot modes and to engage and disengage the autopilot functions, as shown in Figure 9.19. In Figure 9.20, the autopilot heading select (A/P HDG) is engaged.

Figure 9.15 Main menu commands

9.5 Real-time Interaction

The instructor station runs as a single process, executing a number of functions every frame, as shown in Figure 9.21. In particular,

- Incoming packets are stored;
- Instructor touch screen inputs are detected;
- Commands initiated by the instructor or a script are executed;
- The display is redrawn, to provide up-to-date information;
- If a command has been actioned, it is added to the packet to be broadcast.

Recalling that the display is implemented in OpenGL (using double-buffered rendering), the state of the display must be stored to enable the display and any changes to the display to be redrawn in the current frame. For example, the aircraft track may have been extended, or another set of flight data may have been acquired or the instructor may have partially input a command. The instructor station is a real-time system; packets arrive from the other simulator computers every frame and the response of the user interface must not contain any discernible lags or discontinuity.

Two OpenGL Glut procedures are used to maintain the display: glutDisplayFunc and glutIdleFunc. The argument of glutDisplayFunc is a user-defined procedure, which is invoked once per frame and is used to update the display. The argument of glutIdleFunc is a user-defined procedure, which is invoked by Glut when there are no OpenGL actions to be performed. The functions performed by the procedure passed by glutDisplayFunc are as follows:

The Instructor Station

```
Initialise the graphics settings for the current frame
Clear the display
Depending on the IOS mode:
    Display the map, or
    Display the approach mode, or
    Display flight data, or
    Display raw date
Update the GUI interface
Complete the OpenGL frame operations
```

The graphics settings are the general conditions for line drawing, for example, enabling anti-aliasing. OpenGL allows the display to be cleared to a specific colour. If the current instructor station software state is the map display, say, then during the current frame the lines of latitude and longitude, the beacons and runways and any aircraft tracks are drawn. Similarly, if the user is activating a command, the current GUI is drawn together with the current user input, for example, a string of digits entered as a numeric value. Finally, double-buffering is enabled by the OpenGL library procedure *glutSwapBuffers*.

The Glut OpenGL procedure `glutIdleFunc` enables user-defined functions to be activated at the point in the current frame when the graphics has been executed. For the instructor station, this feature enables any changes to the instructor station to be implemented, at least once per frame. Although such changes may result in changes to the display, they do not involve any graphics

Figure 9.16 Master sub-option menu

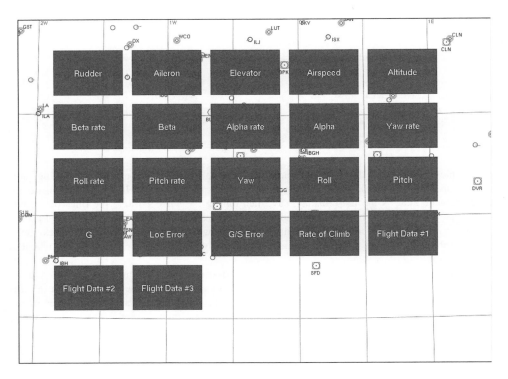

Figure 9.17 Flight data display options

operations. A typical user-defined idle function (idle in the sense that no graphics functions are needed), passed as an argument to `glutIdleFunc`, is summarized below:

```
Check the GUI for new inputs
Detect and copy any incoming packets
If any commands have been initiated, broadcast a packet
Invoke any replay functions (typically copying a packet from a file)
Update any pending Script instructions (if a Script file is active)
Check to see if a shutdown command has been issued.
```

The code to check for GUI user input updates the state of the GUI. For example, the user may enter a new digit, or select the cancel button, or enter a command to re-centre the map display. These inputs alter the specific data structures of the GUI and will be evident in the next frame, when the GUI is redrawn. Of course, to the user, these changes appear instantaneous, as they occur within 20 ms at an update rate of 50 frames per second. The incoming packets may also alter the information displayed, for example, the aircraft position may have changed, extending the display track or, in the flight data mode, another set of data can be plotted. Any functions that must be executed once per frame are likely to be included in the code for the idle functions. For example, if a *Script* file is active, the *Script* module is invoked to execute *Script* instructions; the program may be waiting for the aircraft altitude to exceed 500 ft and the Script program will check the aircraft altitude value in the appropriate incoming packet every frame. Similarly, in the replay mode, a saved packet is transmitted to the other computer systems to override current inputs, in order to

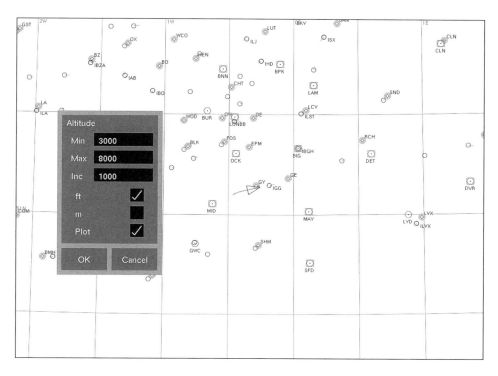

Figure 9.18 Flight data display parameters

replay previous flight conditions. Finally, the simulator can be shut down either by an explicit instructor command or if the simulator key switch is turned off. In practice, as this event needs to be communicated to the other simulator computers, considerable care is needed to schedule these packet transfers so that no computer is left waiting for another packet, when all the other computers have stopped. The instructor station is the natural place to coordinate this action.

As packets are received continuously from the other computers in the flight simulator, it is important that no packets are lost or dropped, particularly in data recording applications. To some degree, this activity is transparent; the incoming packets are buffered by the socket handler in Linux. For example, if it requires 200 ms to read a file, corresponding to 10 frames, the software accessing the incoming packets will catch up immediately, because the packets from the previous 10 frames would be acquired from the sockets buffer with negligible delay. In practice, the display is likely to be updated well within a single frame but significant delays can occur if large amounts of data are written to disk or if the operating system executes a background task.

The instructor station is different from the other simulator processes because it only needs to broadcast data if the instructor has executed a command or if a script is running. The majority of the time, there are no packets to send, and even when a user command is executed this may only amount to a single command and its arguments. Unlike the incoming packets, where the static structure of the packets is defined in header files, the structure of the outgoing packets is best represented as a dynamic structure. The main complication is that the commands have different numbers of arguments and types of arguments. The solution is to encode the data as a serial byte stream, which is encoded at the instructor station and decoded by the other simulator computer systems. A simple encoding method was chosen; each command comprises a single 16-bit integer, followed by the arguments specific to the command. For example, to position the aircraft at 10,000 ft, the

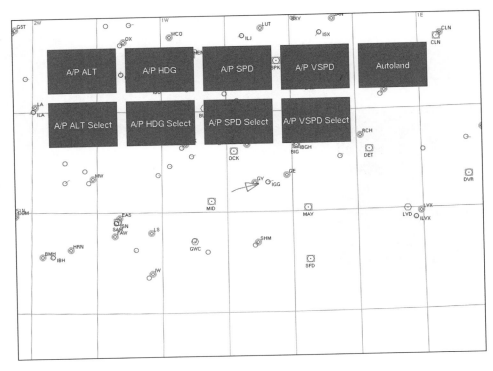

Figure 9.19 Automatic flight mode selection

'set altitude' command (121) is transmitted as 2 bytes followed by the altitude (in metres), defined as a 32-bit floating-point number and transmitted as 4 bytes. Transmitting a floating-point number as 4 bytes is easily achieved by using a C union of a 4 byte floating-point type (float) and an array of type unsigned char, also of 4 bytes. A similar decoding method is used at the receiver to receive the incoming bytes into a 4 byte unsigned char array, which is defined as a union with a floating-point variable.

Although the operating system is able to buffer incoming packets, an alternative approach is to organize the instructor station as a set of concurrent real-time processes, where one process is responsible for reading incoming packets, one process responds to user inputs and another process maintains the GUI and the displays. In this case, the packet reading process must be guaranteed to be executed once per frame, requiring a real-time operating system that allows processes to be prioritized.

While Linux supports many of the features required for an instructor operating station, particularly Ethernet packet handling, OpenGL graphics and mouse or touch screen inputs, it is not strictly a real-time operating system. By careful design of the code, it may be possible for the user software to always execute within a single frame, and, by inhibiting specific system processes, it may be possible to force the instruction station application to be the major system process. However, there is no guarantee of real-time response by the operating system. In practice, such problems are rarely encountered and the benefits afforded by Linux far outweigh the possibility of non-real-time performance. Moreover, any latency occurring in the instructor station software is likely to have a minimal effect on other parts of the simulation. For example, if the display is drawn five frames late (1/10 s), it will not be noticed by the instructor or similarly, a short delay in changing the visibility is unlikely to have an adverse effect on any training task.

The Instructor Station

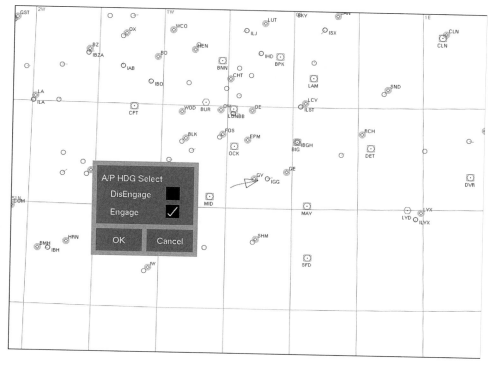

Figure 9.20 Autopilot heading-hold selection

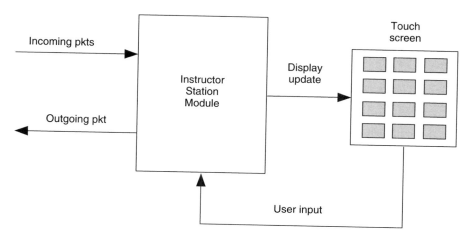

Figure 9.21 Instructor station execution cycle

9.6 Map Displays

The map projections described in Section 6.3 can be adapted to provide navigation charts for an instructor station. Although charts are published for approaches and en route navigation, and these are available nowadays in electronic form, they are mostly produced as graphical images, and consequently the image resolution restricts the level of zooming and moreover it is not straight-forward to align vector graphics with pixel-based images. In addition, detailed charts produced for specific airfield approaches are not appropriate for a continuous map display. There is also a wide assortment of charts. For example, in the United Kingdom, the Civil Aviation Authority (CAA) 1 : 500,000 aeronautical chart for visual navigation is a VFR topographical chart in an ICAO approved format. Although this chart contains much useful information for visual flying, it would provide far too much unnecessary information for use in an instructor station.

The map display in an instructor station is also dependent on the application. For example, many military aircraft have a moving map display in the cockpit and, for such applications, the instructor station would include a replica of the cockpit display. However, it might also be necessary for the instructor to introduce hostile aircraft in a military simulator or conflicting traffic in a civil simulator, possibly 60 nautical miles from the aircraft position shown on the map display. Consequently, there are several advantages in providing a dynamic charting capability in the instructor station:

- It minimizes clutter on the display, allowing the instructor to focus on the training task;
- If the chart information is organized as classes, it is straightforward to remove unnecessary information from the display;
- It gives the instructor freedom to organize the chart display in order to display essential information for the training session;
- Aircraft tracks, distance and times to waypoints can be overlaid on the map display.

The Lambert conformal map projections can be readily adapted to varying the scale factor and centre of a map and these projections support the four basic functions needed for a map display:

- Zooming – this is equivalent to changing the map scale factor;
- Panning – moving the display left or right, up or down;
- Centering – positioning the centre of the map display at a specific location;
- Rotation – aligning the chart to aircraft heading, true north or magnetic north.

Two forms of presentation are possible with map displays: to provide a static map display where the map settings are set explicitly by the instructor or a dynamic map display (also known as a *moving map*) where the map parameters change according to the aircraft track. Of course, if OpenGL is used to render the map display using a double-buffering mode, then the complete map is redrawn every frame and the computational effort to generate a dynamic map display is the same as a static map display.

The content of a map display depends on the application and, in many instructor stations, the instructor is provided with options to include or exclude specific features. For example, the display of towns, rivers or coast lines may be inappropriate in a general aviation simulator used for instrument procedures training, but may be relevant for low-level flying in a military simulator. The information presented on a map display also depends on the content of the navigation database. In a minimal form, the map display is likely to contain lines of latitude, lines of longitude, navigation beacons and airfields. A typical instructor map display is shown in Figure 9.22. The display includes lines of latitude and longitude (at 30′ intervals), ILS, VOR, DME, tactical air navigation (TACAN) and NDB beacons and airfields, using standard chart symbols for navigation beacons and airfields.

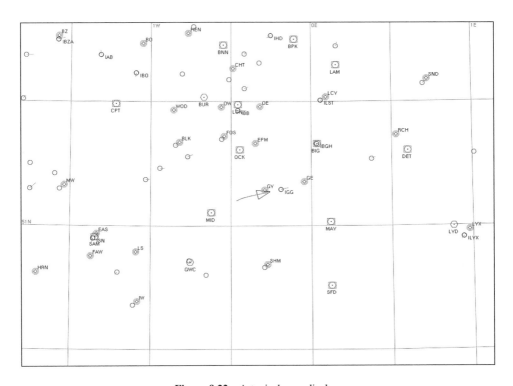

Figure 9.22 A typical map display

In addition, the triangle symbol (near the GY NDB) denotes the aircraft position and heading. The same map display is shown zoomed in (Figure 9.23), where the aircraft track is clearly shown. The grey box in the bottom right-hand corner is a navigation panel, used for zooming and panning. Positioning the cursor over the inner circle, the display zooms in and positioning the cursor over the outer circle, the display zooms out. Similarly, positioning the cursor over the left, right, up and down arrows, the display pans left, right, up and down, respectively.

Figure 9.24 shows the aircraft track overlaid on the map, in this case to display a hold pattern at the Midhurst VOR(MID).

A variant of the chart display is the approach display, shown in Figure 9.25, which displays the glide slope, localizer and airspeed, in this case for a light twin-engine aircraft with the approach flown at 90 kt. This display provides a side view and plan view of the approach, with the ideal approach path (in this case 3°) shown on the plot (not to scale). The glide slope position is derived from the aircraft altitude and distance from the ILS beacon; the localizer position is derived from the aircraft position and the airspeed is obtained from the aerodynamic data.

Lambert conformal projections are combined with OpenGL drawing functions to render the chart display. All entities in the simulation are defined by latitude, longitude and altitude, and these are transformed to the 2D coordinate frame of the screen in the instructor station. The actual display dimensions (in pixels) determine the scale factor. Although there is no requirement to conform to any specific chart format for an instructor station, a standard conformal projection is normally used with predefined co-latitudes.

Each time the screen is centred, the global variables Lambert_N and Lambert_K are recomputed in the procedure Map_SetMapCentre, as shown in Example 9.1.

Example 9.1 Setting the map centre

```
void Map_SetMapCentre(float Latitude, float Longitude)
{
    float Chi;
    float lat1;
    float lat2;
    float r;
    float r1;
    float r2;

    Map_MapLatitude = Latitude;
    Map_MapLongitude = Longitude;
    Chi = PIBY2 - fabs(Latitude);
    Lambert_N = cos(Chi);
    Lambert_K = tan(Chi) * pow(tan(Chi / 2.0), -Lambert_N);
    lat1 = Chi - DEG6;
    lat2 = Chi + DEG6;
    r = Lambert_K * pow(tan(Chi / 2.0), Lambert_N);
    r1 = Lambert_K * pow(tan(lat1 / 2.0), Lambert_N);
    r2 = Lambert_K * pow(tan(lat2 / 2.0), Lambert_N);
    msf = D12 / fabs(r1 - r2);
    MapOffset = r * msf;
}
```

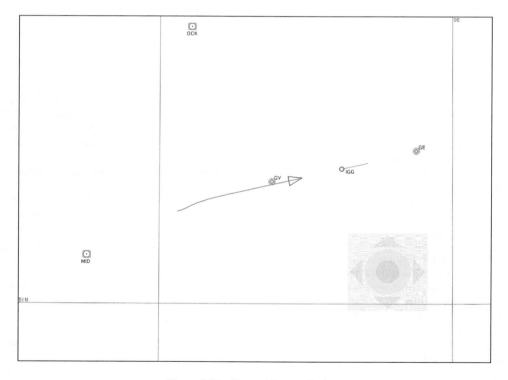

Figure 9.23 Zoomed in map display

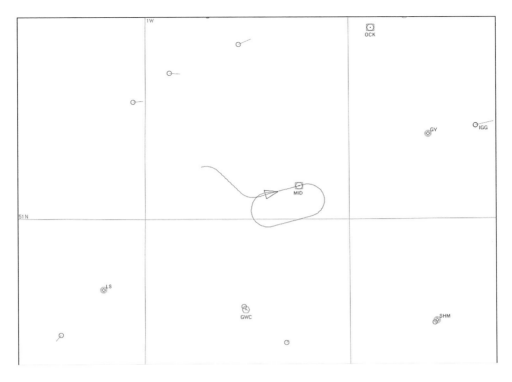

Figure 9.24 Hold pattern

The constant D12 is the distance of 12° of latitude in metres, given by $2\pi R/30$, where R is the radius of the earth. The constant DEG6 is 6° in radians. The integer constants MapCentreX and MapCentreY define the centre of the map display in pixels, typically the centre of the current window. The global variables Map_MapLatitude and Map_MapLongitude are set to the latitude and longitude of the centre of the map, respectively. The co-latitudes of the current map are given by the latitude at the map centre ±6°. The transformation of the latitude and longitude (λ, ϕ) to polar coordinates (r, θ) follows the method outlined in Section 6.3 for the co-latitude ξ (chi)

$$\xi = \frac{\pi}{2} - |\lambda| \tag{9.1}$$

$$N = \cos(\xi) \tag{9.2}$$

$$K = \tan(\xi) \left(\tan\left(\frac{\xi}{2}\right) \right)^{-N} \tag{9.3}$$

where λ is latitude, N is the convergence factor (Lambert_N) and K is a constant (Lambert_K) depending on the co-latitudes. MapOffset is the screen offset from the pole in metres and msf is the local scale factor between distance in metres and pixels.

The following procedure Map_GlobeToScreen transforms a world position (latitude, longitude) to its corresponding screen coordinates (x, y), as shown in Example 9.2.

Example 9.2 Globe to screen transformation

```
void Map_GlobeToScreen(float Latitude, float Longitude, float *x, float *y)
{
  float Chi;
  float Theta;
  float r;
  float t;

  Chi = PIBY2 - fabs(Latitude);
  Theta = Lambert_N * (Longitude - Map_MapLongitude);
  r = Lambert_K * pow(tan(Chi / 2.0), Lambert_N);
  *x = (r * sin(Theta) * msf / Map_MapScaleFactor) + MapCentreX;
  t = (r * cos(Theta) * msf - MapOffset) / Map_MapScaleFactor;
  if (Map_MapLatitude > 0.0){
     *y = MapCentreY - t;
  }
  else {
     *y = MapCentreY + t;
  }
}
```

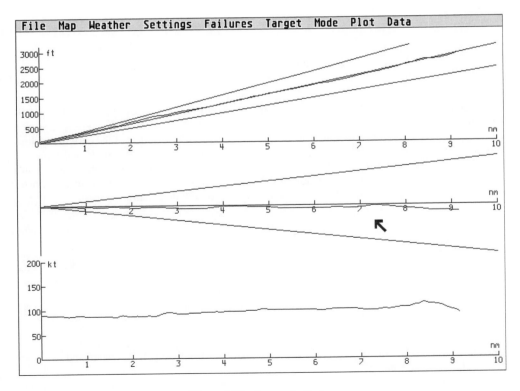

Figure 9.25 Approach display

The parameters Latitude and Longitude are given in radians and the screen coordinates x and y are returned in pixels. The variable Map_MapScaleFactor is a global variable defining the scale factor to be used to display the map. Note the check in the last seven lines of the procedure, which corrects the transformation for the northern or southern hemispheres.

The inverse transformation, mapping a screen coordinate to a latitude and longitude, is given by the following procedure Map_ScreenToGlobe, as shown in Example 9.3.

Example 9.3 Screen to globe transformation

```
void Map_ScreenToGlobe(float x, float y, float *Latitude, float *Longitude)
{
  float Chi;
  float Theta;
  float r;
  float dx;
  float dy;

  dx = (x - MapCentreX) * Map_MapScaleFactor;
  if (Map_MapLatitude > 0.0) {
    dy = (MapCentreY - y) * Map_MapScaleFactor + MapOffset;
  }
  else {
    dy = (y - MapCentreY) * Map_MapScaleFactor + MapOffset;
  }
  r = sqrt(dx * dx + dy * dy) / msf;
  Theta = asin((x - MapCentreX) * Map_MapScaleFactor / (r * msf));
  *Longitude = Theta / Lambert_N + Map_MapLongitude;
  Chi = 2.0 * atan(pow(r / Lambert_K, 1.0 / Lambert_N));
  *Latitude = PIBY2 - Chi;
  if (Map_MapLatitude < 0.0) {
      *Latitude = -(*Latitude);
  }
}
```

It is straightforward to apply these transformations to render the map display. However, three further points need to be considered:

- When the display is zoomed in, most of the entities are outside the border of the map display. Although they would be excluded as a result of clipping, this can be a slow process; it is more efficient (in terms of rendering speed) to transform the extent of the map display to world coordinates, discarding the majority of the objects by their location rather than by clipping during rendering;
- In drawing lines of latitude and longitude, the actual values should be appended to these lines and the position of these labels can be determined from the coordinates of the clipping of the lines of latitude and longitude;
- The display can become cluttered as it is zoomed out; care is needed in adding text to entities to avoid obscuring parts of the display.

9.7 Flight Data Recording

In an engineering flight simulator, as in aircraft flight testing, considerable emphasis is given to flight data recording. Flight data, acquired from any test is often very valuable (in terms of the effort needed to acquire it) and loss of data, either because it was not recorded or cannot subsequently be retrieved, can introduce delays in design programmes. The major problem in flight data recording is the sheer volume of flight data. If a variable is stored as a 32-bit floating point value, it occupies 4 bytes. If there are 500 variables in the simulator, then 2 kB of data is stored per frame, or 100 kB per second (at a 50 Hz frame rate) or 6 MB of data per minute. With modern PCs, this storage requirement is not the issue it might have been 10 years ago, but even so, 30 minutes of flight data requires almost 200 MB. In addition, in writing this data to disk, delays caused by the operating system actually transferring blocks of data to disk could affect the real-time performance of the simulator or introduce discontinuities into the acquisition of data.

One solution is to record only useful data, that is, data that is actually needed for a specific test, in order to reduce the amount of data stored. The main drawback is that, quite frequently, subsequent analysis of the data may highlight a problem where insufficient data has been recorded, necessitating further flight tests. Alternatively, recording the complete set of flight data is subject to the storage capacity.

The most important aspect of flight data recording is the real-time performance of the data acquisition software. If raw data is generated at 50 frames per second, say, data needs to be recorded at this rate, without any loss of data. However, this requirement assumes that, independent of any simulator software, all operating system calls are either completed well within this 20 ms frame rate or, alternatively, that data acquisition processes take precedence over the relatively slow data storage processes. In many operating systems, I/O data can be queued without any loss of data, allowing a data access process to subsequently catch up, despite periods where the operating system may have exclusive use of the processor.

In flight simulation, data acquisition covers both online data analysis and off-line data analysis. Generally, raw data is acquired in the simulator and then processed by data analysis tools, which may apply filtering to the data. However, even for online analysis, the sheer volume of data can pose problems and tools need to be provided to assist in the management of the data (as opposed to analysis of the data, for which there any many packages available). For example, if data is recorded for specific events, there should be some way of marking this event (adding an identifier to the data, which can be accessed at a later time). Although the time base can be compressed to identify events, specific events might be characterized by the relationship between signals or damped cycles or fast changing signals or saturation of a signal. In other words, to be able to search through the data, looking for a magnitude, frequency or phase relationship between variables can be very helpful in processing large amounts of flight data. Similarly, the ability to skip over pages of data or set thresholds in searching can identify events, which might be difficult to detect manually, particularly in cases where flight data recording is used to detect a problem. Such tools are particularly useful in online analysis, where the result of one test may lead to further tests. Time spent analysing data can be expensive if access to a simulator is limited.

If large amounts of data are produced in flight data recording, data compression techniques can be used to reduce the amount of stored data. However, often the data is binary and standard compression methods are unlikely to provide compression ratios better than two to three. In some cases, logic states (e.g. switch positions), can be stored as single bits rather than bytes or words. Similarly, if the range of a variable is ± 28, say, then storing the variable as a 16-bit integer in the range $\pm 28,000$ may reduce the storage from a 4-byte floating-point value to 2-byte integer value, if an accuracy of 0.001 is appropriate for that specific variable. For example, an angle in the range $\pm \pi$ could be represented as a 16-bit integer representing tenths of milliradians ($\pm 31,416$). It would certainly be wasteful to write flight data to a file as ASCII text, where a floating-point number with 5 leading digits, 5 trailing digits, a decimal point and a space occupies 12 bytes.

The flight simulator covered in this book is based on a distributed architecture where packets are transmitted between computers. All packets transmitted are read by the instructor station computer, and, therefore, these packets provide the basis for flight data recording. By assembling a single block from the set of incoming packets per frame, this block can be written to memory (and subsequently to disk) as a binary image of the packets transmitted between the simulator computers. The definition of the structure of the flight data packet given in *iosdefn.h* is shown below; the packet definition is derived from the definitions of the aerodynamic data packet, the input/output packet and the navigation packet, given in the header files *aerodefn.h*, *iodefn.h* and *navdefn.h*, respectively.

```
typedef struct {
    AeroDefn_AeroDataPkt AeroPkt;
    IODefn_IODataPkt IOPkt;
    NavDefn_NavDataPkt NavPkt;
} IosDefn_PlaybackDataPktRecord;
```

This approach affords one further advantage; most operating systems have a block read and a block write function where the application provides the address of the array to be written and the number of bytes to be transferred to the file. Contiguous file transfer is usually much faster than sequential file transfer, particularly if the block size of the flight data is the same as the physical block size of the disk.

There is an additional advantage in organizing recorded flight data as direct images of frame packets. In normal operation, each module receives incoming packets from the other simulator computers. However, if the data is retrieved from a recording, stored as an identical contiguous disk file, these blocks can be retrieved from disk every frame. In other words, to replay recorded data, as each frame is retrieved, the simulator variables are set from the corresponding value in the recorded packet. The primary variables are reset in each frame, for example to reinstate variables normally produced by the equations of motion or the engine model, from recorded variables.

In many applications, playback is an essential aspect of flight training. This is particularly true in military training, where the playback can be viewed by pilots, but the view port can be easily changed, so that a pilot can see the opponent's view or an overview (sometimes referred to as a *God's eye view*) of manoeuvres and tactics. Similarly, in civil training, a pilot can replay a crosswind landing to review their handling of the crosswind in comparison with other recordings of the same flight conditions.

During flight data recording, the data is typically displayed as strip displays, which are similar to chart recorders. These are time traces of variables, recorded at the simulator frame rate but displayed at rates selected by the user. Default values for the time base and amplitude of variables can be changed during simulation, allowing the user to define both the range and intervals used in plotting. Often, data is recorded over a relatively long time and displayed as individual pages. Facilities are provided to move forwards or backwards one page at a time or to skip to a specific page. In addition, certain events may initiate data recording and, in these cases, it is helpful to mark the point where data recording started. This information is appended to the recorded data enabling the user to skip forwards or backwards between marks. The important point to bear in mind is that recorded data has a high value to its owner and that often large amounts of data are recorded so that efficient access to this data is essential; in many cases, being able to quickly inspect data can save valuable time avoiding the need to repeat a test at a later date.

Although it is possible to display flight data using computer graphics and then copy the screen image to a file, an alternative method is to write the raw data to a file and generate a script for a plotting package, exploiting the benefits of proprietary plotting packages. For example, the script in Example 9.4 was generated automatically for gnuplot by the flight data recording software in the instructor station. In this example, a test pilot was asked to fly the aircraft at 3000 ft at 180 kt (flaps 20) and hold the bank angle at 5° intervals from 0° to 30° for 10 s in order to confirm the relationship between bank angle and turn rate at this specific airspeed. Flight data was recorded at 50 Hz for aileron position, yaw rate, roll rate and roll angle for 80 s, in the file *test10.dat*.

Example 9.4 A gnuplot script for data recording

```
set terminal png truecolor font arial 8 size 600,640
set output "test10.png"
set size 1,1
set origin 0,0
set lmargin 10
set multiplot
set grid
set format y "%5g"
set size 1.0, 0.250000
set xr[0.000000:80.000000]
set xtics 5.000000
set origin 0, 0.000000
set ylabel "Aileron deg"
set yr[-15.000000:15.000000]
set ytics 5.000000
plot 'test10.dat' using 1:2 notitle with lines
set origin 0, 0.250000
set ylabel "Yaw rate deg/s"
set yr[0.000000:3.000000]
set ytics 1.000000
plot 'test10.dat' using 1:3 notitle with lines
set origin 0, 0.500000
set ylabel "Roll rate deg/s"
set yr[-3.000000:3.000000]
set ytics 1.000000
plot 'test10.dat' using 1:4 notitle with lines
set origin 0, 0.750000
set ylabel "Roll deg"
set yr[0.000000:35.000000]
set ytics 5.000000plot 'test10.dat' using 1:5 notitle with lines
unset multiplot
reset
```

The flight data is written as columns of 'space-separated' data. In this case, four channels are plotted for roll, roll rate, yaw rate and aileron position, over 80 s. The five columns of the data file *test10.dat* are time (s), aileron ($\pm 15°$), yaw rate (0 to $3°/s$), roll rate ($\pm 3°/s$) and roll angle (0–35°), respectively. The advantage of this approach is that these two files (the plot file and the raw data file) are straightforward to generate from the flight data and the plot command file needed for gnuplot is only 30 lines. The first 10 lines of the data file are given in Example 9.5 and the generated plot is shown in Figure 9.26.

Example 9.5 A sample of recorded flight data

```
0.000000 0.039063 -0.087189 0.017949 -0.896076
0.020000 0.117188 -0.087043 0.018344 -0.895804
0.040000 0.117188 -0.086844 0.018695 -0.895524
0.060000 0.117188 -0.086700 0.019083 -0.895236
0.080000 0.156250 -0.086536 0.019912 -0.894932
0.100000 0.117188 -0.086340 0.020233 -0.894621
0.120000 0.117188 -0.086199 0.020592 -0.894303
0.140000 0.117188 -0.086060 0.020941 -0.893977
0.160000 0.117188 -0.085922 0.021284 -0.893645
```

```
0.180000  0.117188  -0.085785  0.021621  -0.893306
0.200000  0.117188  -0.085594  0.021914  -0.892961
0.220000  0.078125  -0.085425  0.021755  -0.892619
```

9.8 Scripting

A typical exercise in a simulator session might be to fail an engine as soon as the aircraft reaches 100 ft (above ground level). The instructor could monitor the altimeter and then select the fail engine option and monitor the pilot response to the failure. The pilot may detect the flurry of mouse clicks needed to fail an engine; by the time the instructor has initiated the engine failure the aircraft may have passed 300 ft, and thirdly there is no record of the pilot's response to the event other than the instructor's recollection. A more objective way to manage the session would be to enter these events and recording requirements into a script file that is executed concurrently with the real-time simulation. Such a file is often referred to as a *script*, because the schedule of events and actions are set out in a file, in much the same way that an actor's script encapsulates the sequence of events in a play, which are repeated in every performance.

The advantage is that the script is completely unobtrusive; exactly the same conditions and events can be applied with different pilots, and moreover, the instructor can define explicitly what is to be

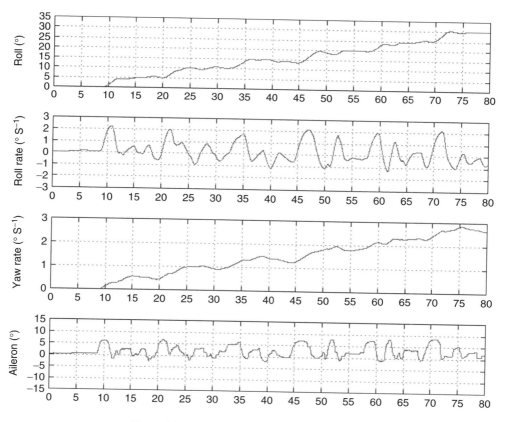

Figure 9.26 Flight data recording – yaw rate test

recorded. Such a system is very demanding. It must execute in parallel with the simulation, detect events that trigger actions, initiate actions, record events and data and moreover it must provide a scripting language for the instructor that is easy to understand and execute.

As an example of a script, during a recurrent check the instructor may introduce an engine failure after take-off and then monitor the airspeed, heading and rate of climb for 50 s after the engine failure. The script would take the form:

Wait until the aircraft passes 100 ft
Then fail the port engine
For 50 seconds
  record airspeed (kt), altitude (ft) and heading (degrees)

The corresponding script program is shown in Example 9.6.

Example 9.6 A simple script program

```
Wait Altitude > 100 ft
Fail Engine1
FOR 50 Secs DO
   Record Airspeed kts
   Record Altitude ft
   Record Heading degs
END
```

A pilot response to the engine failure produced by this script is shown in Figure 9.27, showing the variation in heading (yaw), altitude and airspeed, in addition to pilot rudder input during the simulator exercise, where the data is plotted from raw data captured during the exercise. The script is activated by the instructor, without the student being aware of the instructor's action. The actual engine failure occurs as the aircraft passes 100 ft, where the runway altitude is 210 ft.

The output from a script is produced in a form that can be imported directly into a spreadsheet package or graph plotting package or can be displayed at the instructor station. It can show if the pilot is able to maintain minimum engine-out airspeed, establish the engine-out climb rate and also monitor deviation in heading. If required, the script could be modified to record the pilot's rudder and aileron input, the angle of sideslip or the angle of bank. In addition to assessing pilot performance, a scripting function also allows the instructor to increase pilot workload. For example, a veering wind can be introduced in a navigation exercise, conflicting traffic can be produced to check pilot lookout, visibility can be reduced to assess pilot decision-making and instruments can be failed for limited panel exercises. Scripting also provides a rudimentary form of flight testing, enabling an instructor to measure climb rates, stall speeds or even to display a phugoid. Scripting is therefore a very powerful addition to a flight simulator; it provides objective checking and training and it enables an instructor to develop a library of exercises with minimal intervention during exercises in the simulator.

The scripting language used in this book is referred to as *Script*. In its simplest form, *Script* enables flight variables to be set to a specific value, as shown in Example 9.7.

Example 9.7 Setting flight conditions in a script program

```
set airspeed 120 kts
set altitude 3000 ft
set heading 340 degs
set visibility 10 km
```

The Instructor Station

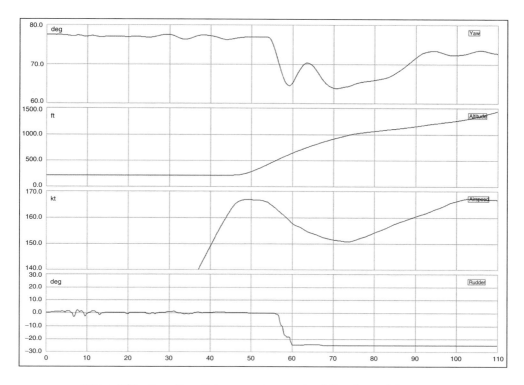

Figure 9.27 Recording pilot response to an engine failure in a script program

The aircraft is positioned at 3000 ft, with an airspeed of 120 kt and a (magnetic) heading of 340° and visibility reduced to 10 km. Although the flight simulator uses SI units in the flight model and engine model, the user can override default values by defining specific units for a variable. Similarly, variables can be recorded during simulation in appropriate units, as shown in Example 9.8.

Example 9.8 Recording flight variables in a script program

```
record airspeed mph
record roll degs
record rudder_trim
record engine3_rpm
```

The airspeed is converted from metres per second to miles per hour, the roll angle is converted from radians to degrees, the rudder trim is recorded in simulator units (± 1.0) and the RPM of engine 3 is recorded as a percentage.

The structure of the *Script* compiler is shown in Figure 9.28. An instructor can write a set of *Script* programs appropriate to flight training sessions in a flight simulator. A specific *Script* program can then be activated as a user command, where the user is prompted for the name of the *Script* file, which is compiled to intermediate code. If an error occurs during compilation, the error is logged and the command is cancelled; otherwise the *Script* program is loaded and activated. In practice, it is rather inefficient to use the instruction station to test programs for compilation errors and an off-line version of the *Script* compiler is provided so that a *Script* program can be

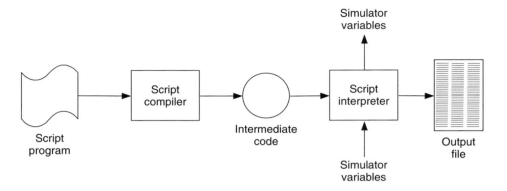

Figure 9.28 Script compiler structure

checked for compilation errors. The *Script* interpreter is triggered once per frame to execute the intermediate code. The interpreter has access to the simulator variables (in the packets transmitted to the instructor station) and can set simulator variables to specific values, in a packet broadcast to the other simulator computers. Any output generated by the intermediate code is written to a *Script* output file.

There are two limitations to be overcome in implementing a real-time scripting language of this form. First, the interpreter must be executed in parallel with the simulation. In practice, it is executed every frame, but if only one instruction is executed per frame, then only 50 instructions would be executed per second, in contrast to the simulator, which executes several million instructions per second. To avoid this constraint, the interpreter could be created as a separate process, which is executed at the same rate as the instructor station process. Such an implementation would also require a message passing mechanism between the two processes to enable the interpreter to access simulator variables. Secondly, the action of writing to disk can introduce delays of several milliseconds. The solution is to write directly to memory, where an area of memory is set aside and treated as an output device. On completion of the *Script* program, the output that has been written to this temporary buffer is copied to a specific disk file.

Although most compilers generate instructions for a specific machine, to ensure portability, instructions can be generated for a general purpose (hypothetical) machine. The instructions of this virtual machine are subsequently fetched and executed by a small program known as an *interpreter*, which can be written in C. Admittedly, there is a potential loss in performance, but the application of a scripting program is likely to be relatively simple and the programs are likely to be small.

A commonly used architecture for machine-independent compilation is a stack machine. The instructions are stored in a program memory and are fetched one at a time. A program counter (PC) points to the next available instruction. Dedicated storage is allocated for program variables and also for a stack, with a stack pointer (SP) pointing to the next available location in the stack. The only access to the stack is via the stack pointer. An item can be pushed onto the stack, or popped from the stack, as shown in Figure 9.29 where the stack contains items a, b and c, and the stack pointer points to the next available location on the stack. The variable x is pushed onto the stack at the location pointed to by stack pointer and then stack pointer is incremented.

In Figure 9.29b, the item c is popped from the stack and then the stack pointer is decremented. Note the symmetry of the push and pop operations and also that there is no actual movement of data on the stack; all access to the stack is via the stack pointer. Of course, sufficient memory is needed for the stack that will expand and contract as data is pushed and popped. In addition, the stack pointer must be initialized for an empty stack.

The Instructor Station

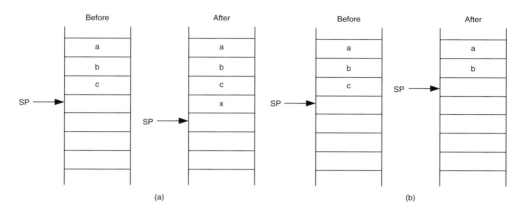

Figure 9.29 (a) Stack push and (b) stack pop operations

Arithmetic instructions operate on the items at the top of the stack, and consequently have no arguments. For example, the *Script* code

```
a = b + c
```

is implemented by the following instructions

```
push b
push c
add
pop a
```

The add instruction adds the top two items on the stack and leaves the result of the addition on the top of the stack, decrementing the stack pointer, as shown in Figure 9.30, which depicts the state of the stack after each instruction.

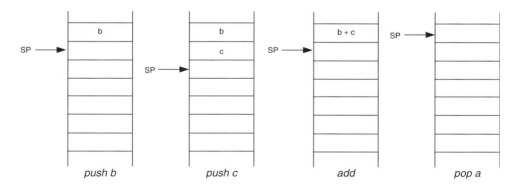

Figure 9.30 Execution of an addition instruction for a stack machine

By way of illustration, consider a simple example to record the deviation from an airspeed of 100 kt. The code fragment and the corresponding intermediate code are shown in Example 9.9.

Example 9.9 Stack machine code

```
record airspeed * 0.51444 - 100 * 0.51444
```

PUSHV 12	*push airspeed*
PUSHC 0.514440	*push conversion factor*
MULT	*convert to kt*
PUSHC 100.000000	*push 100*
PUSHC 0.514440	*conversion factor*
MULT	*convert m/s to kt*
SUB	*relative airspeed*
RECORD	*invoke data recording*

The single *Script* statement is compiled to eight intermediate code instructions. The first line PUSHV 12 pushes variable number 12 (airspeed) onto the stack. The last instruction RECORD pops the item on the top of the stack and writes the value to the output file. Note that, in this case, the user has applied an explicit conversion from metres per second to knots and also that bracketing can be used to ensure the ordering of operations in arithmetic expressions.

Although these operations enable variables to be set or recorded, the primary advantage of a scripting facility is that it enables complex situations to be described in terms of events and actions. These situations may be conditional; they may only occur when certain events occur or they may be repetitive; they may occur while a set of conditions apply or for a specific period of time. Of course, conditional and repetitive operations are provided in conventional programming languages in the form of if-then, if-then-else, while-do and repeat-until expressions. For example, an instrument might be failed above 3000 ft to simulate icing conditions or data may be recorded only when the aircraft is within 10 nm of a beacon. The compiler parses the *Script* code to generate intermediate code, following the simple syntax rules of the scripting language. Two examples are given below to illustrate the power of a scripting language, together with the compiled code to show how the intermediate code can emulate complex events. Comments are added to explain the intermediate code. In Example 9.10, a limited panel exercise, the attitude indicator is failed if the aircraft is above 3000 ft or at least 5 nautical miles from the selected DME station and also if the airspeed is at least 80 mph. If the condition is appropriate to fail the attitude indicator, the instrument is failed and a message is logged to note this failure, together with the altitude at which the failure was initiated. Admittedly, this is a contrived example, but it illustrates the detection of a complex series of events.

Example 9.10 Conditional failing of an instrument

```
IF ((Altitude > 3000 ft) OR (NAV1_DME > 5 nm)) AND (Airspeed > 80 mph) THEN
  fail ai
  log "failing AI"
  record altitude m
END
```

The Instructor Station

PUSHV 11	*get altitude (m)*
PUSHC 914.400024	*3000 ft (m)*
COMP	*comparison of altitude and 3000 ft (boolean)*
PUSHV 79	*get NAV1 DME (m)*
PUSHC 9260.000000	*5 nm (m)*
COMP	*comparison of DME distance and 5 nm (boolean)*
OR	*OR of the two comparisons (boolean result)*
PUSHV 12	*get airspeed (m/s)*
PUSHC 35.759998	*80 mph (m)*
COMP	*comparison of airspeed and 80 mph (Boolean)*
AND	*AND of the two comparisons (boolean result)*
JF L17	*skip next 4 instructions if the test fails*
FAIL 15	*fail the attitude indicator*
LOG "FAILING AI"	*log the message to indicate that the AI has been failed*
PUSHV 11	*get altitude (m)*
RECORD	*output the altitude (no conversion needed for metres)*

In Example 9.11, the airspeed, altitude and elevator position are recorded, but only while the aircraft heading is between 20° and 120°. As soon the heading is outside this range, the recording stops and the next *Script* statement following this fragment is executed. Note the loop repetition at instruction 20 and the possible loop exit test at instruction 9.

Example 9.11 Conditional recording of flight data

```
WHILE (heading > 20 degs) AND (heading < 120 degs) DO
   record airspeed kts
   record altitude ft
   record elevator rads
END
```

PUSHV 4	*get heading (radians)*
PUSHC 0.349060	*20 degrees (radians)*
COMP	*comparison with 20 degrees (boolean)*
PUSHV 4	*get heading (radians)*
PUSHC 2.094360	*120 degrees (radians)*
SWAP	*reverse the order of the stack items for the comparison*
COMP	*compare with 120 degrees (boolean)*
AND	*AND of the comparisons (boolean)*
JF L21	*skip the remainder of the code if this test fails (exit the loop)*
PUSHV 12	*get airspeed (m/s)*
PUSHC 0.514400	*get conversion factor for kt*
DIV	*convert to kt*
RECORD	*output airspeed (kt)*
PUSHV 11	*get altitude (m)*
PUSHC 0.304800	*get conversion factor for ft*
DIV	*convert to ft*
RECORD	*output altitude (ft)*
PUSHV 25	*get elevator position (radians)*
RECORD	*output elevator position*
JUMP L1	*repeat the loop*

In many ways, a scripting language is similar to a programming language. Its instructions are executed sequentially, it supports conditional and repetitive operations and provides input and output of external variables. However, the language is also very much simplified; the variables are all predefined (no user-defined variables are permitted); there are no user-defined types – all variables are scalar values; there are no user-defined procedures – by implication there are built-in procedures to set variables or to write to the output file. These simplifications afford two advantages; they make the language much easier to understand and to write and the implementation of the language is straightforward in terms of the compiler and the interpreter.

These control flow structures enable a *Script* program to synchronize with the simulator. If the *Script* program is waiting for a simulator event, for example, for the bank angle to exceed 15°, then if this condition is not satisfied, there is no point in testing for the condition again until the next frame, when a new packet can be accessed. In other words, if any test fails, the *Script* program is suspended (simply by storing its program counter and stack pointer) to be resumed during the next frame. There are two examples where this situation can occur explicitly in *Script*. First a WAIT statement waits for a condition to occur, for example

```
wait altitude < 2000 ft
```

will wait until the aircraft is below 2000 ft. This condition is tested every frame and the *Script* program will not resume until this condition is valid. A variant of a WAIT statement is the DELAY statement. For example

```
delay 3 mins
```

will introduce a delay of 3 minutes into a *Script* program. This delay starts from the time it is initiated but only applies to the *Script* program, which is effectively suspended. One further timing function is to execute a set of *Script* instructions for a specific period of time and this is provided by the for-loop construction in *Script*. Unlike a conventional programming language, where the for-loop index is a program variable, *Script* treats the index as time, so that a set of actions can be implemented over a specific period. For example, altitude, heading and airspeed are recorded for 50 s in Example 9.6. In executing a for-loop, the time of termination of the loop is checked against the current simulator time (in units of 20 ms ticks) during the update of the *Script* program every frame.

A *Script* program can perform all the actions of an instructor, in terms of controlling and monitoring the simulation. Systems can be failed or reset to normal operation, flight control systems can be engaged and disengaged and the aircraft can be repositioned to a specific latitude/longitude. Finally, *Script* provides a means of logging messages and writing variables to the *Script* file at any point during a Script program. These lines of output can be time-stamped with local time, elapsed time (since the *Script* program started) or the simulator time. In addition, the rate at which data is sampled can also be defined. By default, data is recorded at 50 Hz, but in many applications, much lower sampling rates may be preferable.

Once per frame, packets are read from the other computers, the instructor inputs are checked, the display is updated, a packet is broadcast and the *Script* interpreter is activated, if there is an active *Script* program. As the *Script* language is very basic and the control flow is limited, it is possible to emulate the parallelism needed to implement *Script* because all the conditional branches in *Script* are organized so that a failed test branches backwards. This convention requires only a simple test in the *Script* interpreter; if the instruction does not result in a backward branch, the next instruction is fetched and executed (unless either a STOP instruction is encountered or the end of the program is reached). Strictly, this is not a valid method to implement parallel processing, as any number of *Script* instructions can be executed until a test fails or an event occurs. Nevertheless, given the simple syntax of the language and the fact that a *Script* program is activated once per frame, it works exceedingly well in implementing an apparently parallel process without recourse to complex multiprocessing. Moreover, the implementation is transparent to the user.

A powerful scripting language is a valuable asset in a flight simulator. It enables events to be detected, flight data to be recorded and actions to be initiated that would otherwise require manual intervention. In many areas of flight training, the role of the instructor is often to provide an environment that may be encountered in actual flying. These exercises may not only involve failures, malfunctions, weather events or more subtle changes, but they also provide a means to increase pilot workload in training situations where this may be important and they allow flight crews to make judgements and decisions in real time; this form of training, where two flight crew cooperate to manage both the flight and unexpected situations, is sometimes referred to as *crew resource management* (*CRM*). The combination of a simulator and a powerful instructor station provides a means of accelerating experience and provides flight crew with situations they might encounter rarely. In these situations, a scripting program enables an instructor to construct very elaborate flight situations and to monitor pilot performance and response very closely. It also provides objective output for both the instructor and the training crew to discuss specific outcomes during training.

The capacity of the instructor station to enhance synthetic training cannot be too highly emphasized. Features designed in the instructor station provide opportunities to train and assess flight crews but, with many of the features outlined in this chapter, they also provide a means to extend this training. Nowadays, the experience of modern pilots is a combination of events encountered in flight and events encountered in a flight simulator. In the case of the flight simulator, the pilot's experience can be artificially accelerated, albeit in a controlled way, so that the pilot encounters situations that (hopefully) may not occur during their career. However, if such events are encountered in flight, the pilot will be able to use training for such situations, which has been thoroughly practised in a simulator. Consequently, the instructor station has a pivotal role in flight training, and the design of the instructor station (particularly the software) should be developed with the same analytic approach that is used to develop the other main components of the simulator.

References

Ahn, Y.H. (1997) Advances in instructor operating stations, Royal Aeronautical Society Conference. Flight Simulation – Expanding the Boundaries, London.

Alexandria, V.A. and Dalheimer, M.K. (2002) *Programming with QT*, O'Reilly Media, Inc, USA.

Caro, P.W. (1972) Transfer of instrument training and the synthetic flight training system. Report No. *HumRRO-PP-7-72*. Human Resources Research Organisation, Alexandria.

Fountain, A. and Ferguson, P. (2000) *Motif Reference Manual*, O'Reilly Media, Inc, USA.

Galitz, W.O. (2007a) *The Essential Guide to User Interface Design: An Introduction to GUI Design Principles and Techniques*, John Wiley & Sons, Ltd.

Goode, M. and Evans, D. (1992) An instructor station for the instructor, Royal Aeronautical Society Conference. European Forum – Matching Technology to Training Requirements, London.

Hayes, R.T. and Singer, M.J. (1989) *Simulation Fidelity in Training System Design*, Springer-Verlag, New York.

Krause, A. (2007) *Foundations of GTK+ Development*, APress.

Newman, W. and Sproull, R. (1979) *Principles of Interactive Computer Graphics*, McGraw-Hill.

Raskin, J. (2000) *The Humane Interface: New Directions for Designing Interactive Systems*, Addison-Wesley.

Schneiderman, B. (1992) *Designing the User Interface*, Addison-Wesley.

10

Motion Systems[1]

10.1 Motion or No Motion?

It is said by engineers that if you were to ask five pilots for their views on a technical subject, you would get ten opinions. Certainly, the value of motion in a flight simulator provokes heated debate. The argument for motion is based on the observation that motion in an aircraft has a disorienting effect on pilot perception, which can affect their performance; if pilots are trained in the benign environment of a fixed-base (non-motion) flight simulator, skills learnt in the simulator may not transfer to the aircraft, owing to the disturbance introduced by the accelerations experienced in the aircraft. The counter argument is that the accelerations experienced in an aircraft cannot be accurately replicated in a flight simulator and consequently, no motion is better than wrong motion. Moreover, no aspect of flight training requires pilots to learn to identify motion cues to execute flying tasks. Although most pilots and engineers would add the caveat that the benefit of motion depends on the training application, the case for motion is far from clear.

This debate is exacerbated by our lack of understanding of human cognitive processes. Man has evolved with sensory systems designed for survival. We can run at speed, moving our limbs, head and eyes independently, optimizing both balance (stability) and speed (performance). The brain performs sensor fusion, combining and weighting sensory inputs to make the best assessment of any situation. This form of information fusion is very relevant to pilot training, where visual, motion and tactile cues are combined, as shown in Figure 10.1. Our understanding of the fusion processes performed by the brain is very basic, and given this lack of understanding, as we are not sure how motion cues impact pilot behaviour, it may be a sensible precaution to try to replicate airborne motion cues in a flight simulator. The understandable concern of the regulatory authorities is that if a pilot who has trained in a fixed-base simulator subsequently encounters accelerations in an aircraft, which were not associated with skills learnt in the simulator, it may result in a different (and potentially undesirable) response in the aircraft.

It is very tempting to consider undertaking experiments to compare pilot performance in learning the same task in both a full-motion and a fixed-base simulator. Although a number of studies have been undertaken (Parrish *et al*. 1977; McCauley, 2006; Cardullo, 1991; Anderson and Morrison, 1993), the results are far from conclusive. The effect of motion varies with the specific flying task. For example, the effect of motion is undoubtedly more important in a helicopter hovering close to the ground than it is in operating a flight management system of an airliner in the cruise. It is also difficult to measure the effect of motion objectively, particularly as there is also a wide variation of

---

[1] Part of the material in Sections 10.2–10.5 is based on notes by Gerhard Serapins and Len Allen of CAE, which were delivered in the Cranfield University annual short course in flight simulation.

Principles of Flight Simulation   D. J. Allerton
© 2009, John Wiley & Sons, Ltd

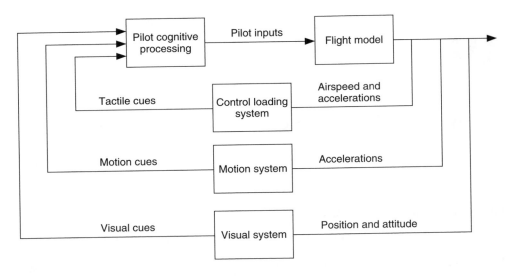

Figure 10.1 Pilot cues

pilot response to motion inputs. For example, pilots with a light aircraft background may be more used to disturbances from turbulence (in terms of filtering out motion disturbance) than fast jet pilots. One additional element of variability is found in flight simulation equipment including the motion actuators and the control loading systems. For example, the linear motion displacement may vary for different flight simulators, so that results obtained for one simulator may not necessarily be applicable to other simulators.

The dilemma of the 'motion versus no motion' argument can be summarized by the following (hypothetical) conversation between the Chief Executive Officer (CEO) of an airline and the Chief Pilot (CP), who is asking the CEO to approve the purchase of a new flight simulator for the airline:

CEO: According to the figures, we can save $2M if we purchase the simulator without motion.

CP: But motion is essential in the training of our pilots, particularly for engine failures, where there is a large side force.

CEO: So you train pilots to detect motion?

CP: Well no, we train them to ignore all motion cues and use the visual cues provided by the flight instruments.

CEO: We provide motion, so that they can ignore it?

CP: Well, it needs to be realistic motion, so that they are ignoring the right sort of cues.

CEO: I can't see the point of adding something so that it's specifically ignored.

CP: The engineers and psychologists would tend to agree but the regulatory authorities use pilots to approve flight simulators.

CEO: Never mind them, let's go for a simulator without motion and save the airline $2M.

CP: If we do that, the simulator will not be qualified for training by the regulator. We will have to do a lot more training on the aircraft. The additional cost of training will bankrupt the airline.

CEO: Where do I sign?

As with all flight training simulators, it is important to match the specific technology to the training requirement (Caro, 1979). In the case of motion platforms, it is not straightforward to assess the contribution of a motion system to training effectiveness. Some studies have been undertaken (Caro, 1977; Martin and Waag, 1978; Lintern, 1987), including comparisons of similar training tasks in moving base simulators with fixed-base simulators. However, there are surprisingly few studies and consequently, there is an overall lack of objective guidelines on the contribution of motion to training effectiveness.

Despite the lobby for fixed-base flight simulators, this chapter examines the physiology of motion detection by humans and the mechanical actuation systems to move a simulator platform in a way that closely resembles actual aircraft motion. Of course, the motion of an unconstrained aircraft is very different from that of a simulator platform with limited motion displacement. Nevertheless, for civil flight simulators, simulator manufacturers have developed actuator systems, which can provide remarkably realistic motion cues, to provide sensations similar to the motion cues experienced in an aircraft. For military trainers, particularly for combat simulation, where high-G cues dominate and the visual systems provide extremely wide-angle projection, motion platforms have been discarded in favour of actuators, which provide tactile forces between the pilot and the seat.

10.2 Physiological Aspects of Motion

The human body has a remarkable set of motion sensors, which are capable of detecting linear and angular accelerations, in addition to the haptic sensors that detect tactile pressure applied to the skin. These sensors enable us to walk up a flight of stairs, jump off a chair or run after a bus, without endangering ourselves. They also provide orientation in three dimensions, for example, an athlete can jump over a 2-m bar or a diver can perform several somersaults, jumping backwards from a diving board. These same sensors detect accelerations during conventional flight and in aerobatic manoeuvres. For the flight simulator designer, an understanding of the human motion sensors is essential; it can identify the limits of the brain to detect motion and possibly exploit this information to reduce or simplify the motion applied to a simulator platform. If the dynamic responses of the human balance sensors are known, filters can be designed to match the combined response of the motion platform and pilot to the response of the aircraft and pilot (Gum, 1973). In attempting to provide realistic motion cues, the motion inputs applied to a flight simulator platform are constrained:

- The structural limitations of a simulator motion platform determine the maximum forces that can be applied to the pilot in a simulator; it is necessary to establish the limits where lack of motion may affect the simulator fidelity;
- Reducing the motion inputs may allow a lighter and less expensive structure to be used in a motion platform; alternatively, the mass of the platform and cabin determines the power needed from the actuators;
- If acceptable platform motion can be achieved with reduced power, considerable savings can be made in the running costs of a flight simulator; the use of electrical actuation may reduce both power requirements and the environmental problems associated with hydraulic systems.

The vestibular system (Howard, 1986) of the human body senses the orientation of the head and dynamic movement of the head. As the head moves, the eyes are stabilized, so that the vision is not blurred by the head movement; the vestibular system also provides signals for the eye muscles to accommodate this movement. In effect, the vestibular system provides a stabilized platform as it is capable of detecting both linear and angular accelerations about three axes. The angular accelerations are detected by semicircular canals and the linear accelerations are detected by the otoliths. One set of these sensors is located in each inner ear.

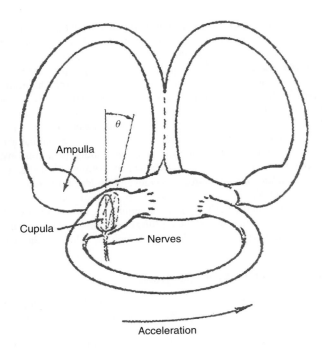

Figure 10.2 Semicircular canals

The semicircular canals (Melville-Jones *et al.* 1964) are arranged as an orthogonal set of canals, in three mutually perpendicular planes, as shown in Figure 10.2. The canals are attached to the skull and filled with a fluid, known as *endolymph*. In each canal, there is an expanded section called the *ampulla*, which is sealed by a flap known as the *cupula*. Angular acceleration of the head about one of the three axes causes the fluid in the canal to move (with a short lag) deflecting the cupula by a small amount. The nerves in the cupula detect this movement, sending signals to both the brain and the oculomotor muscles in the eye (to stabilize eye movement). The response of the semicircular canals is given by Ormsby (1974):

$$\frac{y(s)}{\phi(s)} = \frac{0.07s^3(s+50)}{(s+0.05)(s+0.03)} \qquad (10.1)$$

where ϕ is the angular displacement and y is the perceived displacement.

Generally, a second-order approximation (Hixson, 1961) is used, given by

$$\ddot{\theta} + 2\zeta\omega_n\dot{\theta} + \omega_n\theta = u(t) \qquad (10.2)$$

where
- θ is the angle of the cupula
- ζ is the damping factor
- ω_n is the natural frequency
- $u(t)$ is the input acceleration

Values of ζ lie between 3.6 and 6.7. Values of ω_n lie between 0.75 and 1.9 (Van Egmond *et al.* 1949), giving an over-damped system with two real roots. The short time constant is 0.1 s.

The longer time constant depends on the sensor axis, with 10.0–11.8 s in yaw, 5.3 s in pitch and 6.1–6.8 s in roll (Benson and Bodin, 1965). Under laboratory conditions, it has been shown that angular acceleration cannot be detected below a minimum acceleration, which is between 0.12 and 4.0°/s². Threshold values of 0.5°/s² have been reported for pitch and roll under flight simulator conditions (Meiry, 1965). This is a very important observation; it implies that a pilot in a simulator is unaware of angular accelerations below 0.5°/s² being applied to the platform. Similarly, angular velocities below 0.5t/s cannot be detected, where t is the time the acceleration has been applied. For example, to move the simulator through 10° of pitch, with an acceleration of 0.5°/s², takes 6.3 s (0.25 t^2). In other words, the simulator could be restored by 10° in pitch in about 6 s, without the pilot being aware of this applied motion. For example, although the maximum pitch angle of the platform might be 15°, as the simulator motion approaches this limit, the platform can 'leak' 10° of pitch in 6.3 s, enabling the simulator to achieve a further (perceived) pitch up of 10° in response to subsequent pilot input.

Linear accelerations are detected by the otoliths. The sensor consists of hair cells in a gelatinous fluid containing particles of calcium carbonate, as shown in Figure 10.3. As acceleration is applied to the head, the calcium carbonate particles lag slightly behind the head movement, deflecting the hair cells. Movement of the hair cells is detected by nerve cells which transmit signals to the brain and the oculomotor muscles in the eye. Studies have shown that the otoliths detect the tangent component of applied forces (Miller and Graybiel, 1963; Schone, 1964). Ormsby (1974) gives the following transfer function for the response of the otoliths:

$$\frac{y(s)}{f(s)} = \frac{2.02(s+0.1)}{s+0.2} \qquad (10.3)$$

where f is the applied force and y is the perceived force.

In practice, an equivalent second-order transfer function is generally used with time constants of 0.66 and 10 s. Threshold values of 0.0011 ft/s² have been measured in centrifuge experiments

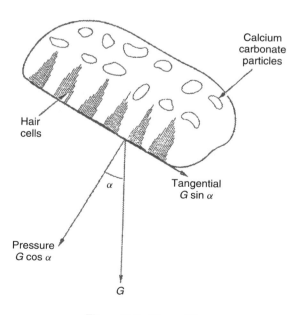

Figure 10.3 The otoliths

(Graybiel, 1948), although higher values of 0.1–0.4 ft/s² were obtained from airborne studies in the USSR (Derevyanko and Mylinikod, 1964). The same studies suggest that the threshold is actually jerk motion (rate of change of acceleration) with values from 0.3 to 0.64 ft/s³.

Clearly, the pilot's vestibular system detects accelerations before the effect of the accelerations are perceived on the aircraft instruments. In particular, attitude and altitude are second integrals of acceleration, introducing a lag before the initial acceleration takes effect. There is, arguably, an inner control loop in which the pilot detects and responds to accelerations, which occur in a full-motion simulator and also in an aircraft, but is omitted in a fixed-base flight simulator. This lack of acceleration cues in a fixed-base simulator is cited by some pilots as a potential cause of negative training transfer in transitioning from a fixed-base simulator to an aircraft. In other words, pilots apply one technique in the simulator and another technique in the aircraft. Certainly, there have been instances during in-flight refuelling exercises and also with vertical take-off aircraft, where there have been noticeable differences between the pilot's performance in the simulator and in the aircraft. In refuelling applications, such differences can be attributed equally to the visual system (potentially the reduction of the vertical field-of-view or the projector focal length with near objects), poor turbulence modelling or incorrect aerodynamic interaction with the tanker aircraft.

10.3 Actuator Configurations

The very early developments of trainers focused attention on motion rather than aircraft dynamics or instrumentation. These trainers, such as the Antionette trainer (Allen, 1993), were rotated mechanically by human operators. Later, the Link trainer was rotated in pitch, roll and yaw by pneumatic actuation. During the 1960s, Link refined their design, using electrical actuation with their General Aviation Trainer, known as the *GAT-1*. However, these trainers only provided rotational accelerations, possibly because instructors felt that experiencing the disorientation of banking was essential in instrument flight training. As these trainers also provided continuous yawing, slip rings were provided to connect the electrical signals to the cockpit systems, adding considerably to the complexity and cost of the trainer. Despite these reservations, it is important to note that the Link Trainer was a very effective trainer; many World War II pilots and subsequently civilian pilots owe their lives to lessons learnt in these primitive simulators.

It was not until 1965, with the development of the Stewart motion platform (Stewart, 1965), that linear motion and angular motion were successfully combined. The platform consists of six independently operated hydraulic legs, to provide six degrees of freedom (6-DOF). This form of motion platform is also referred to as a *synergistic platform* or a *hexapod*. Although there have been major improvements in the performance and smoothness of the actuation systems and some degree of variation in the geometry of these platforms, the basic configuration has remained unchanged for over 40 years.

A modern civil flight simulator cabin contains a flight deck, typically with seating for two pilots and the training instructor. The cabin may also contain computer systems for the instrument displays, the instructor station and the projectors for the visual system. In simple terms, the cabin is a rectangular box mounted on the motion platform, with a single door access at the back of the cabin and an emergency exit through the cabin windows, via a rope ladder. It should be appreciated that the mass of a typical simulator in a training session with three flight crew is probably of the order of 40,000 lb. The centre of mass of the cabin is well above the mounting point of the actuators, which is particularly demanding of any mechanical system. Increasing the structural integrity of the simulator increases the mass and an increase in the mass of the load requires further strength of the structure, particularly for off-axis loads. Raising and lowering the platform over several metres also introduces major safety considerations to ensure that the occupants are not harmed or injured.

A pilot is an aircraft experiences linear motion along the three body axes of the aircraft and angular motion about these three axes (Nahon and Reid, 1988). A synergistic motion platform provides three linear accelerations: heave (vertically), sway (laterally) and surge (longitudinally) and three angular accelerations: pitch, roll and yaw, giving rise to the expression 'six degrees of freedom' or 6-DOF, in describing the motion platform dynamics. These acceleration cues, combining both linear and angular motion, are provided by actuators attached between the platform frame and a solid concrete floor. The actuators require sufficient power to generate these accelerations, but must also operate reliably (to minimize simulator down time) and smoothly (so that no spurious accelerations are applied). Since the 1960s, hydraulic systems have been used in the motion platforms of most airline flight simulators. However, recent advances in electrical drive technology have led to the introduction of electrical motor drives, each controlled by a power amplifier, to replace hydraulic systems. Although the current range of electrical drives do not fully meet the performance of hydraulic actuation systems, it is likely that the advantages afforded by electrical drive technology will see the demise of hydraulic systems in flight simulation, for several reasons:

- There is considerable risk to maintenance staff working in close proximity to high-pressure hydraulic systems;
- Leakage of hydraulic fluid can result in environmental contamination;
- A small electrical motor for each actuator, combined with a mass compensation scheme (typically to support 10–12 tonnes) can reduce energy costs by as much as 80% (Anon, 2008).

The combination of mechanical translation in three axes and rotation about three axes is particularly challenging. Although the Stewart platform is limited in terms of linear and angular displacement, it has become a de facto standard as a motion platform in the simulation industry. The linear actuators are attached to the base of the platform and the floor of the building, where the attachment points to the floor and the platform are universal joints. Figure 10.4 illustrates the translation from linear actuation to rotary motion. In Figure 10.4a, the cabin is rolled by extending the left linear actuator. In Figure 10.4b, the cabin is pitched up by extending the forward linear actuator.

Notice that the angular displacement is limited by the stroke length of the actuators and that rotation about the yaw axis is more restricted than the pitch and roll axes, as a consequence of the actuator geometry. For linear motion, if all actuators are extended simultaneously, the cabin will

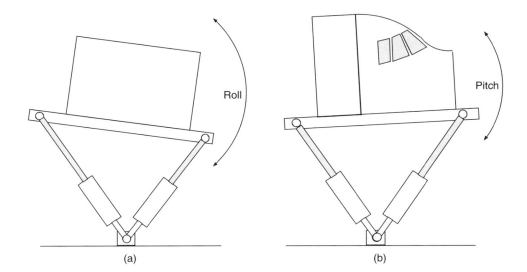

Figure 10.4 Platform angular motion. (a) Rear view. (b) Side view

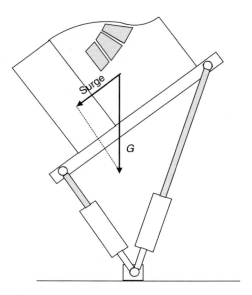

Figure 10.5 Tilting the platform to provide surge acceleration

move up vertically (heave). Surge and sway motion is accomplished by a combination of actuations. For example, in Figure 10.4b, if the rear actuator is retracted the cabin will move forwards (surge), assuming the right actuator is used to maintain the pitch angle. The simulator platform would drop by a small amount in maintaining the pitch angle and consequently, care is needed to avoid introducing a false vertical cue when a longitudinal motion cue was intended.

There is one further consideration with motion actuation. Although accelerations can be applied by the linear actuators, the flight crew also experience gravitational acceleration. For example, if the cabin is tilted upwards in pitch, as shown in Figure 10.5, the gravity component can provide both heave and surge cues for the flight crew, provided that the flight crew are unaware that the cabin has been rotated in pitch. For example, during take-off, the engine thrust will result in an initial surge acceleration, which cannot be provided simply by moving the cabin forwards. However, the surge acceleration is less than 1 G and if the cabin is titled to 45°, say, then the flight crew will experience a continuous surge acceleration of 0.7 G. A similar effect is possible when reverse thrust and braking is applied after touchdown, by tilting the platform downwards, to provide the deceleration associated with braking. A lateral acceleration can be achieved by rolling the platform to provide sway cues, provided that the angular motion is below the vestibular threshold.

A major design decision for the motion platform designer is the configuration of the actuators. There are a number of constraints that influence the geometry of actuator placement:

- The number of degrees of freedom of motion of the simulator platform;
- The maximum platform payload;
- The size of the simulator cabin;
- The actuator dimensions;
- The actuator dynamics;
- Trade-off between the responses in heave, sway and surge.

Most manufacturers have adopted the Stewart platform configuration, with six linear actuators (hydraulic or electrical), although the rectangular platform poses a problem as the six actuators are

Motion Systems

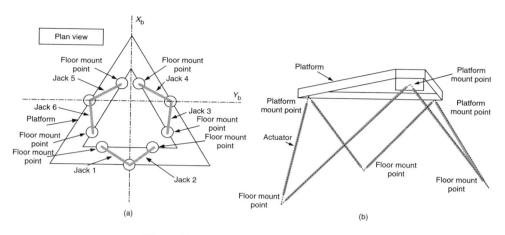

Figure 10.6 Hexapod platform configuration

Table 10.1 Motion requirements

Axis	Displacement	Velocity	Acceleration
Vertical	±34 in.	30 in./s	1 G
Lateral	±60 in.	35 in./s	1 G
Longitudinal	±60 in.	39 in./s	1 G
Pitch	±35°	22°/s	400°/s^2
Roll	±33°	24°/s	400°/s^2
Yaw	±37°	28°/s	400°/s^2

attached to four sides. The solution is to attach two actuators to a single mounting point on the left-hand side of the frame, two actuators to a single mounting point on the right-hand side of the frame and two actuators to a single mounting point on the front side of the frame, as shown in Figure 10.6b. This configuration, which is referred to as a *hexapod configuration*, is in effect two equilateral triangular bases. Three pairs of actuators are attached to the points of an equilateral triangular base on the floor and three pairs are attached to the points of an equilateral triangular base on the platform, as shown in Figure 10.6a.

The typical requirements of a modern motion platform for a Level-D full flight simulator are summarized in Table 10.1. The values in the table are best case. In situations where motion occurs in two channels simultaneously, there is a significant reduction in the performance of the individual channels, in terms of reduced displacement or actuator velocity. However, it is straightforward for the manufacturer to model the platform motion in order to construct motion 'footprints' to establish the limits of motion. For example, plots of vertical displacement versus lateral displacement could be plotted as a function of pitch angle. Having produced these plots, any limitations, in terms of training requirements, can be established prior to construction of the platform. In practice, although the manufacturer can use more powerful actuators, the geometry of the 6-DOF platform, the physical dimensions of the cabin and the performance of hydraulic actuators impose severe constraints on the motion that can be provided.

The dimensions and mass of most flight simulator platforms are similar and the emphasis in platform design focuses on the motion algorithms rather than the platform configuration. By providing strong initial motion cues, the pilot is alerted to a change of acceleration. The subsequent use of wash-out filters can remove accelerations and minimize the resultant platform displacement. The combination of simultaneous linear and angular motion of the platform can emphasize specific motion cues (Cardullo, 1983). In particular, rotation of the platform changes the components of gravitational acceleration felt by the cabin occupants, so that angular motion of the platform is sensed as a linear acceleration by the flight crew.

The motion platform must meet the following functional requirements:

- Respond to the three linear acceleration inputs and the three angular acceleration inputs produced by the flight model (in body axes);
- Provide the acceleration cues sensed at the flight crew position (rather than the aircraft cg);
- Wash out accelerations below the vestibular threshold so that they are not sensed by the flight crew;
- Minimize the time lag between a pilot input and the motion platform response – typically, a maximum value of 150 ms is recommended by the regulatory authorities;
- Synchronize the motion response to the visual system to minimize the relative lag between the motion system and the visual system.

These are demanding engineering requirements and consequently, a modern motion system will include:

- Hydrostatic bearings to reduce the actuator friction and minimize any mechanical 'noise' as the actuator changes direction;
- Non-contacting sensors to detect actuator position and actuator velocity;
- A hydraulic drive system with a fast response in acceleration and velocity;
- A high-speed servo valve controller for accurate control of the actuator;
- A fast processor to compute the equations of motion.

10.4 Equations of Motion

A motion platform comprising six linear actuators can be considered as a set of dynamic systems (Cardullo and Kosut, 1979), where the response of each actuator contributes to the dynamic response of the platform to motion demands to provide appropriate motion cues. In addition, it is also important to optimize the motion of the platform, given its very limited motion envelope (Advani et al., 1999).

The equations of motions of the aircraft are solved every frame, and the linear and angular body accelerations are passed to the motion system computer, which has to determine the position of each actuator to achieve as close a match as possible to the accelerations that should be felt by the flight crew. Although the flight model equations of motion are computed at 50 Hz, much higher computation rates (typically 500–1000 Hz) are required to solve the motion equations for hydraulic actuators (Nahon and Reid, 1990). In addition to the computation of the jack lengths (Nanua et al. 1990), appropriate filtering is applied to the platform motion, for example, to wash out any accelerations or to emphasize a specific motion cue.

Two coordinate frames are used in computation of the equations of motion. The base system is an inertial frame, with axes X_I, Y_I and Z_I, with an origin O_I. The platform is a moving frame with an origin O_P at the centroid of the platform, with an axis system X_P forwards, Y_P right and Z_P downwards, as shown in Figure 10.7, where:

O_I is the origin of the inertial motion base coordinate system and is coincident with the platform at rest

Motion Systems

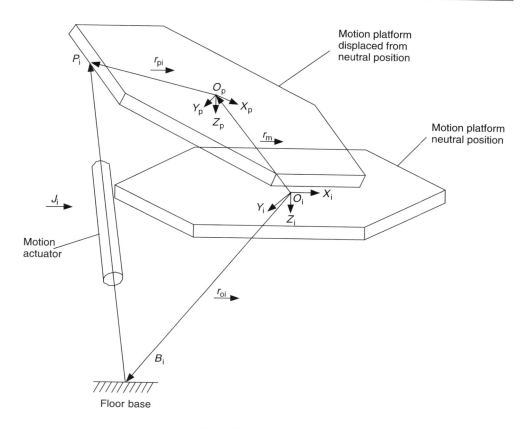

Figure 10.7 Platform axes

O_P is the origin of the platform coordinate system
$\underrightarrow{J_i}$ is a vector defining the length and orientation of the ith jack – this variable is computed for all six jacks to determine the platform position
$\underrightarrow{r_{oi}}$ is the vector from the base axes system to the attachment point
$\underrightarrow{r_m}$ is the vector joining the origins of the two platform coordinate systems
$\underrightarrow{r_{pi}}$ is the vector from the origin of the platform to the jack upper bearing point.

For each of the i jacks,

$$\underrightarrow{J_i} = -\underrightarrow{r_{oi}} + \underrightarrow{r_m} + \underrightarrow{r_{pi}} \qquad (10.4)$$

where
$\underrightarrow{r_{oi}}$ is known – it depends only on the platform dimensions and is fixed for each jack
$\underrightarrow{r_m}$ is given by the three translational positions X_p, Y_p and Z_p
$\underrightarrow{r_{pi}}$ is obtained from the platform attitude ϕ, θ and ψ.

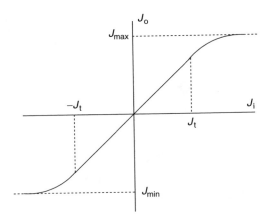

Figure 10.8 Jack soft limits

The motion system equations of motion are given by

$$\begin{bmatrix} J_{xi} \\ J_{yi} \\ J_{zi} \end{bmatrix} = - \begin{bmatrix} X_{oi} \\ Y_{oi} \\ Z_{oi} \end{bmatrix} + \begin{bmatrix} X_m \\ Y_m \\ Z_m \end{bmatrix} + [A] \begin{bmatrix} X_{pi} \\ Y_{pi} \\ Z_{pi} \end{bmatrix} \tag{10.5}$$

The direction cosine matrix (DCM) A is the same as the DCM used to transform from body axes to Euler axes, in Chapter 3. The coordinates X_{pi}, Y_{pi} and Z_{pi} are the coordinates of the upper bearings in platform axes (they are fixed in the platform geometry). The values J_{Xi}, J_{Yi} and J_{Zi} define the position of each jack, enabling the length of each jack to be computed, where

$$J_{Li} = \sqrt{J_{Xi}^2 + J_{Yi}^2 + J_{Zi}^2} \tag{10.6}$$

The value J_{Li} is passed to the actuator controller. To avoid abrupt motion as a jack reaches its limit, soft limits are applied to smooth the jack motion, as illustrated in Figure 10.8. Smoothing is applied when the jack length exceeds J_t.

Although these equations enable the platform to be moved to respond to the accelerations computed in the flight model, the motion is constrained by the lengths of the jacks and also the maximum jack velocity. Consider a continuous longitudinal acceleration of 0.3 g. If this acceleration is applied directly, the platform will reach its maximum velocity (24 in./s) after 0.2 s. If this velocity is then maintained, the jack will reach its limit of travel (56 in., say) within 2 s. If the jack motion is stopped at this position, the sudden (and unexpected) deceleration would be felt by the flight crew. To wash out this motion below the vestibular threshold of 0.015 g, the wash out must be applied after 0.2 s in order not to exceed a jack travel of 56 in., as shown in Figure 10.9, where the upper trace shows acceleration (in./s²), the middle trace shows velocity (in./s) and the lower trace shows position (in.).

The platform is accelerated until the jack reaches 24 in./s, whereupon the jack is decelerated at 0.015 in./s². When the jack velocity reaches 18 in./s (compression), a positive acceleration of 0.015 in./s² is applied. Note that the jack reaches a maximum length of 54 in. and is returned to the neutral position in approximately 10 s. In this example, the flight crew experience an initial surge of 0.3 g, which is washed out over the next 10 s. Having felt an initial acceleration of 0.3 g

Motion Systems

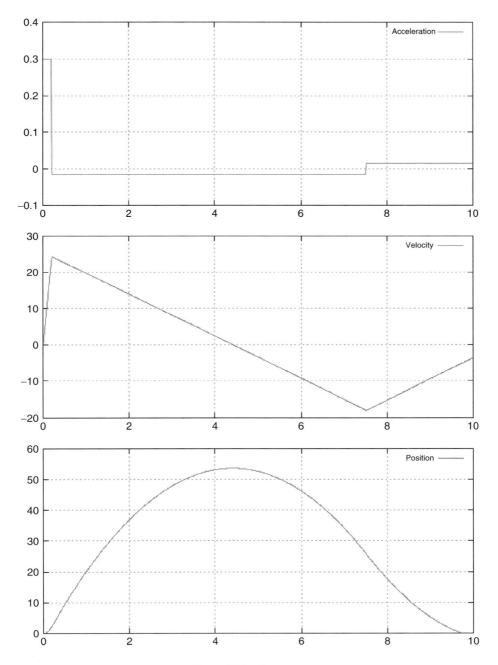

Figure 10.9 Surge cue

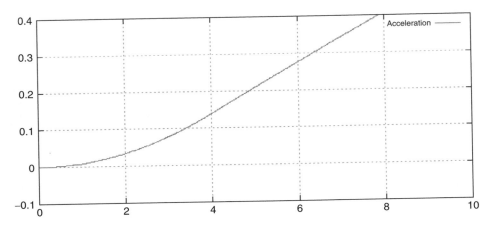

Figure 10.10 Surge cue from platform tilt

and being unaware that it has been removed, the flight crew will feel that the acceleration is still present.

An alternative method to provide a sustained surge of 0.3 g is to tilt the cabin up to a pitch attitude of 17.46°, but with a pitch acceleration below the vestibular threshold, so that no pitching cue is felt. By pitching the cabin at an angular acceleration of $1.0°/s^2$, the flight crew experience an increasing surge acceleration, reaching 0.3 g after approximately 6 s, as shown in Figure 10.10. A similar approach applies to side force. Although sway can be implemented as direct lateral acceleration, the motion must be washed out quickly, whereas side force can be provided by tilting the platform in the roll axis.

10.5 Implementation of a Motion System

The motion system responds to the aircraft linear and angular accelerations in order to compute the most appropriate cabin motion to replicate these accelerations, subject to the displacement limits and the velocity limits of the jacks. Of course, the motion platform computer has no knowledge of the future motion of the simulated aircraft and consequently, it seeks to restore the platform to its neutral position, so that it is positioned to generate new motion cues. The aircraft accelerations are transformed from aircraft cg axes to the pilot position and finally to the inertial frame (the floor of the simulator room) in order to compute the jack lengths. The cabin accelerations are filtered so that the cabin motion never exceeds the mechanical limits of the motion platform, particularly the maximum jack displacements and the maximum jack velocities.

Sustained cues are provided by tilting the cabin to exploit the gravity vector, but where the rotational accelerations of the cabin (to tilt the platform) are below the vestibular thresholds. However, the initial or onset cues can be implemented as translational accelerations of the platform, until the desired tilt angle is obtained. Consequently, it is necessary to combine the translational and rotational motion; on the one hand, to replicate the aircraft motion and the other hand, to avoid introducing any spurious motion cues. If an angular acceleration is applied to the cabin when it is tilted, a false cue may occur because the actual gravity vector is misaligned and an additional translational acceleration is needed to correct for this error.

The computed jack positions are passed to the actuator controllers. Each controller computes the actuator inputs to drive the actuator to its new position, without violating any mechanical limits of the actuator. The overall system is shown in Figure 10.11.

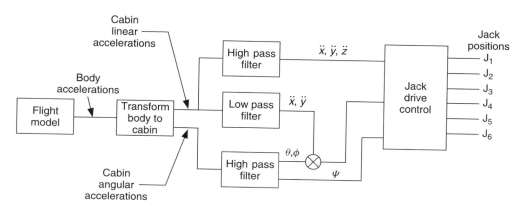

Figure 10.11 Motion algorithm block diagram

The high-pass filter responds to sudden changes to provide onset cues as linear accelerations, but this motion is quickly attenuated to limit the jack displacement. Sustained motion, which can only be represented by tilting the cabin (the gravitational acceleration vector), is achieved by the low-pass filter, providing the relatively slow cabin motion as angular accelerations. The actual filter coefficients vary for the six axes and depend on the jack configurations and also on compromises, which are appropriate to the training application, for example, where take-off and landing cues are considered to be important. In practice, although the simulator manufacturer attempts to match the response of the motion platform to the aircraft motion, this is constrained by the jack displacements and moreover, during the simulator acceptance, the operator may request that specific cues are emphasized or attenuated.

The high-pass filter can be implemented as a second-order system, as shown in Figure 10.12. The input acceleration is limited to 1 G. As the velocity increases, it is fed back to reduce the acceleration input. Similarly, as the position of the actuator increases, this term is fed back to reduce the platform acceleration and velocity to zero. The final integrator term is used to bleed the actuator position back to zero. The effect of the high-pass filter is to generate the onset motion cues, but to wash out these cues before the actuator reaches its end of travel (Dieudonne et al. 1975).

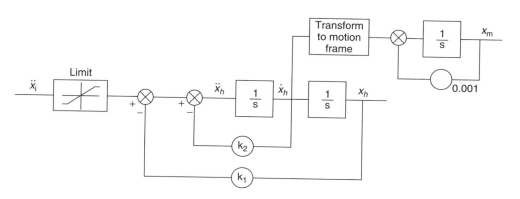

Figure 10.12 Second-order high-pass filter

The second-order form of a high-pass filter is given by

$$\ddot{x}_h = \left(\frac{s^2}{s^2 + 2\delta\omega_n s + \omega_n^2} \right) \ddot{x}_i \tag{10.7}$$

or

$$x_h = \left(\frac{s^2}{s^2 + 2\delta\omega_n s + \omega_n^2} \right) x_i \tag{10.8}$$

where
  x_i is the input position
  x_h is the output position
  δ is the damping ratio
  $2\delta\omega_n$ corresponds to the viscous damping of the actuator
  X_m is the motion platform position (after transformation).

The two feedback coefficients are given by

$$k_1 = \omega_n^2 = \frac{a_{max}}{x_{max}} \tag{10.9}$$

$$k_2 = 2\delta\omega_n = \frac{a_{max}}{v_{max}} \tag{10.10}$$

where, for a specific axis
  a_{max} is the maximum acceleration of the platform
  v_{max} is the maximum velocity of the platform
  x_{max} is the maximum displacement of the platform.

The effect of a high-pass filter on the platform motion is shown in Figure 10.13. The top trace shows a surge acceleration demand increasing to 0.3 g after 2 s. The lower trace shows the platform longitudinal acceleration. The initial acceleration reaches 0.04 g before it is washed out. The platform forward velocity reaches approximately 7 in./s before it is washed out to zero and the platform displacement is approximately 18 in. Notice that the platform does not exceed its limits of 1 G (acceleration), 39 in./s (velocity) or 60 in. (displacement), respectively, in the longitudinal axis. In practice, a higher-order filter is likely to be used to improve the response and in many installations, the gain of the applied signal is varied with the platform displacement (Reid and Nahon, 1986), to provide an adaptive response.

A third-order high-pass filter, with a third pole added at $-\omega_1$, takes the form

$$x_m = \left(\frac{s^3}{(s^2 + 2\delta\omega_0 s + \omega_0^2)(s + \omega_1)} \right) x_i \tag{10.11}$$

The schematic of a third-order high-pass filter is shown in Figure 10.14, where the three feedback coefficients are given by

$$k_1 = \omega_1 + 2\delta\omega_0 \tag{10.12}$$

$$k_2 = 2\delta\omega_0\omega_1 + \omega_0^2 \tag{10.13}$$

$$k_3 = \omega_0^2\omega_1 \tag{10.14}$$

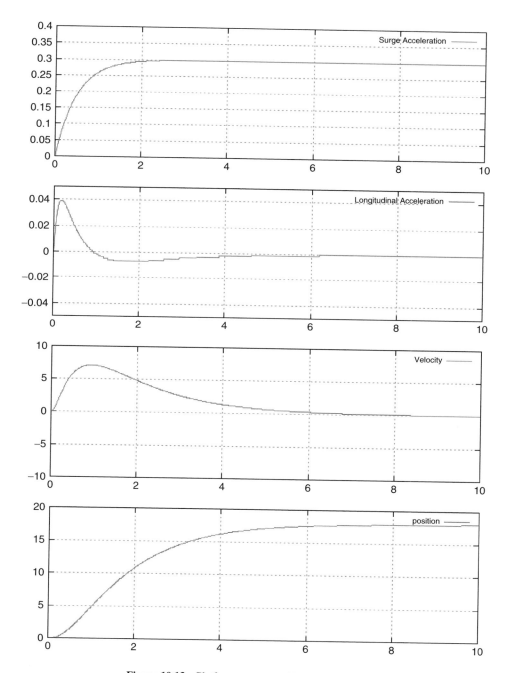

Figure 10.13 Platform response with high-pass filter

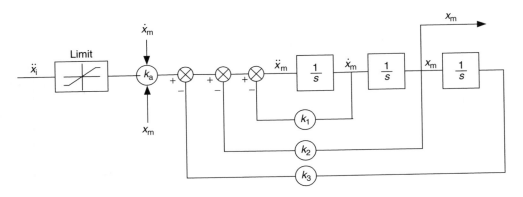

Figure 10.14 Third-order high-pass filter

The gain k_a depends on the velocity of the jack and the displacement of the jack and is continuously computed, providing an adaptive gain.

A low-pass filter is used for pitch and roll motion of the platform. Linear acceleration in the flight model results in increasing angles of pitch or roll to provide longitudinal and lateral accelerations, respectively. The low-pass filter takes the form

$$\theta = \left(\frac{\omega^2}{s^2 + 2\delta\omega s + \omega^2} \right) \tan^{-1} \left(\frac{\ddot{x}_i}{g} \right) \tag{10.15}$$

Note that, in the steady-state, $\tan \theta = \ddot{x}_i/g$, where g is gravitational acceleration. In other words, the cabin is rotated until the desired acceleration is achieved. Of course, this value is limited to 1 g. In addition, there is no transformation to the cabin frame, because the cabin pitch and roll angles correspond to Euler angles.

A typical (second-order) low-pass filter is shown in Figure 10.15. As before, the feedback coefficients k_1 and k_2 are given by

$$k_1 = \omega^2 \tag{10.16}$$

$$k_2 = 2\xi\omega \tag{10.17}$$

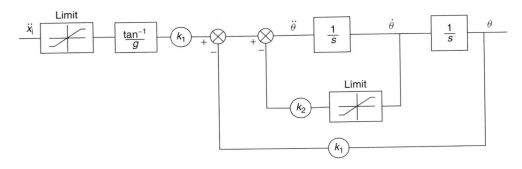

Figure 10.15 Second-order low-pass filter

Motion Systems

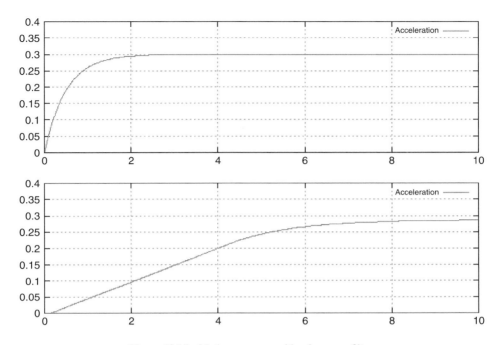

Figure 10.16 Motion response with a low-pass filter

The motion system response with a low-pass filter is shown in Figure 10.16, for an applied acceleration increasing to 0.3 g after approximately 2 s. The upper trace shows the input acceleration and the lower trace shows the resultant platform motion.

These examples are simply provided to illustrate the equations of motion which form the basis of the computations for the motion platform dynamics. In practice, considerable effort is given to the selection of filter coefficients and the design of high-order algorithms to attempt to replicate aircraft cues as accurately as possible. By understanding the dynamics of the motion sensors of the human body and constraining the motion to the physical limits of the actuation system, it is still possible to provide realistic motion cues (Cardullo and Telban, 1999). However, the validity of the motion cues should not be considered in isolation. A wide-angle visual scene provides very powerful vection cues, which can help to enforce the motion cues felt by a pilot, in the sense that the pilot may be less aware of any wash-out motion in the presence of powerful visual cues.

There have been many attempts in the literature to improve motion cues, in particular, the provision of improved wash-out cues (Parrish *et al*. 1975) and adaptive motion algorithms (Dorbolo and Van Sliedregt, 1987). Some of these methods have been driven by a need to reduce the mass of the platform, or to provide less powerful electrical drive systems or to reduce the number of degrees of freedom for applications where the motion is restricted, for example, to limit the heave cues for land vehicles. High-frequency motion cues also occur in aircraft, resulting from vibration (e.g. helicopters), turbulence and ground manoeuvring, which may be difficult to implement with a classical motion platform; high-frequency inputs are demanding in terms of actuator bandwidth and can also induce vibration in other simulator systems (e.g. the projection screen), which are undesirable. In these applications, additional vibration cues are usually provided by a G-seat (Keirl *et al.*, 1995), where vibrations and pressure are applied locally to the pilot seat, enhancing G cues and providing pilot haptic sensations derived from aircraft accelerations. In some cases, and with

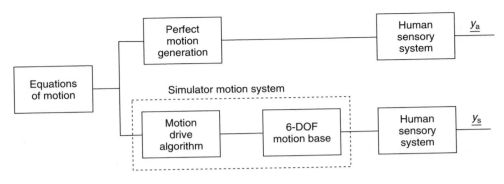

Figure 10.17 Motion cue modelling

the approval of the regulatory authorities, a manufacturer may limit the vertical acceleration cues (particularly felt in heavy landings) to protect both the simulator equipment from impact damage and also to avoid injury to the flight crew.

One method of implementing platform motion that has received considerable interest is to determine the difference between the actual accelerations felt by the flight crew in an aircraft and the accelerations sensed in a flight simulator. These differences can be formulated as a cost function, which is used to adapt the wash-out gain values to attempt to match the simulated accelerations to the airborne accelerations. The cost function takes the following form:

$$J_x = (a_a - a_s)^2 + w_1 v^2 + w_2 x^2 \qquad (10.18)$$

where, for each axis
    J_x is the cost function for the xth axis
    a_a is the aircraft acceleration
    a_s is the simulator acceleration
    v is the velocity of the motion platform
    x is the displacement of the motion platform from the neutral position.

Reid (1984) investigated a motion algorithm based on modelling motion cues, as shown in Figure 10.17. The method attempts to drive the simulator motion vector y_s as closely as possible to the actual motion experienced in an aircraft, y_a. Having derived detailed models of the human sensory systems, the motion drive algorithms (particularly the coefficients of the high-pass and low-pass filters) were adapted to match the ideal response as closely as possible. However, simulator trials with experienced pilots showed a wide variation of acceptance and the weighting factors were adjusted empirically rather than theoretically. There are several useful observations of the outcome of this programme that apply to most motion systems:

- The model of the human vestibular system may not be accurate;
- There is a wide variation in the response of the vestibular systems of individual pilots;
- Pilots may sense motion from other cues as well as the vestibular system;
- Motion in the air is very different from motion on the ground in terms of subjective assessment, particularly for braking and steering on the ground;
- Eventually, a motion trial requires the subjective analysis of pilots, with potential variability in the assessment technique;

- The coupling between pitch and surge and between roll and sway made it very difficult to tune the motion cues and the adaptive gains were limited to the high-pass filters.

Consequently, during acceptance of a simulator, the manufacturer may need to 'tune' the motion platform to the requirements of the airline test pilots. Some pilots prefer strong cues, which are washed out, whereas others prefer to minimize false cues, which may require attenuating some of the motion cues. Airlines, manufacturers and regulators are aware of this problem, and the simulator qualification process does try to minimize the variation that can occur in acceptance tests. In addition, the regulatory teams include pilots with a wide experience of both the aircraft and equivalent flight simulators operated by other airlines. One observation is that pilots undertaking a standard manoeuvre may respond in very different ways. Some pilots are considered to be 'high-gain' pilots, with frequent large inputs, whereas others are considered as 'low-gain' pilots, with relatively small and less frequent inputs. Although both groups of pilots may operate the aircraft safely, each group may be using different cues or have very different perception of motion cues. However, there is the likelihood of much higher variability with assessments by high-gain pilots, possibly because they generate many more motion inputs.

10.6 Hydraulic Actuation

Hydraulic actuators are almost ideal candidates for flight simulation motion systems:

- The technology is mature and well proven – there are many companies manufacturing hydraulic actuators covering a wide range of industrial applications;
- They are capable of generating the forces needed to balance and move a simulator motion platform at relatively fast rates;
- It is straightforward to connect a linear actuator to the platform frame;
- The high-pressure supply of hydraulic fluid can be connected to the actuator by means of flexible hoses;
- They can be controlled by computers.

A hydraulic actuator of the motion platform of a full flight simulator is shown in Figure 10.18. A simple form of hydraulic actuator is shown in Figure 10.19. Hydraulic fluid is pumped at pressure into the cylinder. In Figure 10.19, if fluid is pumped into the cylinder via the left-hand valve, the piston will move to the right. Alternatively, if fluid is pumped into the cylinder via the right-hand valve, the piston will move to the left. As the piston moves, pressure is removed from the non-active side of the piston and hydraulic fluid is returned to a reservoir. The force produced by the piston is given by

$$f = pa \qquad (10.19)$$

where p is the pressure and a is the piston area. One obvious problem with this arrangement is that the area of the piston is much higher at the cap end than at the rod end. Special valves are used, otherwise a different oil flow would be needed to extend or retract the actuator. Moreover, there may be a discontinuity as the piston reverses direction. An early solution to this problem was to provide a symmetric arrangement with a rod on both sides of the piston. The main drawback is that one rod is attached to the platform, but the other rod sticks out of the base of the cylinder, requiring a large pit beneath each actuator, for clearance.

The modern method of actuation is to use hydrostatic bearings, as shown in Figure 10.20, and compensate for the difference in areas of the piston by means of the servo valve used to control the flow of oil into the cylinder. To extend the actuator, hydraulic fluid is pumped at high pressure

Figure 10.18 A motion platform hydraulic actuator (Courtesy: CAE)

into the cylinder at C_1. To retract the actuator, hydraulic fluid is pumped at high pressure into the cylinder at C_2. With a normal hydraulic actuator, there is friction between the piston and the cylinder wall, which introduces a small dead band as the actuator changes direction (i.e. at very low actuator velocity), which results in a noticeable jolt (known as the *turnaround bump*). The hydrostatic bearing, where the piston rod is kept in place by pumping hydraulic fluid at high pressure at P, effectively eliminates friction. The piston is smaller than the cylinder, so that there is always oil (at high pressure) between the piston and the cylinder wall. The piston is machined with a drain hole running the length of the piston, so that oil is forced through the piston enabling this continuous leakage to be drained. Similarly, high-pressure oil is supplied to the hydrostatic bearing; this oil leaks into the cylinder and also into the drain. A low-pressure low-friction seal prevents oil leaking from the actuator assembly.

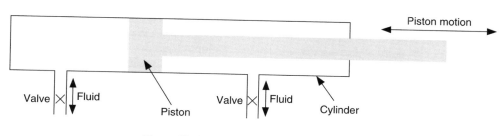

Figure 10.19 Conventional hydraulic actuator

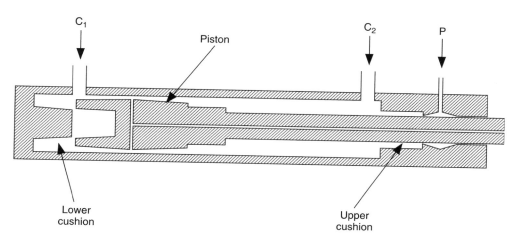

Figure 10.20 Actuator with hydrostatic bearing and hydraulic safety cushions

One further point to note is the shape of the piston. With hydraulic pressures of 1500 psi, the actuator can travel very fast over a relatively small distance and considerable attention must be given to the safety of the flight crew in the simulator cabin. The piston is designed so that there is a cushion of hydraulic fluid at each end of the cylinder. Near the end of the piston travel, the piston enters the cushion region, where the oil is compressed, providing resistance and avoiding the situation where the piston comes into contact with the end of the cylinder. In addition, the actuator control system continuously monitors the actuator position and velocity and can inhibit violent or unintended accelerations. Typically, a sonic sensor is mounted in the cylinder base to provide non-contacting position and velocity measurement. Sound waves are reflected from the piston, where the delay provides a measurement of position and the Doppler shift provides a measurement of velocity.

Typically, the cap end is designed so that it has exactly twice the area of the rod end of the piston. Consider the piston shown in Figure 10.21, where the simulator weighs 31,500 lb and is supported by six actuators. Assume that the hydraulic pressure available is 1500 psi. In the neutral position (all six actuators at their mid-position), 750 psi is applied to both sides of the piston. The upward force $= 14 \times 750 \times 6 = 63,000$ lb. The downward force is $7 \times 750 \times 6 = 31,500$ lb. The actuators can therefore support the 315,000 lb weight of the simulator. If 1500 psi is applied to the cap end of the piston (with 0 psi at the rod end), the maximum up force $= 14 \times 1500 \times 6 - 31,500 = 94,500$ lb. Similarly, if 1500 psi is applied to the rod end of the piston (with 0 psi at the cap end), the maximum down force $= 7 \times 1500 \times 6 + 31,500 = 94,500$ lb.

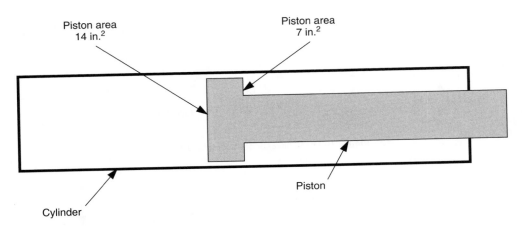

Figure 10.21 Unequal area actuator

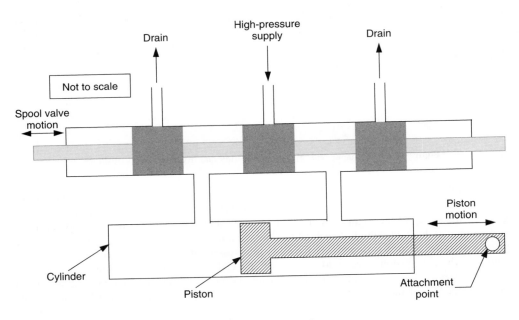

Figure 10.22 Spool valve

The platform motion actuation is computed by a real-time processor, solving the equations of motion for the platform at rates concomitant with the time constants of the hydraulic actuators, typically 500–1000 Hz. At the actuator, the controller applies high pressure from a hydraulic pump to the two inputs, on each side of the piston. This metering of hydraulic fluid is provided by means of a servo valve, which is also known as a spool valve, as shown in Figure 10.22.

In this double-acting arrangement, if the spool valve is moved to the right, the high-pressure supply of hydraulic fluid is applied to the region above the piston, extending the piston rod out of the cylinder. Similarly, if the spool is moved to the left, the high-pressure supply of hydraulic fluid

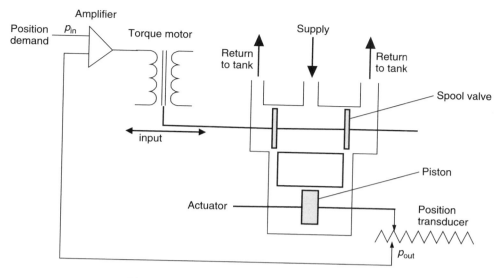

Figure 10.23 Electrical actuation of a spool valve

is applied to the region below the piston, retracting the piston rod. The piston rod is connected to the platform (via a universal joint) at the attachment point shown in the diagram. However, the spool valve also opens and closes the ports to the cylinder, providing a means of controlling the flow rate of oil into the cylinder. Consequently, the spool valve acts as a very high-gain amplifier. A small displacement of the spool valve results in a change of flow rate in the cylinder, causing the piston to move. The precision of the spool valve is critical to the smoothness and accuracy of the control system and not surprisingly, spool valves used in flight simulation are very expensive. The power needed to move the spool valve is small and therefore, an electrical linear actuator (Walters, 1991) can be used for the spool valve. A schematic of an electrical spool valve is shown in Figure 10.23.

Rather than use an open-loop controller, the position transducer signal is fed back to the torque motor amplifier to improve the response and stability. As a further safety precaution, mechanical springs at each end of the spool valve return the spool valve to the mid-position in the presence of an electrical failure.

One further point to note with a double-acting hydraulic actuator is that when the platform is stationary, the piston is a sealed system; the two input ports are covered by the spool valve and the high pressure on both sides of the piston is maintained, leaking away very slowly (over several hours). In other words, loss of pressure at the supply will not result in hazardous motion of the platform. The system is irreversible; the weight of the platform cannot move the piston when the actuator is at rest.

10.7 Modelling Hydraulic Actuators

In much the same way that an electrical engineer will analyse a circuit, using computer-aided circuit simulation tools to model the circuit components before committing to a specific design, a mechanical engineer will analyse the behaviour of an actuator, particularly as the cost of construction of a high-performance hydraulic actuator can be very high. In addition to the physical geometry of the actuators, the performance of an actuator, in terms of its bandwidth and response (Dorf and

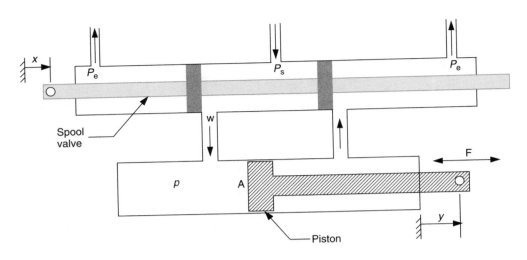

Figure 10.24 Model of a hydraulic actuator

Bishop, 2008), is critical in flight simulation. A slow response of the hydraulic system can reduce the effect of motion onset cues. Similarly, high gains in the system, introduced to improve the system response may lead to instability or produce undesirable motion cues.

Consider the typical spool valve arrangement (McLean, 1990) shown in Figure 10.24. The shaft of the spool valve is moved to close or uncover the inlets to the piston. The simulator motion platform is connected to the end of the piston rod. Note that high-pressure oil can be applied on either side of the piston. In Figure 10.24, the piston is being extended. If oil is fed to the cap end, the piston extends to support the cabin and oil in the rod end of the cylinder is returned to the drain. Similarly, if oil is fed to the rod end, the piston retracts, enabling the cabin to be lowered with an acceleration greater than $1\,g$.

Clearly, if an orifice is completely covered by the spool valve, no oil can flow into the piston. Similarly, if an orifice is completely uncovered, the oil flows at the maximum rate, depending on the pressure of the supply. If the spool valve is moved by an amount x, it will uncover a proportion of the orifice and ideally, the relationship between the spool valve position x and the flow rate into the piston should be linear. In practice, this relationship depends on the shape of the orifice and the shape of the spool valve. Generally, it is assumed that the flow rate is proportional to the displacement

$$q = K_x x \qquad (10.20)$$

where q is the flow rate and K_x is a constant for a specific valve and supply pressure.

At the piston, assuming there is no leakage, the flow rate must equal the rate of change of volume

$$K_x x = A\dot{y} \qquad (10.21)$$

$$\frac{y(s)}{x(s)} = \frac{1}{s}\left(\frac{K_x}{A}\right) \qquad (10.22)$$

In other words, the spool valve acts as an integrator. The piston moves at a velocity proportional to the spool valve displacement, similar to a DC servo motor where speed of rotation is proportional to the applied voltage.

Motion Systems

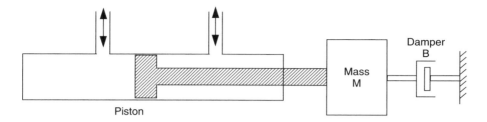

Figure 10.25 Piston with a load

In the case of a motion platform, there is a load on the piston resulting from the mass of the platform (including the cabin, projectors and on-board computers, etc.) and also from any damping in the piston caused by friction. Of course, with hydrostatic bearings, the damping will be very low, although there is also a small contribution from friction in the mounting bearings. The schematic of the piston and its load is shown in Figure 10.25. It is also assumed that the platform is a rigid structure and therefore, there is no spring term associated with the mass/damper load.

Although the flow rate depends on the displacement of the spool valve, it will also vary with the pressure differential, in this case $p - p_s$. Assuming the supply pressure is constant, Equation 10.20 is modified as follows:

$$q = K_x x - K_p p \qquad (10.23)$$

where K_x and K_p are constants taken from the valve characteristics and p is the pressure in the cylinder. The force produced by the piston is pa (Equation 10.19), where A is the area of the piston, which matches the load, giving

$$pA = m\ddot{y} + b\dot{y} \qquad (10.24)$$

or in Laplace terms

$$pA = ms^2 y + bsy = s(ms + b)y \qquad (10.25)$$

Equation 10.20 ignores any compressibility of the oil resulting from the change in pressure in the cylinder and is modified to include this factor as follows:

$$q = A\dot{y} + \frac{V}{\beta}\dot{p} \qquad (10.26)$$

or in Laplace terms

$$q = Asy + \frac{V}{\beta}sp \qquad (10.27)$$

where V is the volume of oil in the cylinder and β is the bulk modulus of the oil. Combining Equations 10.23, 10.25 and 10.27

$$\frac{y(s)}{x(s)} = \frac{K_x}{s\left(s^2 \frac{mV}{A\beta} + s\left(\frac{K_p m}{A} + \frac{bV}{A\beta}\right) + \frac{K_p b + A^2}{A}\right)} \qquad (10.28)$$

In other words, the transfer function represents an integrator ($1/s$) and a second-order system. For hydraulic actuators used with hydrostatic bearings, the damping b is very small and can be neglected, simplifying Equation 10.28.

$$\frac{y(s)}{x(s)} = \frac{K_x}{s\left(s^2 \frac{mV}{A\beta} + s\left(\frac{K_p m}{A}\right) + A\right)} \qquad (10.29)$$

For small actuators, the ratio of the volume to the bulk modulus is small and the first coefficient can also be dropped, reducing the transfer function to a first-order response, plus an integrator. Although a much more detailed derivation of the transfer function, which takes into account additional terms associated with hydraulic actuation, is given by Merritt (1967), several of these terms are non-linear; consequently, they are linearized about an operating point, to enable the system stability and response to be determined.

In a general form, Equation 10.29 can be expressed as

$$\frac{y(s)}{x(s)} = \left(\frac{1}{s}\right) \frac{K\omega_n^2}{s^2 + 2\xi\omega_n s + \omega_n^2} \qquad (10.30)$$

where

$$\omega_n = A\sqrt{\frac{\beta}{mV}} \qquad (10.31)$$

$$\xi = \frac{K_p}{2A^2}\sqrt{\frac{mV}{\beta}} \qquad (10.32)$$

The motion platform control system will compute the desired actuator positions, typically at 400–500 Hz and an internal controller will control each actuator, at rates up to 5000 Hz, as shown in Figure 10.26. The input to the control system is the demanded actuator position p_d, which is computed for each actuator. The controller (typically a PID controller) computes the actuator position based on the current actuator position and the actuator velocity, derived from the actuator sensors measuring position p and velocity v. The actuator position is converted to a signal for the spool valve solenoid (a small torque motor). Although the motor lag is shown in Figure 10.26, spool valve solenoids are designed with a short time constant and in comparison with the actuator time constant, can be neglected.

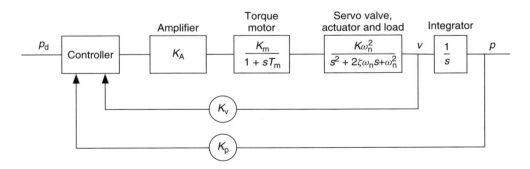

Figure 10.26 Actuator control system

10.8 Limitations of Motion Systems

The major limitation of the conventional motion platform is the lack of sustained acceleration cues. A pilot in an aircraft banked at 60° in a sustained turn will experience a downward acceleration of 2 G (perpendicular to the base of the seat). Although a motion platform can provide initial acceleration cues, these must be washed out quickly, owing to the limited travel of the platform jacks. However, in the steady-state, the motion platform is limited to 1 G. Pilots in an aircraft, banked at 60° for 30 s, say, will notice that their head feels heavy if moved or that their hand and arm feel heavy if a switch is selected. For pilots banked at 60° for 30 s in a flight simulator, the sensation of acceleration will be no different from sitting in a chair in an office.

For airline simulation, the limited acceleration cues are not seen to be important. In normal operations, a civil transport aircraft is unlikely to exceed a bank angle of 45° or accelerations of 1.4 G. The surge acceleration and deceleration on take-off and landing, respectively, are easily accomplished by tilting the platform in pitch. There are three main areas where the conventional motion platform is lacking for the training of civil pilots:

- The provision of high-frequency cues, for example, engine vibrations during engine malfunctions, tyre bursts, structural vibrations or during turbulence;
- In ground handling where the interaction of aerodynamic forces, undercarriage dynamics, steering and braking can produce high lateral accelerations;
- In practising high-attitude manoeuvres or upset recovery, where the aircraft may be subject to accelerations up to 3 or 4 G, and large pitch and roll displacements.

The generation of high-frequency accelerations by six hydraulic jacks would be very demanding of the bandwidth of the actuators (increasing significantly the cost of the motion platform) and would also place additional requirements on on-board systems, to ensure their immunity to vibrations (increasing the cost of electronic equipment installed in the simulator cabin). In recent years, considerable effort has been given to the improvement of ground handling cues, particularly to model the effects of icing, water on the runway and tyre bursts. Recent attention has also been given to upset recovery, following the malfunction of automatic flight control systems and incidents of wake vortex encounters. The accepted view was that civil pilots did not need aerobatic skills (e.g. spin recovery) because they would never encounter such conditions, although basic aerobatics had been included in the flight training syllabus up to the 1980s. As upset recovery techniques are based on visual cues provided by the flight instruments, the need for realistic motion for such training in a flight simulator is debatable. Airlines do practise for such eventualities, but the emphasis is on recovery techniques using flight instruments.

The situation in military training is rather different. Pilots are subject to high-G cues and also to high-frequency vibrations, for example, from the stall buffet or the firing of guns. Realizing the limitations of the conventional 6-DOF platform, military training organizations have elected to install:

- Fixed-base simulators;
- Wide-angle high-resolution projection systems, particularly hemispherical domes;
- G-seats.

On the assumption that incorrect motion cues are worse than no motion cues and that visual cues can dominate in high-rate manoeuvres, few moving platforms have been installed in military flight simulators since the 1980s, and the G-seat is now widely accepted as an effective training aid (Ashworth *et al*. 1984). Actuators in the seat can press the pilot into the harness for positive G cues and relax the harness pressure (by lowering the seat pan) for negative G cues. During high-G manoeuvres, the pilot eye line is also lowered slightly, to simulate the equivalent effect in airborne manoeuvres. Additional surge and sway accelerations are provided by haptic cues, where pressure is applied by appropriate pads in the seat for longitudinal and lateral accelerations. In addition, the seat can be vibrated (by attaching it to an off-centric shaft driven by an electric motor) to simulate high-frequency accelerations, for example, runway rumble, helicopter rotor interactions, weapons release and stall buffet, where the seat is mechanically isolated (by rubber mounts) from the other cockpit systems to minimize the effect of vibration on electronic equipment.

Although some consideration has also been given to the use of centrifuges to simulate high-G accelerations, these devices are mostly limited to experimental rigs used in aeromedical studies, for the following reasons:

- The device is large (50-ft arm) and expensive;
- As the cabin is rotating, electrical signals have to be passed via slip rings, further increasing the cost and complexity;
- While they can provide sustained G cues, fast variation in acceleration can only be achieved by accelerating or decelerating the centrifuge, requiring very powerful motors;
- Once rotating, the minimum G value is about 1.6 G – to simulate 1 G, the centrifuge must be at rest.

The other major problem with motion platforms is latency. In an aircraft, as the pilot moves the control surfaces, there is an instantaneous change in the aircraft accelerations. However, in a flight simulator, the pilot inputs are acquired as analogue signals, the equations of motion are computed to derive the linear and angular accelerations, velocities and positions, which are transmitted to the instrument display computers, the visual system computers and to the motion platform computers. Consequently, there is a lag, referred to as *latency*, between the time a pilot applies an input and the simulator responds to provide both visual and motion cues. Clearly, there is a limit to the amount of latency that can be tolerated, before it is perceived by a pilot or alters the dynamics of the various loops that include the pilot. The maximum value is 150 ms for civil simulators qualified by the FAA. However, the FAA also requires that the motion response occurs before the visual system response. Most military operators require latency to be less than 100 ms.

However, latency is inevitable in flight simulators. There is a finite time to acquire pilot inputs, to compute the equations of motion and to transmit data between computers. For visual systems, there is a further delay as the image is rendered and then projected. For motion platforms, the latency is also increased by the response of the mechanical actuation systems, which include the spool valve response, the hydraulic actuator response and any computation of the platform orientation. As with all dynamic systems, the introduction of a lag in any path can introduce non-linearity, altering the system behaviour and potentially introducing instability. Some simulator designers have attempted to implement delay compensation methods (Merriken, 1988) where the future state of the motion platform is predicted to reduce the perceived latency. However, with a pilot in-the-loop, the future pilot inputs cannot be assured and such systems are prone to introducing adverse cues.

With the use of modern processors and high-speed networks, many of the sources of latency have been reduced. Nevertheless, the time constant of large hydraulic actuators can be as high as tens of milliseconds. The response and power of electrical actuation systems is less than equivalent hydraulic systems, offering very little scope to reduce the actuator latency. Consequently, with flight model iteration rates and visual system update rates of 50 Hz and hydraulic control system update rates in excess of 500 Hz, considerable care is needed to match the latency of the motion platform and the visual system (including projection). In cases where this has not been achieved, cases of pilot nausea (often referred to as *simulator sickness*) and pilot-induced oscillations have been reported (Gower and Lilienthal, 1988; Kennedy *et al.*, 1989; Kolasinski, 1995). Simulator sickness occurs because the brain detects a conflict between the visual sensors and the motion sensors and assumes the effect has resulted from some form of poisoning. Pilot-induced oscillations occurring in a flight simulator are also of concern, because it implies that the pilot behaviour in the simulator is different from the aircraft. In such cases, the effect of potential negative transfer of training needs to be investigated to avoid pilots using techniques learnt in the simulator, which might be dangerous if applied in the aircraft.

One other major consideration with motion platforms is flight crew safety. The simulator cabin can move through 2 m with considerable acceleration, particularly if the actuator is commanded incorrectly, with a risk of major injury to the flight crew. Consequently, two or more parallel channels of computation are used; the outputs of each system are compared continuously and in the event of any discrepancy between these systems, the power is removed in such a way that the platform is restored to a safe state. In addition, each computer monitors the actuators and if any unexpected motion is detected, the platform motion is switched off. As normal access to the cabin is via a door at the back of the simulator and a gangway, which is withdrawn when the motion system is active, an additional emergency exit is provided to enable the flight crew to exit through the simulator windscreen (or floor) via a rope ladder.

For simulators with a high cg, the worst-case loading occurs when the front jacks are fully retracted and the rear jacks are fully extended (the cabin is pitched down). The actuator cushions (shown in Figure 10.20) are designed for such eventualities, albeit that they occur rarely. They are machined so that there is a progressive trapping of oil in the last 3–4 in. of travel, allowing the platform to come to rest with a constant deceleration, but without generating excessive forces or causing structural damage.

10.9 Future Motion Systems

Motion is the most contentious element of flight simulation. It is expensive to implement, it increases the size of the simulator bay and it can be hazardous in the event of system failures. Moreover the case for (or against) motion has not been proven conclusively. Military organizations have discarded 6-DOF platforms for G-seats without any noticeable reduction in the effectiveness of training. At the same time, the regulatory authorities have been insistent that motion is essential in the training of airline flight crews.

However, motion contributes approximately 20% to the cost of a flight simulator and there are considerable advantages in removing it. To some degree, the cost may be ameliorated by the provision of electrical motion platforms, which offer improved safety and reduced costs of both installation and operation. Having been approved for Level-D full flight simulators, electrical actuators are likely to replace hydraulic actuators as there is a compelling safety case and financial benefit. An electrical actuator for the motion platform of a full flight simulator is shown in Figure 10.27.

For military training, the G-seat already fulfils the major requirements. In training for the physiological effects of G, a centrifuge (not necessarily with an aircraft cockpit) enables pilots to experience and cope with high G, where the effects are closely monitored by medical teams. However, there has been surprisingly little research into the relative training benefits of fixed-base platforms, synergistic platforms and G-seats, and there is the possibility that pilots trained in a fixed-base flight simulator might be just as effective as pilots trained with a synergistic platform. This situation is further complicated by the fusion of a wide-angle visual system with high levels of scene detail and high-resolution projection. Many pilots comment on the sensation of motion experienced in fixed-base simulators with wide-angle visual systems, and there have been many instances where pilots fail to notice that the motion system has been switched off.

Despite these reservations, the main thrust is to improve the algorithms of motion platforms, in an attempt to replicate the accelerations experienced in aircraft, in particular, to better understand the vestibular system and the sensor fusion applied by the brain. In an industry that continually strives to improve the underpinning technology, motion platforms are likely to be retained for the

Figure 10.27 A motion platform electrical actuator (Courtesy: CAE)

foreseeable future, with research into motion conducted on the few flight simulators operated by research organizations.

References

Advani, S.K., Nahon, A., Haeck, N. and Albronda, J. (1999) Optimisation of six-degrees-of-freedom motion systems for flight simulators. *AIAA Journal of Aircraft*, **36**(5), 753–763.

Allen, L. (1993) Evolution of flight simulation. AIAA Flight Simulation and Technologies Conference, Monterey.

Anderson, S.B. and Morrison, R.H. (1993) Lessons learnt from a historical review of piloted flight simulators. AIAA Flight Simulation and Technologies Conference, Monterey, CA.

Anon. (2008) *The Impact of Flight Simulation in Aerospace*, Specialist Paper, The Royal Aeronautical Society, London.

Ashworth, W.R., McKissick, B.T. and Parrish, R.V. (1984) Effects of motion base and G-seat cueing on simulator pilot performance, NASA *Tech. Paper No. 2247*.

Benson, A.J. and Bodin, M.A. (1965) Interaction of linear and angular accelerations on vestibular receptors in man, Institute of Aviation Medicine, Report No. 323, Farnborough.

Cardullo, F.M. (1983) Old problems/new solutions – motion cueing algorithms revisited. AIAA Flight Simulation Technologies Conference, Niagra Falls.

Cardullo, F.M. (1991) An assessment of the importance of motion cueing based on the relationship between simulated aircraft dynamics and pilot performance: a review of the literature. AIAA-91-2980-CP, pp. 436–447, New Orleans.

Cardullo, F.M. and Kosut, R.L. (1979) A systems approach to the perception of motion in flight simulation. Royal Aeronautical Society Conference 50 Years of Flight Simulation, London.

Caro, P.W. (1977) Platform motion and simulator training effectiveness. Proceedings 10th NTEC Industry Conference, Orlando, 93–97.

Caro, P.W. (1979) The relationship between flight simulator motion and training requirements. *Human Factors*, **21**, 493–501.

Derevyanko, Y.E. and Mylinikod, V.G. (1964) *Biological Effects of Gravitational Acceleration*, translation U.S. Department of Commerce Joint Publication Research Service, No. JPRS25–929.

Dieudonne, J.E., Parrish, R.V. and Bowles, R.L. (1975) Coordinated adaptive washout for motion simulators. *Journal of Aircraft*, **12**(1), 44–50.

Dorbolo, G. and Van Sliedregt, J.M. (1987) Improvements in motion drive algorithms. Proceedings Summer Computer Simulation Conference, San Diego, 721–723.

Dorf, R.C. and Bishop, R.H. (2008) *Modern Control Systems*, Pearson Education.

Gower, D.W. and Lilienthal, M.G. (1988) Simulator sickness in US army and navy fixed and rotary wing flight simulators. AGARD Conference Motion Cues in *Flight Simulation and Simulation Induced Sickness*, Brussels, Belgium. AGARD-CP-433.

Graybiel, A. (1948) Stimulus thresholds of the semi-circular canals as a function of angular acceleration. *American Journal of Psychology*, **61**, 21–36.

Gum, D.R. (1973) Modelling of the human force and motion-sensing mechanisms. Tech. Report AFHRL-TR-72-54, NASA.

Hixson, W.C. and Niven, J.I. (1961) Application of the system transfer function concept to a mathematical description of the labyrinth steady state nystagmus response to semi-circular canal stimulation by angular acceleration. US Navy School of Aviation Medicine, *Report No. 57*.

Howard, I.P. (1986) The vestibular system, in *Handbook of Perception and Human Performance: Sensor Processes and Perception* (eds K.R. Boff, L. Kaufmann and J.P. Thomas), John Wiley & Sons, Inc., New York.

Keirl, J.M., Cook, R.J. and White, A.D. (1995) *Dynamic Seats–A Replacement for Platform Motion?* DRA Report, HMSO, London.

Kennedy, R.S., Lilienthal, M.G., Berbaum, K.S., Baltzely, K.S. and McCauley, M.E. (1989) Simulator sickness in US Navy flight simulators. *Aviation Space and Environmental Medicine*, **60**, 10–16.

Kolasinski, M. (1995) Simulator sickness in virtual environments. *Tech. Report 1027*, US Army Research Institute for the Behavioural and Social Sciences, Virginia.

Lintern, G. (1987) Flight simulation motion systems revisited. *Human Factors Society Bulletin*, **30**(12), 1–3.

Martin, E.L. and Waag, W.L. (1978) *Contribution of Motion Platform to Simulator Training Effectiveness, Contract AFHRL-TR-78-52*, ADA-064-305, USAF Human Resources Laboratory, Williams Air Force Base.

McCauley, M.E. (2006) Do army helicopter training simulators need motion bases? *Report No. TR-1176*, U.S. Army Research Institute for the Behavioural and Social Sciences.

McLean, D. (1990) *Automatic Flight Control Systems*, Prentice Hall International.

Meiry, J.L. (1965) *The Vestibular System and Human Dynamic Space Orientation*, Massachusetts Institute of Technology, Cambridge.

Melville-Jones, G., Barry, W. and Kowalsky, N. (1964) Dynamics of the semi-circular canals compared in pitch, roll and yaw. *Aerospace Medicine*, **35**, 984–989.

Merriken, M.S. (1988) Time delay compensation using supplementary cues in aircraft simulator systems. AIAA Flight Simulation Technologies Conference, Atlanta, GA, 295–303.

Merritt, H.E. (1967) *Hydraulic Control Systems*, John Wiley & Sons, Ltd.

Miller, E.F. and Graybiel, A. (1963) *Vestibular Mechanisms and Vision*, NASA CR-56080, NASA, Washington.

Nahon, M.A. and Reid, L.D. (1988) Response of airline pilots to variations in flight simulator motion algorithms. *Journal of Aircraft*, **25**(7), 639–646.

Nahon, M.A. and Reid, L.D (1990) Simulator motion drive algorithms: A designer's perspective, AIAA. *Journal of Guidance*, **13**(2), 356–362.

Nanua, P., Waldron, K.J. and Murthy, V. (1990) Direct kinematic solution of a Stewart platform. *IEEE Transactions on Robotics and Automation*, **6**(4), 438–443.

Ormsby, C.C. (1974) *Model of Human Dynamic Orientation*, NASA CR-132537, NASA, Washington.

Parrish, R.V., Dieudonne, J.E. and Bowles, R.L. (1975) Coordinated adaptive washout for motion simulators. *Journal of Aircraft*, **12**(1), 44–50.

Parrish, T.D., Houck, J.A. and Martin, D.J. Jr. (1977) Empirical comparison of a fixed base and a moving base simulation of a helicopter engaged in visually conducted slalom runs. *NASA Tech. Report TND-8424*, Hampton, Virginia.

Reid, L.D. (1984) Computer control of flight simulator motion. Proceedings of Canadian Conference on Industrial Computer Systems, 1.11.7, Ottowa, Canada

Reid, L.D. and Nahon, M.A. (1986) Flight simulation motion-base drive algorithms. Report Nos. UTIAS-296, UTIAS-307 and UTIAS-319, Institute for Space Studies, University of Toronto.

Schone, H. (1964) On the role of gravity in human spatial orientation. *Aerospace Medicine*, **35**(8), 764–772.

Stewart, D. (1965) A platform with six-degrees-of-freedom. *Proceedings of the Institution of Mechanical Engineers*, **180**(Part 1, 5), 371–386.

Telban, R.J., Cardullo, F.M. and Houck, J.A. (1999) Developments in human-centered cueing algorithms for control of flight simulator motion systems. AIAA Conference Modelling and Simulation Technologies, AIAA 99-4328, Portland, Oregon.

Van Egmond, A.A.J., Greon, J.J. and Jongkees, L.B.W. (1949) The mechanics of the semi-circular canals. *Journal of Physiology*, **11**, 1–7.

Walters, R.B. (1991) *Hydraulic and Electro-hydraulic Control Systems*. Elsevier Applied Science, London.

Index

2D
  graphics 23, 205, 214, 216, 219–221, 226
  clipping 337
3D
  clipping 346–347
  graphical operation 332
  graphics 23, 216, 219, 223, 226, 336, 347, 350, 376
6-DOF *See* Six degrees of freedom

A/D *See* Analogue-to-digital conversion
ab initio (training) 12, 34, 38
acceleration cue 428–429, 432, 442, 451
accelerometer 93, 263–264, 266, 303
  bias 268–272
  measurement 264, 266, 268–269
acceptance 7, 9, 14, 285, 332, 437, 442
  tests 27, 32, 96, 443
access time 206, 310
active
  flight control 303
  objects 342, 365, 368
  polygons 370
actuator
  cushion 453
  dynamics 430
acuity 28, 287, 332, 375
Adams-Bashforth (integration) 60, 96, 167
adaptive
  motion algorithm 441
  response 438
ADF *See* Automatic direction finding
advance ratio 79, 131–132, 135
Advisory Circular 285, 329
aerodynamic
  data 13, 16–18, 112–113, 117, 127, 405, 411
  forces 17, 19, 97, 99, 115–117, 140, 143, 145, 150, 152, 264, 451
  model 17–18, 19, 66, 291, 347
aerofoil 93, 96, 97, 101–104, 128
aileron 22, 69, 104, 107–109, 111–114, 169, 172–173, 188–189, 260, 262–263, 297, 312, 323, 325, 328, 397, 411–412, 414
air accident
  investigators 9
  statistics 9
air data computer 249, 255
air density 19, 54, 99, 101, 111–113, 128–131, 135, 170, 185, 249, 255, 286, 291
air pressure 19, 98, 249
air temperature 19, 98, 100, 136, 181, 249, 291
airborne flight trials 13
aircraft
  axes 19, 103, 114
  dynamics 4, 20, 42–43, 53, 65, 93, 96, 97–98, 140, 150, 172, 262, 287, 291, 294, 298–299, 312, 314, 322, 328, 428, 455
  instruments 2, 15, 23, 31, 107, 119, 203, 205, 219, 227–228, 246, 350, 428
  motion 24, 54–55, 97, 108, 111, 118–119, 124, 150–151, 247, 264, 332, 336, 425, 436–437
  performance 19, 96, 106–107, 135, 155, 170, 172, 196, 287, 291–293, 328, 381
airline flight simulator 24, 82, 96, 285, 305, 328, 429
airspeed indicator 1, 228–231, 236, 240, 249, 254–255, 286

Principles of Flight Simulation   D. J. Allerton
© 2009, John Wiley & Sons, Ltd

Airy formula 267
algebraic constraint 124
alignment (of accelerometers) 268
all-weather operations 2
alphanumeric text 205
altimeter 1, 23, 100, 227, 232, 236, 239–240, 243, 245, 249, 254–255, 397, 413
altitude hold 182
amplifier 3–4, 5, 54, 67–72, 204, 332, 372, 429, 447
ampulla 426
analogue
  computer 3–4
  data 18
  input 66, 70–72, 74
  interface 72
analogue-to-digital conversion (A/D) 18, 71–74, 299, 302
anchor point 231
angle
  of attack 16–17, 93, 101–105, 107–108, 109, 112–113, 116–117, 128, 136, 145, 152, 170–171, 181, 227, 254, 291, 312, 322, 328, 378
  of incidence 54, 101, 104, 111, 128, 256
  of sideslip 104–105, 112, 116–117, 172, 189, 291, 323, 414
angular
  acceleration 30, 43, 54, 108, 127, 132, 143, 147, 154, 263, 269, 425–427, 429, 432, 436–437, 452, 454–455
  momentum 127
  motion 22, 43, 54, 82, 108, 150, 263, 428–430, 432
anti-aliasing 20, 23, 32, 204, 210–211, 213, 220, 225–226, 231, 235, 237, 375, 399
Antionette 1, 428
approach display 405
approval test guide (ATG) 283, 285
aptitude testing 36
arctan function 197, 251
aspect ratio 205, 243
ATG *See* Approval test guide
atmosphere 19, 98, 100, 154
atmospheric
  attenuation 29
  effects 6, 257
  model 99–100, 129, 143, 255, 286
attitude indicator 108, 119, 228, 232–234, 236–239, 241, 254, 259, 261, 418–419
auditory sensors 27

aural cues 31
auto-land 196, 298
  system 12, 172, 191, 194, 262
automatic
  direction finding (ADF) 24, 33, 196, 255, 256, 274, 378
  flight control 13, 195, 201, 329, 451, 455
  landing system 19, 194
autopilot 1, 163, 172, 182, 185, 197, 262, 298, 329, 397
  functions 1, 172, 298, 397
auto-throttle 23, 176, 177, 179, 196, 316
auto-trim 294
avionics systems 7, 13, 38, 282, 328
axes
  conventions 124, 348
  systems 51, 53, 55, 114–115, 150, 264
  transformations 116, 119, 263
azimuth 259, 374

back face
  elimination 339
  removal 365
bandwidth 5, 57, 83, 85, 88, 160, 262, 332, 335, 441, 447, 451
bank angle
  controller 185
  pointer 237
barometric pressure 23, 100, 227, 240, 249
baseline 286–287, 297
  model 36, 305
beam-splitter 371
bezel 205, 226–229, 232–234
bias compensation 71
billboard 363
binary weighted value 68, 74
binary-spaced partition 368
binocular vision 28–29
bitmapped
  character 211
  font 212–213
blade angle 79, 128–129, 131, 132, 135
blending 368
  function 231
body
  frame 52, 55, 103, 115, 117, 119–121, 123, 139–143, 145, 147–150, 152–154, 171–172, 249, 255, 264, 266–268, 313, 333, 336, 360
  rate 55, 121–123, 127, 146–147, 154, 264, 266, 268

Index

brake 18–19, 43, 51, 69, 105–106, 139, 141, 227, 288, 291, 294
braking friction 140–141
breakout friction 140–141
Bresenham 207, 210, 340
BSP-tree 369
buffet 8, 102, 451–452
bumpless transfer 168–169

CAA *See* Civil Aviation Authority
calibrated airspeed (CAS) 249
calibration software 373
calligraphic display 6, 204
cancel button 388, 400
Cartesian
 coordinates 205
 frame 358
cathode ray tube (CRT) 5, 203, 242
CBT *See* Computer-based training
centre of gravity 51, 54, 100, 103, 106–108, 110, 115–116, 126, 139–141, 149–150, 264, 359, 432, 453
centre stick 22, 69, 107, 291
centripetal acceleration 148–149, 152
CFD *See* Computational fluid dynamics
character
 generation 211, 213, 226, 235
 set 212–213, 237
characteristic equation 165, 299, 323
checksum 84, 86–88
Civil Aviation Authority (CAA) 7, 404
civil flight training 8, 138, 379
classes 355, 383, 392, 404
clipping 216–217, 219, 239, 241, 243, 334, 337–339, 341–343, 346–347, 350, 358, 409
clock
 bias 275–276, 279–280, 282
 resolution 311
cockpit
 drills 3, 36
 geometry 291
 voice recorder 10
coefficient
 of drag 47, 49, 105–106, 378
 of friction 106, 141
 of lift 16, 100–102, 106, 108, 111, 322
 of power 132
 of side force 105
 of thrust 79, 132

cognitive process 262, 303, 423
collimated projector 5–6, 371
collimating mirror 371
collimation 242, 370
colour palette (IC) 207, 333
command 12, 91–92, 173, 176, 181–182, 185, 188–189, 193, 196–197, 199, 241, 261–263, 357, 382, 384–390, 392–401, 412, 415, 453
commercial flight training 32
common mode rejection ratio 72
compass 1, 23, 108, 119, 216, 227–229, 234–236, 240–241, 245, 249, 254–257, 392, 394
 card 23, 216, 228, 234, 240–241, 256
computational fluid dynamics (CFD) 17, 41, 93, 102, 105
computer graphics 15, 23, 203, 205, 207, 216, 220, 227, 235, 240, 246, 332, 376, 411, 421
computer-based training (CBT) 15–16, 36–37, 378–379
computer-generated ADF display 257
computer-generated display 203
concurrency 64
conformal 242–243, 252–254, 404–405
context switching 65–66
contiguous file 411
control
 column 22, 26, 69–70, 107, 113, 286, 291
 flow analysis 311
 loading 15, 22–23, 113, 287, 297, 424
convergence factor 253, 407
coordinate frame 120, 275, 288, 334, 336, 345, 347–358, 360, 405, 432
coordinated turn 172
Coriolis 24, 53, 96, 148–152, 264
crew
 cooperation 7, 10, 34
 resource management (CRM) 421
cross-cockpit view 373
crosswind landing 297
cubic spline 77
cues 7, 8, 21, 22, 27–31, 35, 242, 286, 303, 332, 347, 350–351, 363, 375, 382, 423–425, 428–432, 436–437, 441, 442–443, 448, 451–452, 455
cupula 426
curve fitting 76

curved
  mirror 5, 373
  screen 21, 372
cushion of hydraulic fluid 445

D/A *See* Digital-to-analogue conversion
damping 22, 113–114, 140, 159, 173, 255, 449–450
  factor 165, 294, 426
  in pitch 286, 305
  ratio 142, 315–316, 321–322, 324–325, 438
data
  acquisition 13, 18–19, 35, 65–67, 69, 81, 310, 410
  compression 410
  package 13, 18, 93, 96, 108, 112, 114, 139, 313, 328
DCM *See* Direction cosine matrix
dead reckoning 248
dead band 305
decision height 191
deflection coil 203–204
Delaunay triangulation 365, 376
delay compensation 305, 329, 375–376, 452, 455
density of air 47, 100–101
derivative control 163, 165
desktop simulator 35
deterministic (transfers) 83–86, 88
diagnostic tests 24
dialogue box 383, 387, 392, 394
difference operator 162
differential equations 3–4, 43, 53–55, 56–59, 158
digital
  computer 4–5
  input 67, 70
digital-to-analogue conversion (D/A) 68–69, 72
dilution of precision 24, 274, 278–279
dimensional derivatives 112, 312–314, 323
Direct memory access (DMA) 86
Direction cosine matrix (DCM) 110, 120, 123, 143, 148, 154, 266, 278, 279, 336–337, 345, 360, 434
directional stability 113
discontinuity 24, 47, 74–76, 210, 218, 276, 302, 305, 398, 443
discrete
  data 18
  event model 42
display
  resolution 20, 204, 206, 210, 227, 235, 332, 335
  systems 38, 203, 235, 245, 376
Distance measuring equipment (DME) 258, 259, 282, 397, 404, 418
distortion 5, 19, 21, 77, 107, 212–213, 217, 230, 235, 237, 247, 252, 254, 291, 358, 371, 372
distributed systems 82
DMA *See* Direct memory access
DME *See* Distance measuring equipment
double
  buffering 207, 220, 223–224, 226, 238, 399, 404
  precision 122
doublet input 287, 294, 316, 325
drag 20, 47, 49, 105–108, 114, 117, 129, 136, 137, 171, 181, 286, 288, 377–378
  polar 96, 105
drawing package 228, 342, 364, 382
drift rate 263, 268, 270, 272
Dryden (turbulence) model 19
Dutch roll 113–114, 173, 323, 324–325
dynamic
  pressure 249
  response 128, 135–137, 262, 286, 321, 425, 432
  stability 109, 113, 314

earth
  axes 114
  rate 154, 264, 266, 272
  rotation vector 266
earth-centred earth-fixed (ECEF) 119, 264, 277–279
eccentricity (of the earth) 266, 277
ECEF *See* Earth-centred earth-fixed
EFIS displays 23, 33, 235–236, 239, 261
EICAS *See* Engine indicating and crew alerting system
eigenvalue 314–315, 321–322, 325
electrical
  actuation 14, 373, 425, 428, 447, 452
  control loading 15, 22
  drive systems 11, 441
  drive technology 429
  linear actuator 447

Index   461

motion platform 453
motor drive 22, 429
electromechanical instrument 203
elevator 22, 26, 69, 93, 107–109, 111–112,
    169–170, 172, 181, 260–263, 286, 288,
    291, 305, 312–314, 316, 397, 419
  effectiveness 113–114, 287
emulation 31, 235
encoder 74
encoding 86, 229, 401
endolymph 426
engine
  controls 3, 31, 169, 312
  dynamics 14, 18
  failure 19, 22, 31, 106, 129, 135, 137, 288,
      291, 297, 323, 390, 413–414, 424
  indicating and crew alerting systems
      (EICAS) 235, 287
  lever 69–70, 129, 170, 176, 179, 291, 294
  model 13, 18, 19, 26, 99, 127, 135–137,
      171, 287, 291, 294, 299, 305, 381,
      411, 415
  performance 98, 128, 131, 138, 197, 291,
      294
  power 128–132, 135–137, 170, 176
  pressure ratio (EPR) 128, 137
  response 138
  stall 136
  surge 106
  thrust 18, 106, 113, 128, 135, 137, 288,
      312, 323, 430
  vibration 451
engineering
  design 8, 41, 160
  flight simulator 13, 35, 38, 387, 394, 409
  simulation 8
ephemeris data 274–276, 279
EPR *See* Engine pressure ratio
equations of motion 4–6, 15, 16, 20, 53–54,
    91, 97, 103, 113, 115, 124, 143,
    148–149, 151, 154, 263–264, 268, 272,
    312–314, 323, 411, 432, 434, 441, 446,
    452
error growth 62–63
Ethernet 84–85, 88, 92, 196, 354
  packet 84, 402
Euler
  angle 119–123, 143, 148, 154, 264, 266,
      336–337, 360, 440
  frame 141, 254, 336, 359–360
Euler's forward method 58, 60, 61

evaluation handbook 286–287, 328, 329
extrapolation 74, 78, 93, 305

FAA *See* Federal Aviation Administration
failures 9, 19, 22, 24, 37, 105–107, 135, 139,
    297, 328, 379, 385, 397, 421, 424, 453
FCU *See* Flight control unit
feathering (propeller) 129
Federal Aviation Administration (FAA) 14,
    285–286, 329, 452
fidelity 4, 6–8, 11–12, 16, 18, 19, 21, 27, 29,
    32, 35–38, 98, 113, 128, 138, 205, 219,
    276, 288, 291, 298, 328, 331–332, 342,
    351, 363–364, 365–368, 375, 377, 421,
    425, 432
field of view 9, 243, 287, 346, 428
file operations 388
financial benefits 10–11
first order lag 137, 255, 263, 299
fixed
  frame rate 298, 303, 375
  pitch propeller 129
fixed-base
  cockpit 8
  simulator 22, 423, 425, 428, 451, 453
fixed-point arithmetic 5
flap 18, 69, 102–103, 105–106, 114, 135,
    137, 143, 169–170, 172, 182, 229, 286,
    291, 294, 305, 314, 316, 324, 397, 426
  selection 69
  setting 102, 105, 169–170
flash conversion 73
flight
  conditions 2–3, 4, 9, 13–15, 20–21, 27,
      56, 93, 107, 171, 182, 291, 294, 354,
      379–380, 384, 389, 401, 411,
      414
  control systems 13, 15, 38, 116–117, 157,
      171–172, 201, 287, 305, 329, 420,
      451, 455
  control unit (FCU) 15, 33, 195, 196–197,
      240, 397
  controls 15, 18, 22, 69, 169, 201, 291,
      303
  data display 400–401
  data packet 411
  data recorder 10, 107
  data recording 104, 388, 394, 397,
      409–411, 413
  director 196, 238, 260–263, 282–283

flight (*continued*)
  management system (FMS) 7, 15, 24, 33, 35–37, 195–196, 200, 235, 248, 249, 251, 287, 379, 423
  model 2, 4, 12–13, 18, 19, 21, 27, 32, 66, 67, 82, 85, 97, 99, 113, 117, 127, 129, 249, 255, 262, 286–288, 299, 305, 328, 332, 354, 360, 375, 377, 380, 415, 432, 434, 440, 452
  modelling 14, 97
  path 4, 54, 116, 118, 141, 148, 151, 170, 181–194, 196–197, 201, 243, 245, 246, 247, 252, 259–262, 264, 312, 314, 381, 385
  path angle 116, 170, 181–182, 193–194, 245, 262, 264
  plan 15, 19, 181, 185, 196–199, 201, 247, 272
  plan segment tracking 198
  safety 9, 20
  simulation training device (FSTD) 7, 285, 329
  test data 286, 298
  training 3, 8–10, 12, 30, 32–34, 36, 38, 102, 109, 138, 142–143, 331, 376, 377–378, 411, 415, 421, 423, 425, 428, 451
  training device 1, 2, 285–286, 329, 379
  trials 8, 13, 18, 35, 42, 76, 93, 104–105, 294, 298
floating-point 5, 60, 67, 77, 82, 85, 122, 221, 336, 339, 402, 410
FMS *See* Flight management system
fog 5, 10, 108, 216, 219, 287, 331, 335, 353, 356–357, 360, 374
fogging 216, 219, 353, 361, 375
  algorithm 335
  technique 357
font 211–212, 226, 228, 382–383, 412
  editor 213, 237
  selection 226
footprint 342, 366, 367, 375, 431
force feedback 22
forces 8, 16–17, 19, 23, 29–30, 34, 43–45, 47, 49–51, 52–55, 77, 97–100, 104, 106–108, 111, 113–118, 124–127, 129, 138–143, 145, 148–150, 152–153, 169–170, 255, 264, 268, 287–288, 291, 312, 379, 425, 427, 443, 451, 453

frame
  buffer 20, 207
  margin 303
  of reference 51, 52
  period 26–27, 64, 66, 223, 245, 303, 305, 310
  rate 5, 20, 23, 26, 56, 58–59, 64, 65, 92, 135, 143, 216, 219–220, 223, 226, 264, 268, 298, 303, 305, 312, 333, 336, 342, 375, 410–411
  time 26, 66, 223, 298, 303, 312
  timing 92
  store 206–207, 211, 213, 217, 219–220, 223, 302, 332–333, 335–336, 339–340, 342, 375
FSTD *See* Flight simulation training device
fuel-to-air mixture 129–130
full flight simulator 7, 12, 15, 16, 18, 27, 32–33, 35, 285–286, 329, 379, 431, 443, 453
fusion process 423

Gauss-Jordan method 280
gear model 19, 139, 286
general aviation 14, 32, 34, 37–38, 155, 394, 404, 428
geocentric latitude 107, 267
geodesic latitude 107
geomagnetic latitude 274–275
geometric
  computation 332, 343, 375
  height 98
  transformation 216, 336, 339, 342, 347, 350
geopotential height 98
glBegin 220–221, 226, 231, 233, 349, 350–351
glBindTexture 230–232, 233, 350–351
glBlendFunc 225, 232, 350–351
glClear 220, 221, 225, 348–349
glClearColor 220–221, 234, 349
glColor 220–221, 232, 349, 350–351
glDisable 231
glEnable 225, 232, 234, 350–351
glEnd 221, 226, 232, 234, 349, 351
glFlush 221, 225, 349
glGenTextures 230
glHint 225
glide
  path 182, 191, 194, 259–261, 381

Index   **463**

slope 24, 176, 191, 193–194, 259–260, 261, 405
glLightModelfv 234
glLightModeli 234
glLightfv 234
glLineWidth 225
glLoadIdentity 221
glMaterialfv 234
glMatrixMode 229
glNormal 233
global positioning system (GPS) 24, 151, 249, 274, 275, 277, 282, 365
globe to screen transformation 408
glOrtho 221
glPixelStorei 230
glPopMatrix 222–223, 225, 232, 234, 238–239, 242, 348, 349
glPushMatrix 222–223, 225, 232, 233, 238, 241–242, 348–349
glRectf 222, 238
glRotatef 222–223, 225, 232, 233, 238, 241–242, 348
glRshapeFunc 225
glShadeModel 234
glTexCoord 232–233, 349, 350–351
glTexEnvi 230–231
glTexImage 230–231
glTexParameteri 230
glTranslatef 222, 225, 232, 233, 236, 238, 348
glutCreateWindow 221, 223
glutDisplayFunc 220–221, 223–224, 348, 398
glutIdleFunc 224, 398–399
glutInitDisplayMode 220–221, 223–224
glutInitWindowPosition 221, 223
glutInitWindowSize 221, 223
glutKeyboardFunc 223–224
glutMainLoop 220–221, 223
glutPostRedisplay 226
glutReshapeFunc 223
glutSwapBuffers 225, 348, 399
glVertex 218–219, 220–221, 224, 226, 229–231, 232–233, 347, 349, 351
GNUplot 411–412
Gosport speaking tube 1
governed rpm 132, 135
GPS *See* Global positioning system
    error 263, 274, 276, 280
    message 274
graphical user interface (GUI) 383–386, 390, 392, 399–400, 402, 421

graphics
    architecture 219
    engine 20, 218, 333, 336, 343
    operations 205, 214, 216, 219–223, 226–227, 336–347, 399
    primitive 221–223, 226, 236, 347, 353
    processor 5, 20, 206–207, 210, 213, 217–218, 333, 335, 342, 349
    systems 14, 207, 216
gravitational
    acceleration 44, 120, 148–152, 172, 185, 313, 430, 432, 437, 440, 455
    force 44, 47, 98, 104, 107, 114–115, 120, 150
    latitude 107
gravity 27, 51, 54, 98, 100, 103, 105–108, 110, 115–116, 122, 126, 139–141, 149–150, 171, 264, 266–267, 357, 430, 436, 456
    vector 152, 268, 436
great circle 118, 197, 251–253
ground
    effect 19, 103, 106
    handling 19, 29, 106, 142, 291, 451
    roll 19, 138, 288–288
    speed 141, 182, 189, 194, 249–250, 291
ground-based trainer 2
g-seat 8, 30, 38–39, 374, 441, 451, 453–454
GUI *See* Graphical user interface
guidance cues 242
gyro 1, 30, 93, 104, 113, 232, 254, 263–264, 266, 269, 271–272
    drift rate 268, 270
gyroscopic effects 127

handling qualities 1, 54, 97, 109, 113–114, 154
haptic
    sensor 425
    system 30
haze 219, 335, 357
heading reference 108, 189
heading-hold 189
head-up display (HUD) 23, 242–243, 245, 246, 264, 287, 303, 357
heave 7, 22, 39, 429–430, 441
helmet-mounted display 374
hemispherical dome 9, 451
hexapod 428, 431

hidden surface
    elimination 216, 335, 341, 360
    removal 365
hierarchy 352–353, 392
high frequency cue 451
high pass filter 437, 438, 440, 443
homogeneous coordinates 336
HUD *See* Head-up display
    format 244, 246
human factors 31, 38, 39, 377, 382, 455
hydraulic 3, 7, 70
    actuation 22, 373, 429, 443, 450
    actuator 7, 21–22, 176, 431–432,
        443–448, 450, 452–453
    control system 452, 456
    fluid 24, 429, 443–446
    system 7, 11, 22, 33, 425, 429, 448, 452
hydrostatic
    bearing 443–445, 449–450
    seal 7

I/O
    card 23, 67
    computer 92
    data 67, 74, 91, 510
    device 74
    system 74
IAS *See* Indicated airspeed
icing 10, 19–20, 103, 106, 418, 451
IG system 343, 347, 350, 352–354
illumination 24, 234, 335, 360
ILS *See* Instrument landing system
    localizer 189, 191, 193, 260
    localizer capture geometry 192
image
    file format 229
    generator 5–6, 11, 20–21, 28, 29, 82, 91,
        206, 216, 243, 299, 332–333, 347,
        365, 368, 370, 375
    generator loading 368
impulse input 54
inceptor 22, 36, 70, 129, 143, 152, 291
incremental transfer effectiveness ratio 12
index page 383
indicated air speed (IAS) 249
induced drag 105
inertial
    axes 118
    frame 53, 118, 148, 150–152, 264, 432,
        436

navigation system (INS) 24, 33, 149, 151,
    263, 273–274
in-fill method 340
input selection 388
INS *See* Inertial navigation system
    error model 268
    errors 271–272
instability 59–61, 92, 135, 163, 299, 302, 323,
    448, 452
instructor
    command 91–92, 394, 401
    station 3, 13–15, 91–92, 252, 377–385,
        387–388, 392, 394, 396–405,
        410–411, 414, 416, 421, 428
instrument
    approach 24, 191
    display 23, 85, 107, 116, 223, 225, 228,
        230, 234, 428, 452
    flight trainer 15
    flight training 2, 33, 428
    flying 1–3, 32–33
    landing system (ILS) 11, 24, 31, 182, 196,
        238, 258–259, 261, 381, 404–405
integral
    control 163, 167, 181
    saturation 167
integration step length 124, 303
integrator 4, 56, 58, 161–162, 167, 437, 448,
    450
    wind-up 167
intermediate code 415–418
international
    qualification test guide (IQTG) 285
    standard atmosphere (ISA) 100
internet protocol (IP) 84
interpolation 74, 77–81, 132, 219
interpreter 416, 420
inter-processor communication 82
interrupt service routine 64–65
interrupts 64–66
ionosphere 256, 274–276, 280
ionospheric time-delay 282
IP address 84–85, 87, 91
iron-bird rig 13–14, 35
isogonal 249
iteration rate 57, 59, 61, 65, 92, 272, 312, 452

jack 432–434, 436–437, 440, 451, 453
jerky motion 7, 25, 82, 92, 122, 428

Index

jet engine 21, 96, 104, 106–107, 113, 115, 128, 136, 165, 291, 313
jitter 303

keypad 381, 389
  entry 382
keystone effect 372
kinaesthetic sensors 27

Lambert conformal projection 251–253, 405
landing
  configuration 314
  gear 138–140, 142–143, 155, 286–288, 305
    dynamics 155, 288, 305
    force 140
    model 139, 286
    response 143, 287
landscape (orientation) 205, 227
Laplace transform 157–161, 201, 301
laser projector 9, 374
latency 20, 83, 85, 227, 286–287, 298–299, 302–303, 305, 374–375, 402, 452
lateral
  acceleration 22, 172, 430, 436, 440, 451–452
  dynamics 54, 171, 312, 323
  flight plan 196
  force 51, 104, 312
  guidance 189, 197, 201, 259, 262
  stability 323
latitude 24, 98, 107, 122, 146, 149–151, 197, 201, 248–254, 257, 263–264, 266–268, 273–278, 358, 392, 394, 399, 404–409, 420
LCD *See* Liquid crystal display
leaf node 368–369
least squares 75–76
lesson plans 15–16, 37
level-of-detail 352, 375
lift force 100–101, 104, 107
line drawing 205, 207–212, 226, 228, 399
linear
  acceleration 30, 127, 153, 263, 425, 427, 429, 432, 437, 440
  model 288, 316, 323–324, 328
  motion 30, 43, 108, 424, 428–429
  system 328
  velocity 124–125
  voltage differential transformer (LVDT) 19, 22

linearity 69, 72–74, 452
linearized model 316, 324
line-fitting methods 75
Link
  Edwin 1–2, 38
  trainer 2–3, 23, 227, 428
liquid crystal display (LCD) 203–205, 235
load
  factor 185
  balancing 375
local
  area network 82, 85
  gravitational vector 107
localizer 176, 191–193, 196, 199, 259–262, 405
  tracking 189, 194, 197–198
logging 37, 420
long period phugoid 113, 176, 314–316
longitude 24, 107, 119, 122, 143, 146, 149–151, 197, 201, 246–252, 257, 263–264, 266–267, 272, 276–278, 358, 392, 399, 404–409
longitudinal
  dynamics 312, 323
  stability 113–114
lookup table 76, 77, 99, 107
low pass filter 437, 440–442
LVDT *See* Linear voltage differential transformer

MAC address 85
Mach number 16, 54, 79, 99, 103, 105–106, 128, 136–137, 240, 249, 255, 291
Maclauren series 301
magnetic variation 248–249, 255, 394
maintenance 10, 11, 21, 23–24, 32, 35, 37, 66, 285, 329, 429
  training 16, 37
malfunction 9, 24, 106–107, 248, 385, 421, 451
manifold pressure 128–129, 135, 227
map
  display 252, 385–386, 392, 399–400, 403–407, 409
  projection 252, 254, 282, 404
marker beacon 255, 259
master-slave transfers 83
mathematical model 41, 96, 154, 283, 288
matrix transformation 120–121, 222
mechanical actuation 425, 452

memory
    access 20, 86, 206, 211, 310, 336
    write cycle 206, 210
menu
    definition file 393
    options 385
    pop-up 382
    selection 379, 386, 397
meridian of longitude 119, 149, 249, 251
military flight training 8, 34, 379
mipmap 218–219
mission rehearsal 9, 34
mixture lever 128–129, 131, 135
model
    board 4
    validation methods 288
moment 16, 22, 54, 96, 97, 104, 106–107, 110–118, 127, 139–140, 142–143, 145, 147, 152, 154, 170, 173, 176, 286, 288, 291, 312
    of inertia 43, 108–109, 124, 126, 132, 137
monitoring 15, 22, 191, 310–311, 379–381, 420
Moore's law 6
motion
    actuation 14, 430, 446
    actuator 30, 424
    cue modelling 442
    cueing 7, 29–30, 39, 455
    onset cue 446
    platform 5–7, 11, 21–22, 24, 27, 29–30, 32, 35, 41, 57, 85, 118, 286, 305, 332, 371, 373–375, 375, 425, 428–432, 436–438, 441, 442–444, 448–455
    systems 4, 7, 9–11, 30, 423, 442–443, 451, 453–455
mouse input 25, 381
moving base simulator 425
multicast data transfers 85
multi-engine aircraft 106, 113, 116
multi-purpose control display unit (MCDU) 196
multi-step integration 57

natural frequency 114, 142, 270, 298, 315, 321–322, 324, 325
navigation
    computation 191, 193, 250, 278
    database 189, 193, 257, 260, 358, 404

equations 91, 263–264, 280
flight display (NFD) 235
frame 118, 148–154, 264, 266–268, 334, 336
systems 3, 23–24, 35, 36, 53, 149, 151, 247, 263, 268, 282, 287
training 23
NDB *See* Non-directional beacon
NED *See* North-east-down
negative transfer of training 143, 328, 381, 453
network 9, 34, 68, 74, 82–88, 92, 94, 157–159, 299, 302, 329, 389, 452
    interface 82, 86, 88
Newton's laws of motion 43, 118–119
node 68, 83–85, 87–88, 91–92, 246, 354–357, 368–369
non-dimensional derivatives 112, 312–314, 323
non-directional beacon (NDB) 12, 33, 201, 255–257, 378, 381, 404
normal vector 340
north-east-down (NED) 146, 150, 264, 268, 272–273, 336
nose-wheel steering 51
numeric entry 389
numerical
    integration 45, 56, 56–58, 61, 64, 92–93, 96, 161–162
    methods 43, 55, 60, 64, 81, 92, 96
Nvidia 376

Octave 315, 325, 329
oculomotor muscle 426–427
offline
    data analysis 410
    simulation 42, 272
oleo 19, 51, 139–143, 288
omni bearing selector (OBS) 257
on-line data analysis 410
Open Systems Interconnection (OSI) 83
OpenFlight 20, 353, 355, 357, 365, 370, 376
OpenGL 216, 218–230, 232, 235–237, 239, 242–244, 246, 343–344, 347–350, 353, 375–376, 384–385, 398–399, 402, 404–405
    programming guide 220, 246, 376
OpenSceneGraph 243, 245, 352–353, 370, 376

operating
    costs 11, 33
    point 168, 450
    system 25–26, 64–66, 82–88, 302–303,
        310–311, 384, 401–402, 410–411
operational amplifier 3, 68–72
optical flow 29
orbital plane 276, 278
orthographic projection 245
otolith 30, 425, 427–428
overlay 23, 34, 245, 303

Padé's approximant 301–302
painter's algorithm 341
panning (function) 404
PAPI *See* Precision approach path indicator
parallel of latitude 248
parasitic drag 105–106
part-task
    trainer 3, 7, 38
    training 33–34, 378
performance
    analysis 66, 305, 312
    test 288
perspective 1, 14, 17, 18, 25, 28–29, 68, 84,
        86, 109, 196, 216–217, 219, 236, 328,
        331, 334–337, 339–340, 342–341,
        345–347, 350, 352, 379, 456
    computation 339–340, 342–343, 345–346
    cues 28
    projection 335, 339, 342
PFD *See* Primary flight display
phugoid 113–114, 176, 294, 314–316,
        322–323, 414
physiology 425, 456
PID controller 163, 165–167, 450
pilot
    cues 27, 424
    perception 423
    response 42, 93, 299, 303, 413–414, 424
    station 51, 359–360
    workload 169, 172, 176, 195–196, 421
pilot-eye
    frame 336–337
    point 5, 20, 218, 286, 332, 336, 355, 370,
        372
    viewport 359
pilot-induced oscillation 452
piston engine 69, 79, 98, 127–130, 136
pitch

ladder 238, 243
lever 128–129, 132, 135
lines 237–239, 243
rate 109, 125, 150, 170, 286, 288, 305,
    312, 316, 322, 397
rate control law 181
response 305, 315–316, 321
pitching moment 96, 108–110, 113–114, 116,
    170, 176, 286, 288
pixel grid 210–214
pixelization 210
playback 411
plotting package 411, 414
pneumatic actuator 3
polar coordinates 407
polygon
    in-fill 216, 335, 340
    rendering 216, 375
polynomial fit 57, 76
portrait (orientation) 205, 227–228, 236
position and attitude transform 356
potentiometer 4, 19, 71, 129
precision approach path indicator (PAPI) 259,
    355
pressure 2, 8, 15, 18–19, 23, 27, 30, 36, 93,
    98–100, 106, 128–130, 135, 137–138,
    179, 227, 240, 249, 255, 275, 291, 373,
    425, 429, 441, 443–449, 451–452
    altitude 129, 135
primary flight display (PFD) 196, 223, 231,
    233, 236–237, 260
principles of mechanics 97
probe effect 310
processor architecture 63–64
programming language 5, 38, 56, 63, 96, 220,
    223, 384, 418–420
projection 15, 21, 220–221, 224, 242–243,
    252–254, 282, 331, 335, 339, 342, 358,
    377, 404–405, 425, 441, 452–453
    system 5–6, 11, 23–24, 28, 30, 34,
        242–243, 332, 335, 370–374, 451
projector 5–6, 9, 21, 24, 28, 29, 242–243,
    332–333, 358, 370, 372, 374, 428, 449
    geometry 286
propeller
    efficiency 128
    performance 128, 135
    thrust 128, 131, 135
proportional control 163–165, 172, 185
propulsive forces 99, 106, 115, 127, 264
protocol 82–88, 91–92, 196, 302

pseudo-range 275, 279–280
pulse input 294, 316
pure time delay 301

QDM 189, 193, 198, 260, 297, 358
QNH 100, 249, 394
qualification of flight simulators 7, 285, 329
quaternions 122–124, 143, 147, 154, 264, 266

R-2R ladder network 68
radio button 390
radio magnetic indicator (RMI) 255, 256, 397
RAE (turbulence) model 19
rate limits 55, 185
rate-one (turn) 185, 255
real-time
　computing 26, 63, 83
　IG 347
　image generation 343
　rendering 235, 352–353, 356, 369, 376
　simulation 4, 24–26, 56, 61, 66, 82,
　　　83–85, 93, 96, 286, 299, 363, 413
recording 18, 21, 93, 104, 280, 291, 294, 379,
　　　381, 385, 388, 390, 394, 397, 401,
　　　409–415, 418–419
recurrent
　checks 285
　training 32, 34
recursive data structure 368
reference speed 176, 179
reflective memory 82
refraction 274, 282
refresh rate 20, 204, 375
regional pressure setting 249
re-hosting 377
relative bearing 189, 193, 197, 250–252,
　　　256–258, 260
　indicator (RBI) 256
rendering
　package 245, 353, 356, 369
　process 245, 335, 347, 375
　rate 20, 28, 216–217, 219, 223, 335, 363
replay 21, 385, 400, 411
　mode 400
reverse thrust 19, 106, 136–138, 297–298,
　　　430
rho-theta navigation 258
rhumb line 251–253
ring laser gyro 263
RMI *See* Radio magnetic indicator

roll
　rate 113–114, 125, 129, 188, 323, 325,
　　　397, 411–412
　subsidence 323–325
rolling
　digits 23, 239–240, 243, 246
　friction 140–141
　moment 109, 114, 173, 312
rounding error 60, 217, 342, 375
rudder 18, 22, 69, 104–105, 107–109,
　　　111–114, 116, 141, 169, 172–173, 227,
　　　288, 291, 294, 297, 312, 323, 328, 379,
　　　395, 414–415
　response 324–325

safety 7, 9–10, 13, 20, 22, 27, 32, 34–35,
　　　105, 131, 377, 428, 445, 447, 453
sampling 18, 56–57, 71–74, 107, 162, 165,
　　　167, 298–299, 302, 420
Sanders trainer 1
satellite elevation 275, 282
scaling 68, 212–213, 215–216, 231, 243, 245,
　　　312, 323, 339
scan line 340
scene
　detail 6, 20, 352, 375, 453
　effects 353
　graph 354–356
Schuler
　frequency 271
　loop 24, 270, 273
screen to globe transformation 409
script
　compiler 415
　file 400, 413, 415, 420
　program 400, 414–416, 420
scripting 413–414, 416, 418–421
　language 413–414, 416, 418–421
semi-circular canal 30, 425–426, 455–456
sensor
　measurement 76, 248, 259, 268
　model 282
sensory input 423
sequential file 411
serial conversion 72
servo motor 15, 23
servomechanism 3
servo-valve controller 432
SGI OpenGL Performer 245, 353, 370, 376
shading 205, 216, 220, 226, 228, 245, 335,
　　　353

shared memory 82, 354
short period phugoid 113–114, 314–316, 322
side force 104–105, 107–108, 117, 140–142, 172, 424, 436
sideslip 104–105, 107, 109, 112, 114, 116–117, 145, 152, 173, 189, 237, 245, 291, 297, 323, 414
  response 324–328
side-slip vane 172
side-stick 107, 291, 303
signal conditioning 18, 67, 70, 72
Silloth trainer 3
simulated instruments 23
simulator
  acceptance 96, 437
  data 34, 96, 106, 313, 390
  qualification 14, 18, 22, 285, 329, 443
  sickness 27, 375, 453, 455
singularity 121–122, 268
six degrees of freedom 39, 124, 150, 428–429, 431, 451, 453, 454, 456
sky
  model 357
  texture 360
slats 102, 105
slider scale 381, 390
slip ball 105, 117, 172, 226, 254–255
small perturbation model 323
socket 85, 87–88, 382, 401
solenoid 450
spatial
  filter 211
  transformation 335
speed
  brake 105, 294
  of conversion 71
  of sound 99
spherical earth 24, 252
spiral
  divergence 323
  mode 113–114, 294, 324–328
spline fitting 76–77
spoiler 105, 114
spool valve 446–450, 452
stability 41, 53, 61, 92, 96, 97, 109–111, 113–114, 154–155, 159–160, 163, 171–172, 280, 299, 301–302, 312–312, 314, 323, 329, 447, 450
  axes 112, 117, 146–147, 170
stack machine 416–418
stall characteristics 155, 297

stalling 102, 105, 136, 238, 378
standard parallel 253–254
starter formula 61, 162
state-space 312, 315, 321–323
static
  pressure 249, 255
  stability 109–111
  test 291
steady-state 47, 109, 111, 159–160, 163–166, 169, 181, 194, 292–294, 297, 440, 451
  conditions 93, 104, 168, 291, 303
  measurement 288
step
  input 61, 113, 158–159, 164, 294, 303, 315, 324–328
  length 46–47, 57–61, 64, 92, 124, 303
stepper motor 15, 23, 203
Stewart platform 7, 428–430, 456
stick fixed neutral point 111
stick free manoeuvre 109
strap-down INS 263, 268
strip display 243, 411
stroke display 204
sub-option menu 399
successive approximation 72, 74
summer 4, 151, 364, 455
surge 7, 18, 106, 136, 429–430, 434, 436, 438, 443, 451–452
  cue 430, 434, 436
sway 7, 429–430, 436, 443, 452
synchronization 83–84, 92
synergistic platform 428, 453
synthetic environment 37, 332
system
  behaviour 14, 42, 53, 66, 299, 452
  clock 65, 311
  dynamics 61, 155, 167, 288, 305, 328
  life cycle 13

tabular data 74
tactile
  cues 8, 303, 423
  sensor 27
target 29, 36, 96, 332, 356–357, 374
Taylor series 59, 60
temperature 19, 27, 36, 42–43, 93, 100, 106, 109, 128, 135–138, 142, 163, 181, 227, 249, 275, 286, 291, 294
  lapse 98–99
temporal effects 375

terrain database 288, 355, 357
test pilot 287, 294, 297, 305, 411, 443
texture
    filter 219, 350, 351
    formats 218
    image 216–219, 342, 350, 363
    map 213, 216–219, 228–229, 232, 246, 342, 350–351
    memory 217, 342, 350
    operations 218, 228, 352
    space 218, 351
    storage 351
textured
    characters 211
    image 216, 219, 342, 350, 363
    surfaces 6, 357
throttle 69, 174–75, 177, 194, 316
    lever 23, 128–129, 135–136
thrust specific fuel consumption (TSFC) 136
time
    constant 30, 59, 61, 92, 164, 173, 176, 179, 242, 262–263, 298, 324, 426–427, 446, 450, 452
    division multiple access (TDMA) 83
    response 159–160, 302, 402
token passing (transfers) 83, 88
torque 16, 22, 43, 132, 263, 269, 447, 450
    motor 263, 447, 450
touch screen 10, 15, 33, 227, 381, 384–385, 388–389, 397, 398, 402
tracker-ball 381
training
    accidents 4, 9
    devices 1, 7, 12, 33, 285–286, 329, 378–379
    effectiveness ratio 12
    needs analysis 31
    organization 7, 9–10, 12, 15, 31, 377, 379, 451
    requirement 2, 30–32, 205, 285, 377, 380, 385, 421, 425, 431, 455
    syllabus 1, 30–31, 36, 378, 451
    task analysis 12
    transfer 11–13, 38, 227, 328, 428
transducer 70, 93, 445
transfer function 30, 158–165, 173, 299, 301–302, 315, 323, 325, 427, 450, 455
translation 16, 27, 214–216, 223, 225, 237, 239, 243, 336–337, 354–355, 429, 433, 436, 455
transparency 245, 352, 365

transport rate 150, 266–267, 269
tree structure 354, 368–369
trim
    controls 113, 169
    tab 109, 113, 169
    wheel 113, 169, 291
trimming 22, 69, 113, 169–170, 286
troposphere 98, 275–276, 280, 282
tropospheric delay 275
true air speed (TAS) 249
truncation error 61, 268
turbulence 10, 12, 19, 38–39, 117, 146, 153, 173, 176, 185, 298, 312, 379, 394, 424, 428, 441, 451
turn
    anticipation 199, 201
    coordinator 172–173
    indicator 254–255
    rate 52, 118, 185, 189, 191, 199, 200, 255, 259–260, 262, 297, 411
turn-around bump 444
Tustin transformation 162
tyre 29, 51, 140, 219, 351
    burst 19, 106, 139, 142, 381, 451
    conversion 12
    track angle 141

UDP *See* User datagram protocol
    packet 87, 354
undercarriage
    assembly 19, 51, 138–139, 370
    forces 19
    model 139, 297
upset recovery 451
usage 24, 312
user
    datagram protocol (UDP) 84
    interface 197, 377, 380–384, 387, 394, 398, 421
    interface design 381, 421

validation 8, 13, 17, 27, 41, 96, 113, 138, 272, 285–288, 291, 303, 305, 328, 329
variable pitch propeller 106, 128
vection 29, 350–351, 441
vector
    generated character 211, 213
    graphics 205, 235, 404
vehicle
    dynamics 52

training simulator 34
velocity vector 108, 243, 312
vertical
   flight path 181, 247, 259
   guidance 197
   speed hold 182
   speed indicator 227, 236, 255
   tape 239
vestibular
   sensors 27
   system 27, 30, 38, 425, 428, 442, 453, 455
VHF omni-directional range (VOR) 23–24, 191, 196, 255–259, 274, 363, 376, 392, 404, 451
video
   amplifier 5, 332, 372
   bandwidth 335
   generator 333, 335
   signal 20, 23, 204, 206–207, 332–333, 335, 368
   signal format 23, 333
viewing
   angle 205, 358
   frustum 334, 337–339, 346, 366
visual
   acquisition 262, 303
   artefacts 217, 219, 350, 357
   cues 7, 22, 27–29, 286, 332, 347, 375, 424, 441, 451
   database 9, 29, 34, 219, 333, 336, 340, 342–343, 348–349, 352–353, 355, 357–358, 364–369, 375
   database management 364, 370
   fidelity 6, 29, 205, 219, 331, 332, 342, 351, 363, 368, 375
   latency 20
   scene 29, 115, 218–219, 243, 260, 331, 368, 439
   streaming 29
   system 4–6, 9–11, 14–15, 20–22, 23–29, 32–35, 38, 41, 65, 82, 85, 116, 120, 122, 217, 219, 243, 286–288, 291, 299, 302–303, 305, 328, 331, 333–334, 336, 341–343, 347, 349, 358, 360–361, 364, 373–375, 381, 425, 428, 432, 452–453
   system pipeline 332, 339, 370
   target 374
visualization 14, 35, 349
VOR *See* VHF omni-directional range
   radial 189, 260

wake vortices 363
Walters 1, 456
washout
   filter 432
   motion 441
waypoint 185, 196–201, 404
weather model 19–20, 27, 31, 136, 291
WGS-84 264, 275
wide angle projection 6, 30, 34, 425
wind
   component 143, 146, 185, 189
   model 19
   tunnel test 17, 42, 101, 105, 108, 117
windmilling (propeller) 105
wind-shear 10
wing profile 100
wire-frame 335, 343, 347, 349, 365
   image 340
   triangle 352

yaw
   damper 172–173, 176, 189, 294, 310, 323, 328
   rate 125, 173, 176, 181, 189, 255, 294, 323, 325–328, 397, 411–412
yawing moment 106, 109, 113–114, 116, 173, 312

Z buffering 335, 342, 360
zero flight-time (ZFT) 9
zooming (function) 382, 404